SPECTROSCOPIC TECHNIQUES FOR SEMICONDUCTOR INDUSTRY

SPECTROSCOPIC TECHNIQUES FOR SEMICONDUCTOR INDUSTRY

Vladimir Protopopov

Samsung Electronics Mechatronics Center, South Korea

World Scientific

NEW JERSEY • LONDON • SINGAPORE • BEIJING • SHANGHAI • HONG KONG • TAIPEI • CHENNAI • TOKYO

Published by

World Scientific Publishing Co. Pte. Ltd.
5 Toh Tuck Link, Singapore 596224
USA office: 27 Warren Street, Suite 401-402, Hackensack, NJ 07601
UK office: 57 Shelton Street, Covent Garden, London WC2H 9HE

Library of Congress Cataloging-in-Publication Data
Names: Protopopov, V. V. (Vladimir Vsevolodovich), author.
Title: Spectroscopic techniques for semiconductor industry / Vladimir Protopopov,
Samsung Electronics Mechatronics Center, South Korea.
Description: Hackensack, New Jersey : World Scientific, [2023] | Includes bibliographical references and index.
Identifiers: LCCN 2022021000 | ISBN 9789811257599 (hardcover) |
ISBN 9789811257605 (ebook for institutions) | ISBN 9789811257612 (ebook for individuals)
Subjects: LCSH: Spectrum analysis--Industrial applications. | Semiconductors--Design and construction.
Classification: LCC QC451.6 .P76 2023 | DDC 621.36/1--dc23/eng20220822
LC record available at https://lccn.loc.gov/2022021000

British Library Cataloguing-in-Publication Data
A catalogue record for this book is available from the British Library.

For any available supplementary material, please visit
https://www.worldscientific.com/worldscibooks/10.1142/12882#t=suppl

Desk Editors: Balasubramanian Shanmugam/Steven Patt

Typeset by Stallion Press
Email: enquiries@stallionpress.com

Preface

It is true to say that the 20th century was the century of science. Spectroscopy is a mature science, with its roots originating deeply in the very first discoveries by Newton. The scientific background of spectroscopy is nowadays a rudimentary course in every university, and almost every laboratory in university's physical department is equipped with one or another type of spectrometers that are freely available on the market. Typical university courses on spectroscopy are focused mostly on the physics of atomic and molecular spectra, e.g. rotational, vibrational, electronic transitions, quantum mechanical principles of radiating transfers. These educational topics are supported by a variety of well-established textbooks, printed in several editions. All these books are mostly about spectra and not about spectroscopic instrumentation. However, we live already in the 21st century, which is commonly referred to as the century of technology, and technological progress transformed canonic spectroscopic techniques into new, very peculiar methods and instrumentation, mostly unknown to wide audience. The present book takes the reader into the new world of spectroscopic applications to semiconductor industry — the engine of modern technology. It presents special principles and techniques of spectroscopic measurements that are used in semiconductor manufacturing. Since industrial applications of spectroscopy are significantly different from those traditionally used in scientific laboratories, the design concepts and characteristics of industrial spectroscopic devices may vary significantly from conventional systems available on the market. These peculiarities have never been summarized before in a form of a textbook or monograph, and are buried deeply within specialized publications in scientific and technical journals, thus being hidden from wide audience of students, engineers, and scientific workers. However, the concepts of industrial spectroscopic methods and instruments contain numerous efficient techniques, which may widen the scope of all scientists and engineers, working in the field of spectroscopy and optical measurements in general, thus helping to achieve new results in their work. Even if the readers do not find direct answers to their needs, the variety of practical solutions, explained in detail in this book, will help to open new horizons in their research.

When it comes to spectral measurements in semiconductor manufacturing, grating spectrometers, particularly compact spectrometers, dominate owing to their reliability and durability. Theoretical and practical basics of these devices are summarized in Chapter 1.

The Fourier-transform spectrometers may be classified into three main types: absorption type, emission type, and Fourier-transform reflectometers. In the first type, the white-light radiation, the source of which is built into the spectrometer, passes through the sample within the spectrometer to a photodetector, without getting out. Such spectrometers measure exclusively absorption of samples in a transmission mode and are widely used in laboratories. The emission-type Fourier-transform spectrometers measure spectra of the outer light. These are more complicated and more expensive devices, rarely found in the market. Finally, the Fourier-transform reflectometers are very special instruments, used primarily in semiconductor industry for testing reflectivity of horizontal objects, like semiconductor wafers, liquids, epoxies. These three types of spectrometers are explained in Chapter 2.

Nano-scale precision of translational motion in semiconductor manufacturing is controlled by laser interferometers. These devices use stabilized He–Ne lasers, whose wavelength drift is minimized to megahertz level. Evaluation of such high stability requires high-resolution spectroscopy tools, described in Chapter 3.

Semiconductor manufacturing always requires control of thickness of layered structures. The best way of non-destructive and non-contact measurement of layers' thickness is spectral interferometry. Therefore, spectrometers must not only provide good spectral resolution but also imaging capabilities in order to discern details of the structure. Chapter 4 describes two types of such devices.

Charge coupled devices are unrivaled photodetectors for spectrometers. However, minimum exposure time that can be achieved with them is only several milliseconds — too slow to resolve nanosecond-scale processes. And even if it were a nanosecond, it would be completely useless because only few photons could be recorded in a single pixel during such a short time. The solution for studying fast processes, which was successfully commercialized and is available on the market, is to use gated image intensifiers: fast optically switching devices with high optical gain. This is the topic of Chapter 5.

There are applications, like plasma-etching machines in semiconductor industry, aircraft jet engines, high-voltage coronas on transmission electrical lines, fluorescence spectroscopy, where optical emission is modulated by frequencies ranging from 50 Hz to several megahertz. How to separate weak modulated spectrum from strong non-modulated background? Traditional spectrometers cannot do that. The solution is modulation-sensitive Fourier spectroscopy — the subject of Chapter 6.

The nano-scale pattern of semiconductor devices is made mostly by plasma-chemical etching in process machines — vacuum chambers with plasma excited by either magnetic or electrical high frequency fields. The neutral gases and products of etching — atoms or molecules — fill the space above the wafer and are excited by electrons of plasma to produce characteristic monochromatic radiation at certain wavelengths. Spectral analysis of this radiation is almost the only way to obtain real-time information about the etch process. Therefore, every process machine in production line is equipped with a spectrometer. Chapter 7 discusses the basic functions of optical diagnostics in plasma machines and relevant experimental techniques.

Thin layers are the inherent elements of electronic devices. Therefore, measuring thickness of thin layers is the routine operation in semiconductor industry. The most appropriate and widely used technique is spectral reflectometry — non-destructive optical method based on the analysis of spectrum of light reflected from thin film. Interference of partial waves reflected from the top and bottom of the film create unique picture of the spectrum that may be considered as a fingerprint of the layer, containing information about its thickness, optical constants, etc. Chapter 8 explains how this technology is adopted to semiconductor manufacturing.

In addition to spectroscopic techniques, semiconductor manufacturing uses variety of non-spectroscopic optical methods. Some of these methods are standard, like optical microscopy, detailed explanation of which can be found in university textbooks, others are non-standard, developed in recent years specifically for semiconductor applications and closely related to the tasks already considered — measurement of thin film thickness, plasma diagnostics, characterization of pattern structures. Some of these complementary non-spectroscopic sophisticated techniques are explained in detail Chapter 9.

The physical background of every chapter is explained in depth with all necessary mathematical and engineering details that are needed to understand the subject. Therefore, the references in each chapter contain only entries to supplemental literature that might be interesting to those readers who wish to obtain a wider vision of the subject.

The work on this book was intense, and I always felt support from the World Scientific team and particularly from senior editor Steven Patt and the head of the production team Balasubramanian Shanmugam, who represent the best traditions of this renowned publisher.

Vladimir Protopopov

About the Author

Vladimir V. Protopopov was born in 1953 in Moscow. He graduated with honors in 1976 from Moscow Institute of Physics and Technology (MIPT) in the field of applied physics and received Ph.D. in 1978 from MIPT and Dr. Sc. in 2004 from Moscow State University. He worked for various companies in Europe, America, and Asia and authored many books in the fields of academic, industrial, and military applications of optics, X-ray optics, lasers, and opto-electronics.

Contents

Preface v

About the Author ix

1. Basics of Grating Spectrometers 1

1.1 Typical Schemes . . . 1
1.2 Diffraction Grating Formula . . . 2
1.3 Spectral Resolution . . . 5
1.4 Spectral Orders . . . 7
1.5 Order-Sorting Filters . . . 8
1.6 The Rowland Circle . . . 12
1.7 Calibration Techniques and Formulas . . . 16
1.8 Photodetectors . . . 20
1.9 Thermo-electrical Cooling . . . 25
1.10 Coupling to Optical Fibers . . . 27
1.11 Typical Industrial Spectrometers . . . 29
1.12 Monochromators . . . 31

2. Basics of Fourier-Transform Spectrometers 35

2.1 Theoretical Background . . . 35
2.2 Absorption-Type Fourier-Transform Spectrometers . . . 37
2.3 Emission-Type Fourier-Transform Spectrometers . . . 40
2.4 Fourier-Transform Reflectometers . . . 41
2.4.1 Infrared Fourier-transform reflectometers . . . 41
2.4.2 Combined IR–VIS reflectometer . . . 41

3. High-Resolution Spectroscopy 53

3.1 Scanning Interferometers Fabry–Perot . . . 53
3.1.1 Scanning interferometers Fabry-Perot with flat mirrors . . . 53
3.1.2 Confocal scanning interferometers Fabry-Perot . . . 57
3.1.3 Practical results . . . 59

3.2 Laser Heterodyne Spectroscopy 64
3.2.1 Interference of optical waves on the photodetector 64
3.2.2 Basic concept of heterodyne spectroscopy 67
3.2.3 Optimal conditions for wavefronts 74
3.2.4 Practical schemes of wavefront matching 79

4. Imaging Spectrometers 83
4.1 One-Dimensional Imaging Spectrometers 83
4.1.1 Lens-based optical scheme 83
4.1.2 Typical performance 84
4.2 Two-Dimensional Imaging Spectrometers 87
4.2.1 Basic considerations for the Michelson interferometer 87
4.2.2 Basic properties of CCR 89
4.2.3 Imaging configuration of the Michelson interferometer with CCR 92
4.2.4 Design concept of the IFS 96
4.2.5 Data acquisition and processing 96
4.2.6 Typical performance 105
4.2.7 Sensitivity 108

5. Gated Intensified Spectrometers 113
5.1 The Principle and Characteristics of Image Intensifiers 113
5.2 Sensitivity and Noise 115
5.3 Additional Advantages of Image Intensifiers 118
5.4 Coupling Image Intensifier to a Sensor 119
5.5 Typical Performance of a Gated Intensified Spectrometer 120
5.5.1 Spectral performance 120
5.5.2 Computer-enhanced spectral resolution 122
5.5.3 Gating capabilities 124
5.5.4 Time-resolved fluorescence spectroscopy 126
5.5.5 Modulation-sensitive spectroscopy 128
5.5.6 Enhanced sensitivity spectroscopy 129
5.5.7 Micro-spectroscopy 132
5.5.8 Laser-induced breakdown spectroscopy 133
5.6 Automatic Gain Control in Gated Intensified Spectrometers 136

6. Modulation-Sensitive and Frequency-Selective Spectroscopy 143
6.1 Modulation-Sensitive Fourier-Spectroscopy 143
6.2 Frequency-Selective Spectroscopy 148
6.2.1 Principle of operation 148
6.2.2 Selectivity of modulated spectra 148
6.2.3 Measurement of decay times 149
6.3 Spectroscopy of Harmonics 153
6.3.1 Introduction 153
6.3.2 Qualitative theory 154
6.3.3 Experimental installation and measurement 158

6.3.4 Experimental results . . . 160
6.3.5 Effect of chamber condition . . . 163
6.3.6 Measurement of modulation depth . . . 163

7. Optical Diagnostics in Plasma Etching Machines 167

7.1 Observation Conditions on Plasma Chambers . . . 167
7.2 Basic Applications . . . 168
7.2.1 Endpoint detection . . . 168
7.2.2 Photomultiplier tubes . . . 177
7.2.3 Interference filters . . . 182
7.2.4 Subtraction of plasma fluctuations . . . 184
7.3 Diagnostics of Pulsed Plasma . . . 186
7.3.1 Plasma ignition sensors . . . 186
7.3.2 Life time measurements . . . 194
7.4 Vertical Distribution of Optical Spectra . . . 197
7.4.1 Optical system . . . 197
7.4.2 Spectroscopic system . . . 200
7.4.3 Spectral analysis of the plasma sheath . . . 202
7.5 Radial Distribution of Optical Emission . . . 204
7.5.1 Theoretical concept . . . 204
7.5.2 Simulation . . . 208
7.5.3 Design concept and experimental results . . . 210
7.6 Two-Dimensional Sensor for Measuring Spatial Non-uniformity of Plasma . . . 212
7.6.1 The WDM WLS concept . . . 212
7.6.2 Theory of measuring plasma density . . . 217
7.6.3 Electro-mechanical design and experimental results . . . 219

8. Spectral Reflectometry 223

8.1 Measurement of Thickness of Dielectric Films . . . 223
8.1.1 Basic phenomenology of spectral reflectometry . . . 223
8.1.2 Theory of reflection from multiple layers . . . 227
8.1.3 Experimental techniques . . . 235
8.2 Patterned Structures . . . 240
8.2.1 Phenomenology of spectral reflectometry on patterned structures . . . 240
8.2.2 Theory of reflection from dielectric sub-wavelength patterned structures . . . 242
8.3 Chemical Mechanical Polishing . . . 249
8.3.1 Principle of optical control of silicon thickness . . . 249
8.3.2 Basic phenomenological differences from visible optics . . . 250
8.3.3 Sinusoidal shape of oscillations . . . 252
8.3.4 The role of roughness . . . 256

9. Related Non-spectroscopic Techniques 259

9.1 Angular Reflectometry . . . 259
9.1.1 The concept of angular reflectometry . . . 259
9.1.2 Theoretical restrictions . . . 262
9.1.3 Experimental examples . . . 265

9.2 Surface Polarimetry . . . 268
9.2.1 Principle of surface polarimetry . . . 268
9.2.2 Principle of measuring critical dimension . . . 269
9.2.3 Experimental results . . . 274
9.3 Phase-Resolved Heterodyne Microscopy . . . 276
9.3.1 Principle of heterodyne microscopy . . . 276
9.3.2 Optical scheme and instrumentation . . . 278
9.3.3 Qualitative theory . . . 280
9.3.4 Experimental results . . . 285
9.3.5 Super-resolution . . . 286
9.3.6 Dark-field and bright-field modes of operation . . . 288
9.3.7 Non-patterned anisotropic surfaces . . . 291
9.3.8 Three-dimensional profiling of opaque and transparent samples . . . 292
9.4 Optical Arbitrary Waveform Generator . . . 297
9.4.1 Problem statement . . . 297
9.4.2 The concept . . . 298
9.4.3 Variants of practical realization . . . 300

Index 307

Chapter 1

Basics of Grating Spectrometers

1.1. Typical Schemes

Most of spectroscopic needs in semiconductor industry can be met by compact grating spectrometers. Sealed in small brick-size boxes, these devices require nothing more than an optical fiber and universal serial bus (USB) cable to the computer (Fig. 1.1). Commonly, they arrive already calibrated at manufacturing site and do not need calibration. Calibration parameters are stored in operating program. Some versions allow changes of these parameters. Inside, there may be three different schemes explained in the text.

Typically, compact grating spectrometers cover spectral range 200–1000 nm with spectral resolution 1 nm and exposure time varying from 5 ms to 5 s. Exposure time — the integration time of the photodetector — is the only parameter that can be altered in order to adjust necessary sensitivity. Sensitivity greatly depends on the width of the slit. Basically, the optical scheme of the spectrometer projects the image of the slit into the spectral plane that coincides with the photodetector. Therefore, the narrower the slit the better spectral resolution. Thus, sensitivity and spectral resolution are the competing factors. Practically, to achieve spectral resolution of about 0.5 nm, the slit must be around 5 μm, and this proportion is roughly a constant for compact wide-range spectrometers. For instance, if your spectrometer shows spectral resolution 1.5 nm, then its slit is about 20 μm.

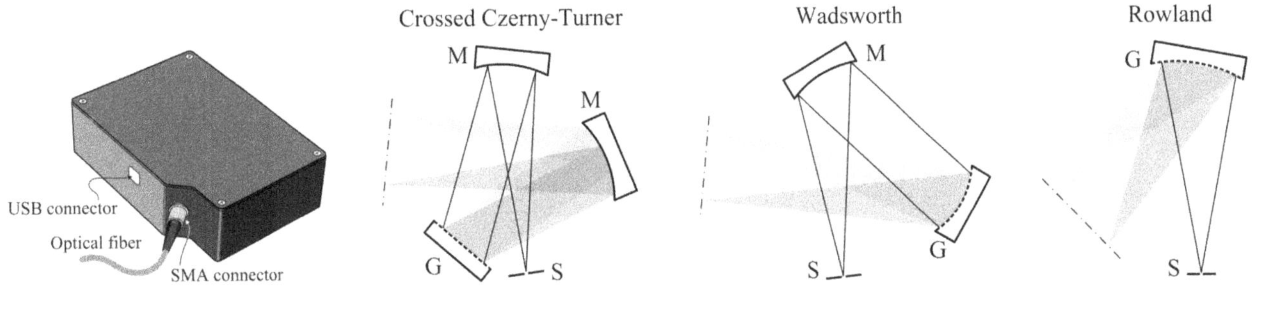

Fig. 1.1. Compact fiber-optic grating spectrometers are ready for use without any additional adjustment. "G" stands for the grating (flat or concave), "M" — for the mirror; "S" — for the slit. Spectrum plane is shown in dashed-dotted line.

1.2. Diffraction Grating Formula

Blazed diffraction grating — flat or curved — is the key part of the spectrometer. The term blazed means that each and everyone groove of the grating is tilted by a certain angle α, which concentrates the diffracted light selectively at a certain angle of reflection (Fig. 1.2). Since, in the spectrometer, each angle of reflection corresponds to a certain wavelength, this feature makes it possible to concentrate energy at a certain part of the spectrum, thus compensating for possible deficiencies of spectral response of the photodetector or reflectivity of the mirrors. Nowadays, blazed gratings are manufactured by deep lithography and subsequent slanted ion etching.

Let $V(\psi, \phi, \varepsilon)$ be angular distribution of the wave with the wavelength λ diffracted on the slanted (blazed) part s of the grating period t. Then the complex amplitude E of the wave diffracted on the entire grating is the sum of $V(\psi, \phi, \varepsilon)$ over all the N grooves taken with respective phases:

$$E = V(\psi, \phi, \varepsilon) \sum_{n=0}^{N-1} e^{-in\delta},$$

with $i = \sqrt{-1}$ and

$$\delta = \frac{2\pi}{\lambda}(\Delta_1 - \Delta_2) = \frac{2\pi}{\lambda}(t \sin \phi - t \sin \psi).$$

We assume that short connecting parts of the profile between adjacent blazed surfaces do not contribute to the wave. Geometrical series sums to

$$1 + q + q^2 + \cdots + q^m = \frac{1 - q^{m+1}}{1 - q},$$

giving

$$E = V(\psi, \phi, \varepsilon)\frac{1 - e^{-iN\delta}}{1 - e^{-i\delta}},$$

and intensity $I = |E|^2$

$$I = |V(\psi, \phi, \varepsilon)|^2 \frac{\sin^2 Nv}{\sin^2 v}, \quad v = \frac{\pi}{\lambda}t(\sin \phi - \sin \psi).$$

The function

$$\frac{\sin^2 Nv}{\sin^2 v}$$

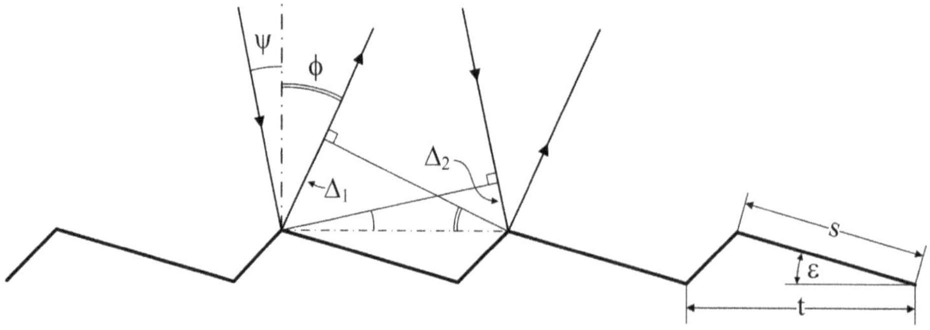

Fig. 1.2. Two processes determine angular profile of the reflected beam: diffraction on the blazed section s and interference of waves spaced periodically by t.

is periodic with narrow peaks of the width $2\pi/N$ as shown in Fig. 1.3. Each maximum corresponds to

$$v_m = \mp m\pi, \quad m = 1, 2, 3, \ldots,$$

which leads to the basic diffraction grating equation:

$$\sin\phi - \sin\psi = \mp m\frac{\lambda}{t}.$$

The values of m are called the orders of diffraction. Thus, the angles ϕ_m, at which the intensity is concentrated, are wavelength-dependent, and this is how diffraction grating decomposes light into its spectral components. Note that upon the convention orders with $\phi < \psi$ are called the positive orders, whereas the negative ones are those with $\phi > \psi$. Positive orders are commonly used.

Angular distribution $V(\psi, \phi, \varepsilon)$ is proportional to the wave-propagation integral over s, which reduces to one dimension because the grating is a one-dimensional object. Omitting insignificant proportionality coefficient, we may write:

$$V(\psi, \phi, \varepsilon) = \int_0^s E(x)e^{-i\frac{2\pi}{\lambda}r(x)}dx,$$

where x is the coordinate along s and r is the distance to the point of observation (Fig. 1.4). In the plane of the reflecting part of the grating, the optical wave is

$$E(x) = Ae^{-i\frac{2\pi}{\lambda}x\sin(\psi+\varepsilon)}.$$

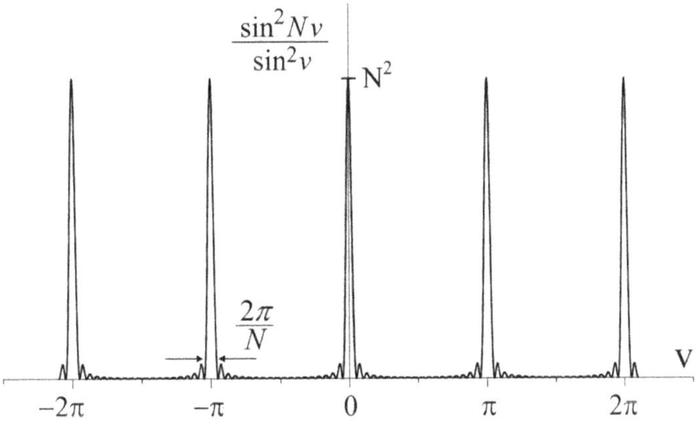

Fig. 1.3. Interference pattern of the diffraction grating.

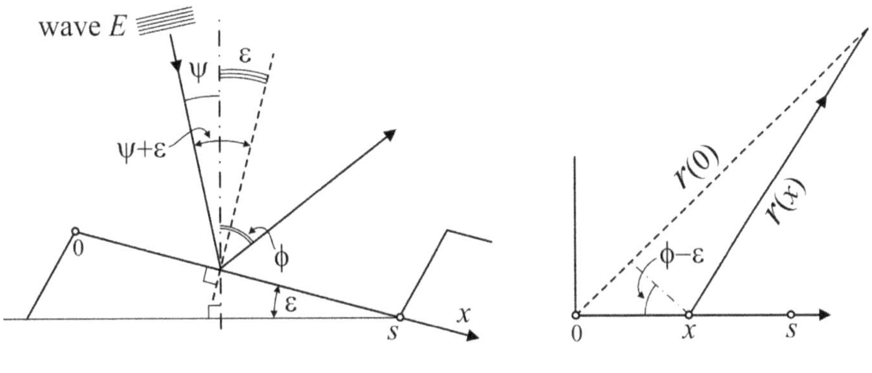

Fig. 1.4. Geometrical considerations for integration.

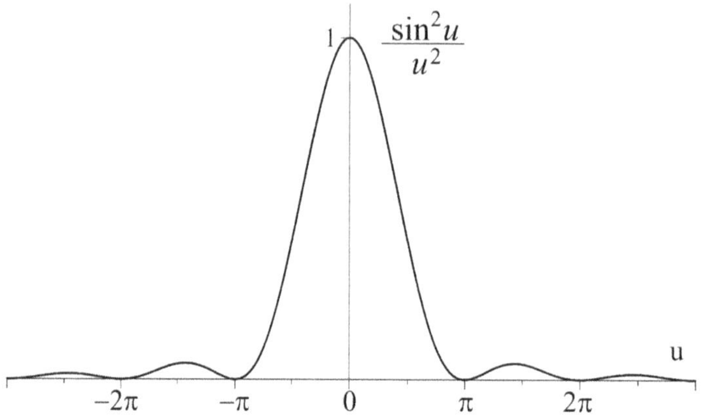

Fig. 1.5. Diffraction pattern of the diffraction grating.

In Fraunhofer diffraction, when only plane waves are considered, the observation point is placed at infinity, making $r(x) = r(0) - x\sin(\phi - \varepsilon)$. Then, disregarding constant phase introduced by $r(0)$ and omitting trivial constant amplitude A of the optical wave, the integral transforms to

$$\int_0^s e^{i\frac{2\pi}{\lambda}x[\sin(\phi-\varepsilon)-\sin(\psi+\varepsilon)]}dx = se^{iu}\left(\frac{\sin u}{u}\right), \quad u = \frac{\pi s}{\lambda}[\sin(\phi-\varepsilon) - \sin(\psi+\varepsilon)],$$

and

$$|V(\psi,\phi,\varepsilon)|^2 = s^2\frac{\sin^2 u}{u^2}.$$

This function maximizes at $u = 0$ and is shown in Fig. 1.5.

With obvious trigonometric manipulations,

$$u = \frac{\pi s}{\lambda}2\sin\frac{\phi-\psi-2\varepsilon}{2}\cdot\cos\frac{\phi+\psi}{2}.$$

To better feel the physical meaning of this result, consider the particular case of $\phi \approx \psi$ and $\varepsilon = 0$–reflection on a narrow stripe of the width s:

$$u \approx \frac{\pi s\cdot\cos\psi}{\lambda}(\phi-\psi).$$

Then, the function $|V|^2$ has the central maximum around $\phi \approx \psi$ (meaning that the angle of incidence is equal to the angle of reflection) with the angular width at half-maximum

$$\approx \frac{\lambda}{s\cdot\cos\psi}.$$

It is the same as in diffraction on a slit of the width $s\cdot\cos\psi$ — projection of the stripe width s on the wavefront of the incident wave.

Returning to the intensity reflected from the blazed grating, we see that it is proportional to the product of the two functions

$$I(\phi) = \frac{\sin^2 u}{u^2}\cdot\frac{\sin^2 Nv}{\sin^2 v}, \quad v = \frac{\pi}{\lambda}t(\sin\phi - \sin\psi), \quad u = \frac{\pi s}{\lambda}2\sin\frac{\phi-\psi-2\varepsilon}{2}\cdot\cos\frac{\phi+\psi}{2}.$$

The overlapping pattern is shown schematically in Fig. 1.6.

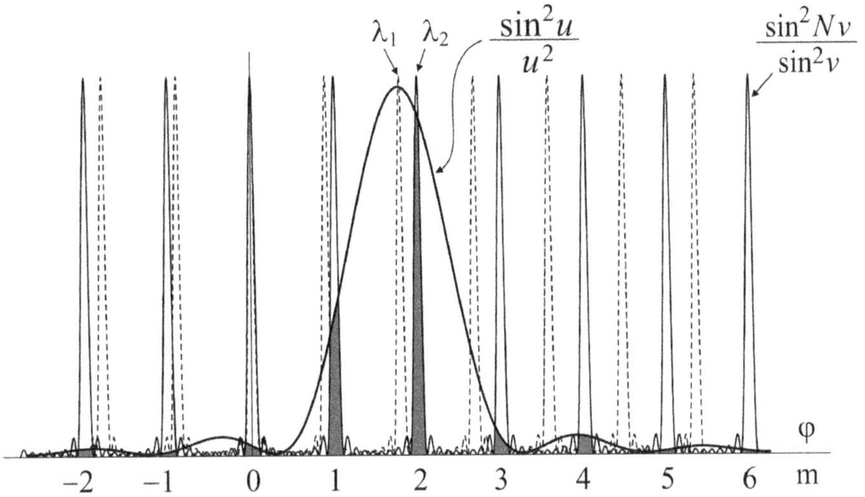

Fig. 1.6. Blazed grating may be designed to support any particular spectral component, for example, wavelength λ_1 in the second order (dashed line). Then other components will be attenuated, like λ_2 in the same order (solid line), or strongly suppressed like other orders.

Thus, choosing the blaze angle ε, it is possible to direct optical power not only to a particular diffraction order of interest, but also to location of the particular spectral component in this order. For that, it is necessary to satisfy the condition $u = 0$ at the specified diffraction angle ψ, which means $\phi - \psi = 2\varepsilon$.

1.3. Spectral Resolution

The most important parameter of a flat diffraction grating is its angular dispersion D_ϕ

$$D_\phi = \frac{d\phi}{d\lambda}.$$

It determines the angle by which the grating separates close wavelengths. Differentiating the equation $v_m = \pm m\pi$ over λ and understanding that $\psi = const$, we obtain

$$D_\phi = \frac{m}{t\cos\phi}.$$

Thus, the smaller grating spacing t is, the bigger angular dispersion is. In catalogs and specifications on gratings, instead of direct values of t another equivalent parameter is used: number of grooves per millimeter, which ranges typically from 100 to 1200 grooves per millimeter. With it, angular dispersion can be easily calculated. Another important conclusion that follows from this formula is that angular dispersion increases with the diffraction order m. Therefore, when very high spectral resolution is needed, researches work in high diffraction orders. However, for general purpose spectrometers even more important parameter is the wavelength range $\Delta\lambda$. This wavelength span, when focused onto photodetector, must fit the length L of its sensitive area that is typically around 30 mm (Fig. 1.7):

$$D_\phi \Delta\lambda \cdot F \approx L \sin\Gamma, \quad \text{or} \quad L \approx \frac{D_\phi F}{\sin\Gamma}\Delta\lambda.$$

The size of the photodetector L and the mirror focal length F are essential for choosing dispersion of the grating. For calculations, focal length F may be taken equal to half of the radius R of the mirror. In this configuration, $\Gamma < 90$o. Red wing of the spectrum goes above the violet one. The dimensionless

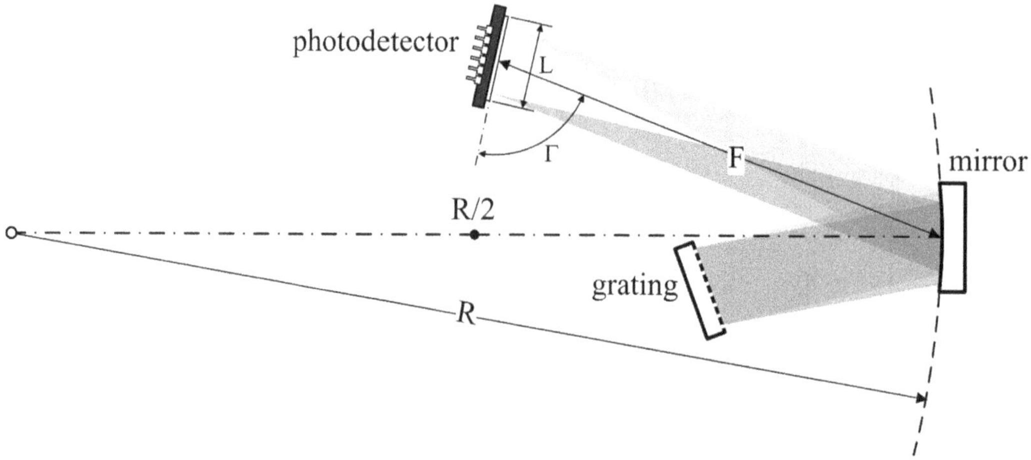

Fig. 1.7. Geometrical considerations in a compact spectrometer.

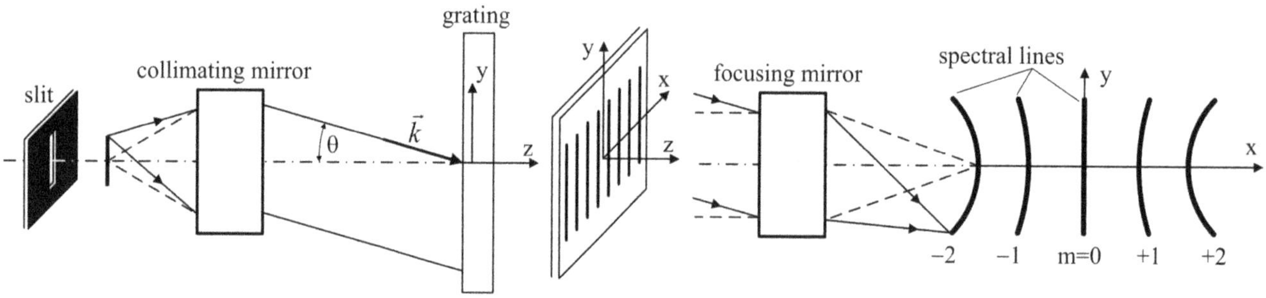

Fig. 1.8. Finite-length slit produces tilted waves, coming to the grating (at left). In the photodetector plane, the images will be curved (at right).

quantity

$$D_l \equiv \frac{L}{\Delta\lambda} = \frac{D_\phi F}{\sin\Gamma}$$

is named linear dispersion, and is usually measured in [mm/nm]. Sometimes, the reciprocal value is used, measured in [nm/mm]. Usually, the acute angle Γ is close to 90°, and the sine in denominator may be considered as unity.

With all the aforementioned, it is easy to choose particular grating needed for particular application. For example, we need to cover wavelength range from 200 nm to 800 nm, i.e. $\Delta\lambda = 600$ nm. Dimensions of the spectrometer give the first estimate for F: typically it is about 80 mm. Photodetector length L is fixed even more precisely: 30 mm is almost a standard. So, linear dispersion of the grating should be 30 mm/600 nm = 0.05 mm/nm, and angular dispersion 0.05/80 = 620 mm^{-1}. Assuming diffraction angle $\phi \approx 30°$ and $\cos\phi = 0.866$, we find that the grating of our choice should have around $620 \times 0.866 \approx 500$ grooves per millimeter.

It would be a mistake to think that the curved mirror focuses diffraction orders exactly into point-like spots located along a straight line — the photodetector plane. Not at all: curved mirror is an aberrated imaging element, especially at off-axial illumination as shown in Fig. 1.7, which may be called geometrical distortion, and diffraction grating produces additional distortion of purely spectral nature. The nature of the latter can be better understood with the scheme in Fig. 1.8.

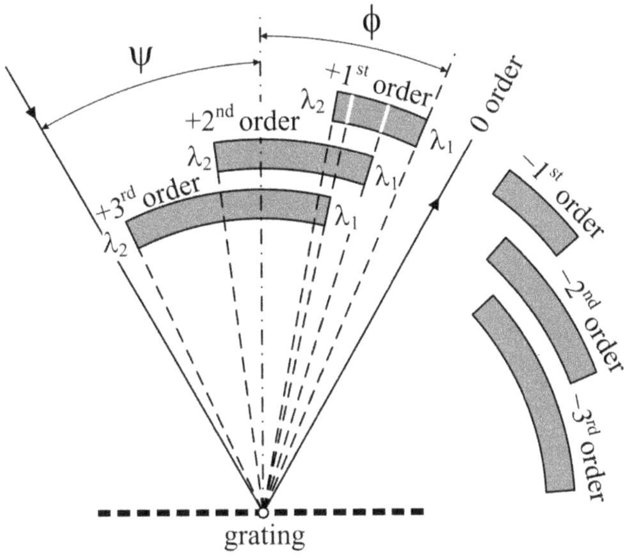

Fig. 1.9. The scheme of diffraction orders in a grating spectrometer.

The ends of the slit produce plane waves, coming at finite angle θ to the grating. It means that normal component of the wave vector $\vec{k}$ decreases to

$$k_z = \frac{2\pi}{\lambda}\cos\theta = \frac{2\pi}{\lambda'}$$

with $\lambda' > \lambda$. As such, these waves will diffract at larger angles ϕ, and will be focused farer from the zeroth-order point. This leads to curviness of spectral lines in the plane of the photodetector as shown in Fig. 1.8. The longer the slit, the more it affects spectral resolution. However, for compact fiber-optic spectrometers with the height of one sensitive element of the photodetector of only 200 μm, this phenomenon is not critical.

1.4. Spectral Orders

What is really critical and causes numerous mistakes, is the overlapping of diffraction orders. The basic diffraction grating equation

$$\sin\phi - \sin\psi = \mp m\frac{\lambda}{t}$$

may be presented schematically in the form shown in Fig. 1.9. In wide-range spectrometers, multiple diffraction orders may angularly overlap. The photodetector size L, angular dispersion of the grating D_ϕ, and focal length of the mirror F (Fig. 1.7) are designed to intercept the first order in between the wavelengths λ_1 and λ_2. Multiple orders may also be focused here, as indicated by white marks on the first-order section.

For small $\phi < 30°$, $\sin\phi \approx \phi$, and approximately for positive orders

$$\phi \approx \psi - m\frac{\lambda}{t}.$$

Thus, when $2\lambda_1 \le \lambda_2$, the second order meddles into the interval of the first-order diffraction. If $3\lambda_1 \le \lambda_2$ then the third order appears in this interval, and so on. For instance, the spectrometer designed to work between 200 nm and 650 nm would show the strongest 253 nm mercury line not only at 253 nm but also at 506 nm, making it impossible to identify real spectrum. Or, in case of a

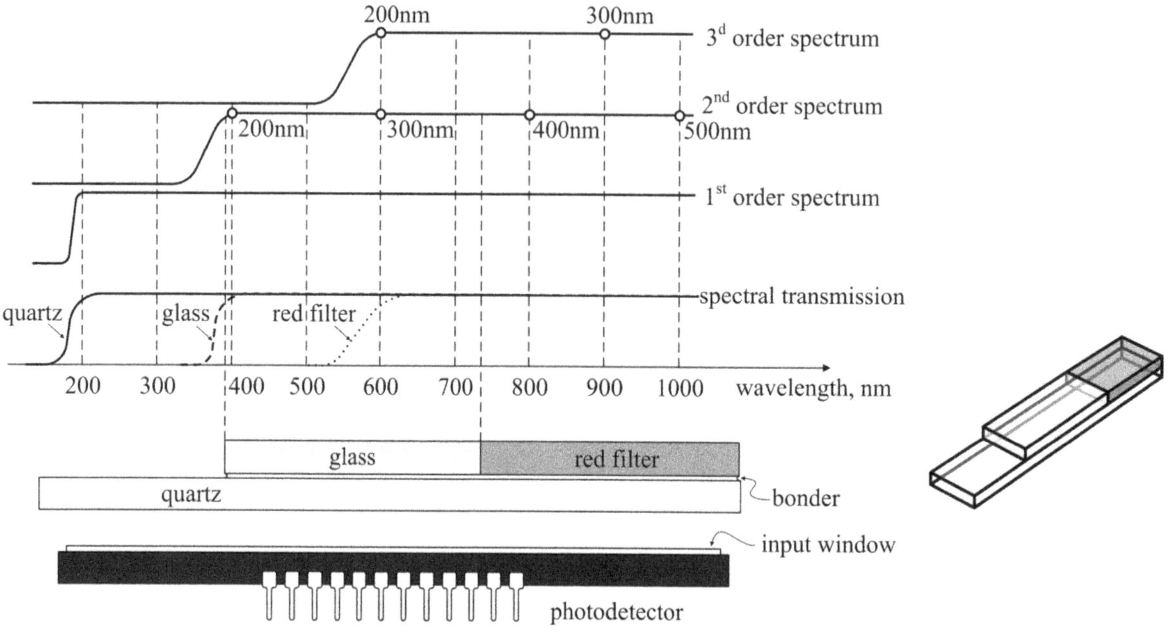

Fig. 1.10. An example of an order-sorting filter for spectral interval 200–1000 nm.

wide-range source with strong ultra-violet (UV) emission, like xenon lamp, the entire UV wing around 300 nm will relocate to visible part of the spectrum around 600 nm, confusing everything. The majority of compact spectrometers, being in use in laboratories, cover very wide spectral interval between 200 nm and 800 nm or even 1100 nm. Then what do they do to eradicate higher-orders intrusion? The solution is the so-called order-sorting filter.

1.5. Order-Sorting Filters

There are two types of order-sorting filters, being used in compact fiber-optic spectrometers: the detachable glass plates and thin-film coating of the photodetector input window. Principle of functionality is the same, but the first type is more versatile. It consists of two, maximum three, glass plates, one on top of another, geometrically and spectrally designed to block multiple orders from reaching the photodetector (Fig. 1.10). Two materials — glass and red filter — sequentially block the second and third orders from reaching the photodetector. Glass blocks all the spectral components below 400 nm, and the red filter does the same below 600 nm. The glass section of a proper length absorbs UV radiation in the interval 200–350 nm that might be able to reach specific area of the photodetector to which all the visible components from 400 nm to 700 nm are focused. The red section shields the photodetector area assigned for 700–1000 nm, blocking both the UV radiation, infiltrating with the third order, and the visible light, coming through the second order. Quartz plate does not block anything but serves as a support for the two other sections. The input spectrum never contains components below 200 nm as they are fully attenuated by a feeding optical fiber.

The thin-film order-sorting filter is more complicated and expensive technique. It requires deposition of thin optical layers in vacuum, mathematical design of multilayer coating, and special technology of depositing layers of spatially varying thickness. A photodetector matrix purchased from the vendor cannot be immediately installed in the spectrometer without depositing the filtering coating on its input window. The best way to realize how such filter works is to measure its spatially-spectral characteristics. From Fig. 1.7 it is clear that, once being installed in the spectrometer, a photodetector is spatially fixed relative to the spectrum of light: each and everyone pixel of the photodetector is connected to only one specific wavelength, and not to another. For example, if the photodetector of 512 pixels must measure the spectrum from 200 nm to 800 nm, then pixel number 1 must always receive light at 200 nm, and pixel number 512 — light at 800 nm. However, due to multiple orders of the diffraction grating, the pixel designed for 400 nm light will receive the light at 200 nm, created by the second harmonic of 200 nm spectral component, the pixel created for 800 nm will receive light at 200 nm from the fourth harmonic of 200 nm and 400 nm light from the second harmonic of the diffraction grating, and so on. The thin-film order-sorting filter must be deposited on the photodetector in such a way that to block unwanted spectral components from reaching the sensitive line of pixels. In other words, each longitudinal point on the window of the photodetector must have its specific spectral transparency. This is the spatially-spectral characteristic of the optical coating.

Measuring spatially-spectral characteristic of optical coating is not a simple task for several reasons. Suppose we opened the spectrometer and took the photodetector from it. The first problem is what to measure: spectral reflection $R(\lambda)$ or spectral transmission $T(\lambda)$? If we assume no reflection from the sensor itself, then $R(\lambda) = 1 - T(\lambda)$ and it looks like there is no difference: measure one parameter and compute another. If one decides to measure reflectivity $R(\lambda)$, then a new problem arises: how to arrange interception of reflected light and block the background? Moreover, the sensing line of pixels also reflects, although very insignificantly. Additionally, the surrounding electrical and protective circuits also reflect, and even much stronger. All that would make measurements unreliable. Therefore, more reliable results can be obtained, measuring spectral transmission $T(\lambda)$. For that, there are two ways: tear the optical window off the photodetector, or use the sensor itself as a photo-receiver. Obviously, the second way is preferable, taking into consideration that all electronic circuitry is already at our hands — from the spectrometer itself.

The second problem, which floats up immediately, is extreme sensitivity of the sensor: it saturates even from very weak optical fluxes, and therefore special well-designed light-protecting chamber is needed. Next, it is necessary to focus monochromatic light of variable wavelength onto the line of pixels and scan it along the photodetector. Since the entire opto-mechanical arrangement must be placed into a closed chamber, the scanning may be controlled only distantly, using motorized mechanical scanning stages, which is also a problem: standard motorized scanning stages include end-switches that use opto-pairs. These opto-pairs emit light, typically in near infrared domain, which is easily sensed by the photodetector, whose maximum spectral sensitivity commonly lies exactly in the same spectral interval. Therefore, these end-switches must be disabled. And it is not the last problem: any focusing optics is spectrally-dependent, which requires additional motion for refocusing at various wavelengths.

The concept of the measuring opto-mechanical installation is outlined in Fig. 1.11, omitting the light-protecting chamber.

For the beginning, the pixel line of the photodetector must be aligned along the scanning direction. This is done by rotational aligner in a series of scans. Next, the focal spot of light, delivered to the photodetector by optical fiber, must be positioned in the middle of the pixel light with the help of manual horizontal aligner. These two alignments may be done manually because they will not be changed during measurements and opening of the light-protecting chamber will not be needed. The vertical aligner, which sets the focusing distance, must be controlled remotely because the focusing depends on the wavelength that is used for measurements. Therefore, it is assembled on motorized vertical aligner.

The optical scheme of measurements is clarified in Fig. 1.12. It uses the argon-mercury calibration source (Section 1.7) in combination with the compact monochromator (Section 1.12). Wavelengths of narrow spectral lines of the light source are not the multiples of each other, therefore the monochromator may be used without appearing of spectral ghosts. In order to obtain fine light spot on the photodetector

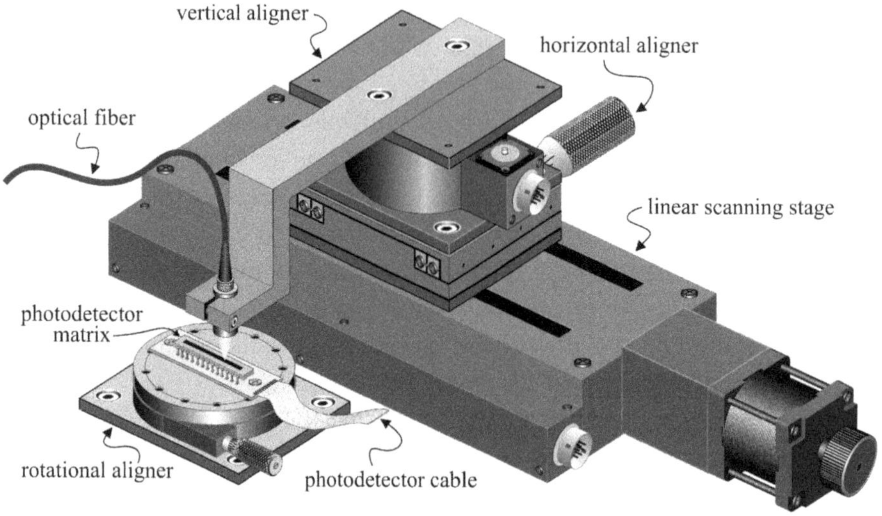

Fig. 1.11. Opto-mechanical arrangement for measuring spectral transmission of order-sorting filters.

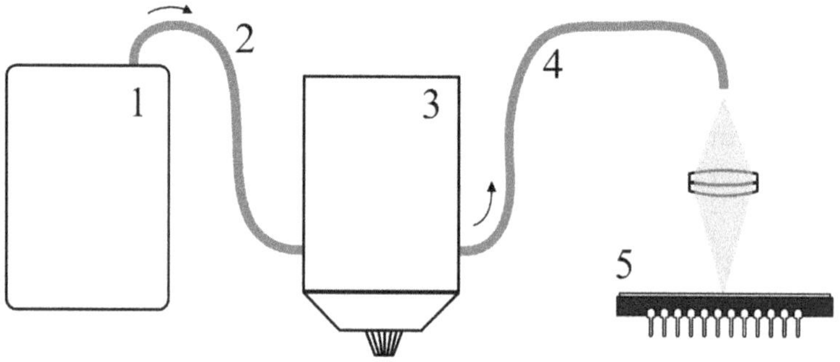

Fig. 1.12. Optical scheme of measurements. 1 — calibration source; 2,4 — optical fibers; 3 — variable monochromator; 5 — photodetector.

fully confined within the pixel line, the diameter of the core of the optical fiber must be small, about 50 μm.

To activate the photodetector during measurements, the software supplied with the spectrometer must be used. Then the electrical signals from pixels are given not as a function of the pixel ordinal number but as a function of wavelength — a very useful feature for creating spatial-spectral characteristic. When the focused light spot of a chosen wavelength scans along the photodetector, quantum efficiency of pixels remains the same, and the electrical output depends only on the transmission of the order-sorting filter at the chosen wavelength. A set of typical curves captured during scanning at different times (different positions of the focused spot on the photodetector) are shown in Fig. 1.13. The amplitudes of the peaks, signifying transmission of the order-sorting filter, were saved as the function of the wavelength, to which the specific pixel is allocated.

Since the pixels are marked in the values of wavelengths that they are adjusted to, it is very convenient to compare actual spectral transmission of the order-sorting filter to the one required for proper performance of the spectrometer. The set of spatial-spectral characteristics for several strongest optical lines of the argon–mercury calibration source, measured on one particular photodetector, is plotted in Fig. 1.14.

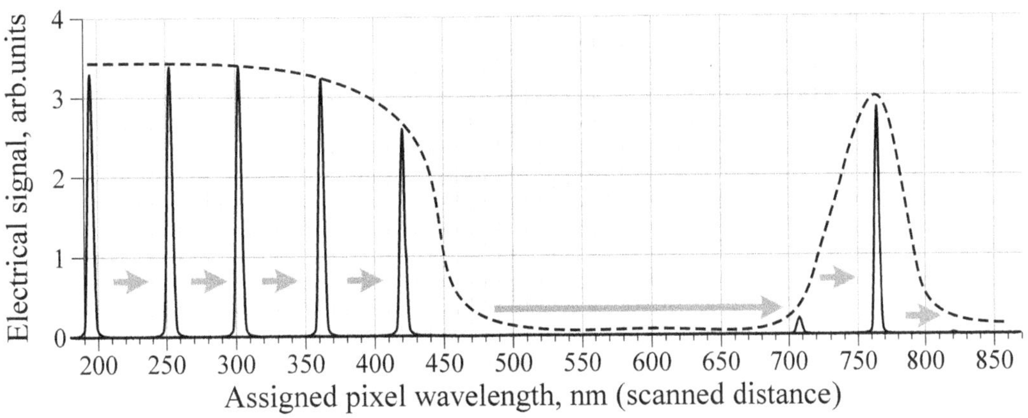

Fig. 1.13. Consecutive electrical response of the photodetector to moving focused light at the wavelength 313 nm. The arrows show direction of scanning. The dashed line shoes tendency of the signal.

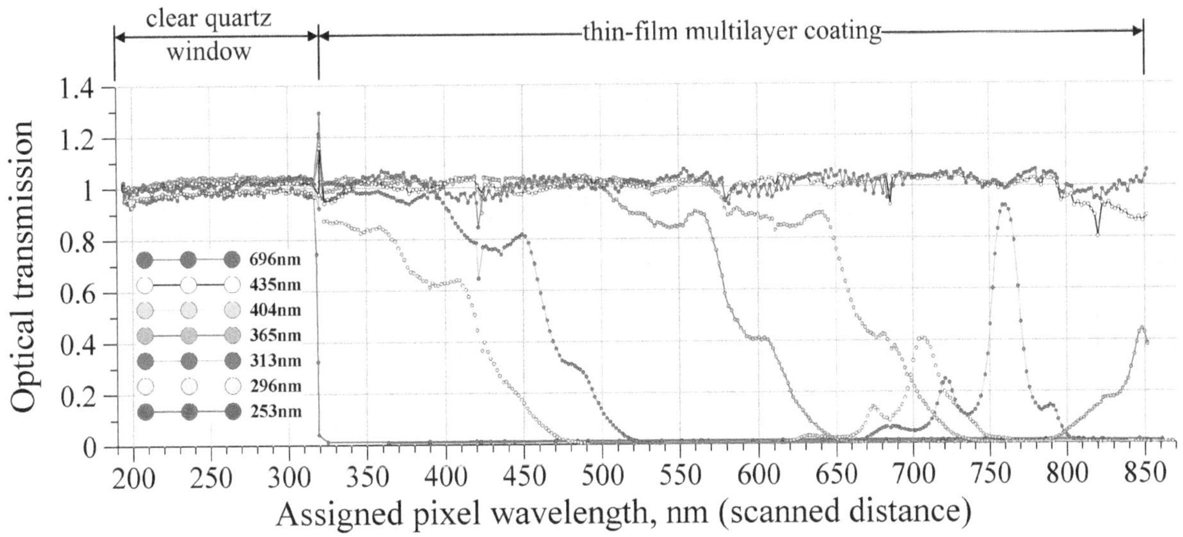

Fig. 1.14. Measured spatial-spectral characteristics of a thin-film order-sorting filter.

The thin-film structure is designed so as to block all possible second and third orders of diffraction within the spectral interval 190–870 nm. For example, the second order of the wavelength 296 nm, which might occur at the position of the wavelength 592 nm, is blocked by zero transmission at this pixel of the photodetector. The enhanced transmission of the filter at 296 nm at the pixels associated with the wavelengths around 700 nm plays no role because no diffraction orders of 296 nm can appear here. The data of the Fig. 1.14 may be presented more explicitly in three-dimensional form, showing the spectral transmission as a function of the pixel number (Fig. 1.15).

Order-sorting filter is a very important point of negotiations when ordering a new spectrometer. Manufacturers offer great variety of diffraction gratings for their spectrometers, listing all the necessary details in their catalogs, but never mention the necessity of the order-sorting filters. This commonly leads to a very pitiful situation when the user calculates parameters of the grating, satisfying his needs in spectral range and blazed wavelength, and then just orders the spectrometer with this grating without even mentioning the order-sorting filter. The aftermaths are dramatic: total confusion of spectral data and the vendor does not accept any responsibility because the filter was not formally ordered.

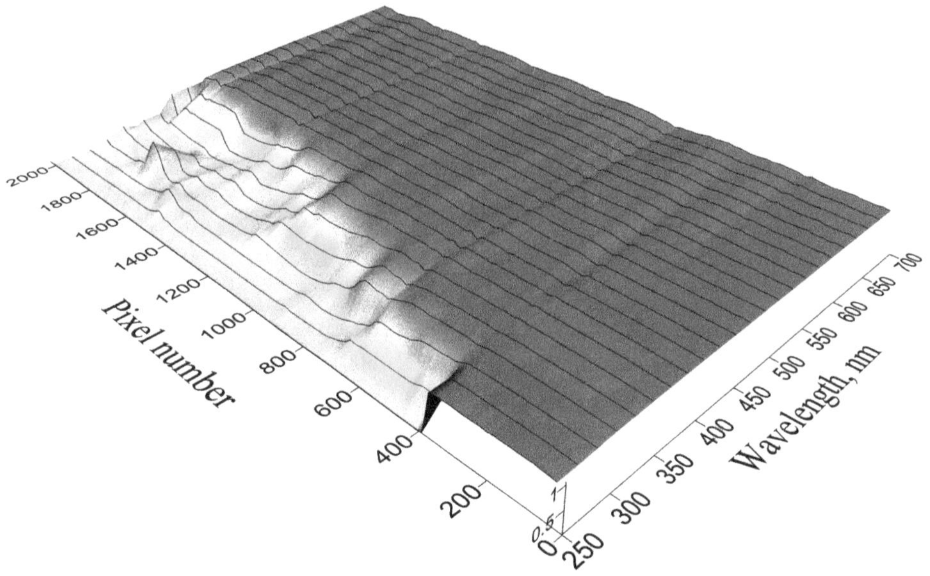

Fig. 1.15. Spatial-spectral characteristics of a thin-film order-sorting filter in three-dimensional presentation.

We discussed the flat type of diffraction gratings — the most easily manufactured one, and the crossed Czerny–Turner scheme shown in Fig. 1.1 is commonly used with them in compact spectrometers. This scheme is named "crossed" because another Czerny–Turner scheme exists, in which optical beams do not intersect. However, the uncrossed Czerny–Turner scheme requires more space, and therefore is not used in compact spectrometers.

1.6. The Rowland Circle

Requirements of small space and least possible number of optical components — the most essential for mass production — dictate usage of curved diffraction gratings that perform two functions simultaneously: dispersion and focusing. It is even more important because reduction of the number of optical components decreases scattered light and aberrations. The basic scheme of that kind is known as the Rowland circle (Fig. 1.16).

Consider a source of monochromatic rays with wavelength λ at the point S. Let the A and B be the two adjacent grooves on the grating and O — the center of the grating curvature with the radius R. The rays SA and SB make the angles ψ and $\psi + \Delta\psi$ with the local normals to the grating (lines OA and OB). Diffracted rays AF and BF come to the point of focusing F at the angles ϕ and $\phi + \Delta\phi$ to the local normals OA and OB. Constructive interference of these rays takes place when

$$(SA + AF) - (SB + BF) = m\lambda, \quad m = 0, \pm 1, \pm 2, \ldots.$$

Continue the sections SB and BF to the points C and D to make their lengths equal to SA and AF. Then

$$(SA + AF) - (SB + BF) = BC + BD.$$

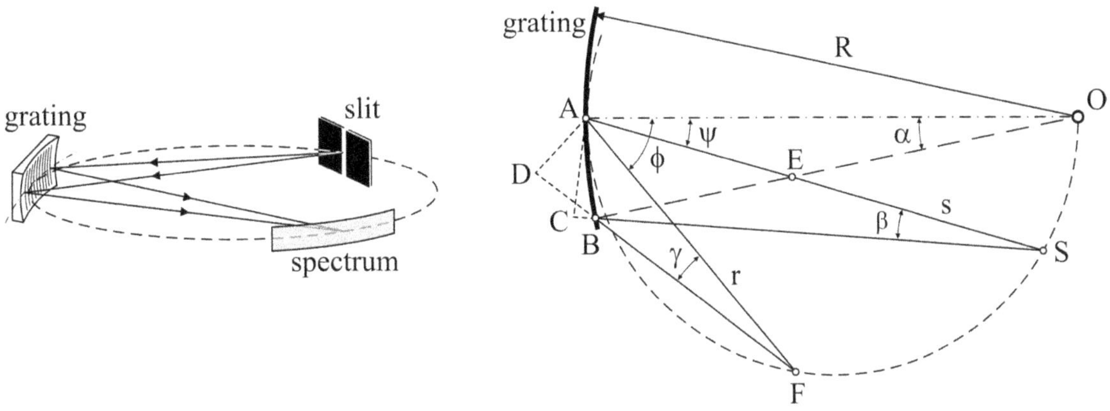

Fig. 1.16. In the Rowland scheme, the entrance slit and the spectrum are positioned on the circle (dashed line), touching the diffraction grating (at left). Its diameter is equal to the radius R of the grating. On this circle, every point S is imaged into another point F — dispersed or not dispersed (at right).

Since the single groove width $t \ll R$, the angles $\beta \ll 1$ and $\gamma \ll 1$, which makes $\angle ACB \approx \pi/2$, $\angle ADB \approx \pi/2$ and $\angle CAB \approx \psi$, $\angle DAB \approx \phi$. As such, $BC \approx t \cdot \sin\psi$ and $BD \approx t \cdot \sin\phi$. Then the basic grating equation transforms to

$$t(\sin\phi + \sin\psi) = m\lambda, \quad m = 0, \pm1, \pm2, \ldots.$$

It is the same equation as for the flat grating, taking into consideration that, according to the previous convention, the angle ϕ is negative (cf. Figs. 1.2 and 1.9).

Next, we are going to show that the Rowland scheme possesses the quality of focusing, i.e. all the rays from S with the same wavelength λ and slightly different angles of incidence $\psi + d\psi$ come to the same point P within the same diffraction order m. If the angle of incidence ψ changes, then the angle of diffraction ϕ also changes to $\phi + d\phi$, according to the basic grating equation. To find $d\phi$ as a function of $d\psi$, we have to take differential of the above equation:

$$\cos\phi \cdot d\phi + \cos\psi \cdot d\psi = 0.$$

We do not need to draw new scheme to proceed with this relation — just look at the Fig. 1.16 and assume that now point B is separated from A by $N \gg 1$ grooves

$$AB = N \cdot t$$

and $\angle OBS = \psi + d\psi$. Consider then triangles AEO and BES. They have the common angle at E, therefore

$$\psi + \alpha = \psi + d\psi + \beta, \text{ or } d\psi = \alpha - \beta.$$

With the same considerations, $d\phi = \alpha - \gamma$. On the other hand,

$$\alpha = \frac{Nt}{R}, \quad AD = Nt \cdot \cos\phi, \quad AC = Nt \cdot \cos\psi, \quad \beta = \frac{AC}{s}; \quad \gamma = \frac{AD}{r}.$$

Altogether gives the differential

$$\cos\phi \cdot \left(\frac{1}{R} - \frac{1}{r}\cos\phi\right) + \cos\psi \cdot \left(\frac{1}{R} - \frac{1}{s}\cos\psi\right) = 0.$$

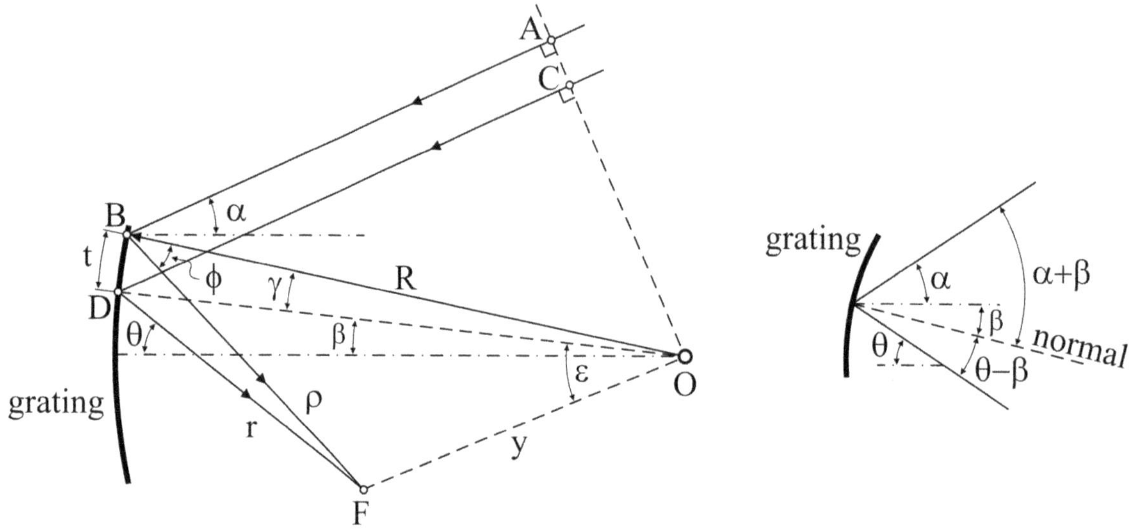

Fig. 1.17. In the Wadsworth scheme, curved diffraction grating is illuminated by a parallel beam of rays, represented by AB and CD.

For this equation to be satisfied for an arbitrary ψ and ϕ, the following result must hold true:

$$r = R\cos\phi \quad \text{and} \quad s = R\cos\psi.$$

This is the equation of a circle with the diameter R, which proves focusing properties of the Rowland circle. It may be useful to understand that not only spectrally dispersed components with $m = \pm 1, \pm 2, \cdots$ are focused on this circle but also zeroth-order wave $m = 0$, which means focusing property of an ordinary spherical mirror, not necessarily the diffraction grating.

The analysis above considers only meridional rays, i.e. the rays in the plane perpendicular to both grating and grooves. The skew rays or sagittal rays, i.e. the rays that propagate at some angle to this plane, do not focus in the same point as the meridional rays. This produces the aberration called astigmatism. Astigmatism of the Rowland scheme is stronger, the bigger the numerical aperture of the grating is. The Wadsworth scheme (Figs. 1.17 and 1.1) produces much smaller astigmatism. Points B and D on the grating are spaced by one groove step t. Then the dispersed rays are focused in F. Point O is the center of grating curvature with the radius R.

In it, the grating is illuminated by a parallel beam of rays. As before, constructive interference of these rays takes place when

$$(CD + r) - (AB + \rho) = m\,\lambda, \quad m = 0, \pm 1, \pm 2, \ldots.$$

Applying standard trigonometric identities, we have

$$CD - AB = t \cdot \sin(\alpha + \beta).$$

The next goal is to determine r and ρ. For that, solve the triangle ODF to find

$$y = \sqrt{R^2 - 2Rr\cos(\theta - \beta) + r^2}, \quad \sin\varepsilon = \frac{r\sin(\theta - \beta)}{\sqrt{R^2 - 2Rr\cos(\theta - \beta) + r^2}}.$$

Similarly, solve the triangle OBF to find

$$\rho = \sqrt{R^2 - 2Ry\cos(\varepsilon + \gamma) + y^2}.$$

Since $t \ll R$,

$$\cos(\varepsilon + \gamma) \approx \cos\varepsilon - \frac{t}{R}\sin\varepsilon.$$

The next sequence of calculations includes the following:

$\rho - r \approx \frac{yt}{\sqrt{p}}\sin\varepsilon$ and $p = R^2 + y^2 - 2Ry\cos\varepsilon$ with the same condition $t \ll R$;

$$p = r^2.$$

With this, the initial equation transforms to

$$\sin(\alpha + \beta) - \sin(\theta - \beta) = \frac{m\lambda}{t},$$

which is again the basic diffraction grating equation as it follows from the right scheme in Fig. 1.17. The angle θ, to where diffracted light is focused, is the function of the wavelength λ.

Now, we are going to find the radius of focusing r and, applying the same formalism as before, assume that the point D is separated from B by a macroscopic space of $N \gg 1$ grooves: Nt. The differential of the above equation over the two independent variables $\theta + d\theta$ and $\beta + d\beta$ with

$$d\beta = \frac{Nt}{R}, \quad d\theta = \frac{Nt}{r}\cos(\theta - \beta)$$

gives

$$r = \frac{R\cos^2(\theta - \beta)}{\cos(\alpha + \beta) + \cos(\theta - \beta)}.$$

Since the radius of focusing r does not depend on rotation of the entire picture around the center of curvature O, we may rotate it until $\beta = 0$. In this configuration, the line OD coincides with horizontal axis, and the angle α becomes the angle of incidence ψ that we introduced in Fig. 1.2. Similarly, the angle θ becomes the angle of diffraction ϕ, and finally we obtain

$$r = \frac{R\cos^2\phi}{\cos\psi + \cos\phi}.$$

This is not a circle like in the Rowland case, and the line determined by this formula is shown in Fig. 1.18.

For practical reasons, it is interesting to realize the scale of curviness that the Wadsworth scheme provides. From Fig. 1.18 follows that its maximum corresponds to $\psi = 0$ and $\phi = 0$. In polar coordinates $r(\phi)$, radius of curvature v is determined by the following formula:

$$v = \frac{r^2(0)}{r(0) - r''(0)} = \frac{R}{5}.$$

Thus, the minimum radius of field curvature in the Wadsworth scheme is 2.5 times less than in the Rowland scheme, which is not a very good news. However, recently developed technology of varied line-space (VLS) gratings minimizes field curvature in the Wadsworth scheme. Well-defined variation of spacing between straight parallel grooves from one side of the grating to another flattens the surface to which the dispersed rays are focused. In our mathematical calculations above, it means that t varies smoothly over the grating. The zeroth-order focus has no relation to t, and therefore the curved mirror would focus on the curved surface as described above. But diffracted (dispersed) rays focus in nearly a plane.

Although fundamental for large-scale spectrometers, the Rowland scheme could not be used in compact versions for a long time because high numerical aperture of the grating produced big astigmatism

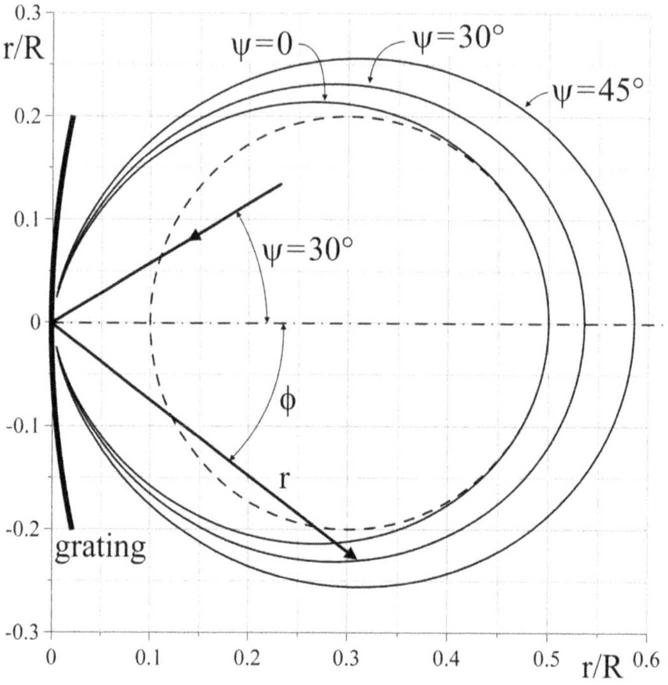

Fig. 1.18. In the Wadsworth scheme, the spectrum is focused to a more complicated curve rather than a circle. The dashed line shows the circle fitted to the most curved section in case of normal incidence $\psi = 0$.

and field curvature. The very compact Rowland scheme in Fig. 1.1 is the product of recent technological developments like independent surface profiling in meridional and sagittal planes and VLS: the first one minimizes astigmatism and the second flattens field of view. For such gratings, only linear dispersion makes sense and it is always carefully specified by manufacturer together with relative positions of slit and focal plane.

1.7. Calibration Techniques and Formulas

Raw spectral information is retrieved from the spectrometer as a one-dimensional array s_j of electrical signals detected in the jth pixel of a line detector array, commonly the charge-coupled device (CCD) photodetector. The number of pixels is typically 1024, 2048, or 3648, and a single-pixel geometry is a rectangle $a \times 200$ μm^2. With the length of such an array being always around 30±1 mm, the width a can be easily calculated. Next, the pixel number j must be associated with the wavelength λ, and for that the calibration procedure should be applied. As a rule, spectrometers are shipped already factory-calibrated, and no additional calibration is needed. However, with time or by any other reason, calibration may be disturbed, and then it should be restored. The best tool for that is an argon–mercury lamp, readily available from the same vendor that supplied the spectrometer itself (Fig. 1.19). Designed to work with the fiber-optic spectrometers, the output is coupled to a fiber by a standard SMA-905 connector. Normally, no special optical interface system is needed to couple the fiber to the lamp.

In it, mercury covers the UV and visible parts of spectrum, whereas argon fills the near-infrared interval (Fig. 1.20). Relative intensities of argon and mercury lines may drift significantly with time as chemical reactions develop inside the bulb. Nevertheless, this process does not affect stability of specific spectral lines of the both gases. Table 1.1 presents wavelengths of clearly seen spectral lines.

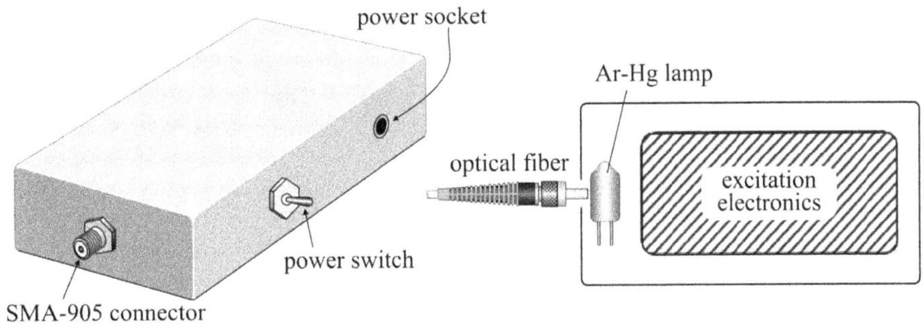

Fig. 1.19. The argon–mercury calibration module contains a miniature lamp and high-voltage excitation circuit powered by any low-voltage source of 12–24 V.

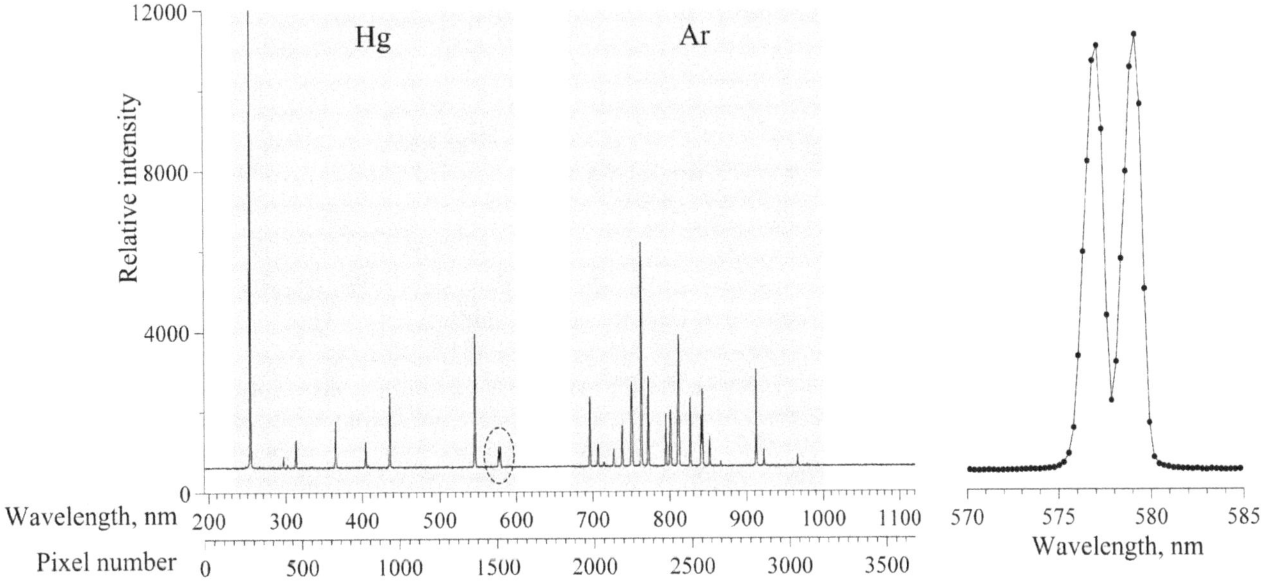

Fig. 1.20. Typical calibration spectrum of the Ar–Hg lamp. The doublet 576.960 nm and 579.066 nm confined within the dashed ellipse is magnified separately at right.

Table 1.1. Most frequently used argon and mercury spectral lines.

Hg	253.652, 296.728, 313.155, 365.015, 404.656, 435.833, 546.074, 576.960, 579.066
Ar	696.543, 706.722, 727.294, 738.398, 750.387, 763.511, 772.376, 794.818, 800.616, 811.531, 826.452, 842.144, 912.297, 922.450

Argon and mercury characteristic lines are exceptionally well calibrated and readily available in the literature. However, do not think that the knowledge of wavelengths alone is enough: the real practical problem, especially for the beginners, is to identify the designated spectral line in the spectrum, especially among the argon lines. These lines are narrowly spaced, and mixing the one for its neighbor is easy if not to take into consideration their relative amplitudes. Therefore, the entire picture of relative intensities like the one presented in Fig. 1.20, is very helpful. Because of possibly different quantum efficiencies of CCDs, blazing efficiencies of gratings, design of order-sorting filters, and other particular

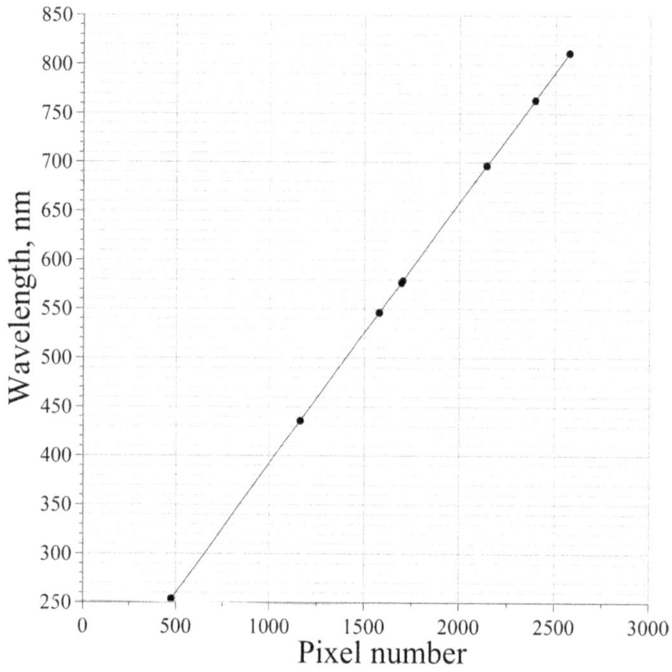

Fig. 1.21. Calibration curve for 100 mm focal length grating and 3648 pixels CCD photodetector. Dots mark the points of the table, the line is the least-square linear fitting. The mercury doublet 577–579 nm is also included in calibration table — these two dots almost coincide.

details each spectrometer has its own relative intensity distribution. That is why the picture shown in Fig. 1.20 should be considered only as a typical one, with significant variations quite possible.

After calibration spectrum is recorded and characteristic lines identified, the function $\lambda(j)$ — the wavelength λ in nanometers as the function of the pixel number j — should be defined. The question is how to know the pixel number, corresponding to specific wavelength. For that, the operating program supplied with the spectrometer always indicates pixel numbers along with the wavelength. It is not necessary to identify all the lines that are visible in the spectrum: 4–5 most distantly located points will suffice. The farer the two end points are located from one another, the better the precision of calibration will be. Therefore, commonly strong 253.652 nm mercury and 763.511 nm argon lines terminate the set of points. The resultant two-column table is used for calibration, which is commonly performed automatically by the operating program, using the calibration option. What this option actually does is the least-square fitting of a polynomial into the set of points of the calibration table.

With all the explanation above about focusing of the spectrum, the function $\lambda(j)$ should be expected nonlinear. And in general it is, although in relatively narrow spectral intervals and for long-focused spectrometers nonlinearity may be hardly noticeable. For example, Fig. 1.21 shows calibration points measured with the diffraction grating, having 100 mm focus.

Simple linear function fits exceptionally well, and if only it were known beforehand then only two points would suffice. However, linearity is never guaranteed, therefore at least three points are needed to fit the second-order polynomial.

Calibration source with argon–mercury lamp is also useful for testing spectral resolution of the spectrometer. Some regretful cases are known when even new spectrometers failed to show specified spectral resolution. The test is simple: turn on the lamp and record the spectrum around the mercury doublet 577–579 nm. Spacing between these lines is almost exactly 2 nm. According to the Rayleigh criterion, two spectral lines are considered resolved if they produce the gap in between of them more

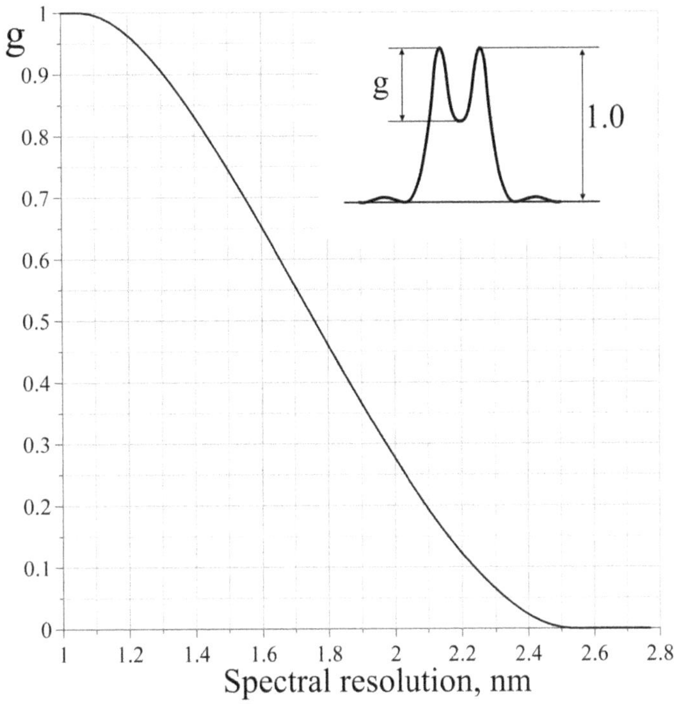

Fig. 1.22. Nomogram for numerical evaluation of spectral resolution on the mercury doublet 577–579 nm.

than 20% of the values in maxima. For example, the gap of the doublet shown in Fig. 1.20 is 80%, meaning that spectral resolution of this spectrometer is significantly better than 2 nm. This simple assessment technique may be converted from the domain of guessing to numerical evaluation if we recall that theoretical shape of a single spectral line with the wavelength λ_0 is defined by diffraction on the entrance slit:

$$\left[\frac{\sin a(\lambda - \lambda_0)}{a(\lambda - \lambda_0)}\right]^2,$$

where λ is the wavelength and a is the unknown scaling parameter. Then, computing the sum of two of these functions nested at λ_0 and $\lambda_0 + \Delta\lambda$ and evaluating the relative gap g in the middle as a function of $\Delta\lambda$, it is easy to find that the 20% Rayleigh criterion satisfies when

$$a = \frac{\pi}{\Delta\lambda}.$$

It means that the spectrometer with spectral resolution $\Delta\lambda$ forms the shape of a single spectral line λ_0 as

$$\left[\frac{\sin \frac{\pi}{\Delta\lambda}(\lambda - \lambda_0)}{\frac{\pi}{\Delta\lambda}(\lambda - \lambda_0)}\right]^2.$$

The next step is to take $\lambda_0 = 576.960$ nm for one line and $\lambda_0 = 579.066$ nm for another and compute g as a function of $\Delta\lambda$. The result in the form of a nomogram is presented in Fig. 1.22. It can be used to quantitatively estimate spectral resolution of spectrometers. For instance, spectral resolution of the spectrometer that was used to obtain the data in Fig. 1.20 may be estimated as 1.4 nm — a very good value for a wide-range compact spectrometer.

1.8. Photodetectors

There are two ways to sense the spectrum formed by diffraction gratings: either mechanical scanning, moving a single-channel photoreceiver along the direction of dispersion, or by electrical commutation inside multi-element photodetector that intercepts the entire length of spectrum. Nowadays, all industrial spectrometers use multi-element charge coupled device (CCD) photodetectors. The principle of these devices is conversion of light into electrical signal by means of generation and separation of carriers in a *pn*-junction — a microscopic area between the *p*- and *n*-type semiconductor areas. Since the technology of single-element *pn*-photodiodes matured in the 1960s, it became clear that multi-element photodiode arrays, capable of detecting images, are feasible. However, the idea remained impractical because traditional photodiodes cannot accumulate light — a feature necessary for sufficiently high sensitivity. The solution was found in the form of a metal-oxide-semiconductor (MOS) capacitor shown in Fig. 1.23(a).

It is formed by a transparent contact — heavily doped poly-crystalline silicon — and silicon dioxide insulator deposited onto *p*-doped silicon. The *n*-doped channel is formed under the contact. Two capacitors — the one on the insulating layer with the capacitance C_{MOS} and the second on the *pn*-junction with the capacitance C_{pn} — are connected in series. Positive voltage is applied to the contact, producing reverse bias on the *pn*-junction. Two switches for reset and read are needed to form a single element signal (Fig. 1.23(b)). In such a scheme, integration is performed in the four-step cycle (Fig. 1.23(c)). The basic cycle, producing electrical signal proportional to the energy of light accumulated during exposure time, looks as follows.

(1) Beginning of the cycle: mechanical shutter is closed, blocking the light from reaching the sensitive structure, and the read switch is open, isolating the element from other circuits. The reset switch is closed, applying reverse bias to the *pn*-junction. At this moment, short burst of current charges both the C_{MOS} and C_{pn} capacitances to the same charge Q_0, making the equation

$$U = \frac{Q_0}{C_{\text{MOS}}} + \frac{Q_0}{C_{pn}}.$$

(2) Mechanical shutter and the reset switch open simultaneously. Photons reach the *pn*-junction, generating carriers in an amount

$$\eta n(t),$$

where η is quantum efficiency and $n(t)$ is the number of photons absorbed by the time t. Separated by the intrinsic electric field inside the *pn*-junction, carriers drift to the opposite electrodes of the

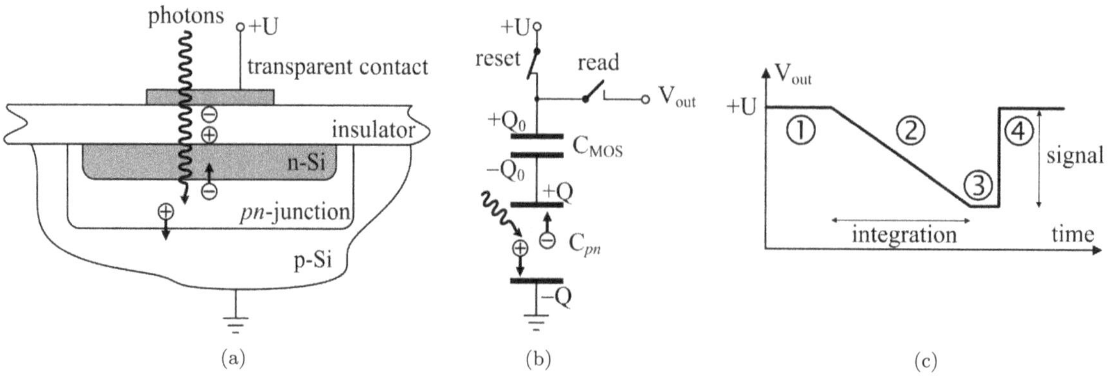

Fig. 1.23. MOS capacitor (a), its equivalent scheme (b), and operation (c).

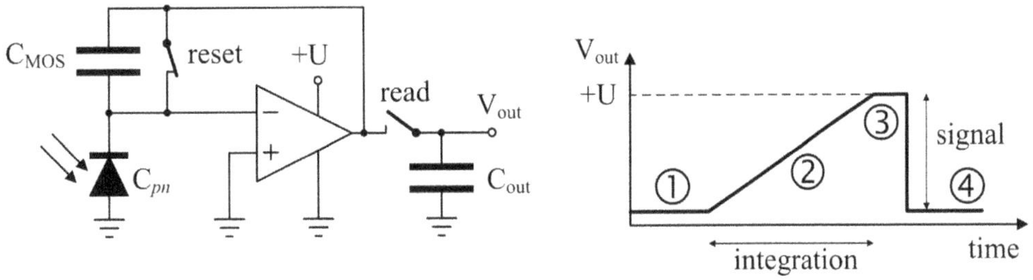

Fig. 1.24. Read-out scheme with inverting amplifier.

C_{pn} capacitor, decreasing the charge to

$$Q(t) = Q_0 - e\eta n(t)$$

with e being the elementary charge. Voltage on the C_{pn} capacitor lowers and so does the voltage on the two capacitors:

$$V(t) = \frac{Q_0}{C_{\mathrm{MOS}}} + \frac{Q_0}{C_{pn}} - \frac{e\eta n(t)}{C_{pn}}.$$

(3) After the exposure time T elapses, both the mechanical shutter and the read switch close. The output voltage

$$V_{\mathrm{out}} = U - \frac{e\eta n(T)}{C_{pn}}$$

connects to the read-out circuitry that accomplishes measurement in some short time.

(4) The read switch opens and the reset switch closes. The circuit returns to initial state.

Although mechanical shutter is quite a realistic option in some scientific applications and especially in photo-cameras, its function can be even better performed by applying positive voltage to the silicon substrate — the technique routinely used in commercial CCD devices.

The read-out scheme in Fig. 1.23 can be further improved by using an inverting amplifier as shown in Fig. 1.24. Inverting amplifier increases the output signal and makes it positive.

In the device, all photo-sensing elements are grouped in line and output capacitors C_{out} are placed adjacently, as shown in Fig. 1.25.

After exposure is finished, switch control logic disconnects amplifiers from output capacitors C_{out}, leaving each of them with their own specific charge, acquired during exposure. Instead of creating read-out electronics around each photo-sensor, the charge propagates through adjacent capacitors, following traveling wave of potential wells (Fig. 1.26).

This wave, being synchronized by the clock, runs from left to right through the shift register, moving accumulated charge from one pixel to another towards the charge amplifier. Thus, the output of the CCD photodetector is the analog signal, reproducing intensity of light as a function of the pixel ordinal number. In reality, more sophisticated algorithms may be used, even including avalanche multiplication during transfer from one element to another.

Switch control logic takes all responsibility for generating necessary sequences of electrical signals and distributing them among sensitive elements. In the simplest case, the user only has to apply the clock signal and start/stop pulses, signifying the end of exposure and the beginning of reading process. The best way to analyze the read-out process in more detail and discern all pulses in figures on a book page is to consider the shortest (in terms of the number of pixels) linear photodetector composed of only 128 elements. A good example is the TAOS TSL1401 photodetector (Fig. 1.27), which can operate

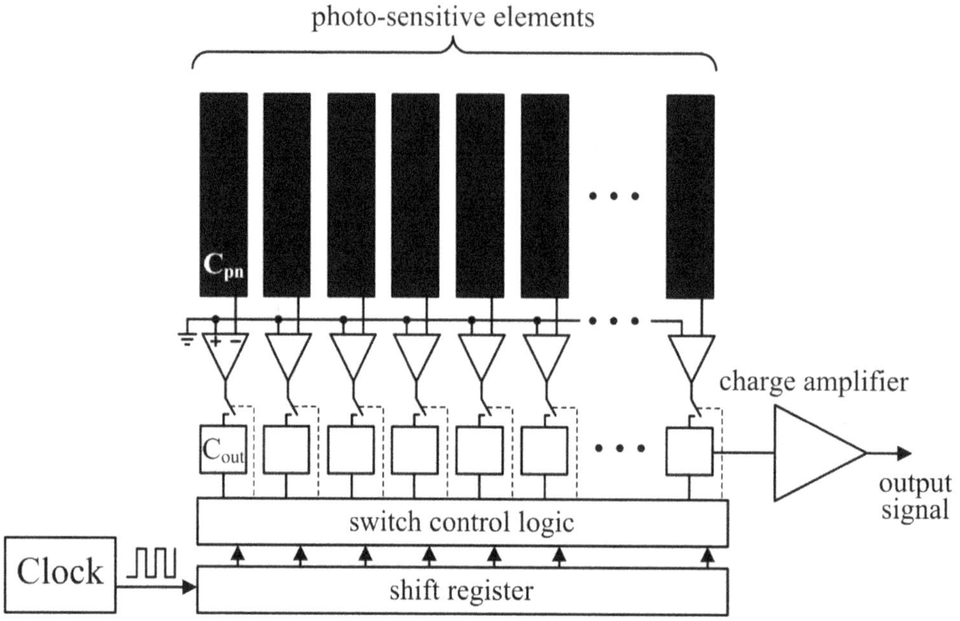

Fig. 1.25. Topological concept of the device.

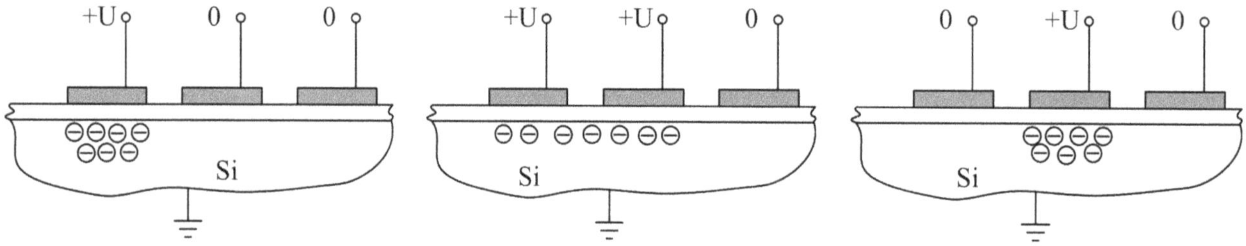

Fig. 1.26. Basic principle of charge transfer in CCDs.

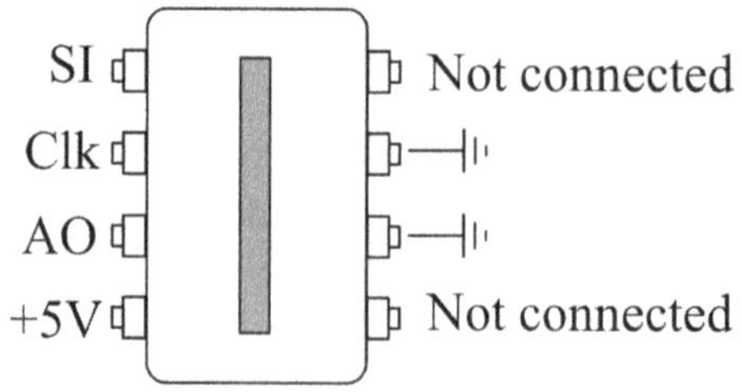

Fig. 1.27. 128-pixels TAOS TSL1401 photodetector.

from only two signals: SI — serial input, defining the start of the data-out sequence, and Clk — a permanent series of rectangular pulses, controlling charge transfer, pixel output, and reset. The result of measurements is the analog signal AO — analog output.

To realize how the output signal depends on the clock sequence, consider expanded oscilloscope traces in Fig. 1.28.

The exposure ends and read-out starts at the positive front of the SI pulse. From this moment, every clock pulse brings electrical charge accumulated in a certain pixel of the sensor to the charge amplifier, forming analog bar-step curve. The insert in the figure shows expanded part of this signal together with

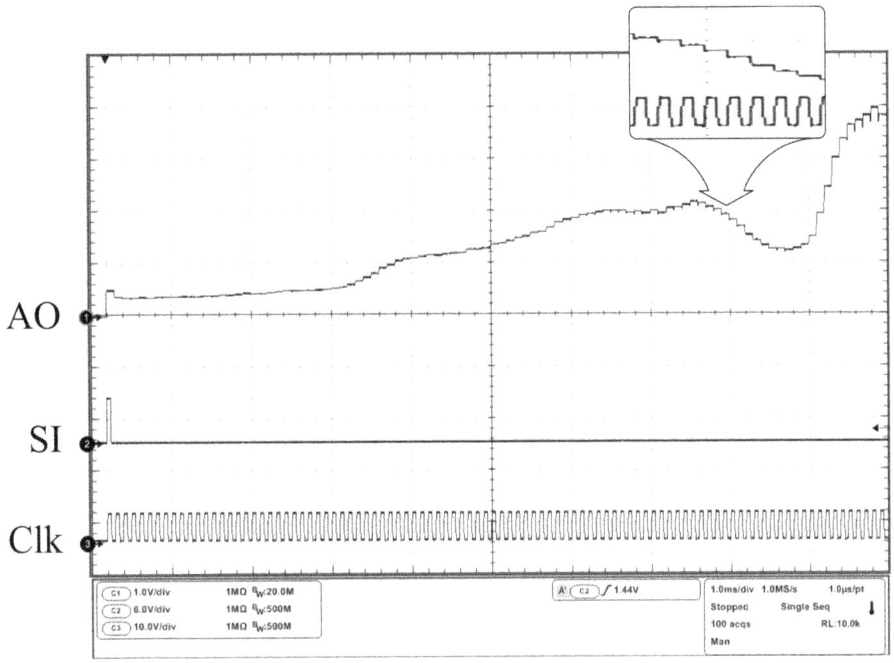

Fig. 1.28. Oscilloscope traces of the driving sequences and the output signal. Oscilloscope settings can be seen in the lower panel of the figure.

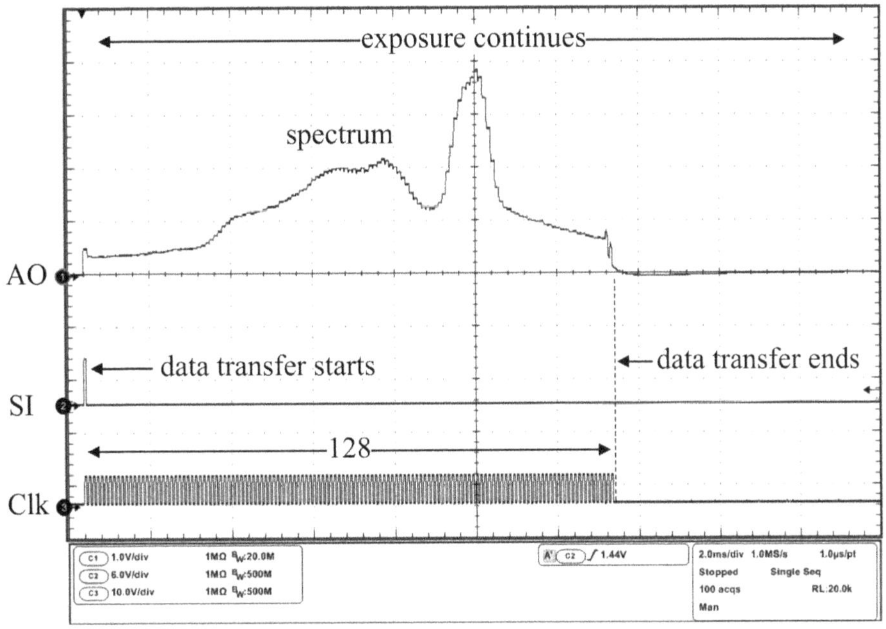

Fig. 1.29. Exposure continues during data transfer.

the relative part of the clock train. If the diffraction grating of the spectrometer is focused onto the photodetector, then the AO signal gives the spectrum of light as the user sees it at the output of the spectrometer.

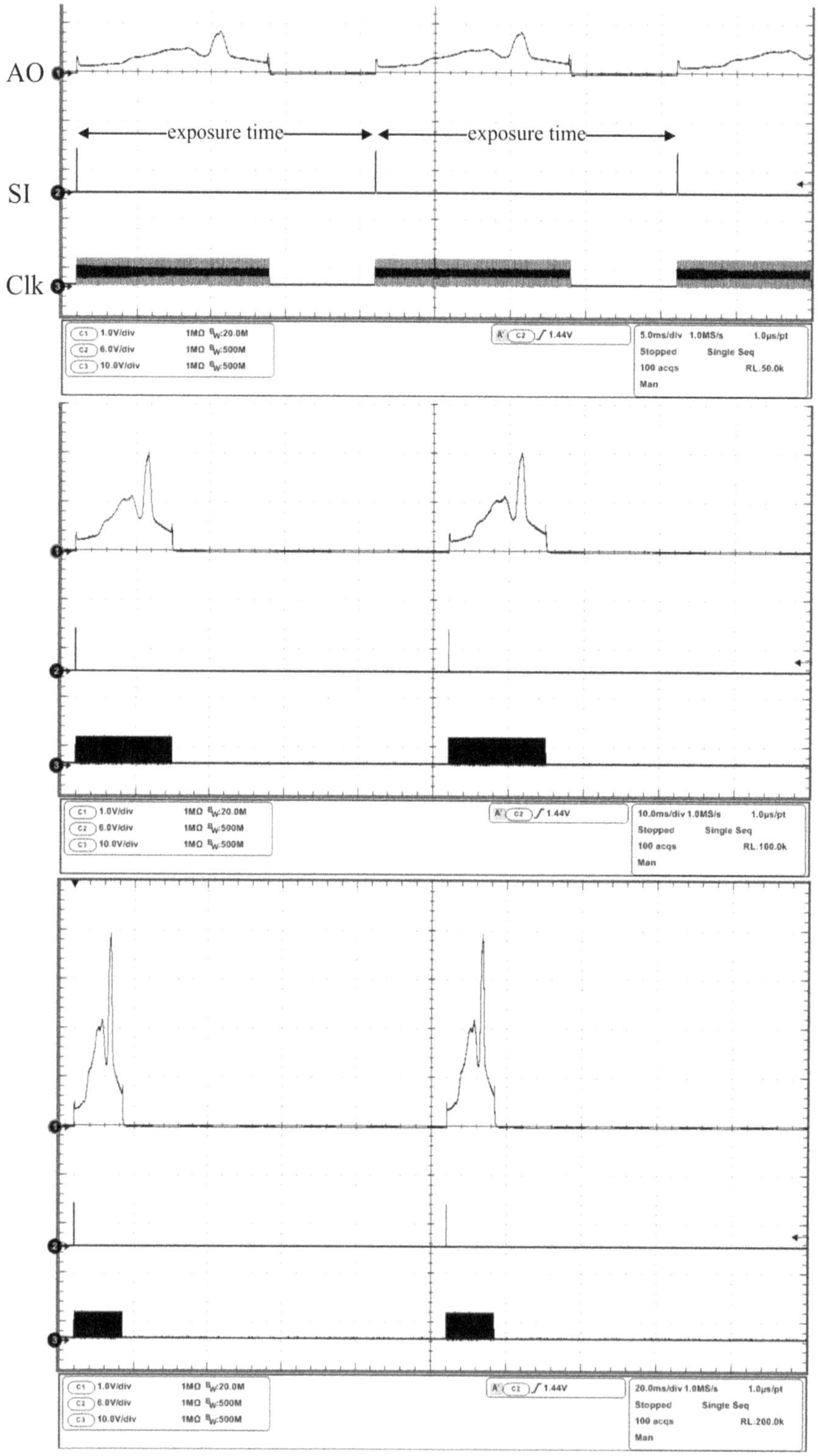

Fig. 1.30. Oscilloscope traces with different exposure time (from top to bottom): 20 ms; 50 ms; 100 ms. Oscilloscope scales respectively: 5, 10, and 20 milliseconds per division.

Figure 1.29 shows how the data transfer stops after all 128 pixels are read. At the moment when exposure ends, the switch control logic connects all 128 amplifiers simultaneously to their output capacitors C_{out}, the amplifiers momentarily charge them all to the voltages proportional to voltages accumulated in corresponding pixels. After this charging is complete (18 clock pulses in TSL1401), the reset switches in each pixel open again and amplifiers are disconnected from the output capacitors. The exposure phase starts again. In the meantime, data transfer continues until all 128 output capacitors have transferred their charges to the charge amplifier. Thus, exposure process does not wait until data transfer is accomplished — an important feature, saving time for measurements, especially for long photodetectors with thousands of pixels.

Figure 1.30 explains how the signals change with exposure time. Note proportional increase of the amplitude of the analog output (AO).

1.9. Thermo-electrical Cooling

In order to reach maximum possible sensitivity, the photodetector matrix should be cooled. In visible domain, the effect of cooling, although noticeable, is not as strong as in infrared domain. The CCD matrices designed for visible domain are always made in silicon substrate with relatively large transition energies, thus being relatively insensitive to thermal generation of carriers. The matrices for infrared domain are made of low-energy semiconductor materials, for which thermal generation of carriers is much more important. The farther the sensitivity of a photodetector goes into infrared domain, the stronger the influence of thermal noise is. Figures 1.31 and 1.32 compare dark currents at different temperatures of two spectrometers designed for different infrared domains. Both sensitive matrices for these spectrometers are made of InGaAs, but of different stoichiometric ratios. Although quantum efficiencies of these two photodetectors are not very different (Fig. 1.33), narrower band gap of the InGaAs optimized for the spectral interval around 2 μm makes it easier for thermally generated electrons to reach the conduction band, thus increasing the dark current.

It is instructive to plot the spectrally averaged dark currents as a function of temperature (Figs.1.34 and 1.35). Obviously, the InGaAs matrix optimized for spectral interval around 2 μm cannot work at room temperature because dark current reaches saturation.

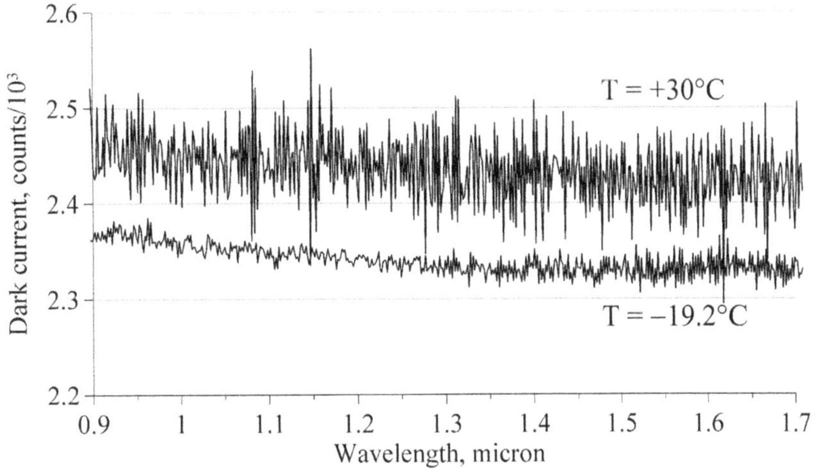

Fig. 1.31. Dark currents of the spectrometer designed for near-infrared domain: at temperatures +30°C and −19.2°C.

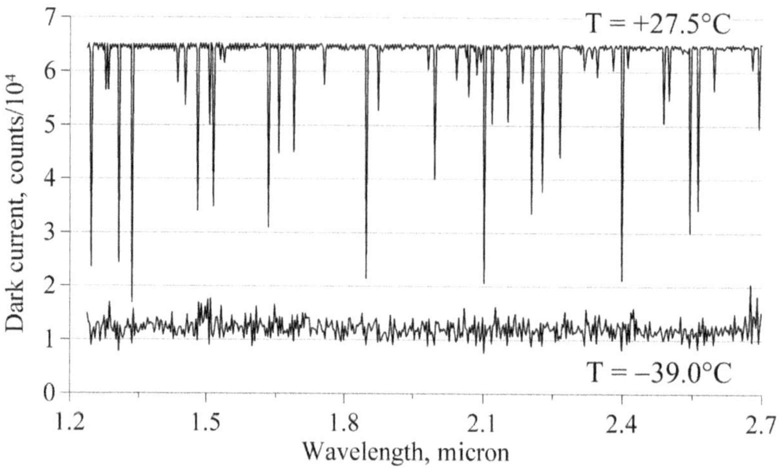

Fig. 1.32. Dark currents of the spectrometer designed for mid-infrared domain: at temperatures +27.5°C and −39.0°C. At temperatures above the room temperature, the dark current already saturates.

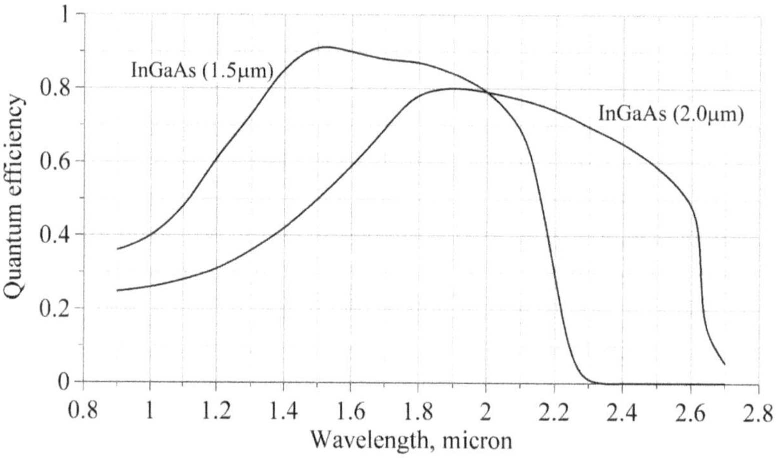

Fig. 1.33. Quantum efficiency of InGaAs photodetectors optimized for two different spectral intervals.

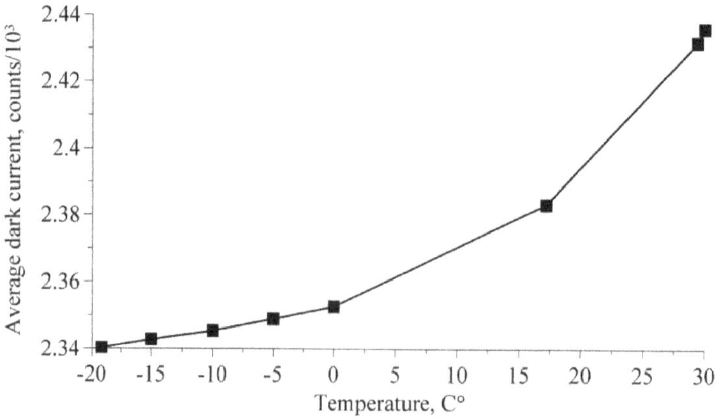

Fig. 1.34. Thermal dependence of spectrally averaged dark current in the spectral domain around 1.3 μm.

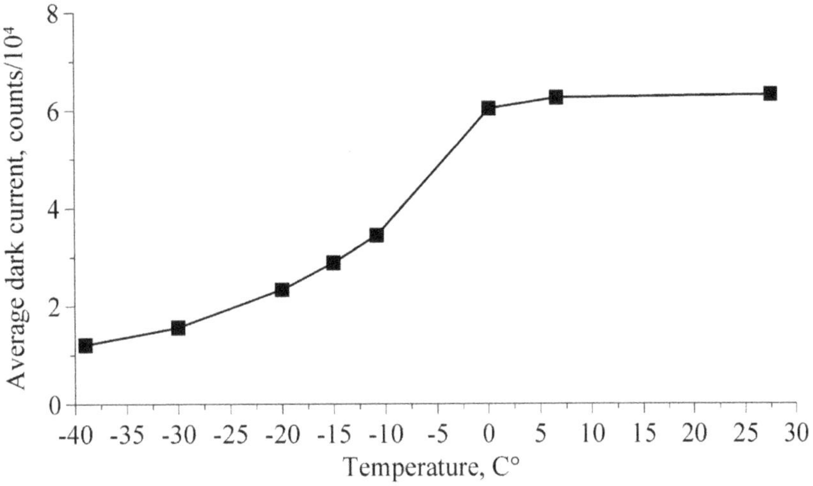

Fig. 1.35. Thermal dependence of spectrally averaged dark current in the spectral domain around 2 μm.

1.10. Coupling to Optical Fibers

Spectral resolution and sensitivity of a spectrometer are two competing parameters. All the three typical optical schemes shown in Fig. 1.1 can be presented in a generalized form as in Fig. 1.36. Since the space is limited and the total length $L+F$ may be considered constant, equal maxima of both L and F, necessary to minimize numerical apertures and minimize aberrations, are reached when $L = F = (L+F)/2$. This corresponds to unity magnification, meaning that the slit is projected onto CCD one-to-one. From the point of view of sensitivity, the more optical flux comes through the slit the better, and thus the wider the slit the more sensitive the spectrometer.

On the other hand, slit width limits spectral resolution since it is projected one-to-one. For example, for a wide-range spectrometer designed to work from 200 nm to 800 nm with spectral resolution 1 nm on 30 mm long CCD, the slit width must be less than 30 mm/(600 nm/1 nm)=50 μm. The priority is always spectral resolution, therefore, in wide-range compact spectrometers slit width never exceeds 30 μm. But even this infinitesimal amount of energy is not used completely because diameter of the fiber core d exceeds transversal dimension of the CCD (Fig. 1.36). In order to ensure axial alignment of the optical fiber to the slit within manufacturing tolerances, typical core diameter must be $d > 400$ μm, whereas the height of the CCD pixel is only 200 μm, wasting half of the usable optical flux. Therefore, it is an illusion to think that thicker optical fiber, like 1 mm core, may improve sensitivity of your spectrometer. However, the thick-core optical fiber is not a mistake, whereas the use of a fiber bundle may be real mistake. This is explained in Fig. 1.37]. Only fiber bundles with central core may work reliably.

Understanding deficiency of sensitivity, manufacturers offer an option that increases efficiency of transferring light to photodetector at the expense of somewhat poorer spectral resolution (Fig. 1.38). The obvious gain in collecting efficiency is plagued by poorer spectral resolution, resulting from stronger aberrations and curviness of spectral lines as explained in Fig. 1.8. Not everyone manufacturer offers this option, and preliminary enquiries should be made prior to ordering.

Another, more efficient but at the same time more expensive, way to increase sensitivity is to use CCD matrix with special read-out technique instead of the line array (Fig. 1.39). Substantially more expensive, this option is used only in high-grade spectrometers.

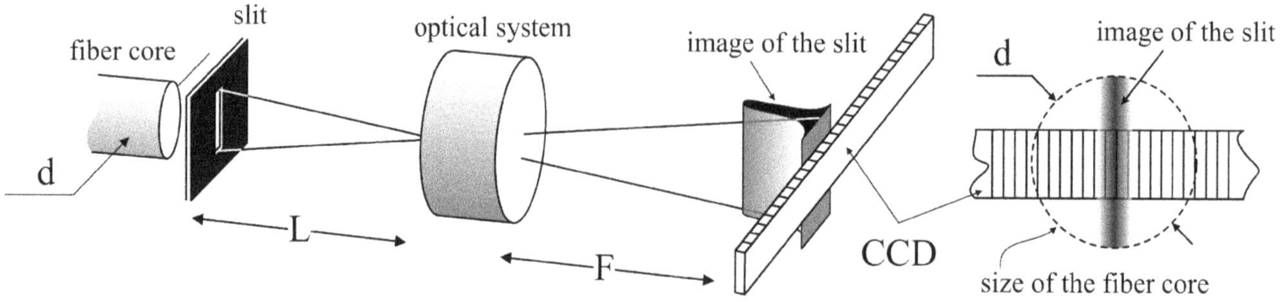

Fig. 1.36. In a spectrometer, photodetector receives the image of the slit magnified by the factor F/L.

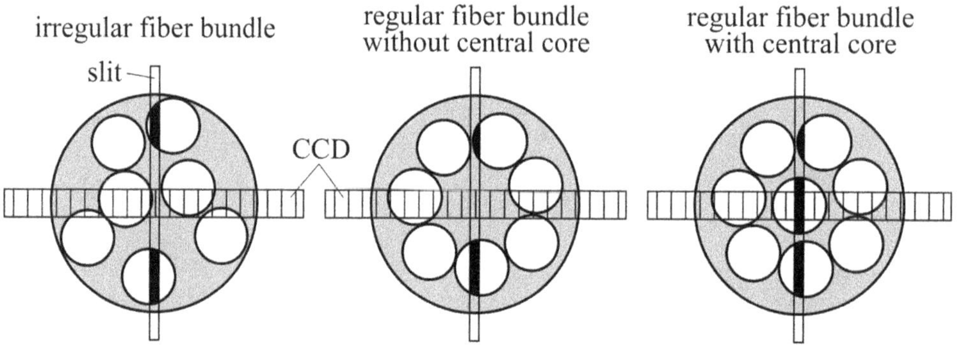

Fig. 1.37. Standard SMA-905 connectors may terminate optical fiber bundles with multiple cores. Only those of them that have a central core deliver light to CCD (charge-coupled device).

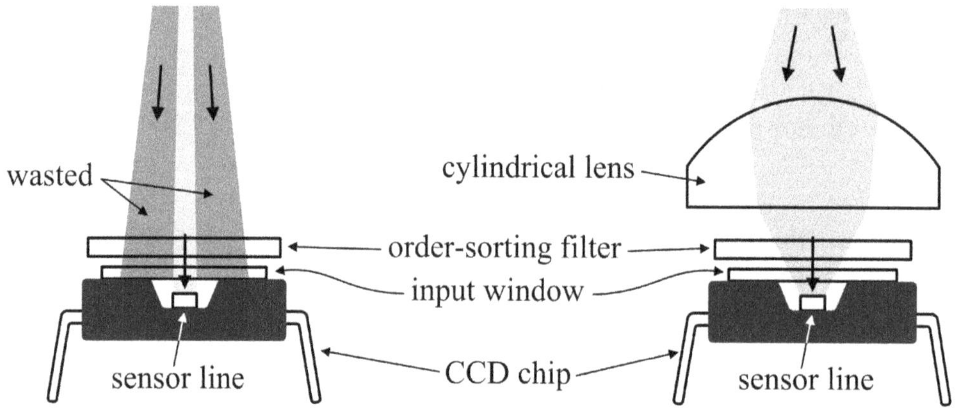

Fig. 1.38. Cylindrical lens installed above the photodetector collects sagittal rays onto the sensor line.

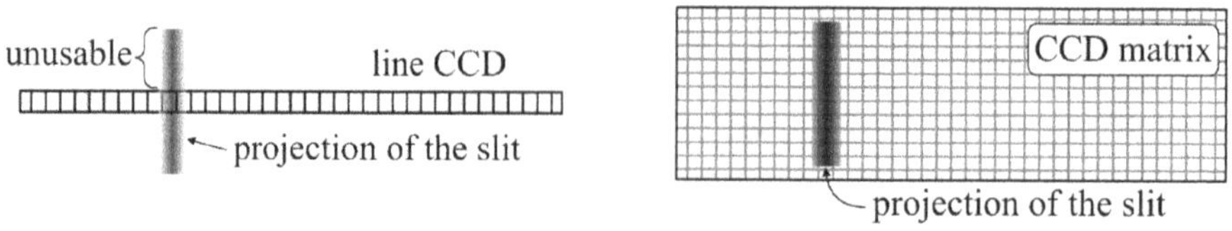

Fig. 1.39. With the CCD matrix, the entire optical flux from the slit may be intercepted.

The read-out protocol combines only vertical elements of the matrix, which is often called "full vertical binning" (FVB). The noise amplitude in the resultant signal also increases as it sums individual noise signals in each vertical pixel. However, they are statistically uncorrelated, producing the increase in signal-to-noise ratio proportionate to the square root of the number of vertical pixels.

1.11. Typical Industrial Spectrometers

In high-volume production facilities, the number of process machines may count hundreds, which makes the size and price of the spectrometers important. In order to minimize both the price and the size, two basic optical schemes are most frequently used (Chapter 1, Fig. 1.1): the crossed Czerny-Turner with less expensive flat diffraction grating but larger number of optical elements, and the Rowland scheme with only one but rather expensive optical element — the curved varied line-spacing diffracting grating. Typical design of the first type is outlined in Fig. 1.40. Angular orientation of the mirrors is adjusted by differential screws, and position of the spectrum on the photodetector is set by manual rotation of the grating.

Figure 1.41 portrays the images of spectra, as they are visible in the plane of the photodetector. With the photodetector and thin-film order-sorting filter on it (Section 1.5 of Chapter 1) removed for observation, the diffraction orders are not filtered out, and multiple-order diffracted harmonics of ultraviolet part of spectra fill the right-hand part of the spectral bar. The upper trace is obtained with the tungsten halogen lamp as the source of light, while the lower one with the argon–mercury calibration source (Section 1.7 of Chapter 1). The mercury doublet 577–579 nm in the green part of the spectrum is clearly resolved. Inclination of the slit is also noticeable — a clear defect of initial adjustment of the spectrometer. Although the width of the photodetector sensitive line is commonly small in this type of spectrometers, typically 0.2 mm, this defect may lead to loss of spectral resolution.

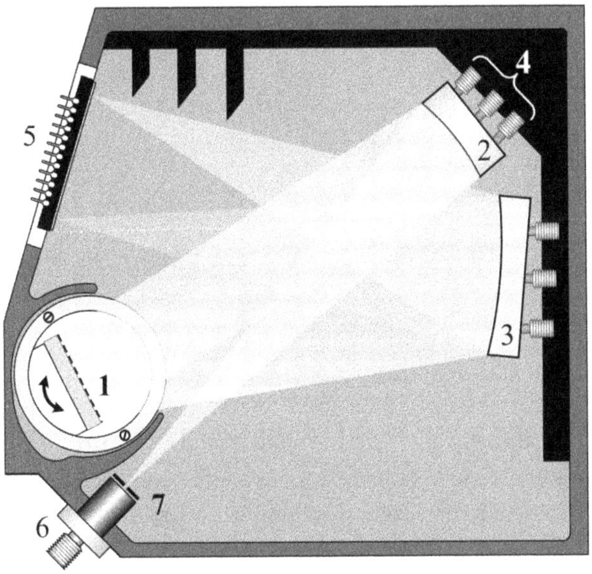

Fig. 1.40. Interior design of the crossed Czerny–Turner compact spectrometer. 1 — flat diffraction grating; 2 — collimating mirror; 3 — focusing mirror; 4 — differential screws; 5 — photodetector matrix; 6 — SMA-type connector for optical fiber; 7 — slit.

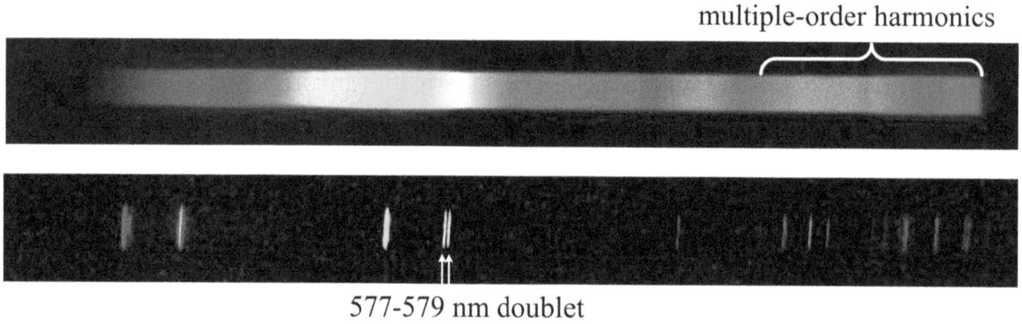

Fig. 1.41. Typical pictures of spectrum in the plane of the photodetector. Upper trace — with the tungsten halogen lamp as the source; lower trace — with the argon–mercury calibration source.

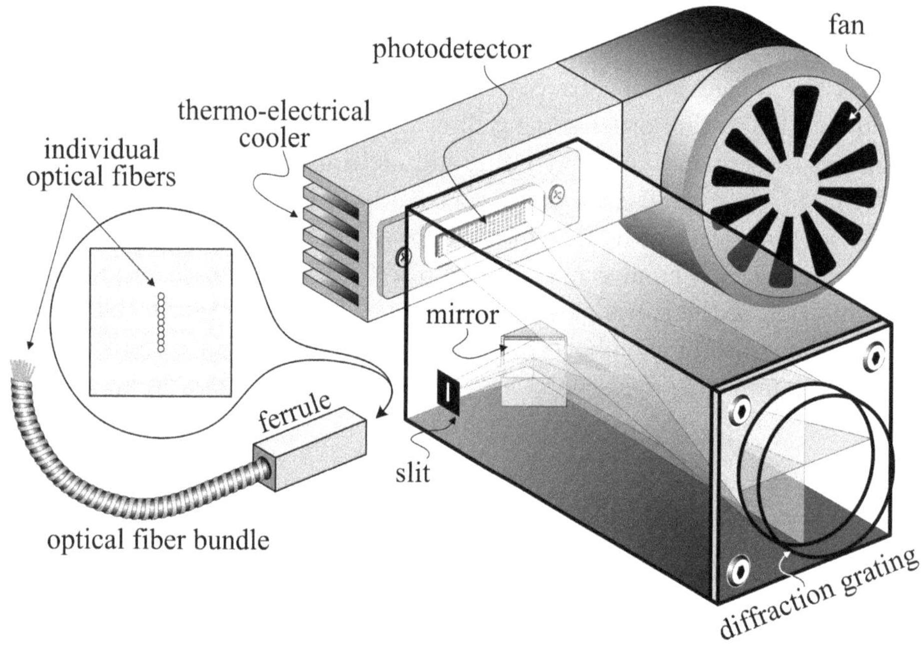

Fig. 1.42. Principal components of the spectrometer with cooled photodetecor and Rowland diffraction grating.

Typical design of more expensive spectrometers based on the Rowland scheme is explained in Fig. 1.42. These spectrometers commonly use thermo-electric cooling of the photodetector and, therefore, require more compact optical design in order to keep the entire size of the spectrometer small enough. The Rowland scheme with curved varied line-spacing diffracting grating makes this possible at the expense of significantly higher price.

Sensitivity of such devices is significantly better than can be obtained with the spectrometers shown in Fig. 1.40 for two main reasons: reduction of noise due to cooling of the photodetector and bigger optical flux due to increased area of the input slit that intercepts light from several individual fibers. The common spectral resolution of this type of spectrometers is around 1.5 nm with 25 μm slit.

1.12. Monochromators

Monochromators are optical devices, selecting narrow spectral components from white light. It can be done either with the help of dispersive optical elements like diffraction gratings or spectrally blocking devices like filters. Grating monochromators can arbitrarily select any spectral line from their working spectral interval whereas filters are designed for a fixed spectral line. Spectrometers do actually the same as grating monochromators but terminate spectrally decomposed beam by a relatively slow photodetector. In some applications, fast response is required at one specific spectral line, and then CCD detectors must be substituted for fast photomultipliers connected to the exit slit of a monochromator. Available in a variety of forms and sizes, and spectral resolution varying from 0.1 nm to 2 nm, grating monochromators basically reproduce the same uncrossed Czerny–Turner scheme (Fig. 1.43). Slits S are positioned at focal distances from spherical mirrors M to convert diverging beam to parallel and in reverse. Flat diffraction grating G turns to select desirable spectral line on the exit slit. Wavelength selection is almost a linear function of the rotation angle α. Such a scheme is known as the Fastie–Ebert configuration. Light may enter through any one of the two slits.

Consider the basic diffraction grating equation (Section 1.2) for the first positive diffraction order:

$$\lambda = t(\sin\psi - \sin\phi)$$

with t being the period of the grating. Since all the optical elements except grating are fixed, the rays coming to the grating and exiting through the exit slit are also fixed, making the angle

$$\varphi = \psi + \phi$$

constant. Practically, $\varphi \sim 30$. In horizontal position of the grating $\psi = \varphi/2$, and being rotated by the angle α grating comes to a new angle of incidence $\psi = \varphi/2 + \alpha$, selecting the wavelength

$$\lambda = t\left[\sin\psi - \sin(\varphi - \psi)\right] = 2\,t\,\cos\frac{\varphi}{2}\cdot\sin\alpha.$$

This function is called the tuning curve and is shown in the left-hand part of Fig. 1.44. It is not linear as a function of rotation angle. The so-called sine-bar mechanism, shown in the center of the figure, provides linear wavelength tuning as a function of the micrometer screw displacement x. With the ball at the end, the lever keeps constant hypotenuse l, making $\alpha = \arcsin(x/l)$ and $\lambda \sim x$.

Grating monochromators do not have order-sorting filters like spectrometers do (Section 1.5) because the selected wavelength is an arbitrary choice of the user. Therefore, working with broad-band sources, always be prepared for higher-order ghosts, emerging from the exit slit along with the selected spectral component. Figure 1.45 exemplifies some typical situations. The lamp spectrum does not contain components below 220 nm, therefore when monochromator is set for 400 nm, no ghosts are visible (upper

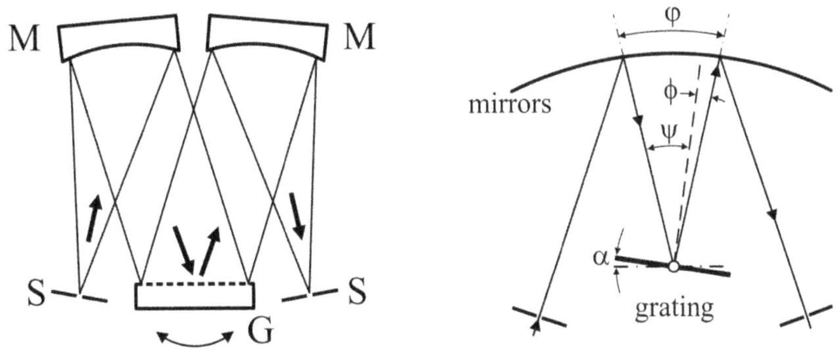

Fig. 1.43. Uncrossed Czerny–Turner monochromator (at left). Some compact designs use a single long mirror instead of two (at right).

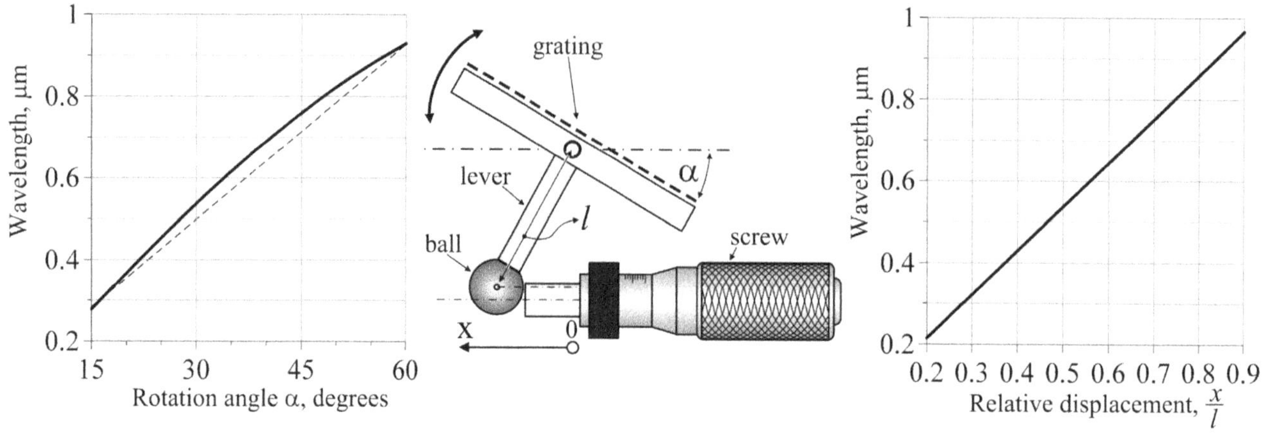

Fig. 1.44. Tuning curve $\lambda(\alpha)$ computed in the range 200–900 nm for the grating with 1800 grooves per millimeter and $\varphi = 30°$. Straight dashed line, connecting the ends of the curve, clearly shows nonlinearity of angular tuning.

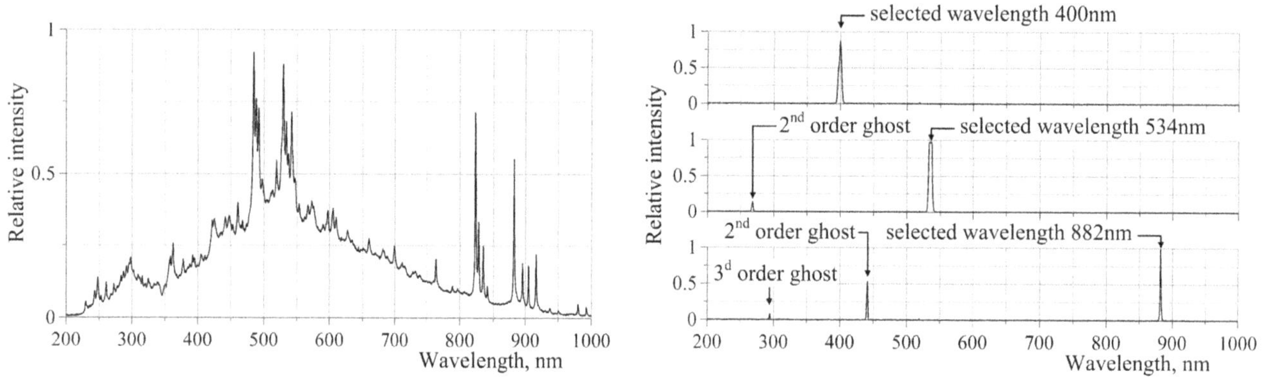

Fig. 1.45. Broad-band spectrum of Xe flash lamp (at left) creates multiple-order ghosts at the output of a monochromator without additional filters (at right).

spectrum at right). Selection of the 534 nm line drags $534/2 = 267$ nm ghost from the xenon spectrum (middle section at right). Even third-order ghost may accompany longer wavelengths (bottom at right).

As well as there are compact fiber-optic spectrometers, there are compact grating monochromators in the market — handheld devices with manual selection of wavelengths and fiber-optic input–output, which can be a good solution for limited space in industrial environment (Fig. 1.46). Belonging in the class of compact devices, they are not supposed to show too high spectral resolution, which is typically about 3 nm with 300 μm slits.

The indispensable feature of grating monochromators is the ability to arbitrarily select the wavelength with potentially high spectral resolution. On the other hand, small area of a slit, not exceeding 5×0.3 mm^2, sets the limit to energy efficiency of grating monochromators. It is almost hundred of times less than a standard photodetector like photodiode or photomultiplier with 10×10 mm^2 sensitive aperture can accept. Interference filters, with clear apertures up to 50 mm and transmission-bandwidth values from 20%-1nm to 70%-40nm at the central wavelength, leave no chance to grating monochromators to compete in delivering optical flux from wide-area sources to photodetectors (Chapter 7).

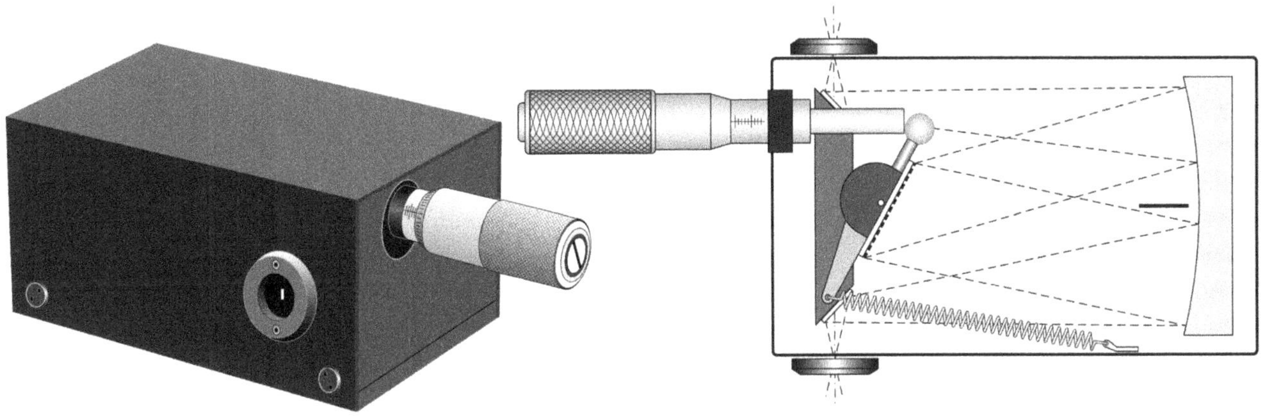

Fig. 1.46. Compact grating monochromators are commonly built on the Fastie–Ebert scheme.

Supplemental Reading

M. Born, E. Wolf, *Principles of Optics*, Cambridge University Press, 7th edn., 1999.

E. G. Loewen, E. Popov, *Diffraction Gratings and Applications*, CRC Press, 1997.

R. Petit, (ed.), *Electromagnetic Theory of Gratings*, Springer-Verlag, Heidelberg, 1980.

M. Johnson, *Photo-detection and Measurement: Maximizing Performance in Optical Systems*, McGraw-Hill, 2003.

Chapter 2

Basics of Fourier-Transform Spectrometers

2.1. Theoretical Background

Fourier-transform spectrometers, in all their plurality, are based on one optical principle — the Michelson interferometer with moving mirror (Fig. 2.1).

Consider spectral intensity I of an input optical flux as a function of wavelength λ. In the photodetector plane of the interferometer with 50% beamsplitter, variable part of the intensity in a quasi-monochromatic (narrow) spectral interval $d\lambda$ is

$$I(\lambda)\cos\Phi \cdot d\lambda,$$

where Φ is the total phase difference between the waves in two shoulders of the interferometer. Splitting the total phase in two components, accounting for spectral dispersion of optics $\psi(\lambda)$ and non-zero initial position of the moving mirror x_0, and summing over the entire spectrum, we may write the photodetector signal as proportional to

$$S(x) = \int I(\lambda)\cos\left[\psi(\lambda) + \frac{4\pi}{\lambda}(x - x_0)\right] d\lambda = \mathrm{Re}\left\{\int I(\lambda)e^{i\psi(\lambda)}e^{i\frac{4\pi}{\lambda}(x-x_0)}d\lambda\right\}, \quad i = \sqrt{-1}.$$

This formula looks similar to Fourier transform of the input spectrum $I(\lambda)$ but it is not the Fourier transform because the wavelength stands in denominator.

Next, we need a practical algorithm to extract $I(\lambda)$ from this measurement. We may introduce complex functions S and F, having in mind that eventually, in order to obtain $I(\lambda)$, we are interested in $|F|$. Denoting

$$u = \frac{2}{\lambda} \quad \text{and} \quad \delta = x - x_0,$$

the relation between the measured signal $S(\delta)$ and spectrum $I(\lambda)$ formalizes to

$$S(\delta) = \int F(u)e^{i2\pi\delta u}du, \quad F(u) = -\frac{2}{u^2}I\left(\frac{2}{u}\right)e^{i\psi(2/u)}.$$

Already, this is the Fourier transform, and $F(u)$ can be recovered by inverse fast Fourier transform (FFT). The FFT is defined on a discrete mesh of M points $l = 1, 2, 3, \ldots, M$ uniformly separated within the scanner maximum displacement Δx:

$$\delta_l = \frac{l-1}{M}\Delta x,$$

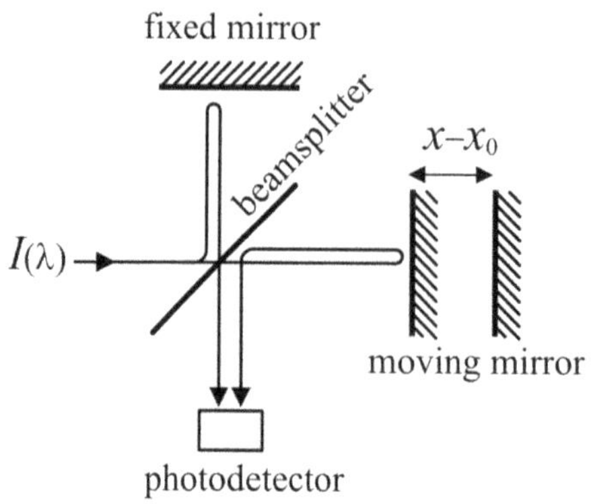

Fig. 2.1. Conceptual view of the Michelson interferometer with moving mirror.

making the recorded array

$$S_l = \int_{-\infty}^{+\infty} F(u)e^{i2\pi u\delta_l}du, \quad i = \sqrt{-1}.$$

The result of the FFT is the array V_j of exactly the same size as S_l:

$$V_j = \sum_{l=1}^{M} S_l \cdot e^{-2\pi i(j-1)(l-1)/M}.$$

Substituting S_l as the integral, bringing summation under the integral, and realizing that only those terms will remain, for which the argument of the complex exponent is zero

$$u\Delta x = j - 1,$$

we obtain

$$V_j = M \cdot F\left(\frac{j-1}{\Delta x}\right).$$

The function $F(u)$ has already been defined, so that it is only a matter of arithmetical manipulations to obtain the final result:

$$I\left(\frac{2\Delta x}{j-1}\right) = -\frac{(j-1)^2}{2M\Delta x^2}|V_j|.$$

There are two fundamental results in one formula. The first result establishes absolute calibration of the final spectrum:

$$\lambda_j = \frac{2\Delta x}{j-1}.$$

It says that if you know exactly the displacement of the mirror Δx (in micrometers, millimeters, or any other units), then the wavelength associated with the number j in the final array V_j depends on nothing but only Δx and the number of samples M that you have recorded. This dependence is nonlinear: wavelengths are ascribed to jth element of the array inversely proportional to $j-1$. From this formula, spectral resolution $\Delta\lambda = \lambda_j - \lambda_{j+1}$ of the FT spectrometer easily follows:

$$\Delta\lambda = \frac{2\Delta x}{j(j-1)} \approx \frac{2\Delta x}{(j-1)^2} = \frac{\lambda_j^2}{2\Delta x}, \quad \text{or} \quad \frac{\Delta\lambda}{\lambda} = \frac{\lambda}{2\Delta x}.$$

Let us see what it means in terms of lateral displacement Δx of the moving mirror. If, for compact grating spectrometers designed for visible domain around 600 nm spectral resolution 1 nm is suitable

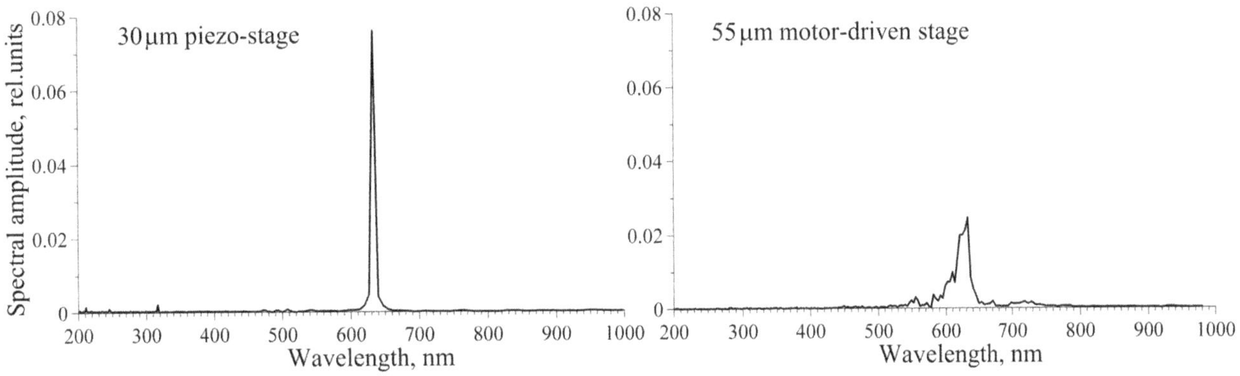

Fig. 2.2. Spectrum of the stabilized He–Ne laser obtained with a piezo-stage and 30 μm stroke (at left) and with the ball-bearing scanning stage driven by a stepping-motor with 55 μm stroke (at right).

for the most industrial applications, then for potential Fourier-transform spectrometer, designed for the same spectral domain with the same spectral resolution, the mirror stroke must be 180 μm — the value readily obtainable with piezo-drives. However, if the same ratio $\Delta\lambda/\lambda$ is needed for far infrared domain around the wavelength say 10 μm, then the mirror displacement must be around 3 mm — the value significantly exceeding potential displacement of piezoelectric actuators. Therefore, in far infrared domain, the Fourier-transform spectrometers implement voice-coil moving actuators. These devices are not as accurate as piezo-drives and require laser control of their translational displacements. The importance of careful control of sampling during mirror motion is exemplified in Fig. 2.2, where the spectrum of a stabilized He–Ne laser is shown as it is reconstructed from the interference pattern obtained on a piezo-driven Michelson interferometer in comparison with the result obtained on the interferometer driven by a precise stepping-motor stage. Even though the motor-driven stage has almost two times longer scan, spectral resolution is much worse: the entire spectrum is torn apart by irregularities of scanning. The area under the curves is the same due to energy conservation principle.

2.2. Absorption-Type Fourier-Transform Spectrometers

The basic opto-mechanical scheme of the absorption-type infrared Fourier-transform spectrometer is outlined in Fig. 2.3.

The beamsplitter 1 is a rather complicated optical device, having several important features: it is non-polarizing, it is compensated and it is semi-transparent in both infrared and visible domains. The latter feature is organized by reserving small specially coated area in the peripheral part of the beamsplitter, through which only the laser beam passes. The infrared beamsplitting coating must be of non-polarizing type and meet stringent requirements on polarization. Any priorities in polarization contribute to lower contrast of fringes. Finally, compensation of the beamsplitter means equality of optical passes from the front surface to the beamsplitting layer and that from the beamsplitting layer to the outer surface on the opposite side. The best way to realize what happens when the beamsplitter is not compensated is to consider a beam-splitting cube (Fig. 2.4). In this figure, vertical thickness of the cube is different from the horizontal one. Since refractive index $n(\lambda)$ is a function of wavelength, optical path difference $s \cdot n(\lambda)$ will be different for short and long wavelengths. Consequently, zero optical path difference will correspond to different displacements x_1 and x_2 of the mirror. If s becomes too big, say 100 μm, then $x_1 - x_2 = s[n(\lambda_1) - n(\lambda_2)]$ may become comparable to the mirror travel range Δx, which causes truncation of useful parts of the interference pattern as shown in the right-hand part of Fig. 2.4. It means that useful spectral information is lost. To prevent this from happening, high-quality

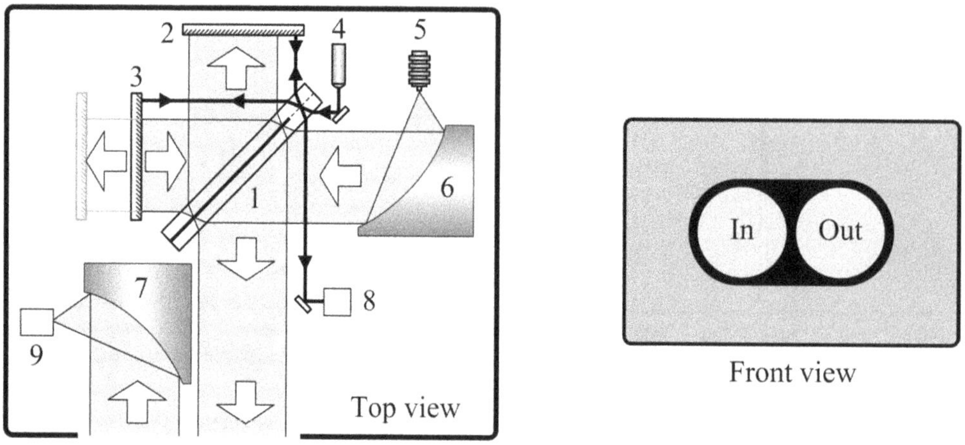

Fig. 2.3. Typical opto-mechanical module of the absorption-type Fourier-transform spectrometer. 1 — compensated dual-wavelength beamsplitter; 2 — fixed mirror; 3 — moving mirror; 4 — laser; 5 — infrared source; 6,7 — parabolic mirrors; 8 — reference photodetector; 9 — infrared photodetector.

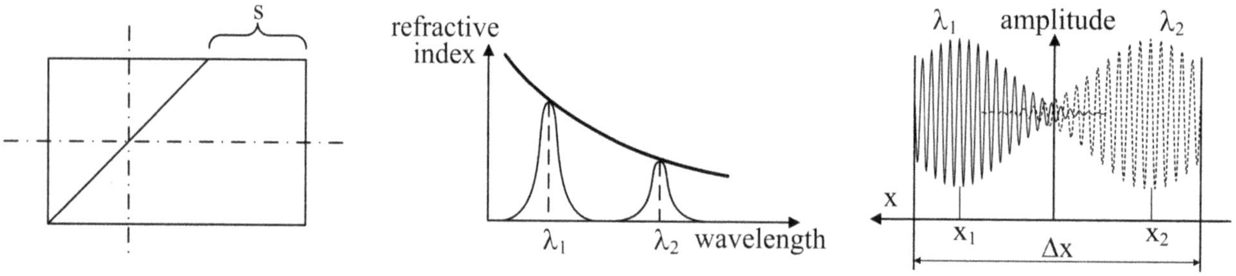

Fig. 2.4. In the beamsplitter, geometrical difference s in two orthogonal directions must be a minimum (at left). Refractive index increases for shorter wavelengths (in the center). As mirror moves, this causes relative displacement of fringe curves for different wavelengths (at right).

beamsplitters are made of two glass plates cut from one precisely parallel substrate and bonded together with semi-transparent coating deposited on one of them. Such beamsplitters are called compensated.

In Fig. 2.3, the reference laser 4 enables sampling of the interferogram signal $S(x)$, coming from the photodetector 9 in the form of a photocurrent, uniformly as a function of the mirror displacement x. Traditional digitizers can sample electrical signals uniformly in time, but if the mirror displacement x increases not uniformly in time due to various irregularities, such sampling is not regular as a function of x. Passing through the interferometer, the laser beam creates oscillations of optical intensity on the photodetector 8, producing the photocurrent synchronized not with time, but exactly with the mirror displacement x (Fig. 2.5). As such, the sampling commands are generated at moments of time when the moving mirror has traveled exactly the same distance δ.

In semiconductor industry, absorption-type Fourier-transform spectrometers are the best candidates for measuring molecular composition of active gases or bi-products during etch, deposition, or cleaning processes. Although different in technology, the common concept of these applications is a large vessel, containing the 300 mm semiconductor wafer, filled with one or several process gases (Fig. 2.6).

The long optical double pass through the vessel provides sufficient absorption by active gases to be reliably recorded. As a simplest example of such an application, Fig. 2.7 shows consecutive absorption spectra, obtained with the compact Fourier-transform spectrometer Cary 630 (Agilent) during ventilation and removal of acetone after cleaning.

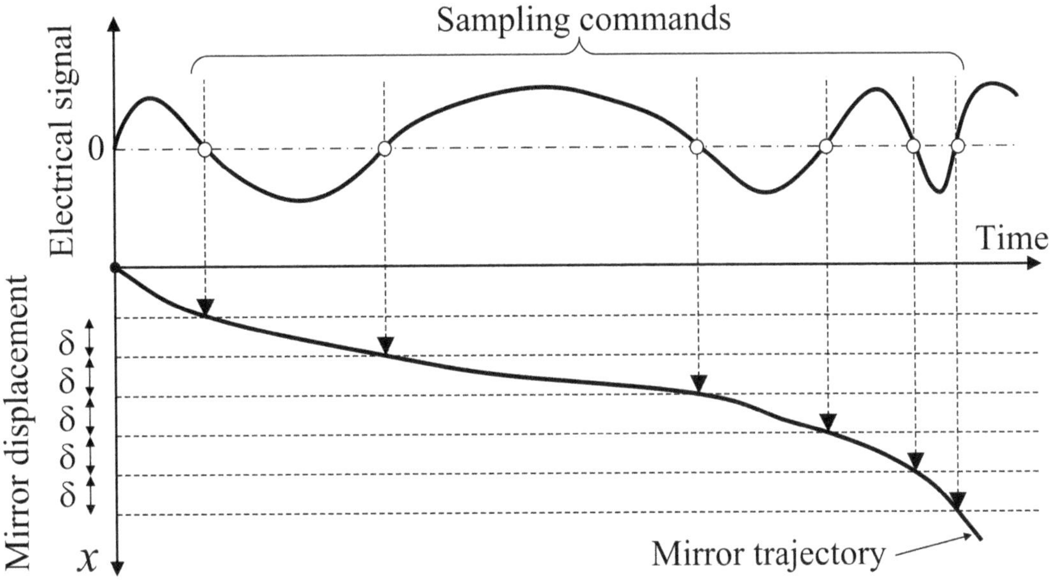

Fig. 2.5. Principle of synchronization of sampling the interferogram.

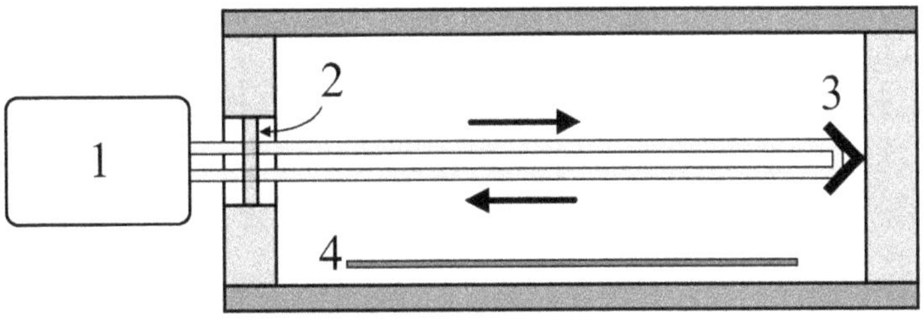

Fig. 2.6. Conceptual application of the absorption-type Fourier-transform spectrometers. 1 — spectrometer; 2 — viewport; 3 — folding retro-reflector; 4 — semiconductor wafer.

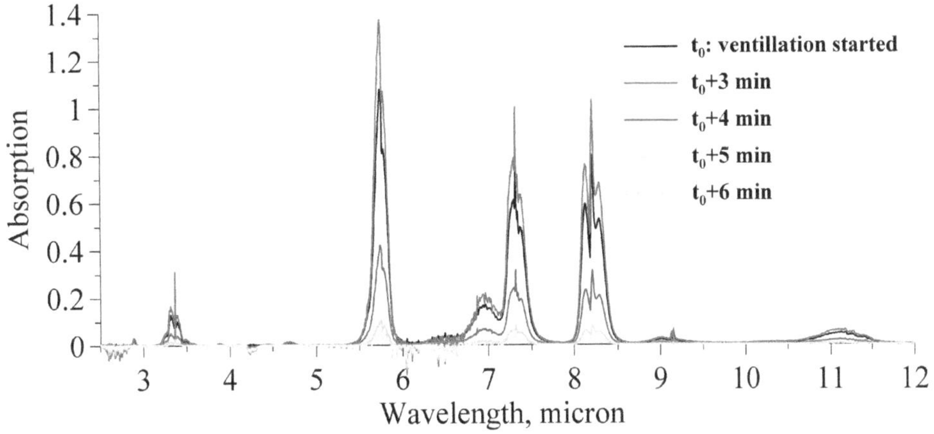

Fig. 2.7. Absorption spectra of acetone vapor during ventilation.

The same scheme in Fig. 2.6 is applicable to detection of leakage in Li-ion batteries: the wafer in this figure must be replaced with a battery pack. Active molecules, leaked from the battery pack, spread in the air, absorbing infrared radiation at very certain spectral lines and clearly identifying the defect.

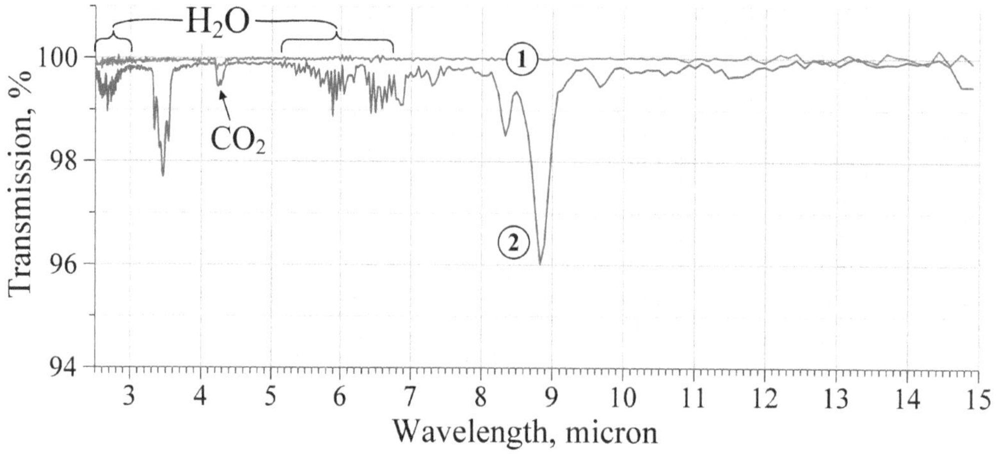

Fig. 2.8. Transmission spectra of the air: (1) normalized background without battery vapor; (2) with the Li-ion battery vapor released. Other air components that strongly contribute to absorption are water (H_2O) and carbon dioxide (CO_2). Obtained using Cary 630 (Agilent) compact Fourier-transform spectrometer with spectral resolution 16 cm^{-1}.

An example of such resonant absorption is clearly visible in Fig. 2.8 that shows transmission spectrum of the air above the opened Li-ion battery. Other components of the air, like CO_2 and water vapor, are also easily seen, and their effect may be substantially decreased by dividing the spectrum by the background obtained before the measurement.

2.3. Emission-Type Fourier-Transform Spectrometers

The emission-type spectrometers measure emission spectra of outer sources, like, for instance, distant objects in environmental applications. Therefore, these devices do not include infrared source in their design (Fig. 2.9). Typical representative of this type of spectrometers is Bruker EM27.

The diaphragm 10 is an essential part of the design. Angular divergence α of the input beam is the second parameter that determines spectral resolution. Even without any additional drawings, it is clear that the ray, coming at the angle α, traverses $\cos^{-1}\alpha$ longer path in the interferometer. As such, the optical path difference increases by approximately $\Delta x \cdot \alpha^2$, after considering the second-order expansion of the cosine of a small angle. For the constructive interference still taking place for these rays, this additional path difference must be smaller than half of the wavelength:

$$\alpha^2 < \frac{\lambda}{2\Delta x}.$$

Combining this with the last formula for spectral resolution, we obtain the requirement for maximum angular divergence of the incoming optical beam:

$$\alpha < \sqrt{\frac{\Delta\lambda}{\lambda}}.$$

Practically, it means that for $\Delta\lambda/\lambda = 10^{-3}$, divergence of the beam must be smaller than 30 mrad or 2°.

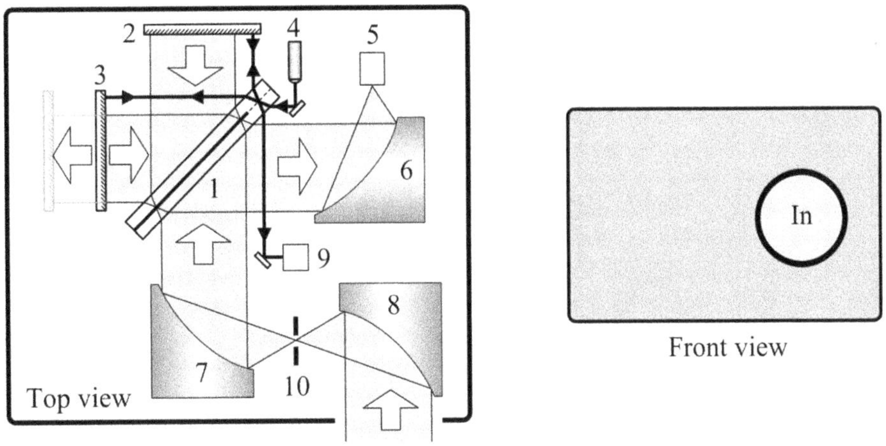

Fig. 2.9. Typical opto-mechanical module of the emission-type Fourier-transform spectrometer. 1 — compensated dual-wavelength beamsplitter; 2 — fixed mirror; 3 — moving mirror; 4 — laser; 5 — infrared photodetector; 6–8 — parabolic mirrors; 9 — reference photodetector; 10 — diaphragm.

2.4. Fourier-Transform Reflectometers

2.4.1. *Infrared Fourier-transform reflectometers*

Reflectometers measure spectra of light reflected from objects. Fourier-transform reflectometers are the devices that can measure infrared spectra of the reflected light, and therefore can be used to monitor the status of photoresist polymerization after the layer of photoresist is formed on a silicon wafer. Transformation of electronic bonds in the photoresist during its curing is clearly mapped on transmission or absorption spectra of infrared light reflected from the photoresist layer. Simultaneously with monitoring chemical transformations, thickness of the layer may also be evaluated, using principle of spectral reflectometry (see Chapter 8 for details). Reflectometers are indispensable instruments in those cases when the material of interest is deposited on opaque substrate or the sample is so large that cannot be inserted into standard spectrometer, like display panels or 300 mm silicon wafers.

The concept of the Fourier-transform reflectometer is clarified in Fig. 2.10, and its practical implementation, using compact Fourier-transform spectrometer Cary 630 (Agilent), is portrayed in Fig. 2.11. The sample is positioned in the focal plane of the parabolic mirror — the feature that makes the entire scheme less vulnerable to tilts of the sample relative to reflectometer.

Figure 2.12 presents the examples of tansmission spectra of photoresist layers on a silicon wafer obtained with this reflectometer. Initially, the two types of photoresists were spinned evenly over the entire surface of two separate wafers. After that, within a small local area on each wafer, the photoresist was removed by acetone to provide two reference spectra reflected from each of the two clean substrates. Finally, the reflected spectra were measured on coated areas of the two wafers and the transmission spectra of the photoresist layers alone were computed by dividing the measured spectra by the reference ones.

2.4.2. *Combined IR–VIS reflectometer*

Spectral reflectometry in visible domain (VIS) is a traditional tool for measuring thickness of transparent thin films. Semiconductor manufacturing requires measurement of thickness of opaque layers, like residual silicon layers after chemical mechanical planarization (CMP process). For that, the reflectometer must work in infrared domain (IR), where silicon is transparent. In an attempt to preserve advantages of both VIS and IR domains, a combined VIS–IR reflectometer may be considered. IR spectroscopy nowadays is based mostly on Fourier-transform IR (FTIR) technology. Therefore, a combined VIS–IR device

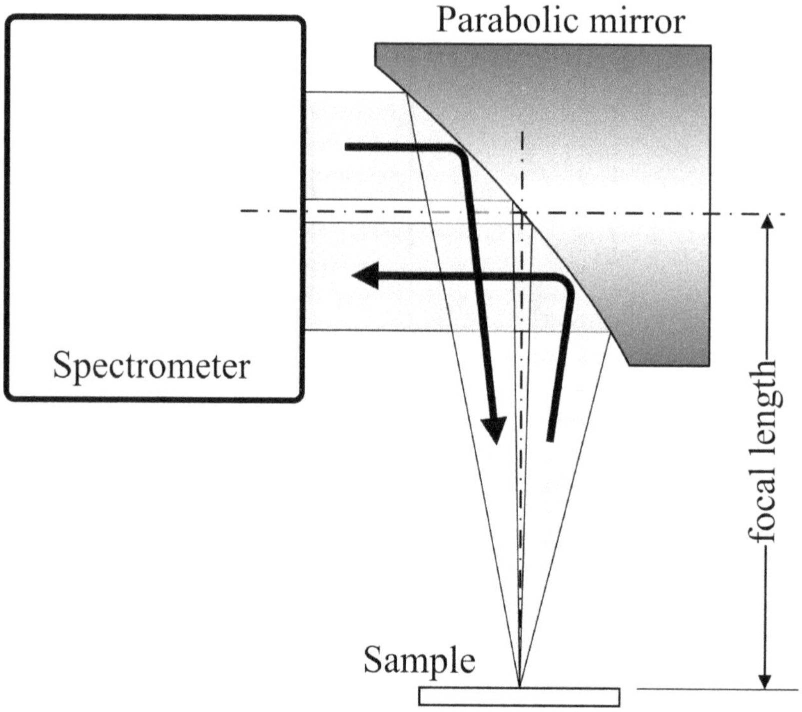

Fig. 2.10. The concept of the Fourier-transform reflectometer.

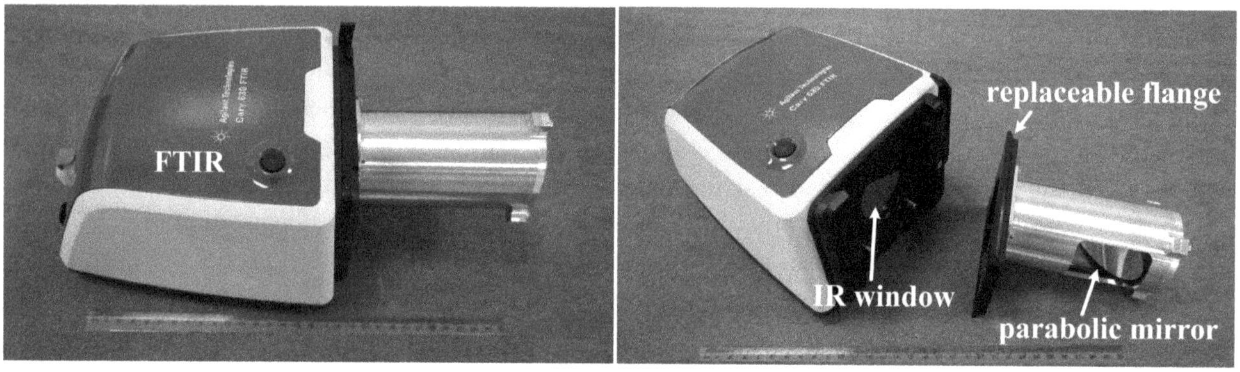

Fig. 2.11. A variant of the infrared Fourier-transform reflectometer, using Cary 630 (Agilent) spectrometer and ThorLabs MPD269 two-inch parabolic mirror with focal length 152.4 mm.

should be designed as a modification of FTIR, incorporating parts of standard VIS spectrometers. Traditional FTIR spectrometers, designed primarily for laboratory applications, are available on the market with horizontal direction of a probe beam. However, in order to satisfy requirements of semiconductor industry, orientation of a sample — silicon wafer — must be horizontal with vertical direction of a probe beam. This only modification, which at the first glance may be considered as insignificant, eventually delivers many useful practical advantages and new applications. In the current section, the prototype of a combined IR–VIS reflectometer is analyzed. The entire list of its features may be summarized as follows:

- ability to measure both visibly transparent and opaque substances, as well as buried structures, such as in semiconductor industry;

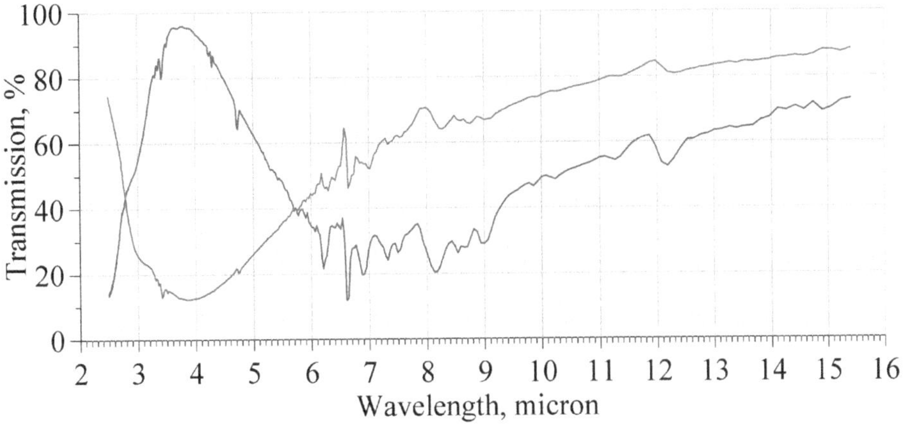

Fig. 2.12. Transmission spectra of the two types of photoresist. Obtained using Cary 630 (Agilent) compact Fourier-transform spectrometer with spectral resolution 16 cm^{-1}.

- horizontal orientation of a sample, convenient for inspection of semiconductor wafers and other flat samples of arbitrary shape without special clamping;
- reflection mode of operation makes 2 times longer optical path inside a sample than in a standard transmission mode;
- spatially resolved measurements with easy manual (automatic in future) adjustment of sample position;
- simultaneous measurement of chemical structure of samples, using IR spectra, and thickness of films, using visible spectra;
- measurement of thickness or, alternatively, refractive indices of thick samples up to 1 cm thick, using high spectral resolution of FTIR, supported by high penetration ability of IR radiation;
- simplicity of measuring liquids from their open surface — no need in side windows in the container;
- insignificant dispersion of most samples in IR;
- position of a sample outside FTIR compartment — convenience for heating or other treatment during experiment.

The prototype of a combined VIS–IR reflectometer is assembled on the platform of Thermo-Fisher Nicolet 6700 FTIR with KBr beamsplitter, using Ocean Optics HR4000CG VIS spectrometer without modifications. Basic optical configuration of FTIR is shown in Fig. 2.13.

Divergence of the beam, entering the interferometer 1, determines spectral resolution, which, in Nicolet 6700, is controlled by setting appropriate diameter of a motorized collimating diaphragm 3. In the modified configuration, described below, sample compartment 9 is not used. Therefore, isolating entry and exit optical windows at opposite sides of it are removed in order to preserve maximum transmission.

Modification was accomplished in two steps: creating vertical probe beam outside FTIR housing, and then installing VIS optics, connected with VIS spectrometer. The first modification is explained in Fig. 2.14.

Two identical 25 mm × 25 mm right angle prism mirrors 1 with reflecting hypotenuses (ThorLabs MRA25-P01, protected silver coating) are combined and glued together, enabling coupling to a right-angle parabolic mirror 2 with focal length 152.4 mm and optical diameter 50 mm (ThorLabs MPD508762-90-P01, protected silver coating). The prisms are installed on two-dimensional linear and two-angle tilt adjustment stages placed one on top of another (Fig. 2.16). When properly aligned, the

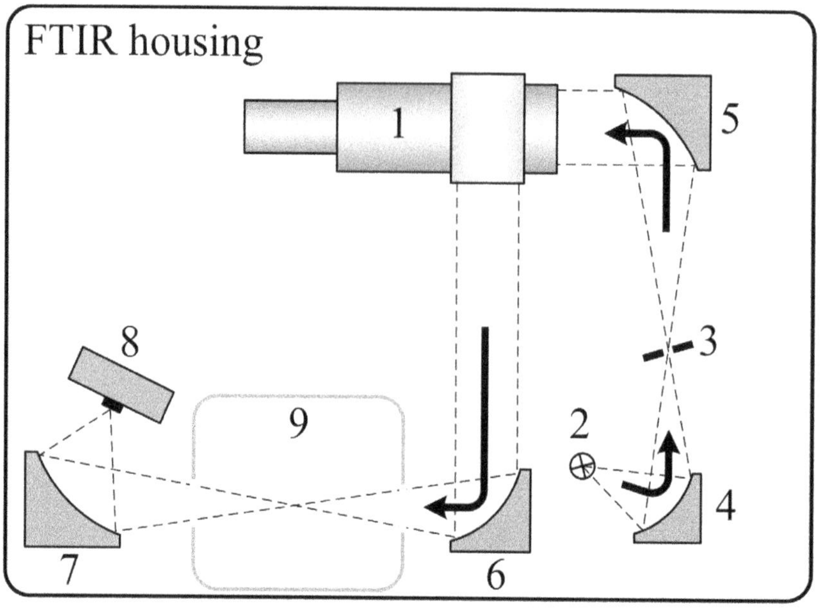

Fig. 2.13. Basic optical configuration of FTIR: 1 — interferometer; 2 — light source; 3 — collimating diaphragm; 4–7 — mirrors; 8 — photodetector; 9 — sample compartment. Arrows show direction of IR beams.

beam focuses on the sample 4, positioned in the focal plane of the parabolic mirror 2, thus making a retro-reflecting optical scheme. Due to symmetry, reflected beam preserves initial collimation and is mirror-shifted around optical axis of the parabolic mirror, being afterwards intercepted by the second right angle prism and returned back to its initial direction. Thus, the exit beam of the interferometer remains coupled to photodetector, not severely decreasing power efficiency of the device. Such a design suffers from three main sources of power loss. The first one is incomplete interception of the exit beam of the interferometer by the first turning prism, whose optical cross section 25 mm $\times$ 25 mm is less than diameter of the exit beam ($\approx$30 mm). The second one is geometrical divergence of the beam reflected from the sample, decreasing interception of it by the second turning prism. And the third source of power loss is longer optical path, which introduces additional absorption in air. Nevertheless, the overall power loss turned out to be not critical, easily compensated for by accumulation option of the Nicolet 6700 software, with which it was possible to average multiple measurements, thus maintaining signal-to-noise ratio at acceptable level. Another problem that appeared from mismatch of sizes of the interferometer exit beam and the first turning mirror was infiltration of the probe beam to photodetector without reflecting from the sample. It was solved by placing a square diaphragm 3 immediately after the second right angle prism (turning mirror). Finally, since most of the samples are transparent in IR, precautions were made to minimize influence of reflection from the sample holder. It turned out to be not as easy as in VIS because, due to much longer wavelength, all machined metal surfaces, even blackened, are good reflectors in IR. Moreover, the aforementioned retro-reflecting property of the optical scheme gathered even scattered (non-specular) component of reflected radiation. To minimize this effect, the opening was made in the sample holder and high-quality absorber placed beneath the focal plane, thus disrupting retro-reflection.

The second step of modification was installation of VIS optics, coupling the VIS spectrometer to the sample. This was done, using optical fibers, as shown in Fig. 2.15, which is almost self-explanatory. The Ocean Optics 74-ACR adjustable fiber collimators, with achromatic doublets corrected for visible

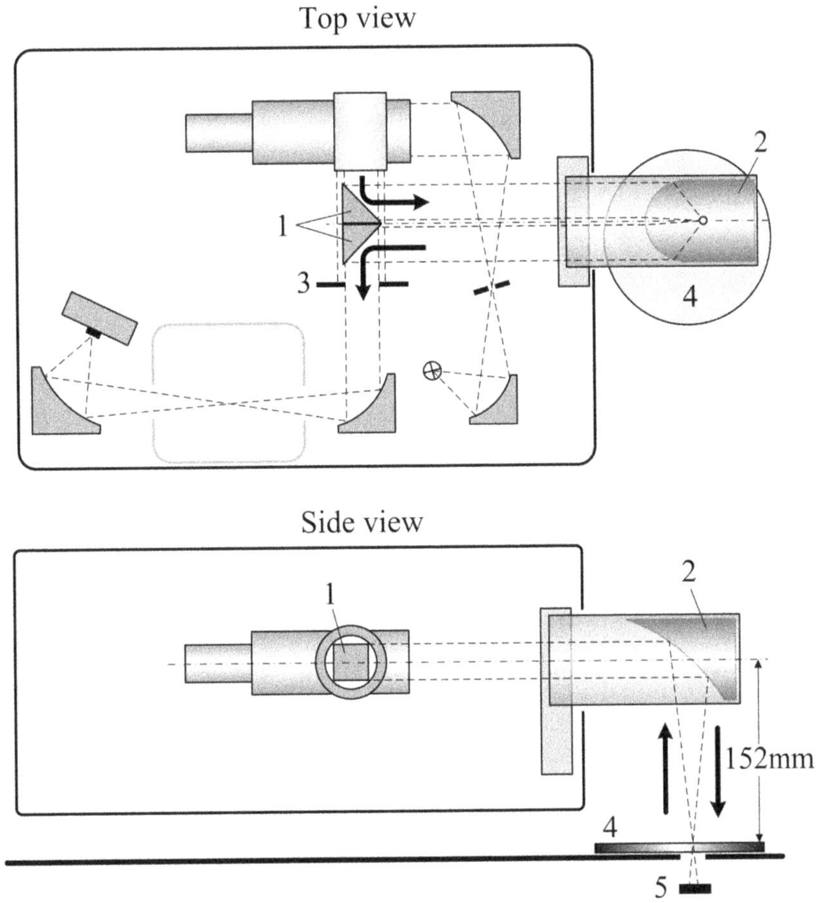

Fig. 2.14. FTIR after modification for vertical probe beam: 1 — right angle prism mirrors; 2 — parabolic mirror with housing; 3 — square diaphragm; 4 — flat sample; 5 — absorber. Arrows show direction of IR beams.

domain, were installed within optical diameter of the parabolic mirror 3 but outside cross section of the probe beam. In order to ensure reliable measurement of the absolute value of reflectivity, which may be affected by tilts of the sample, the fiber collimators should have as big optical apertures as possible, ideally covering the entire free space between the optical diameter of the parabolic mirror and the probe beam of the FTIR. From this point of view, collimators 74-ACR with optical diameters of only 4.5 mm were not the best solution, dictated solely by availability. Requirements to optical fibers, connecting VIS light source and VIS spectrometer, are different. The core of the optical fiber connected to the spectrometer must be wider than 0.2 mm, ensuring full illumination of the detector array inside the spectrometer (the imaging optical scheme is used in HR4000CG). On the contrary, the fiber core, delivering light from the VIS source, must be as narrow as possible to ensure small size of the VIS spot on the sample — the requirement of good spatial resolution of measurements. Specifically, the optical fiber with 50 μm core produced 1mm VIS spot size on the sample. As to the IR spot size, it depends on the collimating diaphragm of Nicolet 6700, and varies from the minimum of approximately 1mm in diameter to the maximum of about 2 mm $\times$ 5 mm. The latter is merely a geometrical projection of a filament of the IR source (fully opened diaphragm) into the sample plane.

Although VIS spectrometer was sensitive up to UV domain (minimum wavelength 200 nm), coating of right angle prism mirrors and parabolic mirror (protected silver P01) was efficient only in VIS and IR

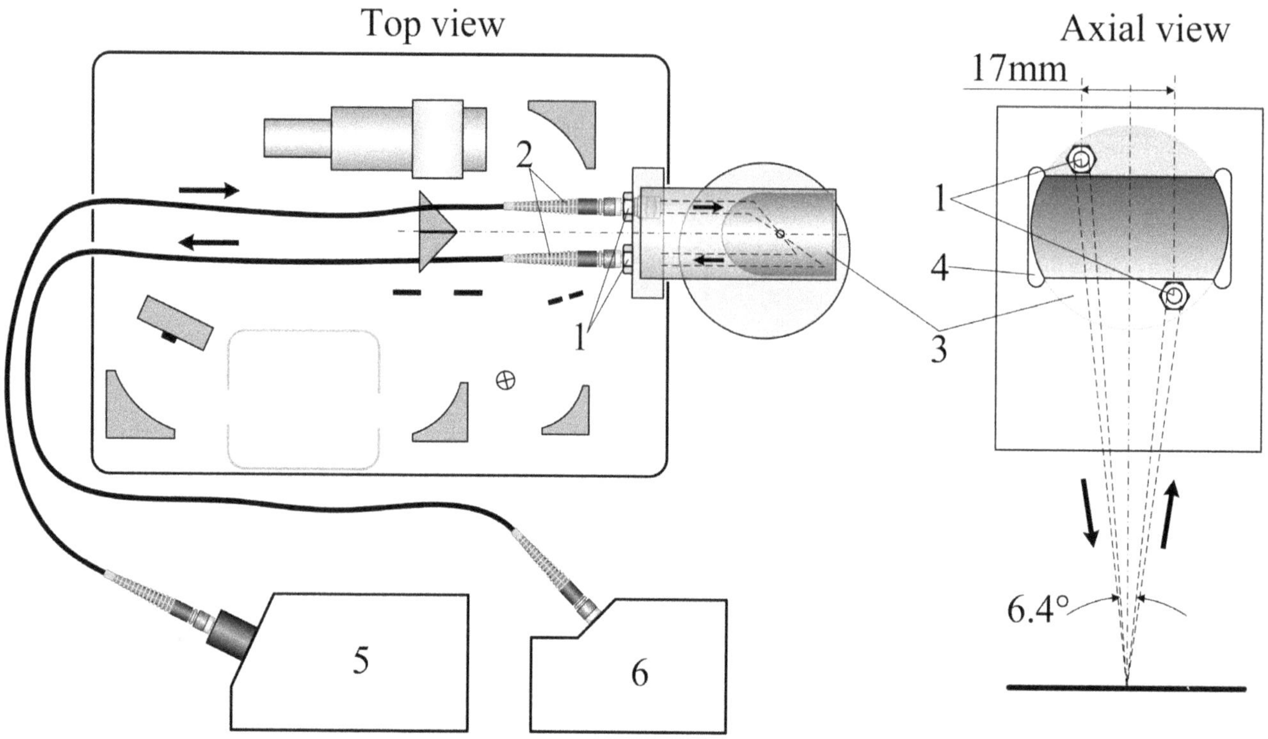

Fig. 2.15. Modification for VIS channel: 1 — fiber-optic collimators; 2 — patch optical fibers; 3 — parabolic mirror; 4 — rectangular hole for IR beam; 5 — VIS light source; 6 — VIS spectrometer. Arrows show direction of VIS light beams.

domains, with minimum wavelength about 400 nm. Therefore, working spectral interval of the combined tool was from 400 nm to 20 μm. The photograph of modified Nicolet 6700 platform is shown in Fig. 2.16.

To begin with, it is instructive to compare combined VIS and IR spectral interferometry on simplest samples, for instance, standard silicon wafer coated with transparent silicon oxide layer of about 5 μm thickness. Typical VIS (25000–10000 cm^{-1}) and IR (12000–1200 cm^{-1}) spectra, obtained within spectral interval 400 nm–8 μm with the same spatial resolution of 1 mm, are shown in Fig. 2.17. Monotonous rise of average reflectivity towards shorter wavelengths in VIS (25000 cm^{-1}, i.e. 0.4 μm) occurs due to smooth increase of refractive index. Abrupt almost two-fold increase of reflectivity towards IR end of spectrum that occurs at 9000 cm^{-1} (1.1 μm) marks the edge of transparency window of silicon, as explained at the end of this section.

In order to properly understand this result, some basic phenomenology of spectral interferometry should be iterated from Chapter 8. Standard silicon wafer of 0.8 mm thickness is almost 90% transparent in IR domain above 2 μm. Therefore, in IR domain, interference is produced by three reflecting surfaces — front surface of SiO_2 layer, interface between SiO_2 and Si, and bottom of the Si wafer (if the wafer is two-side polished). Wavenumber period p of spectral oscillations, produced by two reflecting interfaces spaced by a material of thickness t and refractive index n, is equal to

$$p = \frac{1}{2tn}.$$

In the experiment, the bottom of the wafer and the SiO_2–Si interface are spaced by 0.08 cm with the refractive index of silicon of about 3 in IR (more accurate value will be presented below). Thus, spectral periodicity for this partial interference is $p \sim 2$ cm^{-1}. It means that if spectral resolution of FTIR

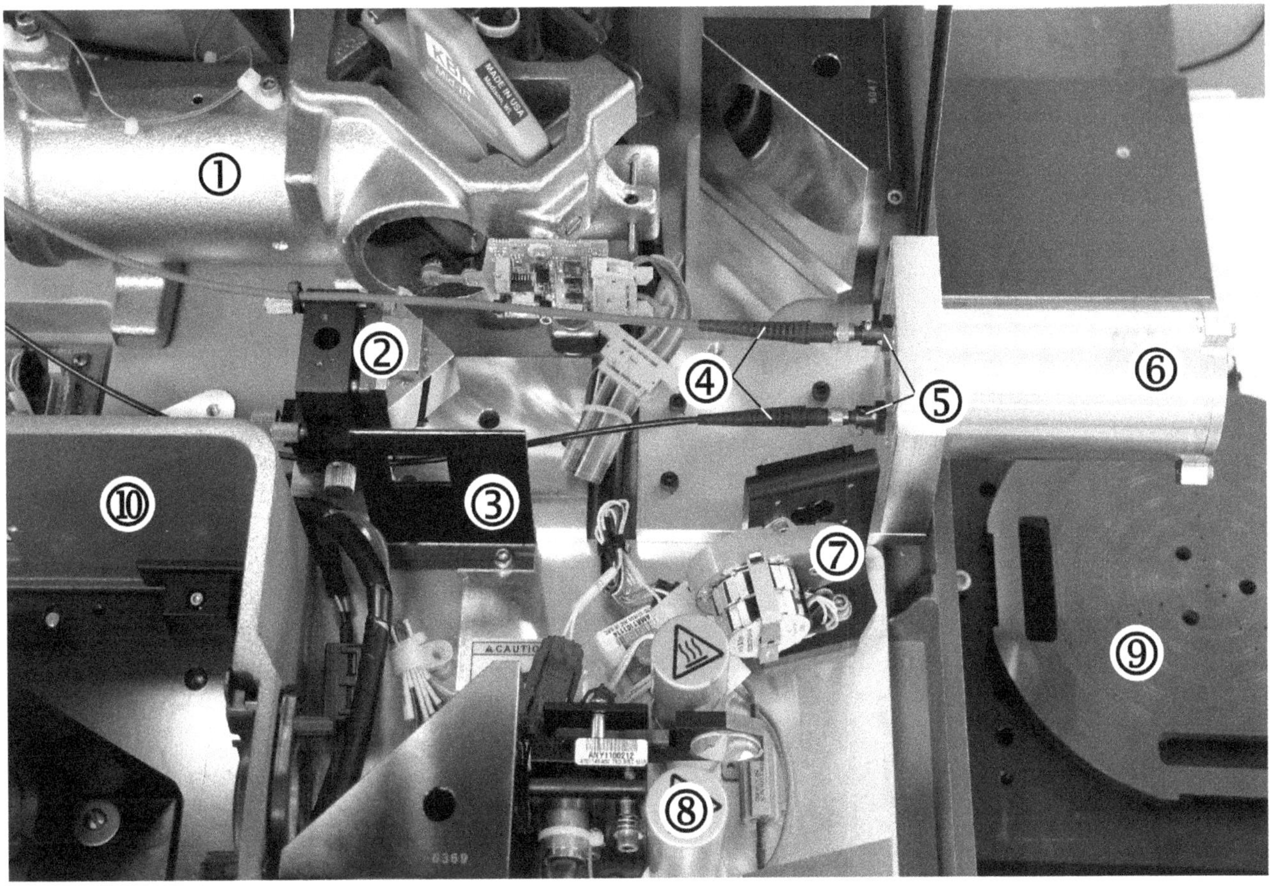

Fig. 2.16. Final optical VIS–IR arrangement: 1 — interferometer; 2 — a pair of right angle prism mirrors; 3 — square diaphragm; 4 — patch optical fibers; 5 — fiber-optic collimators; 6 — parabolic mirror housing; 7 — collimating diaphragm; 8 — light source; 9 — vacuum chuck for samples; 10 — original sample compartment.

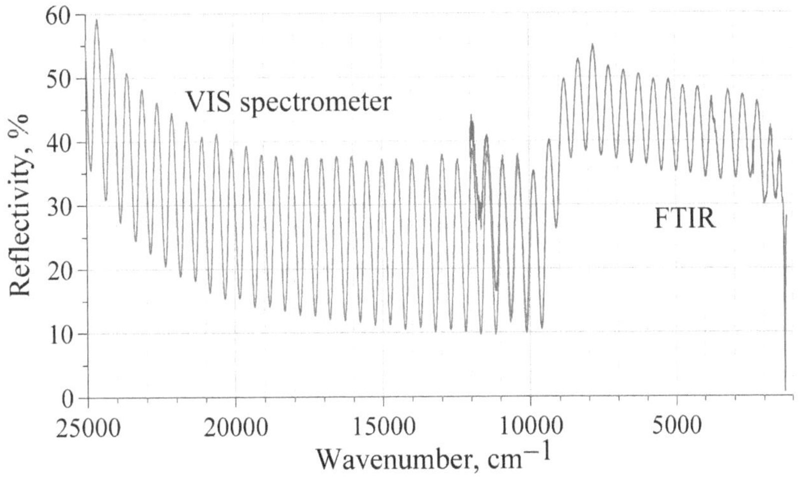

Fig. 2.17. Spectral interference on SiO_2 layer of about 5 μm thick.

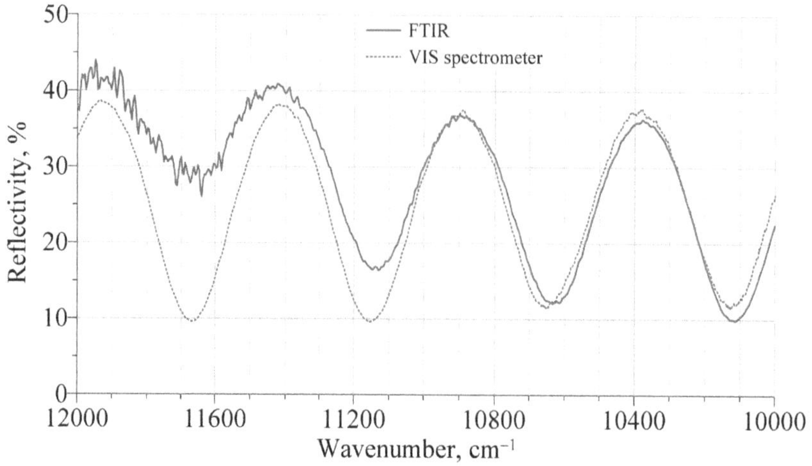

Fig. 2.18. Overlapping part of VIS and IR spectra.

is equal or coarser than this value, then no interference will be observed. In the experiment shown in Fig. 2.17, spectral resolution of FTIR was set to 8 cm^{-1}, which means that no interference on the wafer itself can be seen. As to VIS domain, silicon is opaque and no reflection from the bottom of the wafer can be seen. Therefore, Fig. 2.17 shows only interference on the SiO_2 layer.

VIS and IR spectra shown in Fig. 2.17 overlap within 12000 cm^{-1} and 10000 cm^{-1}, making them hardly discernable. A closer view, showing more details, is presented in Fig. 2.18. Noisier and slightly higher FTIR values near 12000 cm^{-1} are due to quickly falling both optical flux of the FTIR light source and sensitivity of the DTGS photodetector in this part of the spectrum. As to the difference of the phase of modulation, although hardly noticeable, it may be ascribed to slightly different angles of incidence in the VIS and IR beams, as is clear from Fig. 2.15.

Judging on Fig. 2.17, it may be concluded that, for transparent layers, the VIS channel provides more information than the IR one because number of oscillations is bigger. Bigger amplitude of the signal in IR channel may be considered only as insignificant advantage. However, situation changes dramatically for buried structures, which is of the utmost importance for semiconductor industry. A good model of such a structure may be made by placing one silicon wafer upon another: for instance, the one used in the previous experiment and another one with two thin layers on both polished sides of it. The result is predictable (Fig. 2.19): VIS channel gives no information at all, while the IR channel senses the entire 1.6 mm thick structure from top to bottom. In addition to this general conclusion, it is also possible to verify performance of the device as a spectral instrument. Firstly, the VIS and IR channels produce almost identical values of reflectivity in the overlapping region. Secondly, it is possible to compare measured absolute values to theoretical estimates, using VIS part of the spectrum, where 0.8 mm wafer may be considered as infinitely thick bulk. For example, at the wavelength 0.5 μm (20000 cm^{-1}) refractive index of silicon is equal to $4.294 + i \cdot 0.044$. For these values, the Fresnel formulas give reflectivity of silicon at normal incidence 0.387. Measured value, marked by dashed line in Fig. 2.19, is 0.422. Since reflectivity of the sample is computed against the background taken on a mirror, which is supposed to be total reflector, it is possible to evaluate actual reflectivity of the mirror: $R = 0.387/0.422 = 0.92$ — a value very close to standard aluminum-coated mirrors.

Making Fourier transform of IR spectrum, it is possible to identify and analyze in each i-th interface its effective thickness $l_i n_i$, where l_i and n_i are the physical thickness and refractive index respectively. Figure 2.20 shows the example of such computations applied to IR spectrum in Fig. 2.19. If refractive

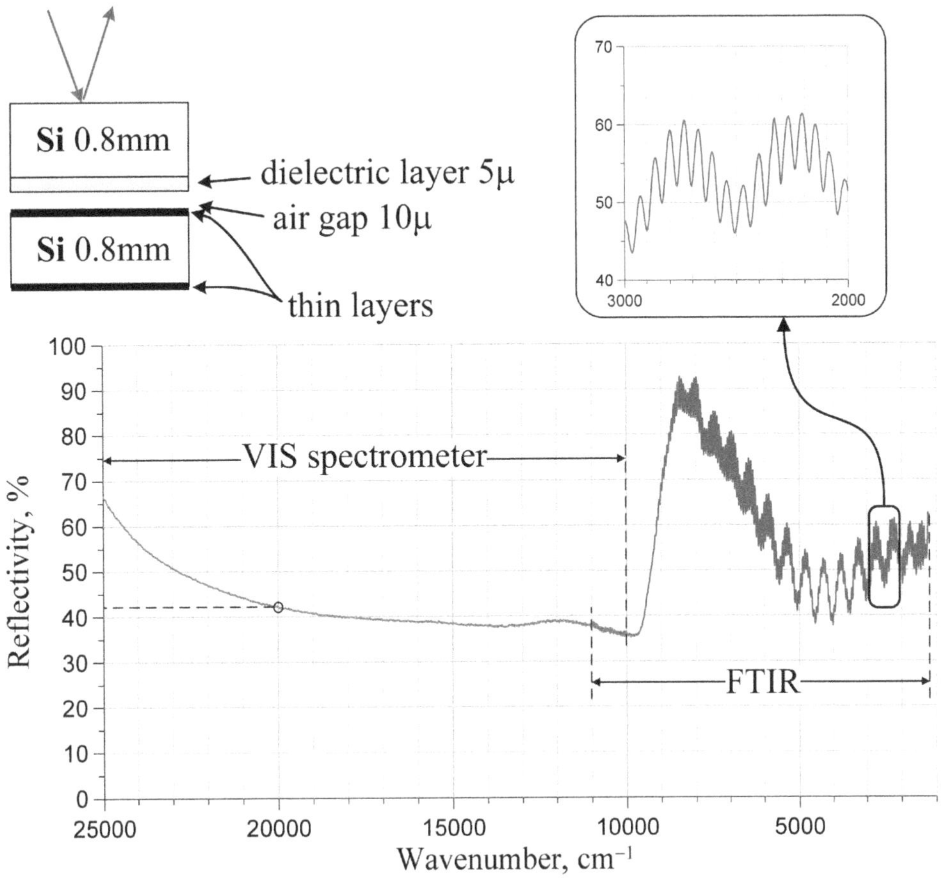

Fig. 2.19. On thick buried structures, like the one shown in the upper left corner, the IR channel senses the entire depth of it, while VIS channel gives no information. Spectral resolution of FTIR 8 cm^{-1}.

indices n_i are known and do not change significantly within the chosen spectral interval, then physical thicknesses l_i can be evaluated separately.

Full potential of a combined VIS–IR scheme is unleashed in those applications where the VIS and IR channels work complementary. As an example, consider curing of a photoresist on silicon wafer. On one hand, it is necessary to know chemical transformations during curing, and on the other — volumetric shrinkage of the material. Chemical transformations can be identified by changes of resonant absorption in IR spectrum, while shrinkage can be directly estimated my measuring thickness of the photoresist layer. The result is shown in Fig. 2.21. Black arrows on the absorption curves mark chemical transformations. On reflection curves, spectral period of oscillations is determined by formula $p = 1/2tn$.

From the formula for period of spectral oscillations, it was computed that thickness of the layer was 1.01 μm before and 0.928 μm after the exposure. Assumption was made that refractive index of the layer was 1.5.

A very elegant and simple application of combined VIS–IR reflectometer is measuring refractive indices of IR materials. From the formula for period p of spectral oscillations, it follows that refractive index n_λ at a certain wavelength λ may be calculated as

$$n_\lambda = \frac{1}{2tp}.$$

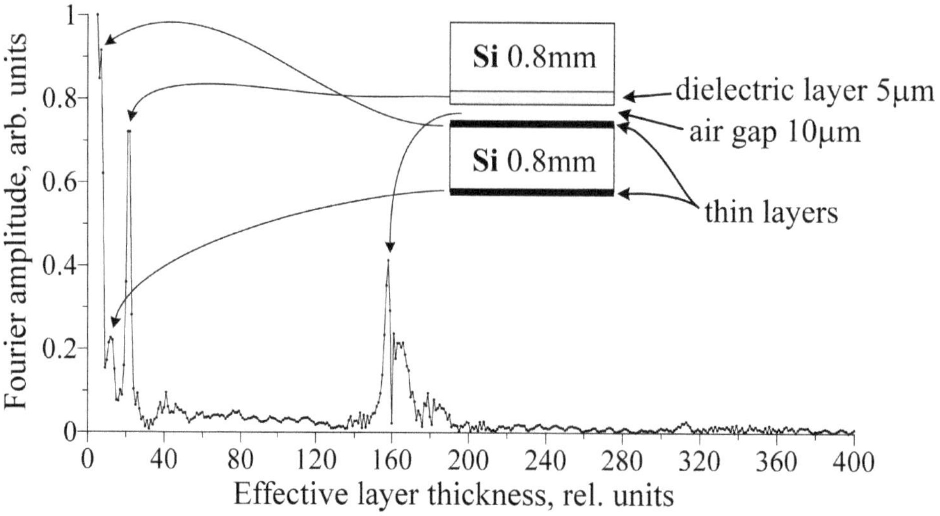

Fig. 2.20. Fourier transform of IR spectra in Fig. 2.19.

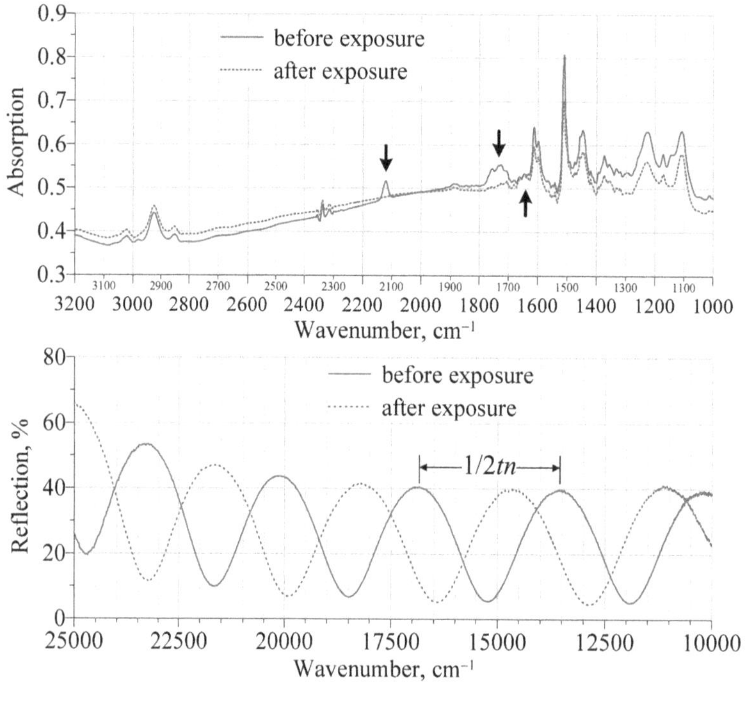

Fig. 2.21. Absorption and reflection spectra are measured, using IR and VIS channels, respectively.

Here wavenumber period p must be measured in the spectral interval of interest, i.e. around specified wavelength λ. Assume, for example, that it is necessary to measure refractive index of a standard 300 mm silicon wafer at $\lambda = 3.0$ μm, i.e. at the wavenumber 3333.3 cm^{-1}. Then spectral interval of FTIR may be chosen from 3383 cm^{-1} to 3283 cm^{-1}, and spectral resolution may be set to the best of Nicolet 6700: 0.125 cm^{-1}. Thickness t of the wafer may be measured by a micrometer, and in this example it was equal to 0.78 mm. The entire 300 mm silicon wafer does not fit into any commercially available FTIR,

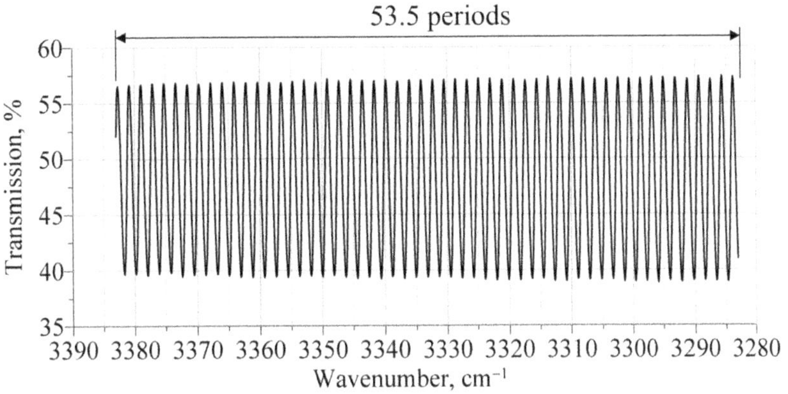

Fig. 2.22. Spectral interferogram of a standard 300 mm silicon wafer around wavelength 3 μm. Spectral resolution 0.125 cm^{-1}.

but with modifications explained in this section it can be easily done. Spectral interferogram is shown in Fig. 2.22.

Here are 53.5 periods of spectral oscillations, which means the average value $p = 100\ \mathrm{cm}^{-1}/53.5 = 1.87\ \mathrm{cm}^{-1}$. Then the above formula gives $n = 3.43$.

Finally, let us explain the sudden increase of average reflectivity in Fig. 2.17, when silicon becomes transparent in IR domain. Consider a bare optical flat in air, reflecting only due to Fresnel reflection on its two interfaces (no reflecting coatings). Then energy reflection coefficient R on both sides is the same. Inside the material, optical losses attenuate intensity by a factor q on each round trip. After first reflection of the incoming ray with intensity I on the front surface, reflected intensity becomes $I \cdot R$. This is the zero partial reflected wave. The ray then penetrates into the flat, reflects from the rear surface, comes back to the front surface, and emerges in air again with the intensity $I(1-R)^2 Rq$. This is the first partial reflected wave. At this point of the front surface, the next reflected ray propagates back into the flat, reflects from the rear surface and again comes to the front with intensity $I(1-R)^2 R^3 q^2$. This is the second partial reflected wave, and so on. Thus, to obtain total intensity of the reflected wave, the following sum must be calculated:

$$IR + I(1-R)^2 \sum_{n=0}^{\infty} R^{2n+1} q^{n+1}.$$

Using the formula for the infinite sum of geometrical progression

$$\sum_{n=0}^{\infty} a^n = \frac{1}{1-a}; \quad a < 1,$$

we obtain the Stokes formula for reflectivity of a semi-transparent optical flat:

$$R\left[1 + \frac{(1-R)^2 q}{1-R^2 q}\right].$$

For fully transparent material, $q = 1$, and reflectivity becomes $2R/(1+R)$, which is approximately the value in the IR domain, where silicon is highly transparent. Alternatively, in VIS, $q = 0$, and reflectivity becomes smaller: R.

Supplemental Reading

R.J. Bell, *Introductory Fourier Transform Spectroscopy*, Academic Press, New York, 1972.

G.G. Stokes, On the intensity of the light reflected from or transmitted through a pile of plates, *Proc. R. Soc.* **11**, pp. 545–556 (1862).

R. Bracewell, *The Fourier Transform and Its Applications*, McGraw-Hill, New York, 1965.

L. Mertz, Auxiliary computation for Fourier spectrometry, *Infrared Phys.* **7**(1), pp. 17–23 (1967).

J. Connes, Recherches sur la spectroscopie par transformation de Fourier, *Rev. d'Optique* **40**, pp. 45–265 (1961).

P.B. Fellgett, The nature and origin of multiplex Fourier spectrometry, *Notes Rec. R. Soc.* **60**(1), pp. 91–93 (2006).

Chapter 3

High-Resolution Spectroscopy

3.1. Scanning Interferometers Fabry–Perot

3.1.1. *Scanning interferometers Fabry-Perot with flat mirrors*

Widely known interferometer Fabry–Perot (IFP) in its scanning modification is used to analyze mode structure of laser beams, and as such is frequently called the laser mode analyzer. It is actually a spectroscopic tool designed to resolve fine structure of laser beams with gigahertz or even megahertz spectral resolution far beyond capabilities of grating spectrometers. However, it must be understood from the very beginning that scanning IFP can never measure absolute values of wavelengths rather than by comparison with another laser source already calibrated with required precision. Therefore, it is better to say that scanning IFP is an indicator of spectral structure of lasers.

Commercially available scanning IFPs, being very simple and reliable in design, are at the same time far more theoretically complicated devices than traditional IFP known from university courses. Therefore, theoretical overview of basic multi-beam interference phenomena is necessary to realize how they work. To simplify mathematics, consider for the beginning a simplified model of a flat-mirror IFP (Fig. 3.1), which is also relevant to interference filters explained in Section 7.2.

Let t and r be the amplitude transmission and reflection coefficients at the mirrors, respectively. Then the plane monochromatic wave with unity amplitude and the wavelength λ, coming at the angle θ, will reflect many times inside the IFP cavity, making the outgoing wave an infinite sum of partial components with phase differences δ:

$$t^2 + t^2r^2e^{-i\delta} + t^2r^4e^{-i2\delta} + \cdots + t^2r^{2(j-1)}e^{-i(j-1)\delta} + \cdots, \quad \delta = \frac{2\pi}{\lambda}p\cos\theta,$$

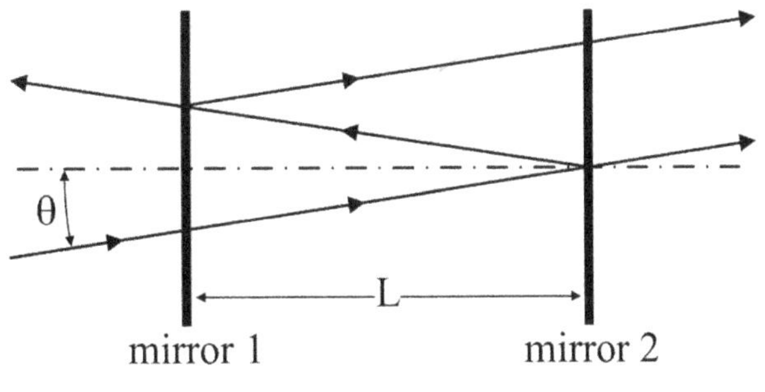

Fig. 3.1. Generalized scheme of the interferometer Fabry–Perot with flat mirrors.

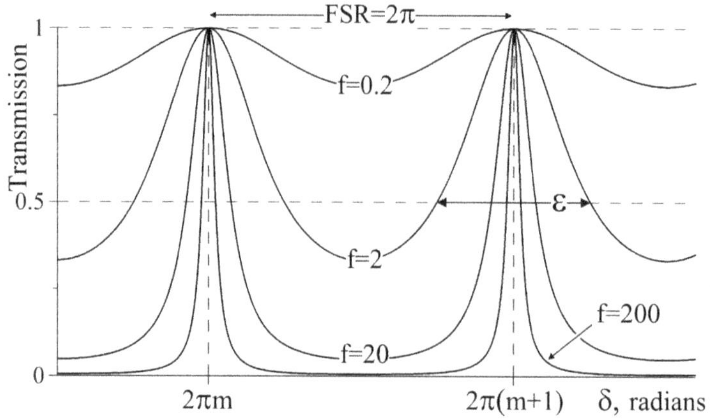

Fig. 3.2. The Airy formula gives periodical function for transmission of the interferometer Fabry–Perot (IFP). With reflectivity of the mirrors $R \to 1$ the shape factor $f \to \infty$, making the peaks narrower.

$i = \sqrt{-1}, j = 1, 2, 3, \ldots,$ and $p = 2L$ — optical path after one period of reflections. In the limit of $j \to \infty$, this geometrical progression converges to

$$a\sum_{j=0}^{\infty} q^j = \frac{a}{1-q},$$

giving the resultant amplitude of the transmitted wave

$$\frac{t^2}{1 - r^2 e^{-i\delta}}.$$

In order to minimize unnecessary details, we allowed some inexactness in this derivation, assuming that transmission coefficients on the entry into the IFP and on the exit of it are the same t. Nevertheless, if this is followed by another minor inexactness in the form of writing energy transmission and reflection coefficients as $T = t^2$ and $R = r^2$, then the final result is exact:

$$\frac{T}{1 - Re^{-i\delta}}.$$

Transmission coefficient of the entire IFP is the square modulus of it, which is better be written in the following form:

$$\frac{T^2}{(1-R)^2 + 4R\sin^2\frac{\delta}{2}}.$$

This formula is known as the Airy formula, which must not be mistaken with the famous Airy function that relates to diffraction on a round hole. In the ideal case of the absence of absorbance, i.e. $T = 1 - R$, the formula transforms to

$$\frac{1}{1 + f\sin^2\frac{\delta}{2}}, \quad f = \frac{4R}{(1-R)^2}.$$

The basic result that follows from here is that transmission of IFP is a periodical function of δ with transmission peaks the narrower, the higher the reflection coefficient R is (Fig. 3.2). The shape reproduces itself every $\delta = 2\pi m$, $m = 0, \pm 1, \pm 2, \ldots$.

Recalling that

$$\delta = \frac{2\pi}{\lambda} p \cos\theta = \frac{4\pi}{\lambda} L \cos\theta,$$

it means that interferometer is transparent for wavelengths λ_m that satisfy condition

$$\lambda_m = \frac{p}{m}$$

at $\theta = 0$. Integer m is a very big number. For example, for $L = 20$ mm and visible domain $\lambda \sim 500$ nm $m \sim 4 \times 10^4$. Spectral separation $\Delta\lambda = \lambda_m - \lambda_{m+1}$ for the particular optical path p is called the free spectral range (FSR) and is equal to

$$\Delta\lambda = \frac{p}{m} - \frac{p}{m+1} = \frac{p}{m(m+1)} \approx \frac{p}{m^2},$$

meaning that only within this spectral interval unambiguous classification of laser modes is possible. The reason for that will be explained below. FSR is typically much smaller than the wavelength, therefore it is usually measured in frequency units $\nu = c/\lambda$:

$$\Delta\nu = \frac{c}{p} = \frac{c}{2L}$$

with c being the speed of light. For the same example with $L = 50$ mm, FSR is 3 GHz — a typical value for commercially available IFPs.

Spectral resolution of the interferometer Fabry–Perot (IFP) is determined by instrumental width ε — the width of the resonant peak in the Airy curve at half the maximum. The number of spectral peaks that can be resolved within the FSR characterizes the quality of an IFP and is called finesse:

$$F = \frac{\text{FSR}}{\varepsilon}.$$

In the domain of phase differences, the FSR $= 2\pi$, and in phase units ε can be derived by applying Taylor expansion for $\varepsilon \ll 1$ to the equation

$$\frac{1}{1 + f \sin^2 \frac{\varepsilon}{4}} = \frac{1}{2}.$$

Then

$$\varepsilon = \frac{4}{\sqrt{f}} \quad \text{and} \quad F = \frac{\pi}{2}\sqrt{f} = \frac{\pi\sqrt{R}}{1-R} \approx \frac{\pi}{1-R}.$$

This very important formula shows that spectral resolution of the flat-mirror IFP increases as reflection coefficient of the mirrors approaches unity. For the finesse to be in the range 100–200, the reflection coefficient of flat mirrors must be as high as 0.98, and it really is that high in the specified spectral interval. Finesse and free spectral range (FSR) are the two basic parameter always included in specifications of scanning IFPs.

Now we are ready to understand how the scanning IFP works (Fig. 3.3). Initially, highly reflective mirrors totally block the laser beam from reaching the photodetector. At the moments when the cavity becomes transparent for a specific wavelength, photodetector (PD) excites short bursts at the screen of the oscilloscope synchronized to the generator. Stable picture of laser modes displays on the screen.

Probability of incidental resonant matching $2L = m\lambda$ between the cavity spacing L and laser wavelength λ is negligible. Fine changes of L are needed to reach full transparency of the IFP. This is done by scanning one of the mirrors. From Fig. 3.2 it is clear that when L changes more than by $\lambda_m/2$ the same wavelength λ_m will be marked for the second time, and may overlap with the wavelength of an adjacent order $\lambda_{m\pm1}$. This happens when spectral width of the laser mode structure is wider than FSR of the IFP. Since longitudinal modes of a laser are spaced in frequency domain by $c/2Z$ where Z is the length of the laser cavity, the full mode structure of a particular laser may be covered by IFP with $L < Z$. However, it does not mean that small spacing L is always better: when super-fine mode structure

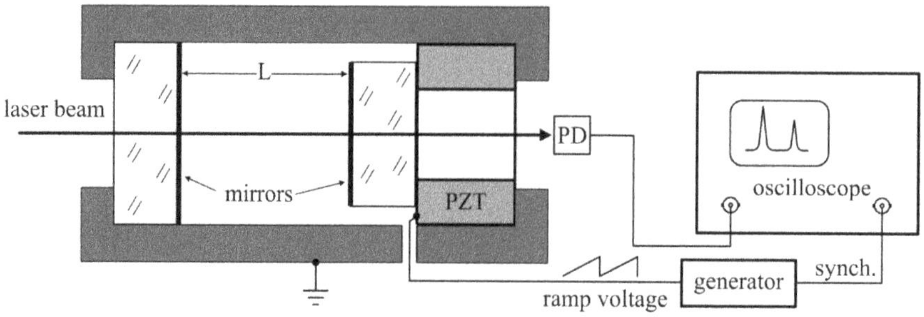

Fig. 3.3. Piezo-transducer (PZT) changes cavity spacing L according to ramp voltage of about 100 V from electrical generator.

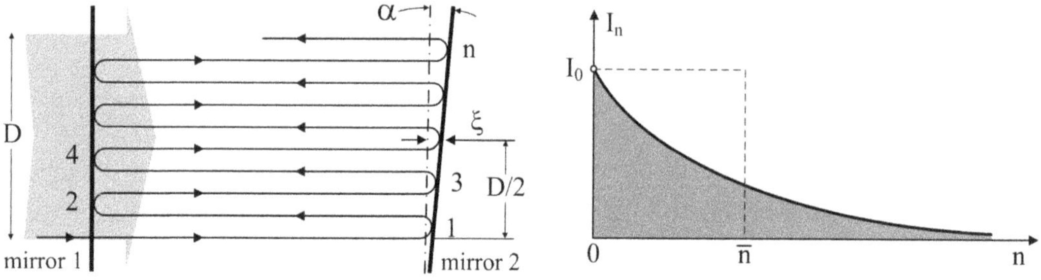

Fig. 3.4. Laser beam of the diameter D comes to an IFP with flat mirrors tilted by the angle α. Average variation of optical paths is 2ξ. Each ray exhibits multiple reflections 1, 2, 3,..., n, each time loosing intensity by the factor R. Intensity of the ray after nth reflection decreases as $I_n = I_0 \cdot R^n$.

is to be analyzed, then longer spacing makes better magnification of the frequency interval of interest. Spectral resolution determined by finesse F does not depend on spacing in our approximation.

Now it is time to make a shocking statement: scanning IFP with plane mirrors never works in laser mode analysis. There are two reasons for that. First, high finesse can be achieved only at normal incidence ($\theta = 0$), and this alignment creates strong reflected wave directed back into the laser cavity, disturbing its operation. Second, tolerance on parallelism of the mirrors is so high that makes volume manufacturing impractical even on the scale of small scientific production. While the first reason is obvious, the second one requires explanation. Figure 3.4 helps to grasp the idea.

With the tilt α, average dephasing within the beam diameter D is

$$\varphi \sim 2\frac{2\pi}{\lambda}\xi = \frac{2\pi D\alpha}{\lambda}.$$

Each ray undergoes on the average $\overline{n}$ reflections until total extinction, and accumulates total dephasing $\overline{n}\varphi$. This total dephasing must not exceed π:

$$\overline{n}\varphi < \pi.$$

Average number of reflections $\overline{n}$ is closely related to finesse F and may be estimated as the integral average (Fig. 3.4):

$$\overline{n} \cdot I_0 = \sum_{n=0}^{\infty} I_n = \frac{I_0}{1-R},$$

giving

$$\overline{n} = \frac{1}{1-R} = \frac{1}{\pi}F.$$

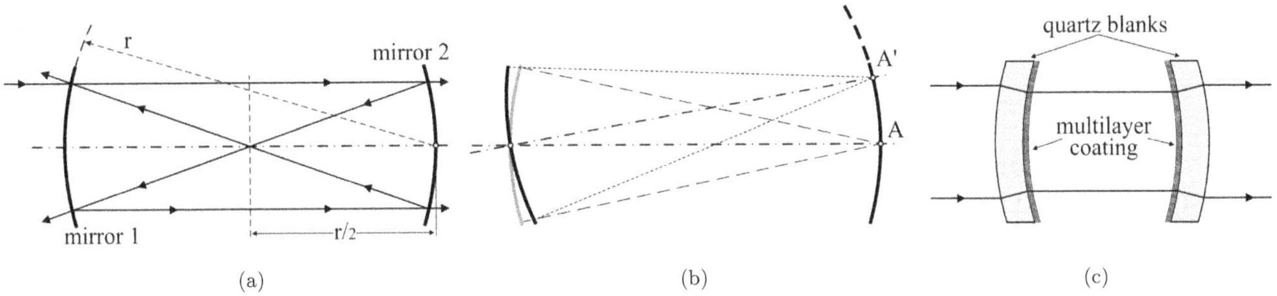

Fig. 3.5. Basic scheme of reflections in confocal IFP.

This gives the estimate for the maximum tilt angle

$$\alpha < \frac{\pi}{2} \cdot \frac{\lambda}{DF} \sim \frac{\lambda}{DF}.$$

Thus, the tilt angle must be F times smaller than the diffraction divergence of the laser beam — a very stringent requirement that cannot be reliably maintained in practice even if the beam is focused inside the cavity to a micrometer-scale size D.

3.1.2. *Confocal scanning interferometers Fabry-Perot*

The solution is the so-called confocal or spherical interferometer Fabry–Perot formed by two spherical mirrors placed at doubled focal length from one another (Fig. 3.5). Before recombining in wavefronts, each ray passes four times through the cavity, exiting also four times but never in the initial direction (Fig. 3.5(a)). Tilting of mirrors does not change properties of the cavity: the point A only slips to a new position A$'$ on the same mirror, reproducing initial configuration (Fig. 3.5(b)). Highly reflecting multilayer coatings are deposited on thin mirror blanks of zero optical power, i.e. concentric surfaces, thus preserving collimation at the output of the IFP (Fig. 3.5(c)).

The greatest advantage of this scheme is insensitivity to tilts of the mirrors. It makes fine angular adjustment redundant, dramatically simplifying manufacturing. The only requirement left is longitudinal adjustment of mirrors to ensure confocality, which can be easily accomplished by fine screws. The second problem of the flat-mirror IFP is also solved: reflected rays do not propagate backwards, making no disturbance to the laser.

Theoretical results obtained for the flat-mirror IFP hold true for the confocal IFP with only very simple modification: since rays reflect four times instead of two, R in the formulas above must be replaced with R^2. Thus, finesse

$$F = \frac{\pi R}{1 - R^2} \approx \frac{\pi}{2(1 - R)};$$

Airy formula

$$\frac{T^2}{(1 - R^2)^2 + 4R^2 \sin^2 \frac{\delta}{2}};$$

shape factor

$$f = \frac{4R^2}{(1 - R^2)^2}.$$

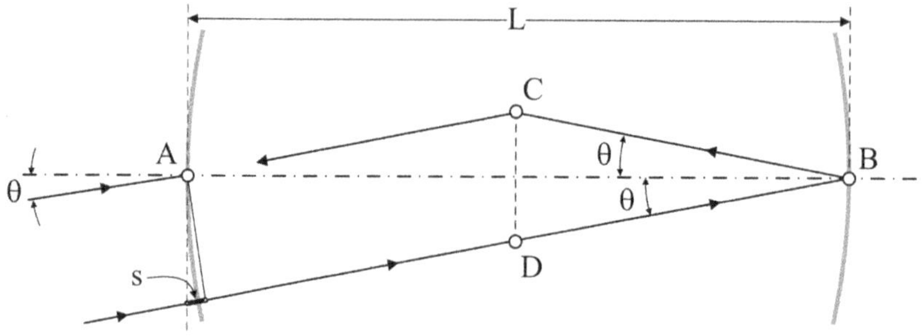

Fig. 3.6. In a confocal IFP, all the rays coming at an angle θ focus first time in the focal point C, then in the focal point D, and then in the focal point C again.

Note that now finesse F is two times smaller than in the case of a flat-mirror IFP, and it brings complications: reflection coefficient of the mirrors must be higher — around 0.993 for $F = 200$.

Another important feature of the confocal IFP is that the phase difference δ does not depend on the angle of incidence θ like in the flat-mirror IFP, dramatically increasing stability of measurements. This is explained in Fig. 3.6.

In order to determine phase difference δ after the round-trip circulation in the cavity, consider its first half from the point of entrance A on the left mirror to the right mirror, then to first focal point C, and then to A again. Instead of exploring cumbersome ray traces from point A to the right mirror and then to point A back, we can use the Descartes principle, saying that optical paths to the focus are the same for all the rays starting at one wavefront surface. As such, the path from A to C is the same as that along the ray passing through D–B–C:

$$\frac{L}{\cos\theta} - s + \frac{L}{2\cos\theta}; \quad s = L\,\tan\theta \cdot \sin\theta.$$

From C the diverging cone of rays passes to the left mirror with the section C–A being

$$\frac{L}{2\cos\theta}.$$

Adding, we obtain the first half of the optical path p

$$\frac{1}{2}p = L\cos\theta + \frac{L}{\cos\theta}$$

and the full optical path after one period of reflections

$$p = 2L\left(\cos\theta + \frac{1}{\cos\theta}\right).$$

For paraxial rays with $\theta \ll 1$ use Taylor expansion of cosine to obtain to the accuracy of the fourth order in θ

$$p = 4L \quad \text{and} \quad \delta = \frac{8\pi}{\lambda}L.$$

Thus, unlike the IFP with flat mirrors, the phase difference in confocal IFP does not depend on the angle of incidence θ and is two times bigger than in the flat-mirror IFP. Accordingly, the FSR for the confocal IFP is two times smaller:

$$\Delta\nu = \frac{c}{p} = \frac{c}{4L}.$$

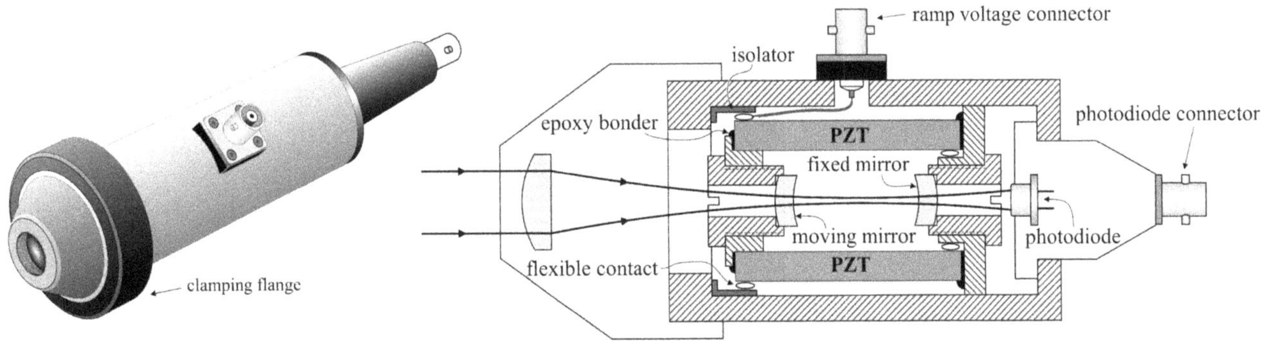

Fig. 3.7. Typical design of a scanning interferometer Fabry–Perot (IFP).

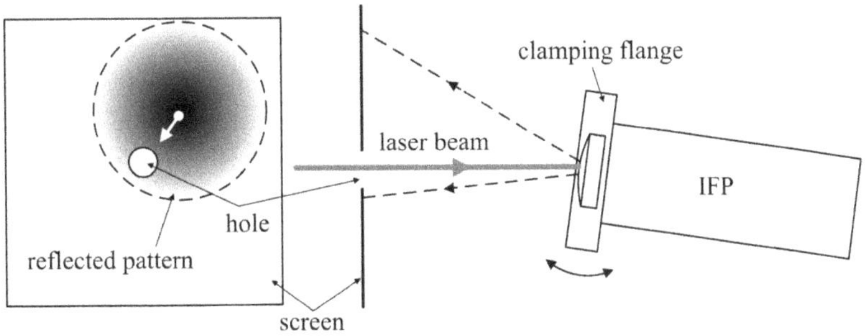

Fig. 3.8. Adjust IFP angularly so that the reflected pattern is roughly centered around the laser beam.

From another point of view, paraxial approximation $\theta \ll 1$ guarantees the absence of spherical aberrations on spherical mirrors. If, however, beam diameter D increases, spherical aberration becomes noticeable, adding corrections to the above formula for the path difference δ. This effect displays itself in a form of circular interference fringes that are formed in the middle plane of the cavity where rays are focused into the image of the source: wide central circle with ever tighter concentric circles around it. In spectroscopic applications, which are the subject of this section, fringes are only a hindrance because they decrease spectral resolution, and as such they should be eliminated. Therefore, the narrower the beam inside the cavity is, the better. The best solution is to focus the beam in the middle of the cavity. Then rays will go exactly the same way as in the case of a collimated beam (Fig. 3.5) but in a reverse order.

3.1.3. *Practical results*

All the commercially available scanning IFPs are of the confocal type (Fig. 3.7). Inner and outer cylindrical surfaces of the piezo-transducer (PZT) are metallized and connected to the ramp voltage through flexible contacts. The photodiode tail is usually removable in order to observe interference fringes when the device is illuminated from the right side. Then the input lens serves as a microscope. Mirrors are fixed in threaded nuts for adjusting the length of the cavity.

Narrow focused laser beam can miss the photodetector if not properly aligned. For that, the IFP must be clamped in an adjustable mount with the laser beam approximately in the middle of the entrance lens (Fig. 3.8).

Then adjust angularly the IFP to bring the reflected beam cone coaxially with the laser beam. Clamping flange is always designed to be roughly in the plane of the lens in order to avoid lateral

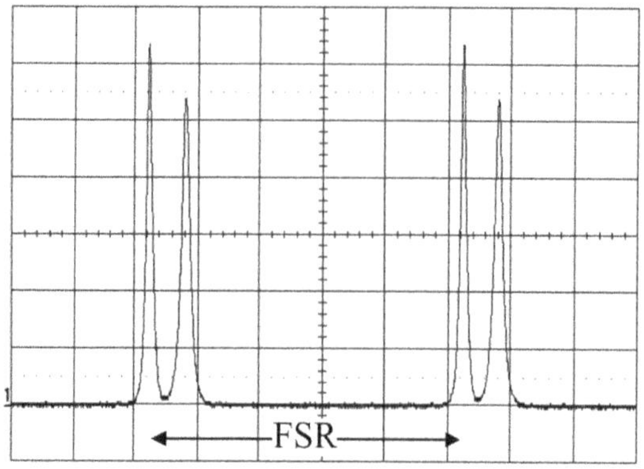

Fig. 3.9. Two laser modes displayed twice during scan over approximately 2 FSR range. The finesse is about 50.

displacements during angular alignment. The IFP is ready for use. If the reflected beam is hardly visible, then another option may be used: remove the photodetector tail and entrance lens; pass the laser beam through the IFP onto the screen; reinstall the lens and adjust the IFP angularly to superimpose the centers of the two beams. Connect the entire system as in Fig. 3.3 and observe the result like the one shown in Fig. 3.9.

Although absolute value of the laser wavelength cannot be measured, using IFP, these devices are routinely used to measure temporal variations of the wavelength of stabilized He–Ne lasers in semiconductor manufacturing. A simple estimate shows that with the reflection coefficient of IFP mirrors being 0.99, its finesse may be $F \sim 100$–200. With the ordinary FSR $\sim$ 1–2 GHz, such a device may produce spectral resolution

$$\varepsilon = \frac{FSR}{F} \sim 5\text{–}15 \text{ MHz}$$

that is a value unattainable by grating or Fourier spectrometers.

As an example, consider variations of optical frequency of the Zygo two-frequency He–Ne laser (0.63 μm wavelength), commonly used in semiconductor industry (Fig. 3.10). Its output beam is combined of two superimposed orthogonally polarized TEM_{00} beams with diameters approximately 5 mm, separated in frequency by exactly 20.0 MHz by means of acousto-optical modulation — sufficiently large frequency separation to be measured by the IFP. Hence, on one hand, the Zygo laser may be used to calibrate the interferometer itself and to compare the data with factory specifications. On the other hand, the IFP may be used to monitor temporal instability of the laser frequency.

Frequency stabilization is achieved by adjusting the length of the laser cavity to produce maximum optical power, i.e. by adjusting the laser line to the center of the neon spectral line. This is done by establishing negative feedback between the signal from the photodetector 16 and the control current, heating the laser tube, causing it to expand or contract in a proper way. The laser beam at the output coupler 13 is not polarized. The polarizer 14 serves in two ways: as a sampler, directing a portion of the laser beam onto the photodetector 16 ("o" polarization state), and as a linear polarizer for the beam that comes to the acousto-optical modulator (17) ("e" polarization state). Polarization axis of the "e" beam makes 45° with the direction of acoustic wave in the modulator (17), and the latter is designed so that to split the output beam into two: the one with vertical polarization (relative to the plane of the figure) and the second one with horizontal polarization. The wavefronts of these two orthogonally linearly polarized beams coincide, while their frequencies are shifted by exactly 20 MHz — the frequency of modulation. While the cross section of the laser beam at the output of the laser

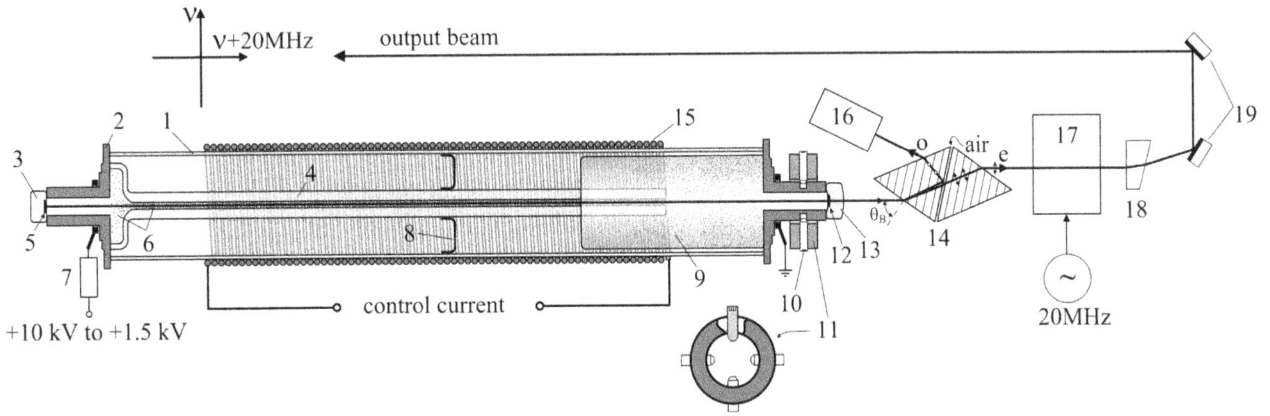

Fig. 3.10. Optical layout of the Zygo laser. 1 — glass tube; 2 — anode; 3 — highly reflecting flat mirror; 4 — capillary tube (bore); 5 — reflecting coating; 6 — gas discharge; 7 — ballast resistor; 8 — supporting spacer; 9 — cathode; 10 — set screws; 11 — fine alignment yoke; 12 — monochromatic reflecting coating; 13 — combined output concave mirror and collimating lens; 14 — Glan-Taylor polarizing prism with Brewster angle; 15 — heating coil; 16 — photodetector; 17 — acousto-optical modulator; 18 — correcting prism; 19 — folding mirrors.

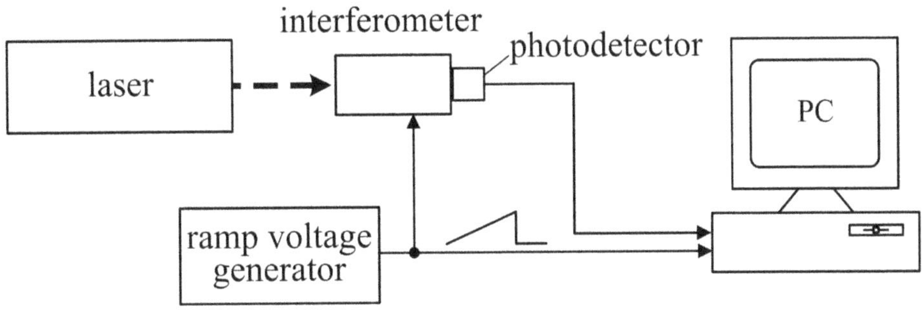

Fig. 3.11. Calibration of the scanning interferometer Fabry–Perot (IFP) on the Zygo laser.

tube is circular, subsequent propagation through polarizer 14 and modulator 17 makes it elliptical. In order to restore axial symmetry of beam, the correcting prism 18 is installed at the output. Unlike other two-frequency heterodyne lasers, being used in semiconductor manufacturing and based on the Zeeman magnetic splitting of spectral lines, the Zygo laser has definite and stable frequency shift between its orthogonally polarized components, which does not depend on the strength of magnetic field and, therefore, is much more stable relative to Zeeman lasers.

Consider first how to calibrate the IFP, using Zygo laser. The scheme of the experiment is outlined in Fig. 3.11. The laser should be turned on in advance and given several minutes to heat up in order to lock on the center of neon line for final stability, as explained in the paragraph above.

The experiments start when the ramp voltage from the generator is applied to the IFP. The ThorLabs SA200 interferometer with specified FSR = 1.5 GHz and 7.5 MHz resolution was used in the experiments. Typically, the amplitude of the linearly rising voltage, corresponding to FSR, is about 5 V. With this amplitude, the output signal of the photodetector is supposed to reach its maximum only once. However, since actual position of the first transparency peak of the interferometer relative to the applied voltage is unknown, the amplitude of the ramp signal should cover at least two FSR in order to guarantee observation of at least two transparency peaks and to be able to compare the 20 MHz split to FSR.

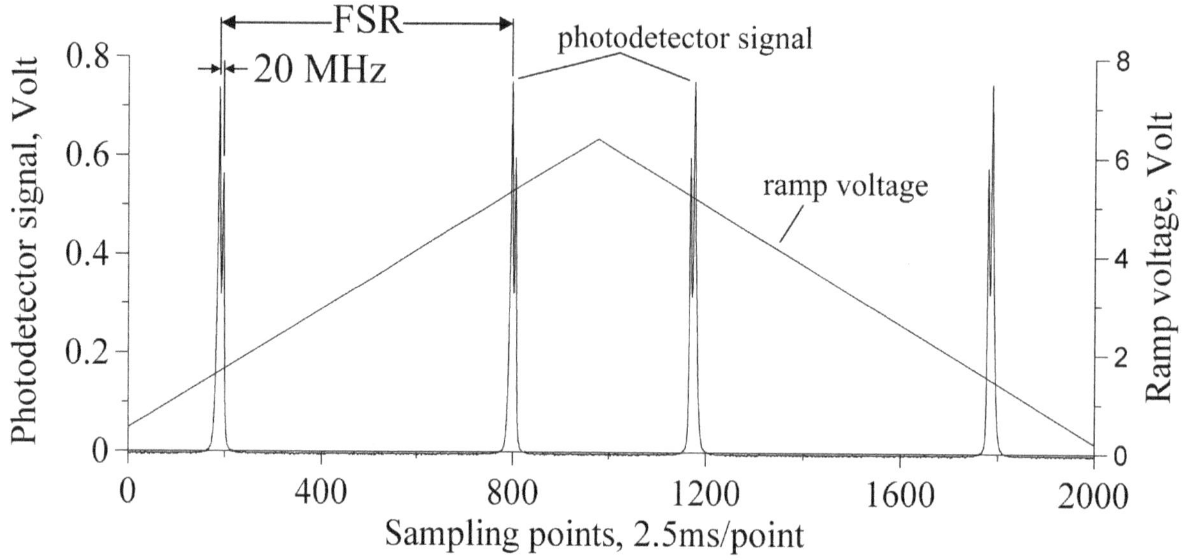

Fig. 3.12. Spectral composition of the Zygo laser in stabilized mode.

Since actual time constant of electro-mechanical response of the IFP is unknown, it is necessary to be sure that electro-mechanical delays do not affect the result. For that, the triangular voltage was applied to the IFP, and the shapes of the curves on the rising and falling parts were compared: their mirror-like identity would guarantee that electro-mechanical delay is negligible. The result is presented in Fig. 3.12, clearly showing such a mirror-like symmetry around the center of the ramp voltage.

The second conclusion is that 20 MHz split of the laser mode structure is well resolved (Fig. 3.13). We may doubt that the values of the FSR and spectral resolution, specified in the passport of the interferometer, are correct, but the 20 MHz frequency split of the Zygo laser modes is a well-defined constant, guaranteed by quartz crystal of its generator. Taking it as a reference, it is possible to accurately estimate actual values of other parameters.

The actual FSR and spectral resolution of the particular IFP can be computed in units of sampling points: if 20 MHz corresponds to 8 sampling points, then 610 points of the FSR correspond to 1525 MHz, i.e. within practical precision FSR = 1.5 GHz. Spectral resolution may be measured as the number of sampling points between the first maximum and its half-value: three points. Within the same proportionality, spectral resolution is measured to be 7.5 MHz. Thus, within practical accuracy of measurements, the values specified by the manufacturer are exactly same as measured.

The next application of the IFP is measuring spectral drifts of stabilized lasers. But even more interesting experiment that can be done, using the same optical scheme in Fig. 3.11, is exploring the process of initial mode locking that precedes the final stabilized operation. In this experiment, the recording process may start immediately after turning the laser on. According to the physics of stabilization explained above, the control electronics of the laser begins oscillating process of heating and cooling the laser cavity, thus causing its resonant frequency to scan across the narrow neon spectral line and the output optical power to oscillate accordingly. The neon gain curve has full width at half-maximum 1500 MHz at room temperature and typical pressure ~5 mTorr, i.e. approximately same as the FSR of the interferometer in our experiments. Consequently, it is possible to expect that, during stabilization, spectral variations of the laser output will cover the entire FSR of the interferometer. Indeed, after

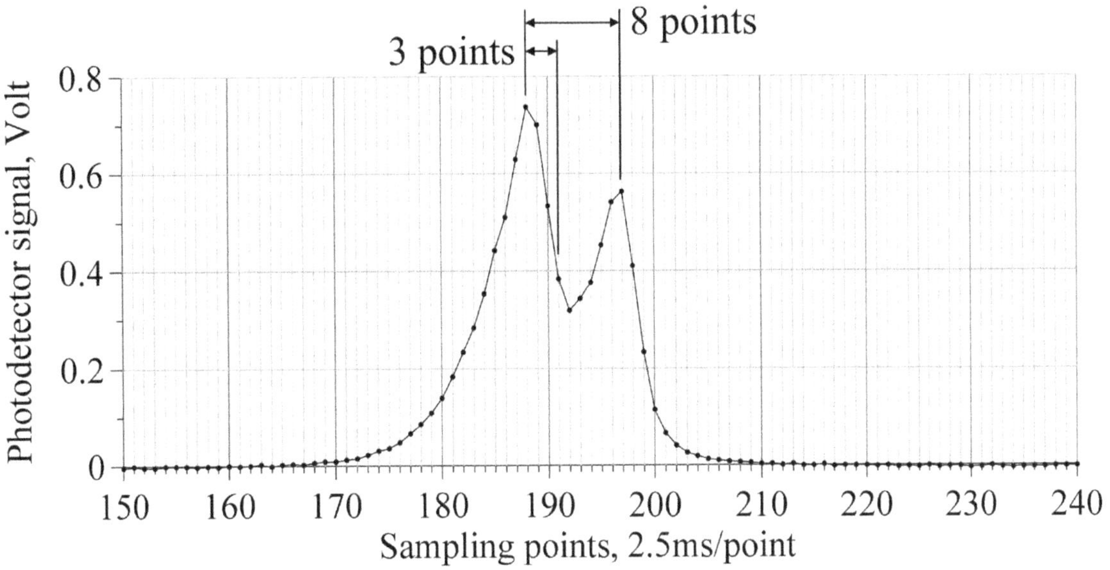

Fig. 3.13. Magnified first peak of the Fig. 3.12.

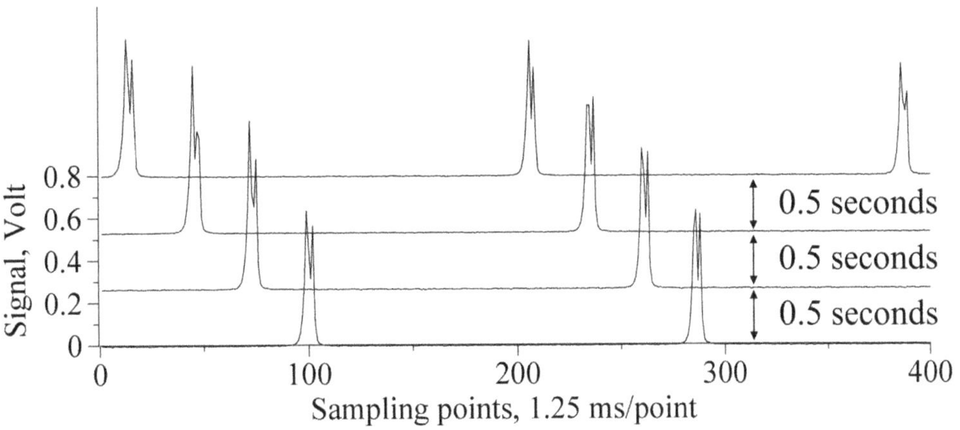

Fig. 3.14. Consecutive spectral traces of the IFP, separated by 0.5 seconds, are combined vertically, showing increasing spectral displacements, as time elapses after turning the Zygo laser on.

turning the laser on, the frequency of the fundamental laser mode begins to change within the working spectral interval of the IFP (Fig. 3.14). Since the IFP mechanical scan is not infinitely short in time, the spectral positions of the transparency peaks may change during one scan due to stabilization process, making the distance between these peaks unequal from trace to trace.

This process can be more conveniently analyzed in two- and three-dimensional pictures, in which one coordinate is formed by the IFP sampling points and the other — by the time elapsed from the turn-on moment (Fig. 3.15).

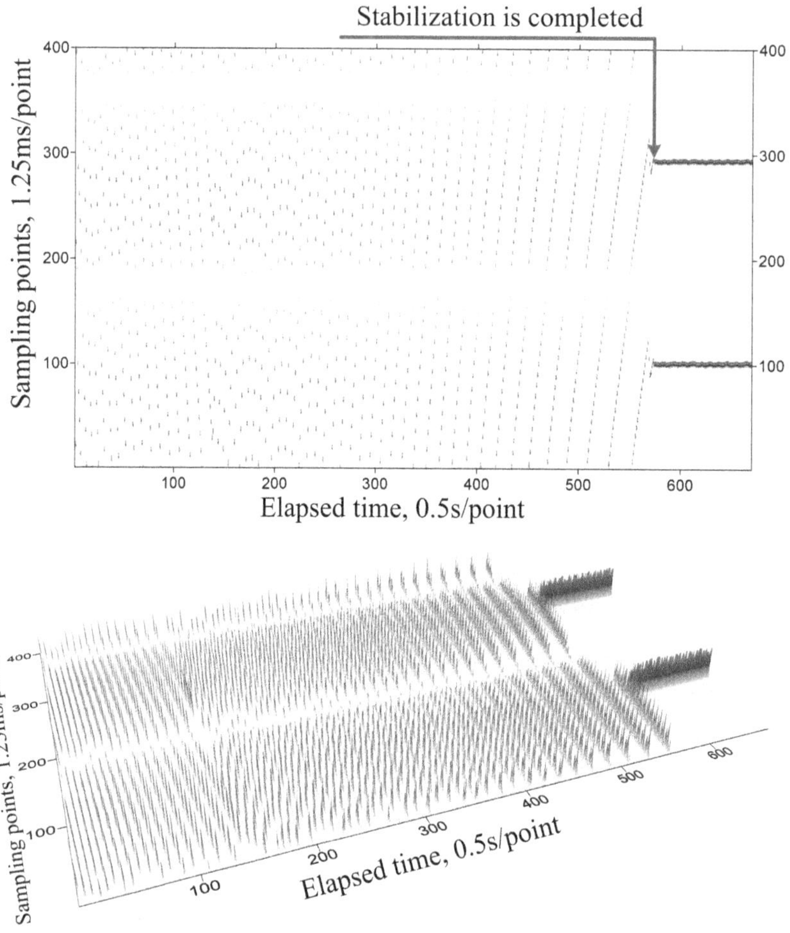

Fig. 3.15. 2D and 3D pictures of the Zygo laser stabilization process.

3.2. Laser Heterodyne Spectroscopy

3.2.1. *Interference of optical waves on the photodetector*

Even higher spectral resolution may be achieved, using the heterodyne spectroscopy. Optical heterodyning and all its phenomenology are based on the interference of two optical waves on the sensitive area of a photo-detector (Fig. 3.16).

Throughout this chapter, we shall be using bold characters as notation of vectors. With this comment, the interfering electrical fields of the waves can be presented in a form of complex exponential functions:

$$\mathbf{E}_1(\mathbf{r}, t) = \mathbf{e}_1 A_1(\mathbf{r}) \exp(i\omega_1 t),$$
$$\mathbf{E}_2(\mathbf{r}, t) = \mathbf{e}_2 A_2(\mathbf{r}) \exp(i\omega_2 t).$$

where $\mathbf{e}_1$ and $\mathbf{e}_2$ — unity vectors of polarization (consider them independent of spatial coordinate $\mathbf{r}$), $A_1(\mathbf{r})$ and $A_2(\mathbf{r})$ — complex amplitudes, determining intensity and phase spatial distribution of the waves, ω_1 and ω_2 — their angular frequencies. Interaction of the resultant field

$$\mathbf{E}(\mathbf{r}, t) = \mathbf{E}_1(\mathbf{r}, t) + \mathbf{E}_2(\mathbf{r}, t)$$

with the material of photo-detector sensitive area originates the photocurrent $j(t)$. Photo-electrons, emerging from a small element $\delta\sigma$ at some point $\mathbf{r}$ of the detector sensitive area, generate pulses of

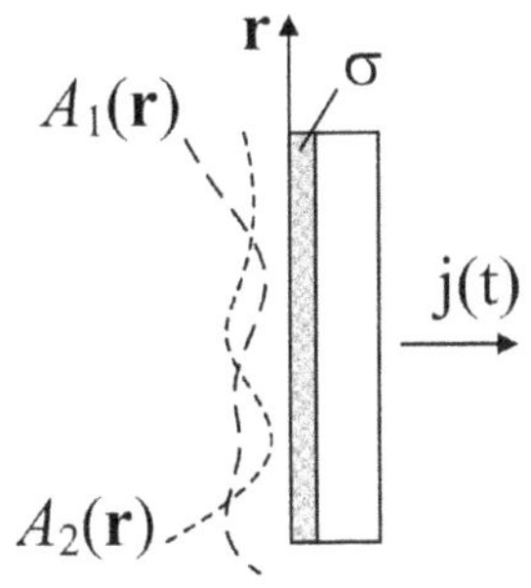

Fig. 3.16. Interference of two fields with complex amplitudes $A_1(\mathbf{r})$ and $A_2(\mathbf{r})$ on the detector sensitive area σ.

electrical current in the recording circuit. These short pulses, merging with each other, contribute to regular continuous component of the current with the amplitude

$$\delta j(\mathbf{r},t) = \frac{q}{h\nu}\eta(\mathbf{r})I(\mathbf{r},t)\delta\sigma.$$

Here q is the electron charge, $\eta(\mathbf{r})$ — quantum efficiency of the detector material at the point $\mathbf{r}$, $h\nu$ — mean photon energy of the interfering waves, $I(\mathbf{r},t) = |\mathbf{E}(\mathbf{r},t)|^2$ — intensity of the resultant field.

Since the nature of the photo-effect is essentially random, the photocurrent from the element $\delta\sigma$ contains both the regular component and the random one (noise) with the amplitude $\delta n(\mathbf{r},t)$. The total photocurrent at the output of the detector is determined by the integral over the detector sensitive area:

$$j(t) = \frac{q}{h\nu}\int_\sigma \eta(\mathbf{r})I(\mathbf{r},t)d^2r + n(t),$$

where the random component is

$$n(t) = \int_\sigma \delta n(\mathbf{r},t)d^2r.$$

Consider first the regular component of the photocurrent $j(t)$, which can be written in the following form:

$$j_1 + j_2 + \frac{q}{h\nu}\mathbf{e}_1\mathbf{e}_2 e^{i(\omega_1-\omega_2)t}\int_\sigma \eta(\mathbf{r})A_1(\mathbf{r})A_2^*(\mathbf{r})d^2r + \frac{q}{h\nu}\mathbf{e}_1\mathbf{e}_2 e^{-i(\omega_1-\omega_2)t}\int_\sigma \eta(\mathbf{r})A_1^*(\mathbf{r})A_2(\mathbf{r})d^2r.$$

Here the asterisk denotes the complex conjugate. The two first terms present independent of time constant photocurrent components, generated separately by $\mathbf{E}_1(\mathbf{r},t)$ and $\mathbf{E}_2(\mathbf{r},t)$:

$$j_1 = \frac{q}{h\nu}\int_\sigma \eta(\mathbf{r})|A_1(\mathbf{r})|^2d^2r; \quad j_2 = \frac{q}{h\nu}\int_\sigma \eta(\mathbf{r})|A_2(\mathbf{r})|^2d^2r.$$

The third and the forth complex conjugated terms present the interference of the fields $\mathbf{E}_1(\mathbf{r},t)$ and $\mathbf{E}_2(\mathbf{r},t)$, and they determine the real oscillating component of the photocurrent that varies harmonically with the angular frequency $|\omega_1 - \omega_2|$. These integrals are commonly called the interference integrals. Consider for the beginning the simplest case of constant η, A_1, and A_2 independent of lateral coordinate $\mathbf{r}$. Then the sum of the third and forth terms transforms to

$$j_3 = 2\frac{\eta q}{h\nu}\sigma\mathbf{e}_1\mathbf{e}_2|A_1||A_2|\cos[(\omega_1-\omega_2)t + \phi],$$

where σ is the area of the photo-detector, ϕ — constant phase term equal to the difference between the phases of the complex amplitudes $A_1(\mathbf{r})$ and $A_2(\mathbf{r})$. This harmonic component, containing information about both the amplitude and phase of the interfering fields, plays the key role in laser heterodyning. Consider its main properties.

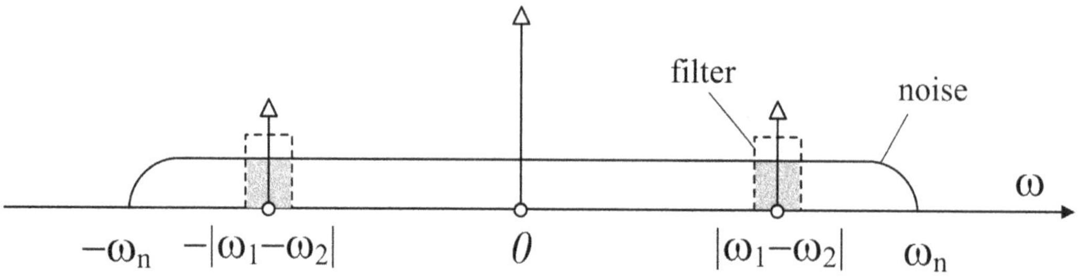

Fig. 3.17. Spectrum of the heterodyne photocurrent.

First of all, when polarization vectors $\mathbf{e}_1$ and $\mathbf{e}_2$ of the interfering waves are orthogonal, no oscillating signal appears at the output of the photodetector. Therefore, since actual polarization of the interfering waves is often unknown, it may be considered as a rule to install a polarizer in front of the photodetector to guarantee oscillations in the output photocurrent.

Secondly, it may be noted that oscillating component j_3, which may also be called "useful", allows easy separation from the two other components j_1 and j_2 by means of spectral filtering. Indeed, consider the spectrum of the total photocurrent $j(t)$. Applying the Fourier transform to $j(t)$, one finds that the two components j_1 and j_2 give two narrow spectral components at zero frequency, while the useful component j_3 — two narrow spectral components at frequencies $\pm|\omega_1 - \omega_2|$, situated symmetrically around zero frequency.

Next, the spectrum of the noise component $n(t)$ is determined by its auto-correlation function. If the time constant of the output electronic circuits were infinitely small, then the auto-correlation function of $n(t)$ would represent the Dirac's delta-function since the single photo-electron pulses are mutually uncorrelated in time. In this approximation, the noise spectrum is merely a constant that does not depend on frequency. In practice, however, the photo-detector time constant is always finite, so that the noise spectrum is always limited by its upper frequency ω_n.

As a result, the spectrum of the output photocurrent may be approximated by the distribution shown schematically in Fig. 3.17.

The useful signal j_3 at the frequency $|\omega_1 - \omega_2|$, which is usually called the intermediate frequency, can be extracted by means of a band-pass filter. Spectral characteristic of the filter shown in dashed line in Fig. 3.17 has always a finite width, so that some portion of noise inevitably transpires to the output. The noise power in the output signal is roughly determined by the shadowed area in Fig. 3.17.

High sensitivity of the heterodyne technique, which is the one of its basic advantages over the direct detection techniques, is determined by the so-called intrinsic amplification of optical signal. Intrinsic amplification means the following. Suppose the field $\mathbf{E}_1$ is under investigation and its amplitude A_1 is small. The amplitude of the useful electrical signal at the output of the heterodyne receiver is proportional to the product $|A_1||A_2|$ where A_2 is the amplitude of optical field from another source which is usually called the reference source. Thus, no matter how small is A_1, the amplitude of the electrical signal can be made as large as necessary by increasing the intensity of the reference source. It means that we can amplify small optical signal before it is combined with the photo-detector noise, thus preserving high signal-to-noise ratio. This is, of course, an idealization of the real situation, but the idea is clear: we can amplify small optical signal by a factor of A_2. In other words, the reference source plays the role of optical amplifier. The limit to the infinite increase of the gain is imposed by the shot noise, being generated in the photo-detector by the reference wave.

3.2.2. *Basic concept of heterodyne spectroscopy*

From the beginning of Section 3.1.1, we considered only regular signals. However, when it comes to spectral measurements, the optical wave to be analyzed is always a random one, generated by scattered or thermal radiation. Such a field may be presented for the analysis in the form of a complex wave

$$\mathbf{E}(\mathbf{r}, t) = \mathbf{e}(\mathbf{r}, t)A(\mathbf{r}, t)\exp(i\omega t),$$

in which the unity polarization vector $\mathbf{e}(\mathbf{r}, t)$ and the complex amplitude $A(\mathbf{r}, t)$ are random functions of the coordinate $\mathbf{r}$ and time t. The output electrical signal of the heterodyne receiver is also a random function of time.

We are interested in finding the relation between statistical characteristics of optical fields, interfering at the sensitive area of the photo-detector, and its output electrical signal. The total photocurrent is equal to

$$j(t) = \frac{\eta q}{h\nu}\int_\sigma |\mathbf{E}_1(\mathbf{r}, t) + \mathbf{E}_2(\mathbf{r}, t)|^2 d^2r + n(t),$$

where, using previous notations, $\mathbf{E}_1(\mathbf{r}, t)$ is the field to be analyzed, and $\mathbf{E}_2(\mathbf{r}, t)$ — the reference field. We shall call $\mathbf{E}_1$ "the signal". For generality, the reference field $\mathbf{E}_2(\mathbf{r}, t)$ will also be assumed random. Then the total photocurrent then may be presented in the form, explicitly showing temporal variations of polarizations and amplitudes:

$$\begin{aligned} j(t) = {} & \frac{\eta q}{h\nu}\int_\sigma |\mathbf{E}_1(\mathbf{r}, t)|^2 d^2r + \frac{\eta q}{h\nu}\int_\sigma |\mathbf{E}_2(\mathbf{r}, t)|^2 d^2r \\ & + \frac{\eta q}{h\nu}\exp[i(\omega_1 - \omega_2)t]\int_\sigma \mathbf{e}_1(\mathbf{r}, t)\mathbf{e}_2(\mathbf{r}, t)A_1(\mathbf{r}, t)A_2^*(\mathbf{r}, t)d^2r \\ & + \frac{\eta q}{h\nu}\exp[-i(\omega_1 - \omega_2)t]\int_\sigma \mathbf{e}_1(\mathbf{r}, t)\mathbf{e}_2(\mathbf{r}, t)A_1^*(\mathbf{r}, t)A_2(\mathbf{r}, t)d^2r + n(t). \end{aligned}$$

Prior to considering statistical characteristics of the photocurrent, it is worth mentioning that in most cases the fields $\mathbf{E}_1(\mathbf{r}, t)$ and $\mathbf{E}_2(\mathbf{r}, t)$ possess the feature of ergodicity, which means that the variables averaged over the time domain are equivalent to averaging over the ensemble of realizations. Angle brackets will denote mathematical operation of averaging over the ensemble of realizations. Besides, as a rule, the random function $\mathbf{e}(\mathbf{r}, t)$, describing polarization state, and the complex amplitude $A(\mathbf{r}, t)$ are statistically independent functions. Therefore, averaging of their product splits into independent averaging of the polarization and amplitude random functions. Thus, the average signal may be written as

$$\begin{aligned} \langle j(t)\rangle = {} & j_1 + j_2 \\ & + \frac{\eta q}{h\nu}\exp[i(\omega_1 - \omega_2)t]\int_\sigma \langle \mathbf{e}_1(\mathbf{r}, t)\mathbf{e}_2(\mathbf{r}, t)\rangle\langle A_1(\mathbf{r}, t)A_2^*(\mathbf{r}, t)\rangle d^2r \\ & + \frac{\eta q}{h\nu}\exp[-i(\omega_1 - \omega_2)t]\int_\sigma \langle \mathbf{e}_1(\mathbf{r}, t)\mathbf{e}_2(\mathbf{r}, t)\rangle\langle A_1^*(\mathbf{r}, t)A_2(\mathbf{r}, t)\rangle d^2r + \langle n(t)\rangle. \end{aligned}$$

Here

$$j_1 = \left\langle \frac{\eta q}{h\nu}\int_\sigma |\mathbf{E}_1(\mathbf{r}, t)|^2 d^2r \right\rangle, \quad j_2 = \left\langle \frac{\eta q}{h\nu}\int_\sigma |\mathbf{E}_2(\mathbf{r}, t)|^2 d^2r \right\rangle$$

are the average photocurrents produced by the fields $\mathbf{E}_1$ and $\mathbf{E}_2$ independently. These fields are assumed to be statistically independent, so that $\langle A_1(\mathbf{r},t)A_2^*(\mathbf{r},t)\rangle = \langle A_1(\mathbf{r},t)\rangle\langle A_2^*(\mathbf{r},t)\rangle$. If the signal field is the result of scattering, then its phase is distributed uniformly in the interval $[0, 2\pi]$. Consequently, in this case $\langle A_1(\mathbf{r},t)\rangle = 0$, and both the second and the third terms vanish. Thus,

$$\langle j(t)\rangle = j_1 + j_2 + \langle n(t)\rangle.$$

This formula does not contain the oscillating component at the intermediate frequency $\omega_1 - \omega_2$. This, however, does not mean that the amplitude of the useful signal at the output of the heterodyne receiver is equal to zero. In order to understand this we have to consider the photocurrent correlation function and to analyze its spectrum.

If the phase of the signal field lies with equal probability within the interval $[0, 2\pi]$ then the photocurrent correlation function is composed of the following components:

$$\begin{aligned} R(t_1,t_2) = \langle j(t_1)j(t_2)\rangle = 2j_1j_2 &+ \left(\frac{\eta q}{h\nu}\right)^2 \int_\sigma\int_\sigma \langle I_1(\mathbf{r}_1,t_1)I_1(\mathbf{r}_2,t_2)\rangle d^2r_1d^2r_2 \\ &+ \left(\frac{\eta q}{h\nu}\right)^2 \int_\sigma\int_\sigma \langle I_2(\mathbf{r}_1,t_1)I_2(\mathbf{r}_2,t_2)\rangle d^2r_1d^2r_2 \\ &+ 2\langle n(t)\rangle j_1 + 2\langle n(t)\rangle j_2 + \langle n(t_1)\,n(t_2)\rangle \\ &+ 2\left(\frac{\eta q}{h\nu}\right)^2 Re\left\{\exp[i(\omega_1-\omega_2)(t_1-t_2)]\right. \\ &\times \int_\sigma\int_\sigma \langle(\mathbf{e}_1(\mathbf{r}_1,t_1)\,\mathbf{e}_2(\mathbf{r}_1,t_1)) \times (\mathbf{e}_1(\mathbf{r}_2,t_2)\mathbf{e}_2(\mathbf{r}_2,t_2))\rangle \\ &\times \left.\langle A_1(\mathbf{r}_1,t_1)A_1^*(\mathbf{r}_2,t_2)\rangle\langle A_2^*(\mathbf{r}_1,t_1)A_2(\mathbf{r}_2,t_2)\rangle d^2r_1d^2r_2\right\}, \end{aligned}$$

where the intensity of the field is introduced: $I(\mathbf{r},t) = |\mathbf{E}(\mathbf{r},t)|^2$.

In order to come to a final result suitable for analysis, it is necessary to make several simplifications which, however, do not impose substantial restrictions on generality of the result. First of all, we may consider all random variables to be stationary in time. Furthermore, polarization of the reference wave, produced commonly by stabilized lasers, may be considered independent of the spatial coordinate and time. Also, intensity of the reference wave is always much stronger than that of the signal: $I_2 \gg I_1$. With these assumptions, one can neglect the second and the forth terms in the above formula. Finally, we may assume for the present analysis that both the signal and reference waves are spatially coherent within the photo-detector sensitive area. The opposite case of partial spatial coherence will be considered in the next section. Then the photocurrent correlation function reduces to

$$\begin{aligned} R(t_1-t_2) = R(\tau) = 2j_1j_2 + 2\langle n\rangle j_2 &+ \left(\frac{\eta q\sigma}{h\nu}\right)^2 R_I(\tau) + R_n(\tau) \\ &+ 2\left(\frac{\eta q\sigma}{h\nu}\right)^2 \mathrm{Re}[e^{i(\omega_1-\omega_2)\tau}R_\theta(\tau)R_1(\tau)R_2^*(\tau)], \end{aligned}$$

where the following correlation functions are introduced: correlation function of the reference wave intensity

$$R_I(\tau) = \langle I_2(t)I_2(t+\tau)\rangle,$$

correlation function of the noise

$$R_n(\tau) = \langle n(t)n(t+\tau)\rangle,$$

correlation function of the cosine of the angle between the polarization vectors of the interfering waves

$$R_\theta(\tau) = \langle(\mathbf{e}_1(t)\mathbf{e}_2)(\mathbf{e}_1(t+\tau)\mathbf{e}_2)\rangle = \langle\cos[\theta(t)]\cos[\theta(t+\tau)]\rangle,$$

and the correlation functions of the complex field amplitudes

$$R_1(\tau) = \langle A_1(t)A_1^*(t+\tau)\rangle, \quad R_2(\tau) = \langle A_2(t)A_2^*(t+\tau)\rangle.$$

When $\tau \to \infty$, the correlation function of intensity reduces to $R_I(\tau) \approx \langle I_2\rangle^2$, so that it is fruitful to introduce the correlation function of the intensity fluctuations $\rho(\tau)$, which tends to zero with infinite τ:

$$R_I(\tau) = \rho(\tau) + \langle I_2\rangle^2.$$

Then

$$\left(\frac{\eta q\sigma}{h\nu}\right)^2 R_I(\tau) = j_2^2 + \left(\frac{\eta q\sigma}{h\nu}\right)^2 \rho(\tau),$$

where $\rho(\tau) \to 0$ when $\tau \to \infty$. Correlation function of noise may be approximated by the delta-function:

$$R_n(\tau) = 2\pi\,\mathsf{N}\,\delta(\tau),$$

where N is the spectral density of noise. Since the phase of the signal wave $\mathbf{E}_1$ is distributed uniformly within $[0,\ 2\pi]$, the correlation function $R_1(\tau) \to 0$ when $\tau \to \infty$. Without limitation of generality, it is possible to assume that phase of the reference field $\mathbf{E}_2$ is also distributed uniformly within $[0, 2\pi]$ independently of its correlation features. Then also $R_2(\tau) \to 0$ when $\tau \to \infty$. Eventually,

$$\begin{aligned} R(\tau) &= 2j_1j_2 + 2\langle n\rangle j_2 + j_2^2 + 2\pi N\delta(\tau) \\ &\quad + \left(\frac{\eta q\sigma}{h\nu}\right)^2 \rho(\tau) + 2\left(\frac{\eta q\sigma}{h\nu}\right)^2 \mathrm{Re}[e^{i(\omega_1-\omega_2)\tau} R_\theta(\tau)R_1(\tau)R_2^*(\tau)]. \end{aligned}$$

It is well known that the real mean square value of a random variable is equal to the value of its correlation function taken at the origin. Thus, the real mean square amplitude of the useful oscillating signal, determined by the last term in this formula, is non-zero and is equal to

$$2\left(\frac{\eta q\sigma}{h\nu}\right)^2 R_\theta(0)R_1(0)R_2(0).$$

Spectral density of the photocurrent is defined as the Fourier transform of $R(\tau)$:

$$S(\omega) = \frac{1}{2\pi}\int_{-\infty}^{+\infty} R(\tau)\exp(-i\omega\tau)d\tau.$$

Omitting the intermediate manipulations, we get

$$\begin{aligned} S(\omega) &= (2j_1j_2 + 2\langle n\rangle j_2 + j_2^2)\delta(\omega) + N + \left(\frac{\eta q\sigma}{h\nu}\right)^2 s(\omega) \\ &\quad + \left(\frac{\eta q\sigma}{h\nu}\right)^2 \int_{-\infty}^{+\infty}\int_{-\infty}^{+\infty} S_\theta(\omega \pm |\omega_1 - \omega_2| - u)S_1(u-v)S_2(v)dudv. \end{aligned}$$

The sign "$-$" should be taken for positive frequencies, and "$+$" for negative frequencies. In this formula, we introduced new notations for spectral densities of fluctuations of the following physical quantities: reference wave intensity

$$s(\omega) = \frac{1}{2\pi}\int_{-\infty}^{+\infty} \rho(\tau)\exp(-i\omega\tau)d\tau,$$

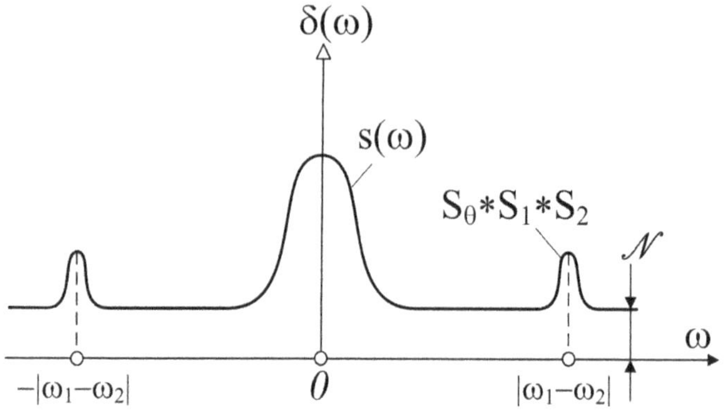

Fig. 3.18. Spectral density of the heterodyne photocurrent in case of random quasi-monochromatic waves.

cosine of the angle between the polarization vectors of the interfering waves

$$S_\theta(\omega) = \frac{1}{2\pi}\int_{-\infty}^{+\infty} R_\theta(\tau)\exp(-i\omega\tau)d\tau,$$

and complex amplitudes of the waves

$$S_{1,2}(\omega) = \frac{1}{2\pi}\int_{-\infty}^{+\infty} R_{1,2}(\tau)\exp(-i\omega\tau)d\tau.$$

In the formula for $S(\omega)$, the last term presents the double convolution of spectral densities of fluctuations and determines spectral properties of the useful signal at the output of the heterodyne receiver. The total spectral density of the photocurrent fluctuations is shown schematically in Fig. 3.18, where the asterisk denotes the convolution operation.

In experiments, the source of the reference wave is always a laser, so that the spectral density $s(\omega)$ is determined by intensity fluctuations of the laser output. These fluctuations are usually caused by such technical features as power supply instability, variations of the temperature of cooling liquid, and also fluctuations of the gas discharge itself. Therefore, the spectral width of $s(\omega)$ does not exceed several kilohertz, maximum — several tens of kilohertz. In order that these fluctuations do not affect the measurements, they have to be filtered away. Efficient filtering requires large spectral separation between $s(\omega)$ and the useful signal, i.e. the intermediate frequency $|\omega_1 - \omega_2|$ must significantly exceed the width of $s(\omega)$.

We are interested now in the spectral width of the useful signal. Note that the convolution of S_θ with S_1 physically means the spectral density of fluctuations of the projection of the vector of the complex amplitude of the signal onto the direction of the reference field vector. This projection is equivalent to the vector of some linearly polarized optical wave $\mathbf{E}'_1$ whose polarization coincides with that of the reference field. This equivalency holds true because the orthogonal component of the signal wave $\mathbf{E}_1$ does not contribute to the useful signal, and also its intensity is negligible to cause any noticeable additional noise in the output photocurrent. In this representation, amplitude fluctuations of the output electrical signal caused by fluctuations of the polarization of actually existing signal wave $\mathbf{E}_1$ should be taken into account by corresponding enhancement of the amplitude fluctuations of the equivalent input field $\mathbf{E}'_1$.

Therefore, we may write down

$$\mathbf{E'}_1(t) = \mathbf{e}_2 A'_1(t)\exp(i\omega t),$$
$$R'_1(\tau) = \langle A'_1(t)A'^*_1(t+\tau)\rangle = R_\theta(\tau)R_1(\tau),$$
$$S'_1(\omega) = \int_{-\infty}^{+\infty} S_\theta(\omega - u)S_1(u)du.$$

In the formula for $S(\omega)$, spectral density of fluctuations of the useful component of the heterodyne signal is defined by the term

$$\left(\frac{\eta q\sigma}{h\nu}\right)^2 \int_{-\infty}^{+\infty}\int_{-\infty}^{+\infty} S_\theta(\omega \pm |\omega_1 - \omega_2| - u)S_1(u - v)S_2(v)dudv,$$

which can be finally named "the signal" and written in the following form:

$$S_{\text{signal}}(\omega) = \left(\frac{\eta q\sigma}{h\nu}\right)^2 \int_{-\infty}^{+\infty} S'_1(\omega \pm |\omega_1 - \omega_2| - v)S_2(v)dv.$$

Suppose that the spectra S'_1 and S_2 are both the Lorentzian with the widths $\Delta\omega_1$ and $\Delta\omega_2$, respectively. It is well known that correlation functions of Lorentz spectra are exponential, and that the convolution can be calculated as the inverse Fourier transform of the product of correlation functions. Then, multiplying the corresponding exponential correlation functions and performing the inverse Fourier transform, one can obtain the simple relation for the spectral width of the signal:

$$\Delta\omega_{\text{signal}} = \Delta\omega_1 + \Delta\omega_2.$$

In particular case of monochromatic reference wave with $\Delta\omega_2 = 0$,

$$S_2(\omega) = c\delta(\omega),$$

where c is the constant coefficient. Substitution of this into (1.40) gives

$$S_{\text{signal}}(\omega) = c' S'_1(\omega \mp |\omega_1 - \omega_2|)$$

and

$$\Delta\omega_{\text{signal}} = \Delta\omega_1.$$

For spectral density over positive frequencies, we have

$$S_{\text{signal}}(\omega) = c' S'_1(\omega - |\omega_1 - \omega_2|),$$

which means that the spectrum of the electrical signal of the heterodyne receiver completely represents spectral density of fluctuations of the equivalent optical field $\mathbf{E'}_1$. This is the basic concept of heterodyne spectroscopy.

In practice, spectral resolution of a heterodyne spectrometer is determined by spectral width of the reference laser. As it was experimentally demonstrated in Section 3.1.3, the industrial stabilized lasers, working in visible domain, may have spectral width within several megahertz. How to measure such a narrow spectrum? It can be done, measuring spectral width of interference beatings between the two frequency-split waves of a heterodyne laser, i.e. by means of self-heterodyning. One type of a heterodyne laser, in which the frequency difference between the output waves is created by means of acousto-optical modulation, was described in Section 3.1.3. However, for the current task it is better to use a laser with smaller frequency difference in order to mitigate requirements to high-speed sampling electronics. Commercially available Zeeman two-frequency He–Ne stabilized lasers provide an order of

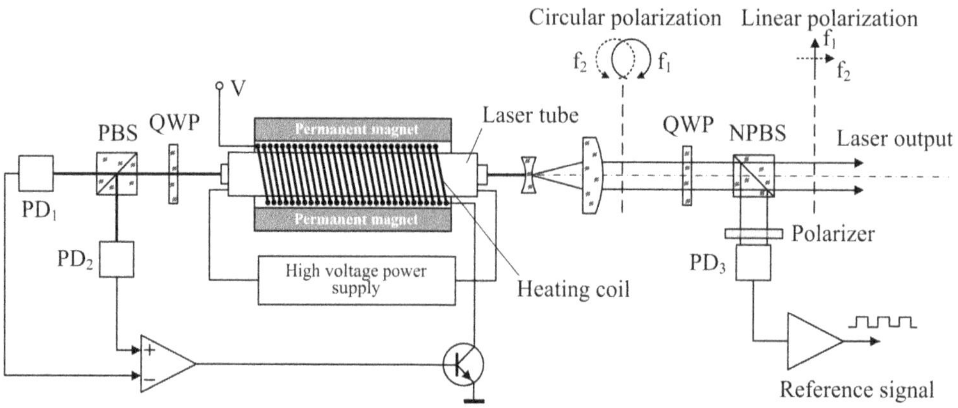

Fig. 3.19. Two-frequency stabilized Zeeman laser. f_1 and f_2 are optical frequencies of the two orthogonally polarized waves.

magnitude lower frequency split, of about 2 MHz, which is quite suitable for the experiment, using ordinary digital oscilloscopes. The scheme of the Zeeman laser is outlined in Fig. 3.19.

Phenomenologically, the Zeeman laser is a low-power He–Ne laser with axially applied magnetic field and a feedback for frequency stabilization. In this type of lasers, Zeeman magnetic splitting of emission line creates two independent orthogonally polarized output waves at the wavelength of 633 nm with frequency split of about several megahertz. It is of a primary importance that the two independent waves are of the same mode structure and travel same paths, experiencing same optical heterogeneities, and therefore, are supposed to have identical wave fronts. Frequency shift between them originates from tiny difference in refraction indices of active medium for the left- and right-hand circularly polarized waves due to Zeeman effect. Both frequency and amplitude stabilization is performed by comparing intensities of the two waves, traveling inside the laser cavity. These waves, originally circularly polarized inside the cavity, are transformed into linearly polarized waves with the help of a quarter wave plate (QWP) and split into two at the polarizing beam splitter (PBS). Photo-detectors PD_1 and PD_2 measure intensities of the two waves, and the differential signal controls the current through the heating coil, maintaining the length of the laser cavity so as to equalize intensities of the two components. Thus, the frequency stabilization principle is the same as in the Zygo laser discussed above in Section 3.1.3. Equal intensities correspond to constant frequency shift between the waves if only magnetic field is constant. Stability of magnetic field is very important for frequency stability. For example, any massive magnetic parts on the optical table, positioned close to the laser, may significantly change its frequency shift. To provide a reference for phase measurements, a small portion of the two output linearly polarized beams is directed to the third photo-detector PD_3 by means of a non-polarizing beam splitter (NPBS). Being orthogonally polarized, these two waves do not produce any interference effect on the photo-detector until they are coupled by a polarizer.

According to physics of operation, the two waves of the Zeeman laser with angular frequencies ω_1 and ω_2 have the same spectral characteristics and stability. Therefore, if these waves produce heterodyne beatings on the photodetector, then spectral width of the photocurrent

$$\Delta\omega_{\text{signal}} = \Delta\omega_1 + \Delta\omega_2 = 2\Delta\omega,$$

where $\Delta\omega_1 = \Delta\omega_2 = \Delta\omega$ is the spectral width of each of the two laser waves. From here, the principle of measurement is clear: measure spectral width of the heterodyne beatings between the two laser waves

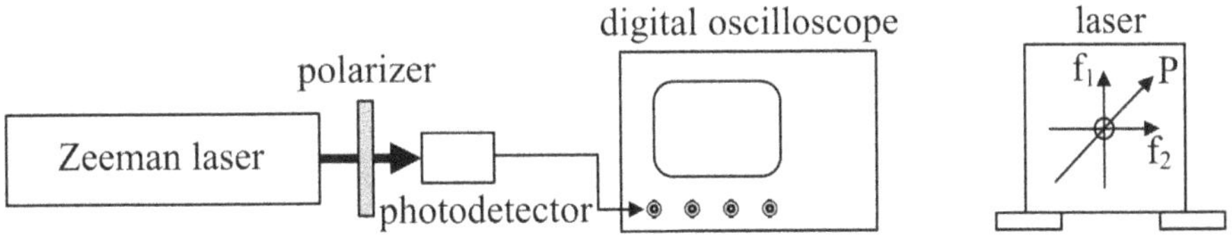

Fig. 3.20. Zeeman laser Agilent (Hewlett-Packard) 5517B. Oscilloscope Tektronix DPO 7104C. Photodetector Hamamatsu C-5460. P — direction of the polarizer that couples orthogonal polarizations of the laser waves.

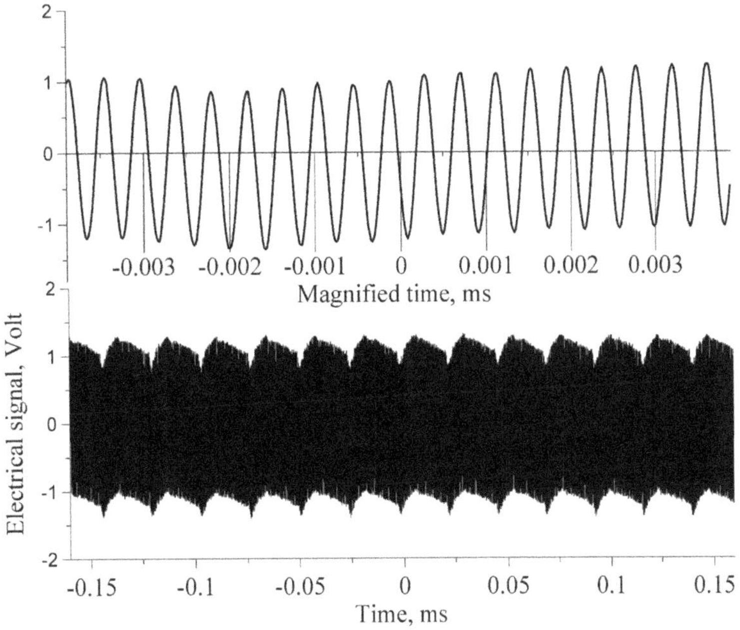

Fig. 3.21. The stacked entire trace of heterodyne oscillations and its magnified part above.

and divide the result by two:

$$\Delta\omega = \frac{\Delta\omega_{\text{signal}}}{2}.$$

The scheme of the measurement is clear from Fig. 3.20.

In order to measure spectral width of the heterodyne beatings, electrical signal from the photodetector must be sampled to a digital form and then subjected to the Fourier transform. According to the Whittaker–Kotelnikov sampling theorem, the sampling rate must be at least 2 times higher than the frequency of oscillations. The frequency measurement option of the oscilloscope gave the average value of oscillations frequency 2.4 MHz. With significant reserve, the sampling rate of the oscilloscope was chosen to be 62.5 MHz, i.e. 26 samples on the period. The Fourier transform is an integral operation over time that averages fluctuations of the signal and, therefore, computing the average spectrum. For such an averaging being correct, the integration time must be long enough compared to the period of oscillations. From the practical point of view, integration of 10^3 periods may be considered sufficient, thus making the total time of measurement 0.4 ms. A typical recorded trace is shown in Fig. 3.21.

Periodical amplitude changes due to work of a stabilization system of the laser are clearly seen. Besides, phase oscillations are also present but not readily visible in the figure. Both the phase and

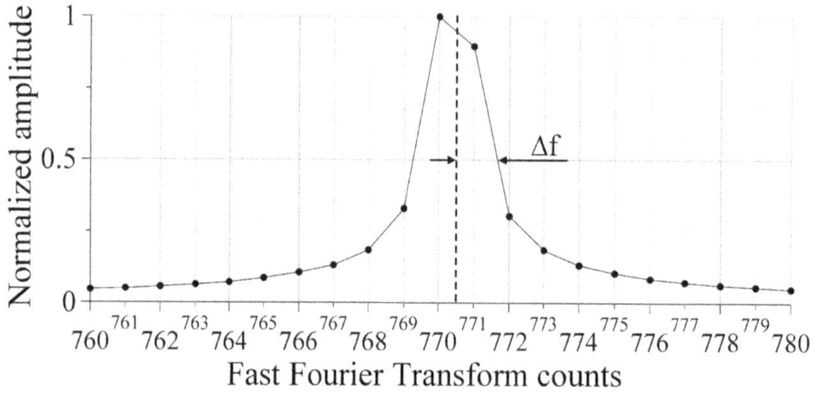

Fig. 3.22. Computed spectrum of the heterodyne signal with the frequency 2.4 MHz.

amplitude fluctuations, caused by both stabilization system and natural instabilities, contribute to finite spectral width of the heterodyne signal, shown in Fig. 3.22.

Spectral width at half maximum of the heterodyne signal Δf_{signal} may be computed, using proportion: the frequency of oscillations (intermediate frequency) is 2.4 MHz and it corresponds to 770.5 counts in Fig. 3.22, then spectral width Δf_{signal}, measured as approximately 1.2 counts in Fig. 3.22, corresponds to 2.4 MHz/770.5∗1.2 $\approx$ 3.7 kHz. Accordingly, the spectral width of a stabilized laser wave is $\Delta f = 0.5 \cdot \Delta f_{\text{signal}} \approx$ 1.9 kHz. Thus, we came to a very important conclusion of this section: spectral resolution of the heterodyne spectrometer may be as small, as several kilohertz or $\sim 10^{-9}$ nm in visible domain — a value completely unattainable for grating spectrometers, whose resolution is commonly of the scale 10^{-2} nm.

However, heterodyne spectrometers suffer from a very important limitation that makes them a very rare guest in the laboratory: poor efficiency of collecting diffuse light. This is explained in the next section.

3.2.3. *Optimal conditions for wavefronts*

There are two most important practical problems on the way of creating the heterodyne spectrometers: adjustment of optical wavefronts and speed of the photodetectors. We address them one by one, starting from the basic considerations on how to combine the optical waves on the sensitive area of a photodetector.

We are going to analyze only wavefronts, therefore there will be no loss of generality to assume that the both waves are linearly polarized in the same direction, i.e. $\mathbf{e}_1\mathbf{e}_2 = 1$, and quantum efficiency is constant over the entire detector surface. Then the complex amplitude of the output signal may be written as

$$u = 2\frac{\eta q}{h\nu}\int_\sigma A_1(\mathbf{r})A_2^*(\mathbf{r})d^2r,$$

where the factor "2" ensures that the amplitude of the real photocurrent is equal to $\mathrm{Re}(u)$. The complex amplitudes of optical fields $A_1(\mathbf{r})$ and $A_2(\mathbf{r})$ determine spatial distributions of both their intensities $|A_1(\mathbf{r})|^2$, $|A_2(\mathbf{r})|^2$ and phases $\arg(A_1(\mathbf{r}))$, $\arg(A_2(\mathbf{r}))$. Therefore, the output signal depends on spatial distributions of both the intensities and phases of the interfering fields within the detector sensitive area. In practice, the rule for optimal adjustment of intensities looks obvious: the two beams must spatially overlap within the detector sensitive area. As to the phases, the optimal condition is non-trivial, and it will be analyzed below.

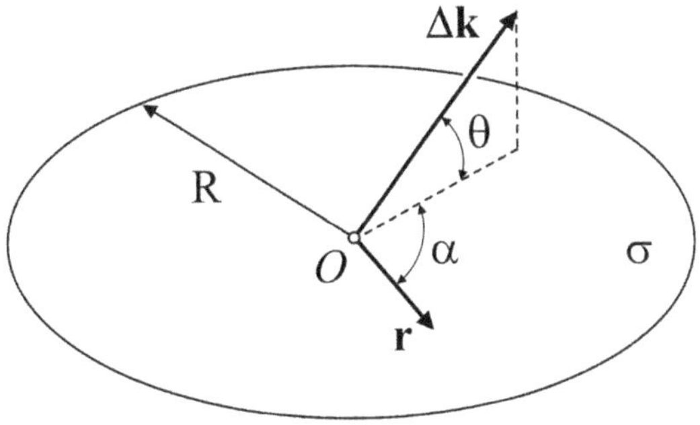

Fig. 3.23. System of coordinates with the origin at O.

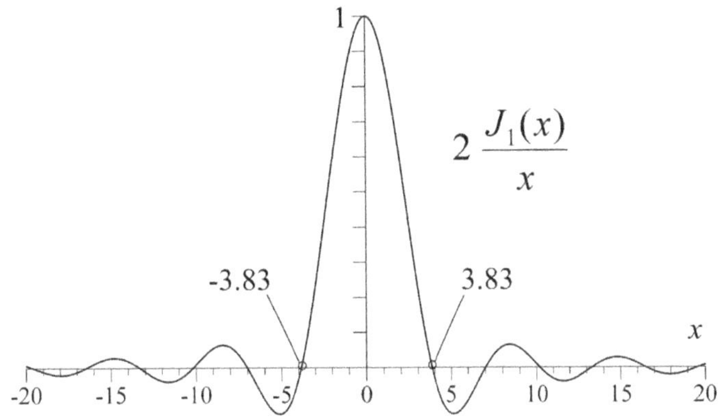

Fig. 3.24. The Airy function.

Suppose for the beginning that $\mathbf{E}_1$ and $\mathbf{E}_2$ are plane waves:

$$A_1(\mathbf{r}) = |A_1| \exp(i\mathbf{k}_1\mathbf{r}), \quad A_2(\mathbf{r}) = |A_2| \exp(i\mathbf{k}_2\mathbf{r}),$$

where $\mathbf{k}_1$, $\mathbf{k}_2$ are the wave vectors. We want to find out how the signal amplitude depends on the directions of the plane waves. Substitution of $A_1(\mathbf{r})$ and $A_2(\mathbf{r})$ gives

$$u = 2\frac{\eta q}{h\nu}|A_1||A_2| \int_\sigma \exp[i(\mathbf{k}_1 - \mathbf{k}_2)\mathbf{r}]d^2r.$$

Now, it is necessary to specify the shape of the detector sensitive area in order to accomplish the integration. Commonly it is a circle.

Then, with the notations of Fig. 3.23, the above formula transforms to

$$u = 2\frac{\eta q}{h\nu}|A_1||A_2| \int_0^R \int_0^{2\pi} \exp(i|\Delta\mathbf{k}|r\cos\theta\cos\alpha)rdr\,d\alpha = 2\frac{\eta q}{h\nu}|A_1||A_2|\sigma\left[2\frac{J_1(R|\Delta\mathbf{k}|\cos\theta)}{R|\Delta\mathbf{k}|\cos\theta}\right],$$

with $\Delta\mathbf{k} = \mathbf{k}_1 - \mathbf{k}_2$, J_1 — Bessel function of the first order, and $\sigma = \pi R^2$ — area of the detector. Square brackets mark the Airy function $2J_1(x)/x$ which is plotted in Fig. 3.24.

The argument of this function is $R|\Delta\mathbf{k}|\cos\theta$. The Airy function reaches its maximum at $x = 0$. Consequently, the signal amplitude reaches its maximum in two cases: when $\Delta\mathbf{k} = 0$ and (or) when $\theta = 90°$. In real situation, one cannot expect any of the two waves coming from the bottom of the detector. Therefore, any of the above two conditions can be fulfilled only if the wavefronts are parallel to each other. The condition $\theta = 90^\circ$ additionally requires the normal incidence of the waves on the detector surface.

The moduli of the vectors $\mathbf{k}_1$, $\mathbf{k}_2$ can be considered equal to each other because the difference δk between them due to frequency difference between the fields $\mathbf{E}_1$ and $\mathbf{E}_2$ is negligible. Indeed, the maximum frequency difference, attainable in the experiments, does not exceed, as a rule, 10^{10} Hz while the optical frequencies themselves are of the order of $3{\cdot}10^{13}$ Hz. Thus,

$$\frac{\delta k}{k} < \frac{10^{10}}{3 \times 10^{13}} \approx 3 \times 10^{-4} \ll 1,$$

which makes it possible to assume $|\mathbf{k}_1| = |\mathbf{k}_2| = k$. Therefore,

$$|\Delta\mathbf{k}| \approx k \sin\beta$$

where β is the angle between the wave vectors $\mathbf{k}_1$ and $\mathbf{k}_2$. This angle is commonly called the mismatch angle between the wavefronts.

It is possible to formulate the condition that must be imposed on the mismatch angle in order to maintain the output signal amplitude close to its maximum. From Fig. 3.24, it is clear that the argument of the Airy function must be smaller than the first zero: $x \leq 3.83$. For small angles β and θ, this transforms to

$$Rk\beta \leq 3.83,$$

or

$$\beta \leq 1.22\frac{\lambda}{D},$$

where λ is the mean wavelength and $D = 2R$ is the diameter of the detector sensitive area. Note that the angle $1.22\lambda/D$ also determines the angle of diffractional divergence of the laser beam on a round hole of the diameter D. It means that the mismatch angle must not exceed the angle of diffractional divergence on the detector aperture. Addressing the real laboratory situation, assume $\lambda = 0.63$ μm, $D = 1$ mm. Then, the mismatch angle must be less than $2.5'$.

Consider now the waves $\mathbf{E}_1$ and $\mathbf{E}_2$ with arbitrary wavefronts $\phi_1(\mathbf{r})$ and $\phi_2(\mathbf{r})$:

$$A_1(\mathbf{r}) = |A_1| \exp[i\phi_1(r)], \quad A_2(\mathbf{r}) = |A_2| \exp[i\phi_2(\mathbf{r})].$$

Then, the modulus of the complex output signal becomes equal to

$$|u| = 2\frac{\eta q}{h\nu} \int_\sigma |A_1(\mathbf{r})||A_2(\mathbf{r})| \cos[\delta\phi(\mathbf{r})] d^2 r,$$

where $\delta\phi(\mathbf{r}) = \phi_1(\mathbf{r}) - \phi_2(\mathbf{r})$. Since the product $|A_1(\mathbf{r})||A_2(\mathbf{r})|$ is non-negative, the following inequality holds true:

$$\left|\int_\sigma |A_1(\mathbf{r})||A_2(\mathbf{r})| \cos[\delta\phi(\mathbf{r})] d^2 r\right| \leq \int_\sigma |A_1(\mathbf{r})||A_2(\mathbf{r})| d^2 r,$$

with the exact equality being reached when $\cos[\delta\phi(\mathbf{r})] = \pm 1$ or $\delta\phi(\mathbf{r}) = \pi n (n = 0, \pm 1, \pm 2, \ldots)$. It means that in general case the output signal of a heterodyne receiver reaches its maximum when

$$\phi_1(\mathbf{r}) = \phi_2(\mathbf{r}) + \pi n (n = 0, \pm 1, \pm 2, \ldots),$$

which is commonly referred to as the wavefronts matching condition. It must take place in each point of the photo-detector sensitive area, and it physically means that the surfaces that represent the wavefronts of $\mathbf{E}_1$ and $\mathbf{E}_2$ must be identical.

The wavefront matching condition defines the unique configurations of the two interfering waves. If one wave can be considered having some fixed configuration, i.e. the reference wave, then the matching condition cannot simultaneously hold true for two or more other waves, coming to the detector at different angles. Therefore, if it is necessary to receive the incoming waves in some finite cone of angles (finite angular field of view) then the amplitude of the heterodyne signal averaged over all possible directions within the field of view is always smaller than the maximum one, corresponding to the matching condition. This decrease in efficiency can be formally ascribed to the decrease in the area of the input optical aperture through which radiation comes to the photo-detector. In other words, it is possible to introduce the so-called effective input aperture. Then a very simple and straightforward relation can be established between the field of view and the effective aperture of the heterodyne receiver. This relation is known as the Siegman's antenna theorem. Consider it in more details.

According to previous notations, the complex amplitude of the useful signal is

$$u = 2\frac{q}{h\nu}\mathbf{e}_1\mathbf{e}_2 \int_\sigma \eta(\mathbf{r})A_1(\mathbf{r})A_2^*(\mathbf{r})d^2r.$$

Assume again that the field $\mathbf{E}_1$ is a plane wave

$$A_1(\mathbf{r}) = A_1 \exp(i\mathbf{kr}).$$

Then

$$u = 2A_1\frac{q}{h\nu}\mathbf{e}_1\mathbf{e}_2 \int_\sigma \eta(\mathbf{r})A_2^*(\mathbf{r})\exp(i\mathbf{kr})d^2r.$$

We can introduce the average quantum efficiency of the photo-detector by averaging the local quantum efficiency over the entire photo-detector sensitive area:

$$\bar{\eta} = \frac{\int_\sigma \eta(\mathbf{r})|A_2(\mathbf{r})|^2 d^2r}{\int_\sigma |A_2(\mathbf{r})|^2 d^2r}.$$

From now on, the line over the top of a letter means spatial averaging. With this definition, it is possible to write the square modulus of the signal complex amplitude as

$$|u|^2 = 4(\mathbf{e}_1\mathbf{e}_2)^2 j_2 \left[\frac{q}{h\nu}|A_1|^2\bar{\eta}S_k\right],$$

where

$$j_2 = \frac{q}{h\nu}\int_\sigma \eta(\mathbf{r})|A_2(\mathbf{r})|^2 d^2r$$

is the average photocurrent generated by the reference wave, and

$$S_k = \frac{|\int_\sigma \eta(\mathbf{r})A_2^*(\mathbf{r})\exp(i\mathbf{kr})d^2r|^2}{\bar{\eta}^2\int_\sigma |A_2(\mathbf{r})|^2 d^2r}$$

is the newly introduced function of the wave vector $\mathbf{k}$, having the dimension of an area. In the formula for $|u|^2$, the quantity $q\bar{\eta}|A_1|^2/h\nu$ is nothing more but the photocurrent from the unity area of the photo-detector caused by the input wave. As such, the expression in the square brackets can be treated as some effective photocurrent amplitude caused by direct detection of the incoming wave, which contributes to the useful output signal at the intermediate frequency.

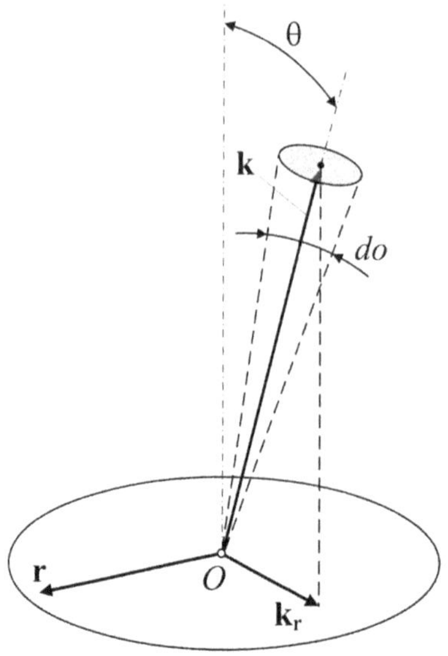

Fig. 3.25. System of coordinates.

In general, not only one plane wave but many plane waves, presenting spatial Fourier components of the incoming field, may come to the detector. Therefore, one should use the effective input aperture surface S_{eff} determined by averaging all S_k over the incoming beam cone solid angle O:

$$S_{\text{eff}} = O^{-1} \int S_k do.$$

The variable of integration is the solid angle o. To proceed with computing the integral in the right-hand side of this formula one has to establish the relation between the differential of the solid angle do and the vector $\mathbf{k}_r$ (Fig. 3.25) that is a projection of the wave vector $\mathbf{k}$ onto the detector surface:

$$do = d^2k_r/(k^2 \cos\theta).$$

In practice, the field of view of the heterodyne receiver is so small that within this angle $\cos\theta \approx 1$. On the other hand, outside the field of view S_k quickly tends to zero. Therefore, we can write down

$$\int S_k do = \frac{1}{k^2} \int S_k d^2k_r.$$

Substitution of S_k into this formula gives:

$$\frac{1}{k^2} \int S_k d^2k_r = \frac{1}{k^2} \left(\overline{\eta}^2 \int_\sigma |A_2(\mathbf{r})|^2 d^2r\right)^{-1}$$
$$\times \int d^2k_r \int_\sigma \int_\sigma \eta(\mathbf{r}_1)\eta(\mathbf{r}_2) A_2^*(\mathbf{r}_1) A_2(\mathbf{r}_2) \exp[i\mathbf{k}_r(\mathbf{r}_1 - \mathbf{r}_2)] d^2r_1 d^2r_2.$$

The limits of integration over $\mathbf{k}_r$ can be formally set to infinity as $S_k \approx 0$ outside the field of view. Changing the order of integration, and using the identity for the Dirac delta-function

$$\int_{-\infty}^{+\infty} \exp(ixy)dx = 2\pi\delta(y),$$

we find:

$$\frac{1}{k^2} \int S_k d^2k_r = \left(\frac{2\pi}{k}\right)^2 \frac{\int_\sigma \eta^2(\mathbf{r})|A_2(\mathbf{r})|^2 d^2r}{\overline{\eta}^2 \int_\sigma |A_2(\mathbf{r})|^2 d^2r}.$$

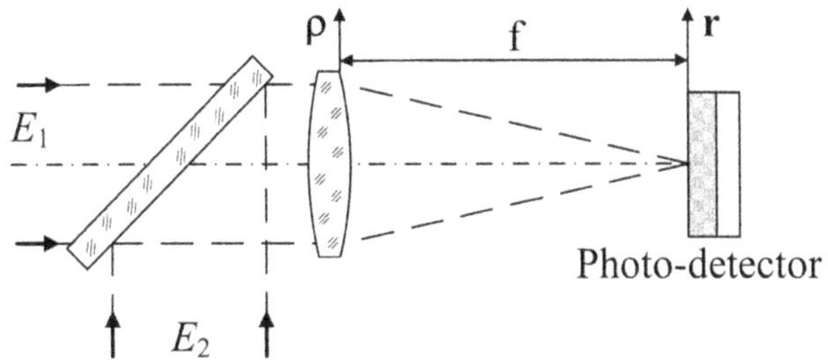

Fig. 3.26. Wavefronts matching scheme with the detector in the focus of a lens.

According to the definition of the spatial average quantum efficiency, which we used above, the ratio of integrals in the right-hand side of this relation is exactly the spatial average value of the square of the quantum efficiency $\overline{\eta^2}$. Thus, the last formula transforms to

$$\frac{1}{k^2}\int S_k d^2k_r = \left(\frac{2\pi}{k}\right)^2 \frac{\overline{\eta^2}}{\overline{\eta}^2} = \lambda^2 \frac{\overline{\eta^2}}{\overline{\eta}^2}.$$

In practice, spatial variation of quantum efficiency is not very large, about several percent. Therefore, it is possible to neglect the difference between the $\overline{\eta^2}$ and $\overline{\eta}^2$ and write down:

$$S_{\text{eff}}O \approx \lambda^2.$$

This identity is known as the Siegman's antenna theorem, which reads that, in the heterodyne receiver, the product of the effective input aperture by the solid angle of the field of view equals the square of the wavelength.

3.2.4. *Practical schemes of wavefront matching*

Consider some frequently used optical schemes for matching wavefronts in heterodyne receivers. Figure 3.26 presents a scheme in which the detector is placed in the focal plane of a lens.

The advantage of this scheme is that the photo-detector may have small sensitive area which means low noise and high speed of operation, and large input aperture, collecting enough input optical flux. It is known that in paraxial approximation the complex field amplitude $A(\mathbf{r})$ in the focal plane of a lens can be presented as the Fourier transform of the complex field amplitude $A(\rho)$ at its front aperture:

$$A(\mathbf{r}) = c\exp\left(i\frac{k}{2f}\mathbf{r}^2\right)\int A(\boldsymbol{\rho})\exp\left(-i\frac{k}{f}\mathbf{r}\rho\right)d^2\rho.$$

Here c is the constant factor, f is the focal length of a lens, and integration is performed over the plane $\boldsymbol{\rho}$. The complex amplitude of the useful (oscillating) signal u at the output of the photo-detector is given by integration of the product $A_1(\mathbf{r})A_2^*(\mathbf{r})$ over the plane $\mathbf{r}$ of the sensitive area of the photodetector:

$$u = B|c|^2\int A_1(\boldsymbol{\rho}_1)A_2^*(\boldsymbol{\rho}_2)\exp\left[-i\frac{k}{f}\mathbf{r}(\boldsymbol{\rho}_1-\boldsymbol{\rho}_2)\right]d^2rd^2\rho_1d^2\rho_2,$$

where $B = 2q\eta/h\nu$. Integration over d^2r gives delta-function with appropriate normalization factor:

$$u = B|c|^2\left(2\pi\frac{f}{k}\right)^2\int A_1(\boldsymbol{\rho}_1)A_2^*(\boldsymbol{\rho}_2)\delta(\boldsymbol{\rho}_1-\boldsymbol{\rho}_2)d^2\rho_1d^2\rho_2.$$

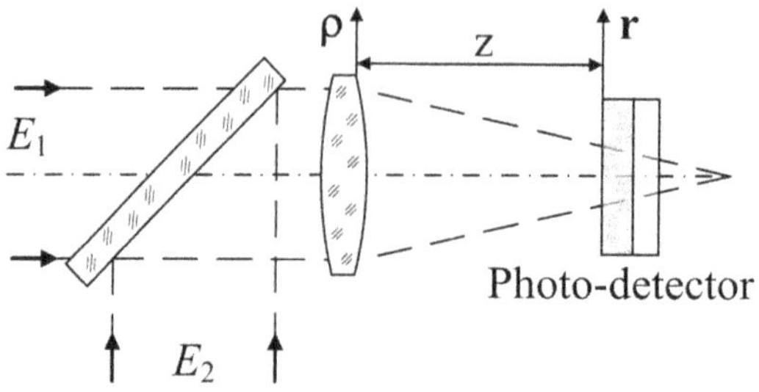

Fig. 3.27. Wavefront matching scheme with shifted photo-detector.

Accomplishing integration over $d^2\rho_1$, we obtain

$$u = B|c|^2\lambda^2 f^2 \int A_1(\boldsymbol{\rho})A_2^*(\boldsymbol{\rho})d^2\rho.$$

Now we have to determine the factor c. For that we shall write the expression for the power W_1 of the field with the complex amplitude $A(\mathbf{r})$, coming onto the photo-detector:

$$W_1 = \int |A(\mathbf{r})|^2 \, d^2r.$$

Here the integration is performed over the sensitive area of the photo-detector. Transforming $A(\mathbf{r})$ to $A(\boldsymbol{\rho})$, integrating over d^2r to obtain delta-function again, and integrating for second time over $d^2\rho_1$, we obtain

$$W_1 = \int |A(\mathbf{r})|^2 d^2r = |c|^2\lambda^2 f^2 \int |A(\boldsymbol{\rho})|^2 \, d^2\rho.$$

On the other hand, the power, passing through the lens aperture, is equal to

$$W_2 = \int |A(\rho)|^2 d^2\rho.$$

The energy conservation law gives $W_1 = W_2$, so that the constant factor equals $|c| = (\lambda f)^{-1}$, and the complex amplitude of the useful (oscillating) signal u transforms to

$$u = B \int A_1(\rho)A_2^*(\rho)\, d^2\rho,$$

where integration is performed over the input aperture of the lens. This is exactly the very first formula of Section 3.2.3 obtained for the case without the focusing lens, showing that the focusing lens does not introduce any changes in the electrical signal.

Suppose now the photo-detector is moved from its position in the focal plane to some other intermediate position located at distance z from the lens (Fig. 3.27).

If the wavefronts are matched then it is rather obvious that moving the photo-detector along the optical axis does not cause any differences in the output signal until only the sensitive area of the photo-detector intercepts the entire flux. However, if there is a mismatch between the wavefronts of the fields E_1 and E_2 then the situations becomes non-trivial, and additional analysis is needed. We shall show that even in case of a wavefront mismatch, shifting the photo-detector along the optical axis does not produce any changes in the output electrical signal.

From the theory of linear optical systems it is known that the complex optical amplitudes in the planes $\mathbf{r}$ and $\boldsymbol{\rho}$ satisfy the Fresnel transform:

$$A(\mathbf{r}) = c\exp(i\alpha\mathbf{r}^2) \int A(\boldsymbol{\rho}) \exp(i\beta\boldsymbol{\rho}^2 + i\gamma\boldsymbol{\rho}\mathbf{r})\, d^2\rho,$$

where c is a constant coefficient, α, β, γ are the parameters, depending on the wavelength and distances z and f. The integration is performed in the plane of the lens.

Consider the interference integral in the plane $\mathbf{r}$:

$$I = \int_\sigma A_1(\mathbf{r})A_2^*(\mathbf{r})d^2r.$$

Using the Fresnel transform, this integral can be rearranged to a form

$$|c|^2 \int A_1(\boldsymbol{\rho}_1)A_2^*(\boldsymbol{\rho}_2)\exp[i\beta(\boldsymbol{\rho}_1^2 - \rho_2^2) + i\gamma(\boldsymbol{\rho}_1 - \boldsymbol{\rho}_2)\mathbf{r}]d^2\rho_1 d^2\rho_2 d^2r.$$

Integration over $\mathbf{r}$ reduces to a delta-function with the new coefficient c':

$$c' \int A_1(\boldsymbol{\rho}_1)A_2^*(\boldsymbol{\rho}_2)\exp[i\beta(\boldsymbol{\rho}_1^2 - \boldsymbol{\rho}_2^2)]\delta(\boldsymbol{\rho}_1 - \rho_2)d^2\rho_1 d^2\rho_2,$$

and, integrating over $\boldsymbol{\rho}_1$, we obtain

$$I = c' \int A_1(\boldsymbol{\rho})A_2^*(\boldsymbol{\rho})d^2\rho.$$

Applying the energy conservation law as it was shown above, one can prove that $c' = 1$. As a result

$$\int A_1(\mathbf{r})A_2^*(\mathbf{r})d^2r = \int A_1(\boldsymbol{\rho})A_2^*(\boldsymbol{\rho})d^2\rho,$$

and, consequently, the complex amplitude u of the output electrical signal remains constant. This proof relies on the Fresnel transform, which is valid only for the aberration-free lens. Nevertheless, it is easy to see that for the thin lens the entire situation does not change even with aberrations, because after integration with delta-function all possible phase terms compensate each other to zero.

Supplemental Reading

M. Born, E. Wolf, *Principles of Optics*, Cambridge University Press, 7th edn., 1999.

R.L. Fork, D.R. Herriott, H. Kogelnik, A scanning spherical mirror interferometer for spectral analysis of laser radiation, *Appl. Opt.*, **7**(12), pp. 1471–1484 (1964).

M.Hercher, The spherical mirror Fabry–Perot interferometer, *Appl. Opt.*, **7**(5), pp. 951–966 (1968).

W.H. Steel, *Interferometry*, Cambridge University Press, 2nd edn., 2009.

J.R. Klauder, E.C.G. Sudarshan, *Fundamentals of Quantum Optics*, 3rd edn., Dover Publications, Mineola, 2006.

S.M. Rytov, Yu.A. Kravzov, V.I. Tatarskii, *Introduction to Statistical Radio Physics, Part 2: Stochastical Fields*, Nauka, Moscow, 1978.

H.Z. Cummins, H.L. Swinney, Light beating spectroscopy. In: *Progress in Optics*, Vol. VIII, E. Wolf, (ed.) North-Holland, Amsterdam, 1970.

Chapter 4

Imaging Spectrometers

4.1. One-Dimensional Imaging Spectrometers

4.1.1. *Lens-based optical scheme*

From the previous chapter, we know that optical system of the spectrometer projects image of the slit in the plane of the photodetector. If the CCD matrix is installed in this plane, then the one-dimensional image of the slit will be obtained along vertical rows at each horizontal pixel of the matrix. As to the horizontal rows, they will be showing spectral intensity at each vertical pixel of the matrix. Thus, the mixed spatial-spectral picture will be the result of the measurement. Does anybody need it? Definitely yes. There are numerous applications where spectral radiance of the source is not spatially uniform and this spatial non-uniformity is the purpose of the research. Imaging spectrometers cover special niche of the market, being much more expensive than ordinary compact spectrometers primarily due to two factors: cheap line CCD photodetector is substituted for high-quality CCD matrix and aberrations of optics are reduced to a minimum. It is well known that aberrations of reflecting optical elements — mirrors — are always stronger than those of the refracting ones — lenses — because reflection does not leave much room for optimization except for only the profile of reflecting surface. In lens design, not only refractive index is an optimization parameter but also a combination of multiple optical components that form a lens. Therefore, if high spatial resolution is needed then the priority should be given to lenses. But typical glass does not work below 400 nm and the red wing of visible domain ends at 800 nm. It means that spectral orders do not overlap and the order-sorting filter is not necessary. As such, lens-based imaging spectrometers may be simplified without loss of quality to a typical scheme shown in Fig. 4.1. This design uses standard high-quality camera lenses and transparent holographic diffraction grating. When design considerations prevail, a folding mirror may be installed to make 90° angle between the input and output axes. The focused image is most symmetrical at all the wavelengths when the grating is oriented parallel to the input plane of the focusing lens. It also makes the shortest design. Anamorphism of the grating, when the diffracted beam is wider than the input one, requires bigger focusing lens with respect to the collimating one. TV lens projects the image of a radiating object into the plane of the entrance slit. Photographic camera lenses are always highly corrected for any type of aberrations, chromaticity, and field curvature. Being designed for standard 36 mm photographic films, their corrected field of view in the focal plane far exceeds dimensions of any CCD matrix, thus leaving no doubts about image quality.

The transmission-type holographic diffraction grating is working in parallel beams. Functionally, it is full analogue of the flat reflection-type diffraction grating already discussed in the previous chapter.

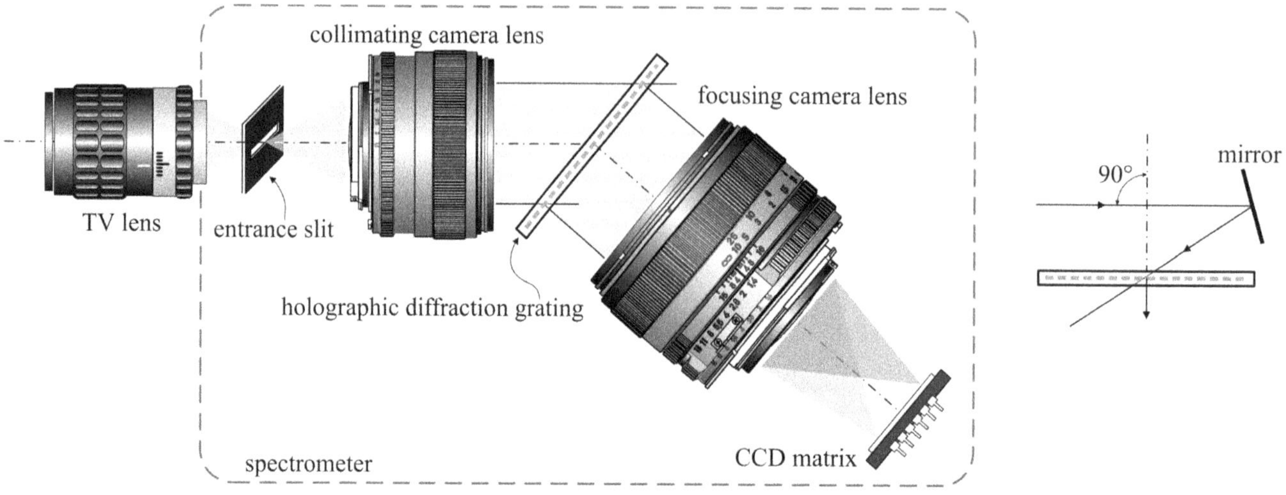

Fig. 4.1. Imaging spectrometer for visible domain.

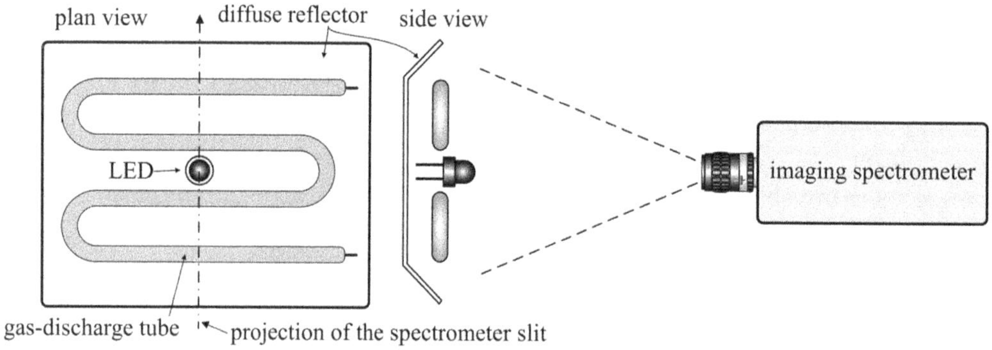

Fig. 4.2. Test source is combined of broad-band gas-discharge tube and narrow-band light-emitting diode (LED).

4.1.2. *Typical performance*

Peculiarity of pictures that imaging spectrometers provide requires some accommodation. As a simplified example, consider a test source shown in Fig. 4.2. Imaging spectrometer, better to say its entrance slit, is horizontally precisely aimed at the center of the light-emitting diode (LED).

Its picture recorded by an imaging spectrometer is shown in Fig. 4.3. Its 1024×256 matrix is composed of 1024 columns, each representing single wavelength, and 256 rows, representing vertical spatial coordinates. Each horizontal section gives a spectrum radiated by a single point on the source. Each vertical section gives vertical distribution of spectral intensity at a single wavelength. Two horizontal sections are made along the LED position and along the axis of third horizontal part of the gas-discharge tube. Two vertical sections are made along 640 nm — maximum spectral intensity of the LED and along 622 nm — red spectral maximum of the gas-discharge tube. In the picture, diffuse component returned from the reflector smoothes sharp physical edges of the tubes.

The best impression gives the three-dimensional picture (Fig. 4.4).

Imaging spectrometers are often used in combination with spectral interferometry to measure thickness profile of thin films and, when equipped with interference objectives — vertical profile of micropatterned surfaces. The advantage is speed: entire profile is measured in one frame of the CCD, without

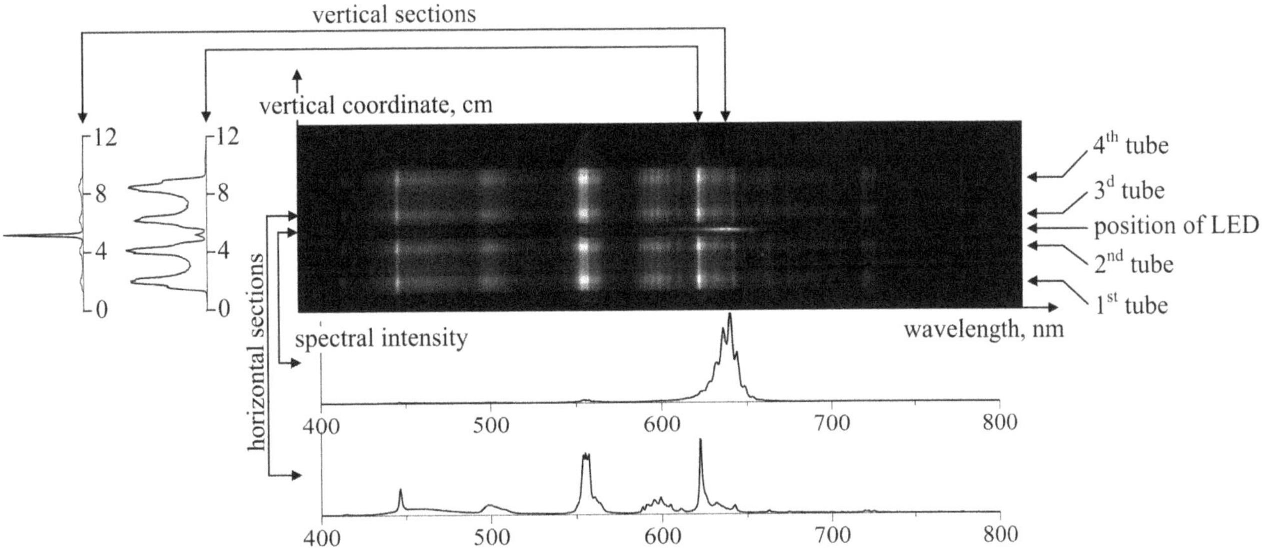

Fig. 4.3. Spatial-spectral picture of the test source shown in Fig. 4.2.

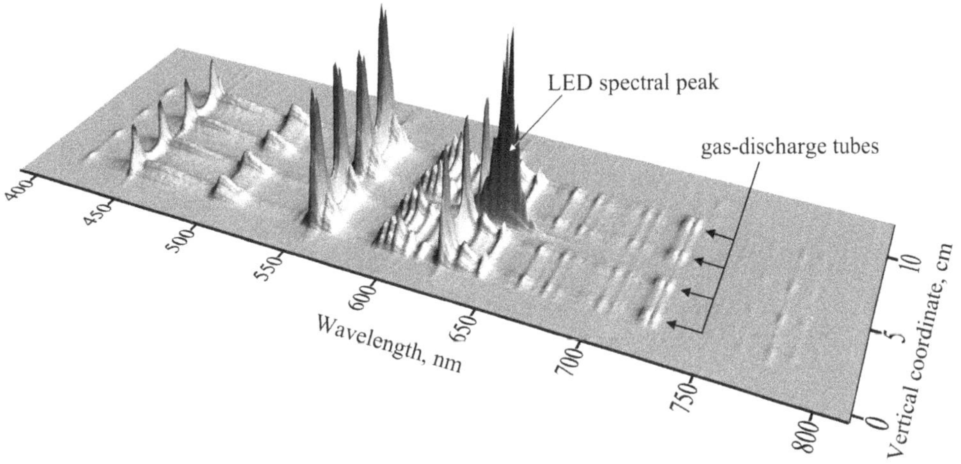

Fig. 4.4. Spatial-spectral map recorded by an imaging spectrometer can be better analyzed in three-dimensional presentation.

mechanical point-by-point scanning. To quickly make spectral interferometer from an imaging spectrometer, TV lens should be replaced by an inspection objective with C-mount flange and illumination port as shown in Fig. 4.5. When there is no pattern on the film, the resultant picture looks very much like an ordinary interference pattern, but it is not: one axis of it represents wavelength. One such spectrum along the white dashed line is portrayed below the picture. Oscillations can be converted to the film thickness in this particular point, using algorithm described in Chapter 6. The spectrum of an ordinary tungsten-halogen lamp is not uniform, with lower spectral emission in the short-wavelength part of the spectrum. This explains black area in the left part of the picture. Maximum emission falls into the infrared domain, which is blocked by multilayer coating of the lamp reflector and additional filters in order to protect outer optical elements from overheating. This explains black area in the right part of the picture. It is clearly seen that the source is optimized for visible domain.

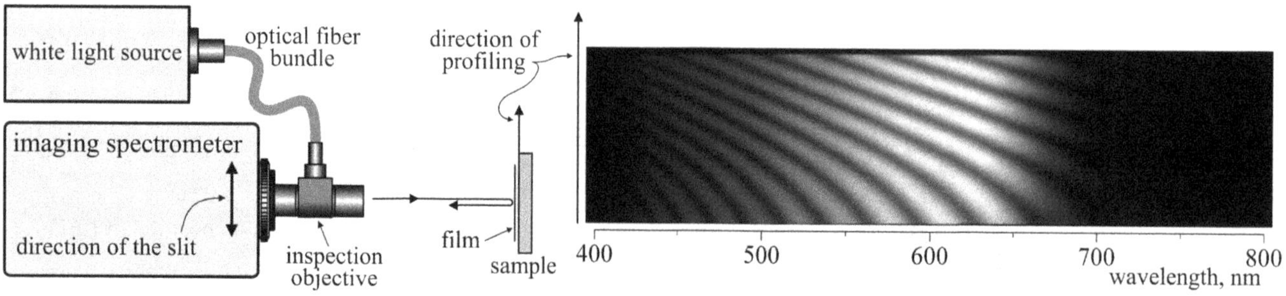

Fig. 4.5. In spectral interferometry mode, imaging spectrometer displays oscillating but not periodical spectra in every point of the slit along vertical axis.

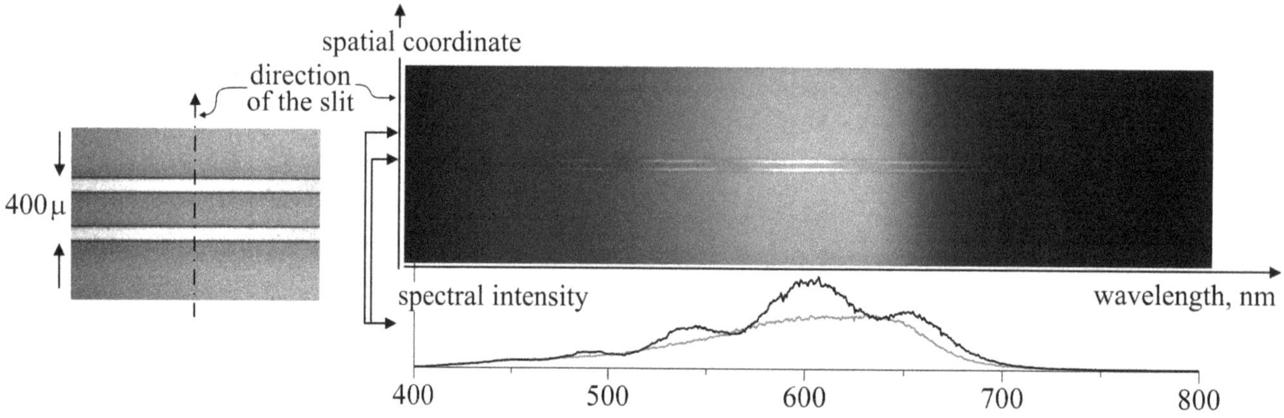

Fig. 4.6. Two thin-film electrodes of 400 μm total width, as they are seen in a microscope, are shown at left. Their spatial-spectral picture is at right. Spectral intensity along the upper strip is shown in black line, spectral intensity along non-patterned area — in gray. Spectral oscillation is clearly visible.

A clear composition of the picture can be seen on regularly patterned films, like the one presented in Fig. 4.6. The two strips are the thin-film leads on the glass of a mobile phone display.

Information about film thickness is encoded in spectral oscillations, and the theory of extracting this information is explained in Chapter 8.

The most confusing thing with the device described above is aiming: exact position of the projection of the spectrometer slit onto the sample, and thus the line along which the profile is being measured, are unknown. One practical solution is to insert temporarily an opaque screen in a form of a small sheet of paper or metal foil with straight vertical edge in front of the sample where you want to measure, and move the sample with such a screen horizontally until the signal is interrupted. Then remove the screen — the spectrometer slit is aimed exactly where you wanted it. The screen may be substituted for a thin wire stretched across the sample along the desired direction. Of course, more fundamental solutions are possible like, for instance, permanent TV monitor coupled through a beamsplitter.

Being connected to a microscope with interference objective of the Mireau or Michelson type, imaging spectrometers can measure not only thickness of thin films but also the profile of bare vertically patterned surfaces (Chapter 8). Also, this technique can be effectively extended even to vibrating surfaces, like in industrial environment, if fast gating is used. If the measurement is performed in much shorter time than the period of vibrations, say in one microsecond, then vibration does not produce any noticeable effect. Gated spectrometers are considered in Chapter 5.

4.2. Two-Dimensional Imaging Spectrometers

4.2.1. *Basic considerations for the Michelson interferometer*

In visible and near-infrared domains, spectroscopic measurements in two dimensions are based on the concept of the Imaging Fourier Spectrometer (IFS) with corner-cube reflectors (CCR). The Fourier spectrometer is based on the Michelson interferometer, which is traditionally understood as a combination of two plane mirrors and a beamsplitter between them (Fig. 4.7). Assume for the beginning that perfectly plane wave comes to the interferometer. If mirror 1 is tilted relatively to mirror 2, then interference fringes are observed in the image plane.

In the Fourier spectroscopy, modulation of light in the image plane is measured by a photodetector and then processed mathematically to reconstruct the spectrum of light. Therefore, amplitude of the electrical signal is important. In order to maximize this signal, the number of fringes within the sensitive area of the photodetector should be as small as possible. Indeed, consider a photodetector with circular sensitive area of radius R, at which two plane waves come with the angle θ between them (Fig. 4.8).

One of the waves has variable phase ϕ, accounting for the displacement of one of the mirrors in the interferometer. The interference pattern is then the oscillating intensity with constant pedestal:

$$\left|e^{ikx\theta/2} + e^{-ikx\theta/2 - i\phi}\right|^2 = 2[1 + \cos(k\theta x + \phi)],$$

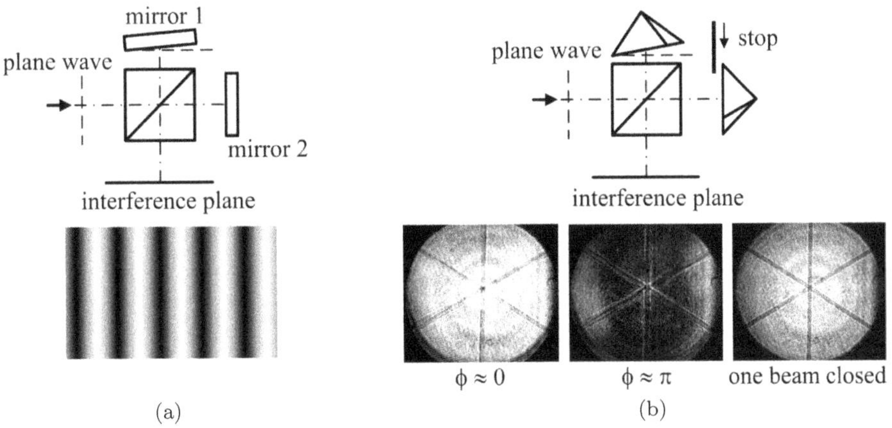

Fig. 4.7. Interference of plane waves. (a) In a standard Michelson interferometer, tilting of any of the two mirrors produces a set of interference fringes. (b) When mirrors a replaced by corner-cube reflectors, no fringes can be observed, although intensity of the image changes as the phase ϕ between the waves changes from 0 to π.

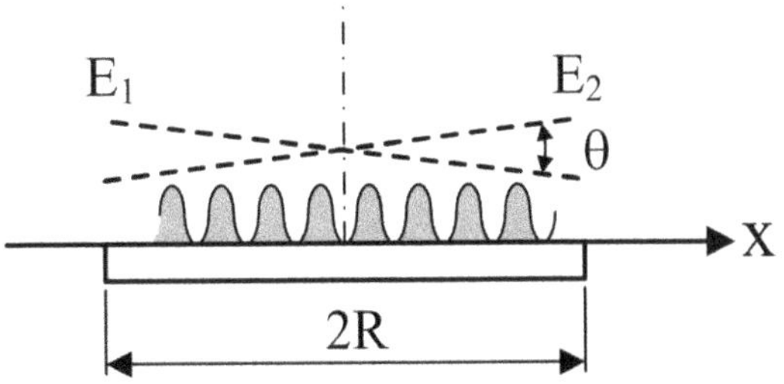

Fig. 4.8. Interference fringes on the detector.

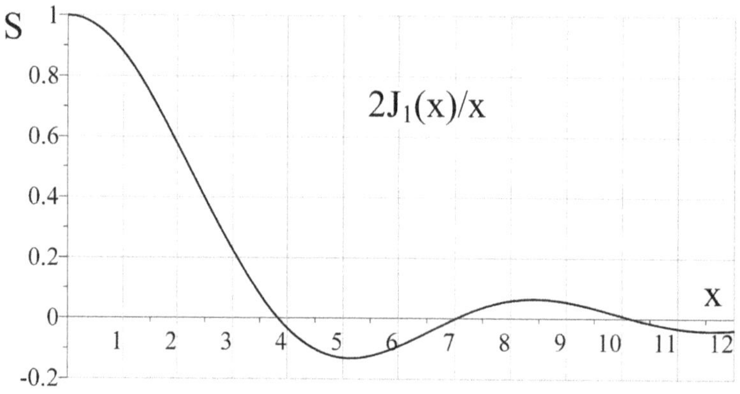

Fig. 4.9. The Airy function.

with $k = 2\pi/\lambda$ being the wavenumber. Only the second oscillating term in the right-hand side of (1) contributes to the useful electrical signal of the photodetector, and we are interested in its amplitude A. Integration over the circular area with polar coordinates r and α gives the electrical signal:

$$S = \int_0^R \int_0^{2\pi} \cos(k\theta r \cos\alpha + \phi) r dr\, d\alpha = \cos\phi \cdot \int_0^R \int_0^{2\pi} \cos(k\theta r \cos\alpha) r dr d\alpha,$$

where the term with $\sin\phi$ vanishes as the integral of the odd function over symmetrical area. Thus, the amplitude of modulation is

$$A = \int_0^R \int_0^{2\pi} \cos(k\theta r \cos\alpha) r dr d\alpha = 2\frac{J_1(k\theta R)}{k\theta R},$$

where J_1 is the Bessel function of the first order. The number of fringes N over the photodetector can be easily expressed through the phase increment over the diameter:

$$N \cdot 2\pi = k\theta \cdot 2R,$$

leading to a simple formula:

$$A(N) = 2\frac{J_1(\pi N)}{\pi N}.$$

This function, known as the Airy function, is shown in Fig. 4.9.

Negative values indicate the change of phase in the output electrical signal. If the signal must not fall below 0.8 of its maximum, then the number of fringes must be less than 0.4. With one fringe on the diameter, the amplitude of the signal falls to 0.2 of the maximum. Therefore, a very accurate angular alignment of mirrors must be done, with accuracy of about angular minutes, which is a big disadvantage of the mirror-type Michelson interferometer. Moreover, in visible domain, which is our priority for the reasons explained in the introduction section and where the wavelength is short, alignment tolerances of this type of the interferometer become so severe that make the mirror-type interferometer impractical. To overcome this drawback, corner-cube reflectors (CCR) may be used. A CCR reflects incoming plane wave back in exactly the same direction, regardless any tilts relative to the direction of the coming wave. Therefore, all spatial parts of interfering plane waves are equally dephased, producing no interference fringes and, hence, maximizing the signal of the photodetector (Fig. 4.7b). Thus, CCR greatly improve stability of the Michelson interferometer, at least for perfectly plane waves.

4.2.2. *Basic properties of CCR*

Plane waves do not carry information about the image: a plane wave produces a point-like spot in the focal plane of a lens — not the image. Therefore, if we want to relay an image through the interferometer, we have to consider a conical wave and to understand the requirements that conical waves impose on the alignment procedure of the interferometer. For that, it is sufficient to analyze a simplified model: a cone beam produced by a point source (Fig. 4.10). It will be shown that a CCR acts exactly as a flat mirror, producing a virtual source on the axis, connecting its apex and the source, at the distance equal to the distance between the source and the apex. Regardless orientation of the CCR.

Let the source be positioned in the origin O and the apex at the point determined by the vector $\mathbf{R}$ (Fig. 4.11). Reflection from the first face of the CCR creates a virtual source O'. The unity normal vector to this face is $\mathbf{n}_1$. Taking the first virtual source O' as the origin, the apex of the CCR is positioned at the point defined by vector $\mathbf{R}'$:

$$2\mathbf{n}_1(\mathbf{R}, \mathbf{n}_1) + \mathbf{R}' = \mathbf{R},$$

where round brackets denote a scalar product. From here

$$\mathbf{R}' = \mathbf{R} - 2\mathbf{n}_1(\mathbf{R}, \mathbf{n}_1).$$

The second face of the CCR is defined by the unity normal vector $\mathbf{n}_2$, and reflection from the second face creates the second virtual source O'', which is connected to the apex by the vector $\mathbf{R}''$:

$$\mathbf{R}'' = \mathbf{R}' - 2\mathbf{n}_2(\mathbf{R}', \mathbf{n}_2).$$

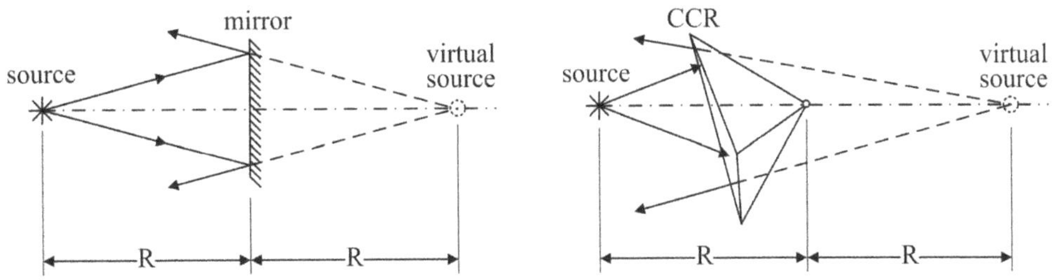

Fig. 4.10. In a CCR, a cone of rays, after multiple reflections, produce the virtual source exactly at the same position, as a flat mirror placed in the apex of the CCR.

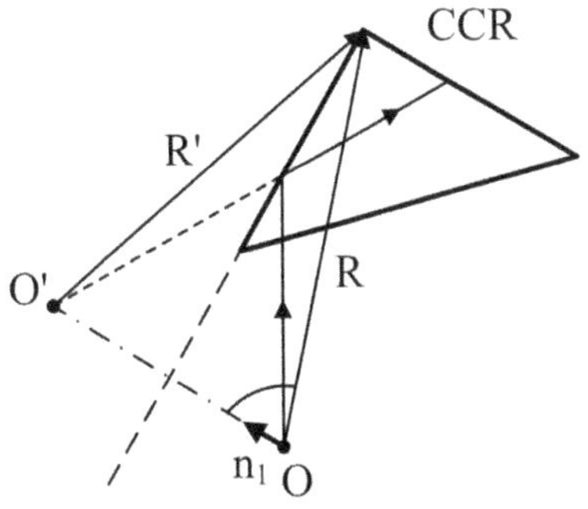

Fig. 4.11. The source O at the origin is mirrored as a virtual source O', when the ray reflects from the first face of the CCR.

The third and the final reflection from the face defined by the unity normal vector $\mathbf{n}_3$ creates virtual source O''' with the vector $\mathbf{R}'''$, connecting it with the apex:

$$\mathbf{R}''' = \mathbf{R}'' - 2\mathbf{n}_3(\mathbf{R}'', \mathbf{n}_3).$$

Then position of the final virtual source, relative to the physical source located at O, is defined by the vector $\mathbf{R} - \mathbf{R}'''$. Computing from (6)–(8), using orthogonality of the normal vectors $(\mathbf{n}_i, \mathbf{n}_j) = 0$ for $i \neq j$, we obtain:

$$\mathbf{R} - \mathbf{R}''' = 2\mathbf{n}_1(\mathbf{R}, \mathbf{n}_1) + 2\mathbf{n}_2(\mathbf{R}, \mathbf{n}_2) + 2\mathbf{n}_3(\mathbf{R}, \mathbf{n}_3).$$

Vectors $\mathbf{n}_1, \mathbf{n}_2, \mathbf{n}_3$ form the Cartesian system of coordinates with the origin at the apex of the CCR, in which the scalar products $(\mathbf{R}, \mathbf{n}_{1,2,3})$ are none others but directional cosines of the vector $\mathbf{R}$. Therefore,

$$\mathbf{n}_1(\mathbf{R}, \mathbf{n}_1) + \mathbf{n}_2(\mathbf{R}, \mathbf{n}_2) + \mathbf{n}_3(\mathbf{R}, \mathbf{n}_3) = \mathbf{R}$$

and

$$\mathbf{R} - \mathbf{R}''' = 2\mathbf{R},$$

which means that, after triple reflection from the CCR, the virtual source is positioned on the axis, connecting the physical source and the apex, at the distance $2R$ from the physical source, as shown in Fig. 4.10(b). This result holds true for any tilts of the CCR relative to the source: only position of the apex is important.

In Section 4.2.1, interference of plane waves in the Michelson interferometer was outlined and it was emphasized that, for the plane waves only, the interferometer with CCR does not produce any fringes. Now, consider what happens when conical beams interfere in the Michelson interferometer with CCR (Fig. 4.12).

Interference pattern in the plane x, y is defined by the intensity of the sum of two spherical waves with the wavelength λ, coming from virtual sources located at the distances r_1, r_2 from the point of observation:

$$|e^{ikr_1} + e^{ikr_2}|^2 = 2 + 2\cos[k(r_1 - r_2)],$$

where $k = 2\pi/\lambda$. Considering, as usual, long $R_{1,2} >> \varepsilon, x, y$, and using the first-order expansion of $r_1 - r_2$, we obtain

$$|e^{ikr_1} + e^{ikr_2}|^2 = 2(1 + \cos\phi); \quad \phi = k\left[R_1 - R_2 + \frac{1}{2}(x^2 + y^2) \cdot \left(\frac{1}{R_1} - \frac{1}{R_2}\right) + \frac{x\varepsilon}{R_2} - \frac{\varepsilon^2}{2R_2}\right].$$

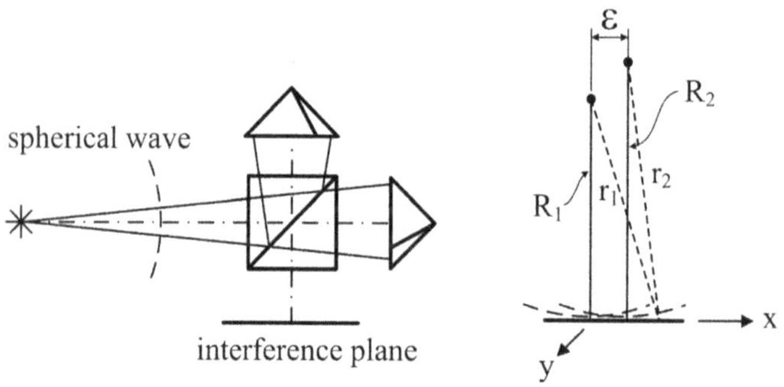

Fig. 4.12. In the interference plane, the two CCR form two spherical waves, coming from virtual sources located at the distances R_1 and R_2 from the interference plane and laterally shifted by ε.

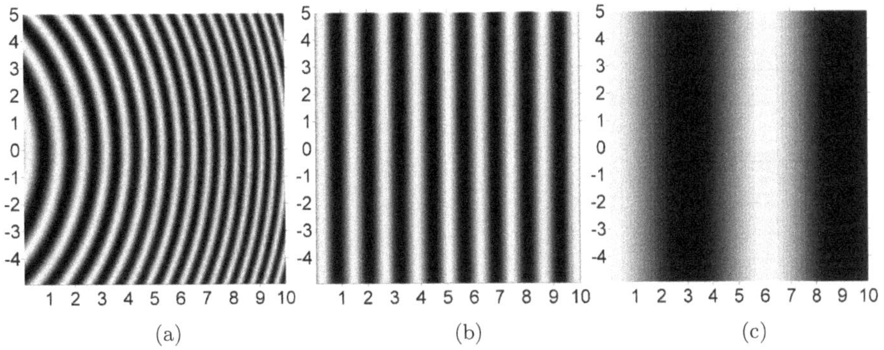

Fig. 4.13. In the interference plane, the two CCR form two spherical waves, coming from virtual sources located at the distances R_1 and R_2 from the interference plane and laterally shifted by ε.

The origin of the system of coordinates in the interference plane was chosen in the projection of the first virtual source. The phase ϕ determines interference pattern: quadratic terms in x, y produce curvature of fringes, while linear term in x is responsible for the straight fringes. When $R_1 = R_2$, i.e. optical paths to the CCR are equal, quadratic term vanishes. When apexes of the CCR are located on one line, i.e. $\varepsilon = 0$, linear term vanishes. A set of three typical simulated pictures is shown in Fig. 4.13. It gives us a very important hint on the alignment procedure for the Michelson interferometer with CCR: (A) if the interference fringes display curvature, then equalize optical paths in the shoulders of the interferometer until straight lines appear; (B) after that, align lateral displacement of the one of the CCR until uniform illumination replaces fringes. Thus, unlike the case of the interferometer with mirrors, requiring angular alignment of the mirrors, alignment of the interferometer with CCR requires lateral displacements of the CCR. This feature makes the Michelson interferometer with CCR far more stable relative to its mirror-based analog.

The alignment sequence is outlined in Fig. 4.14. It starts with connecting the interferometer to a He–Ne laser in order to obtain initial picture with interference fringes, shown schematically at right. Coherence length of the He–Ne laser radiation is on the scale of kilometers, making the two waves interfere at any initial path difference that occurs in the interferometer without preliminary equalization of optical paths. An important feature of the interferometer with CCR is that the curvature of fringes indicates the optical path difference between the two shoulders: the more the curvature is, the bigger the path difference is. Therefore, at the next step, the second CCR installed on a piezo-scanner is manually moved to and fro so that to minimize the curvature of fringes. Although this operation does not guarantee complete equality of optical paths in the shoulders, it minimizes the optical path difference to a value when interference fringes can be visible with interference filter and white light source. Next, the laser is substituted with the white light source (tungsten-halogen lamp) and interference filter at its output. Coherence length shortens to a scale of hundreds of microns, and it is not very difficult to maximize interference contrast to further minimize the optical path difference between the shoulders. Only axial positioner is activated at this phase, leaving interference fringes visible as a mark for estimating the contrast. After that, interference filter is removed and fine axial alignment is done, maximizing fringe contrast again. Finally, the X–Y positioner of the first CCR is used to completely remove fringes from the image.

As additional explanation, Fig. 4.15 portrays the images obtained with laser illumination in four different cases of alignment. Note that the image with concentric circles cannot be achieved in white light illumination because in this case the optical path difference greatly exceeds coherence length of light.

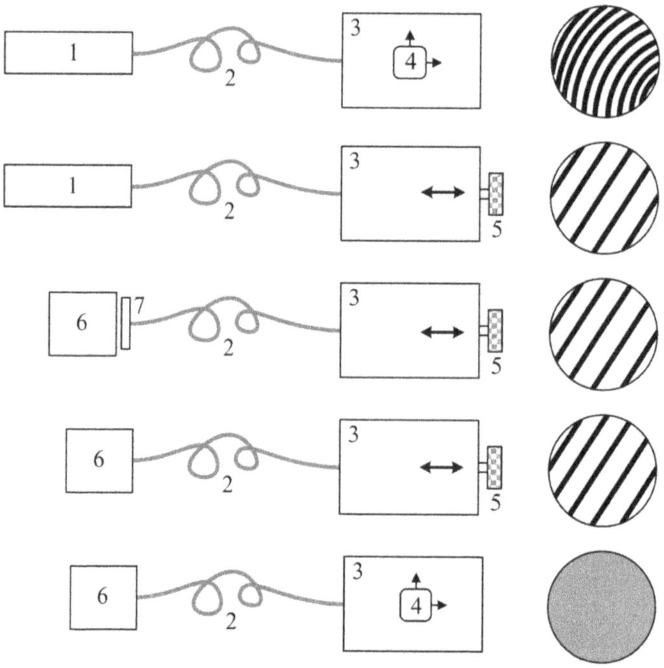

Fig. 4.14. Alignment sequence. (1) He–Ne laser; (2) optical fiber; (3) interferometer; (4) X–Y stage with CCR; (5) translational alignment; (6) white light source; (7) interference filter.

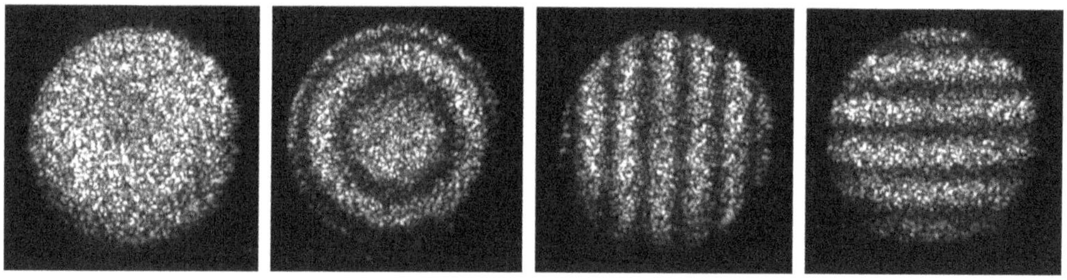

(a) (b) (c) (d)

Fig. 4.15. Possible interference patterns: (a) complete alignment; (b) axial misalignment — inequality of optical paths; (c) horizontal misalignment; (d) vertical misalignment.

4.2.3. *Imaging configuration of the Michelson interferometer with CCR*

Generalized optical scheme of the imaging Michelson interferometer with CCR is shown in Fig. 4.16. According to the Fermat principle, all the rays, emerging from any point in the object plane and converging to its image point (conjugated point) in the image plane, have the same traveling time through their paths. It means that all the conjugated points in the object and image planes are temporally coherent, i.e. have the same phases. Therefore, the waves, forming images of these points, interfere in the image plane. It does not matter, whether the object plane coincides with the focal plane of the lens 1 or not: temporal coherence between all the rays from a single point in the object plane will be preserved in the conjugated point in the image plane. As such, the electrical signal from every pixel of an image sensor, placed in the image plane, will be accordingly modulated when one of the reflectors is being moved along the axis, exactly as in a non-imaging Fourier spectrometer. Thus, making standard mathematical

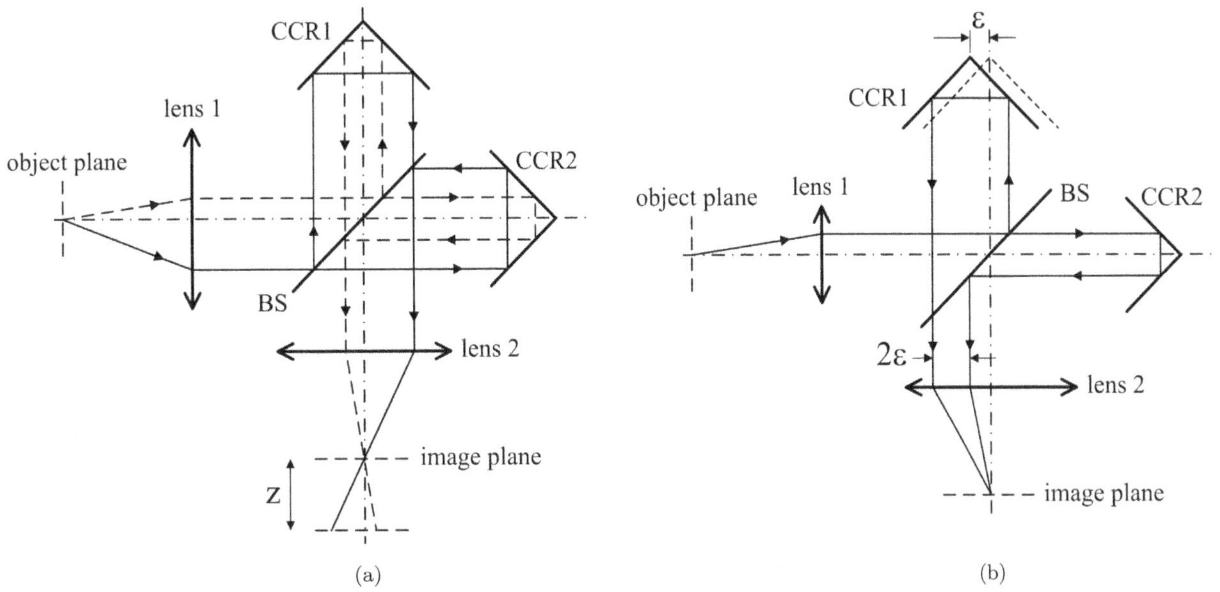

Fig. 4.16. Generalized scheme of an imaging Michelson interferometer with CCR: (a) perfectly aligned interferometer, when apexes of the CCR are mirror-reflected against the beamsplitter BS; (b) misaligned interferometer: interfering rays are shifted by 2ε.

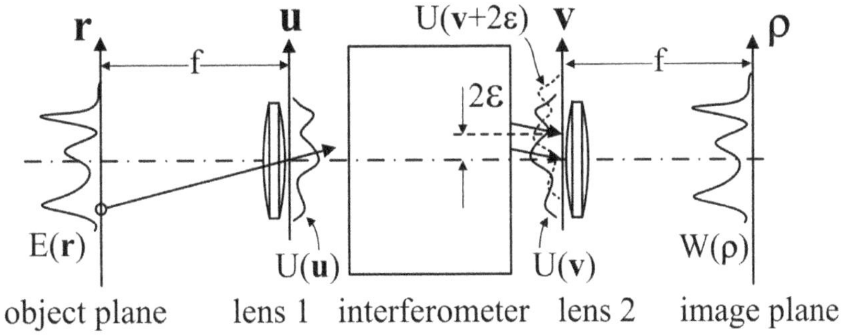

Fig. 4.17. Equivalent optical scheme for the Michelson interferometer with CCR. Bold fonts denote vectors.

computations on the signals from every pixel of the image sensor, it is possible to reconstruct spectrum of light in every pixel. This is the idea of an imaging Fourier spectroscopy.

In a perfectly aligned interferometer, the two rays, in which the beamsplitter splits any single ray coming from the object, recombine in a single ray again. In Fig. 4.16(a), the two arbitrary chosen object rays are shown in solid and dashed lines, splitting at 90° at the beamsplitter and recombining again before they are directed to the lens 2. Since the rays coincide, they propagate with constant phase difference and interfere at any point on their trace. It means that in a perfectly aligned interferometer, interference pattern is not localized in the image plane only and may be observed at any displacement z from it. The image is distorted outside the image plane, but interference pattern is observable.

Consider now what happens when one of the CCR is shifted by a distance ε: the rays, being split by the beamsplitter, do not recombine again and are shifted by 2ε (Fig. 4.16(b)). The equivalent optical scheme in Fig. 4.17 is helpful to address this problem mathematically. For the sake of compactness of formulas, we may assume equal focal lengths f of the lenses 1 and 2. If they are not equal, then a trivial scale coefficient will appear in the final result. Let $E(\mathbf{r})$ be spatially incoherent complex field with the wavelength λ in the object plane, and its intensity being $|E(\mathbf{r})|^2 = I(\mathbf{r})$. Then, according to the theory of linear optical systems, the field at the output pupil of the lens 1 is proportional to the Fourier transform of $E(\mathbf{r})$:

$$U(\mathbf{u}) = \int E(\mathbf{r}) \exp\left(-ik\frac{\mathbf{ru}}{f}\right) d^2r$$

with integration over the entire plane $\mathbf{r}$, and $k = 2\pi/\lambda$ being the wavenumber. The interferometer with CCR creates the second wave and directs both the original and the new waves in the mirrored direction to the second lens. Dropping the proportionality coefficient and assuming 50% splitting of the original wave, the two interfering waves at the input pupil of the lens 2 are $U(\mathbf{v})$ and $U(\mathbf{v} + 2\varepsilon)$. In the image plane $\boldsymbol{\rho}$, the first and second interfering waves create complex fields $W(\boldsymbol{\rho})$ and $W'(\boldsymbol{\rho})$, so that the average intensity is

$$|W(\boldsymbol{\rho}) + W'(\boldsymbol{\rho})|^2 = |W(\boldsymbol{\rho})|^2 + |W'(\boldsymbol{\rho})|^2 + 2\mathrm{Re}[W(\boldsymbol{\rho}) \cdot W'^{*}(\boldsymbol{\rho})],$$

where the asterisk denotes complex conjugate. Since any waves at the input pupil of a lens and in its image plane are Fourier-inversed, it is possible to write down:

$$W(\boldsymbol{\rho}) = \int U(\mathbf{v}) \exp\left(ik\frac{\mathbf{v}\boldsymbol{\rho}}{f}\right) d^2v \approx E(\boldsymbol{\rho}),$$

$$W'(\boldsymbol{\rho}) = \int U(\mathbf{v} + 2\varepsilon) \exp\left(ik\frac{\mathbf{v}\boldsymbol{\rho}}{f}\right) d^2v \approx e^{-ik2\varepsilon\rho/f} \cdot E(\boldsymbol{\rho}).$$

These formulas are approximate to the extent to which the integral over the lens pupil may be assumed the delta-function:

$$\int e^{ikv\rho}/fd^2v \sim \delta(\boldsymbol{\rho}),$$

which means high spatial resolution of the lenses. Then, the intensity in the image plane is given by

$$|W(\boldsymbol{\rho}) + W'(\boldsymbol{\rho})|^2 = I(\boldsymbol{\rho}) \left[1 + \cos\left(k\frac{2\varepsilon\boldsymbol{\rho}}{f}\right)\right].$$

The factor "2" vanishes because the beamsplitter relays only half of the original intensity in each shoulder of the interferometer: $|W(\rho)|^2 = 0.5\, I(\boldsymbol{\rho})$. Thus, the image plane of the misaligned Michelson interferometer with CCR reproduces the intensity in its object plane with fringes $\cos(k2\varepsilon\boldsymbol{\rho}/f)$. Note, that the mirror-based Michelson interferometer would give the inverse intensity distribution, proportional to $I(-\boldsymbol{\rho})$.

Next, we are going to discuss localization of fringes — the feature of the interferometer that determines how accurately the photodetector may be positioned in axial direction relative to the output image plane. According to the principle of fringe localization, "the fringe viewing surface for an interferometer should be the surface that is the locus of points of intersection of the rays which originate from one incident ray coming from the source"[a]. Figure 4.16(b) indicates that, in our case, such locus is the output image plane, where the contrast of fringes is a maximum regardless relative displacement of CCRs ε. It means that placing the photodetector exactly in the output image plane, the maximum sensitivity of the spectrometer is guaranteed. But how accurately it must be positioned and how big the fall of sensitivity will be if actual position of the photodetector is shifted by value z from the image

[a]J.C. Wyant, Fringe localization, *Appl. Opt.*, **17**(12), p. 1853 (1978).

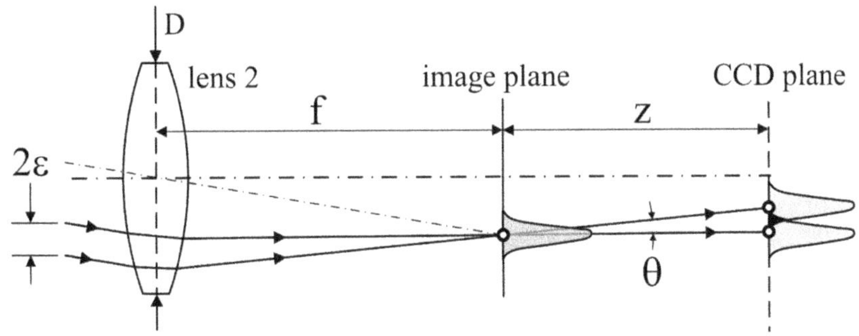

Fig. 4.18. Interference takes place in the plane z only when point-spread functions of the two rays intersect.

plane (Fig. 4.16(a))? Geometrical optics cannot answer this question, therefore, we have to consider diffraction on the output lens (Fig. 4.18). In geometrical optics approximation, the two parallel rays shifted by ε intersect only in the image plane located at the focal distance f from the output lens. In the wave optics approximation, however, these two waves still overlap even outside the image plane because the spots that they form in the focal plane are not infinitely narrow but have some final width of the order of $\sim \lambda f/D$ — the width of the point-spread function of the lens. In reality, this width is even bigger due to aberrations, which are unknown in practice.

Thus, with all the unknowns, it is impossible to make exact mathematical calculations of the tolerance on z, but still very coarse estimate is possible. If we assume that the two rays diverge from the image plane at the angle

$$\theta = \frac{2\varepsilon}{f},$$

then their lateral separation in the plane z is $z\theta$. Assuming further the diffraction-limited width of the point-spread function $\lambda f/D$, the limiting condition for interference my be written as

$$z\theta < \frac{\lambda f}{D}, \quad \text{or} \quad z < \frac{\lambda f}{\theta D} = \frac{\lambda}{2\theta NA},$$

where $NA \approx D/2f$ is the numerical aperture of the lens. In practice, the alignment is always judged visually by the number of fringes N in the image plane. Therefore, we have to rewrite the last formula in terms of the number of fringes N over the photodetector diameter d. Recalling that the fringes are distributed as $\cos(2\pi\theta\rho/\lambda)$, their number is equal to

$$N = \theta d/\lambda,$$

giving the coarse estimate for the axial tolerance as

$$z < \frac{d}{2N \cdot NA}.$$

With typical values $d = 6\,\text{mm}$, $N = 1$, and $\text{NA} = 0.3$, this formula suggests localization depth $z \sim 1\,\text{cm}$. Although this is a rather coarse estimate, it is very important for the beginning of opto-mechanical design to realize, how accurate the detector must be positioned relative to the image plane: with the micrometer, or millimeter, or centimeter precision. For instance, in the prototype of the IFS, described in the next section, localization depth is about 1 cm with the number of fringes up to $N = 10$.

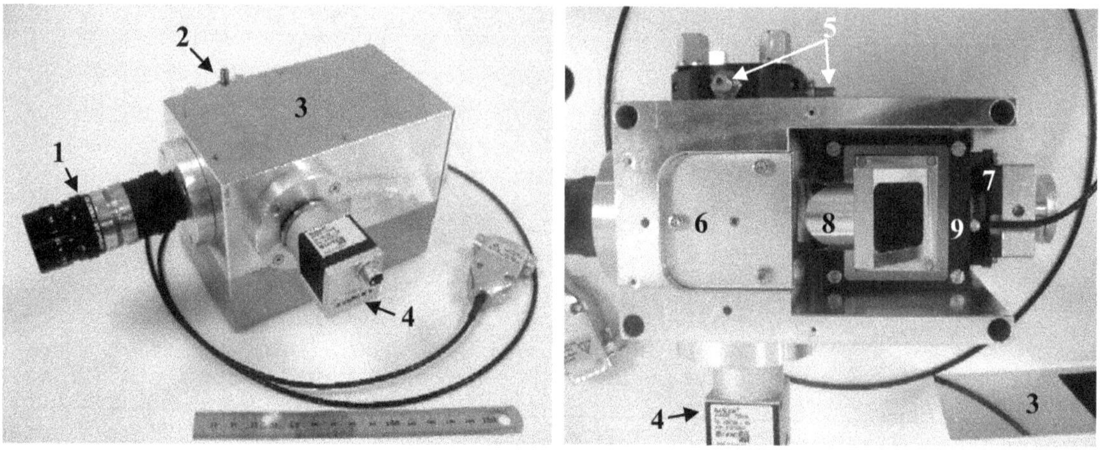

Fig. 4.19. Opto-mechanical module of the reflectometer. 1 — imaging optics (various types); 2 — lateral adjustment stage; 3 — cover plate; 4 — high-speed CCD camera; 5 — X–Y adjustment screws; 6 — beamsplitter compartment; 7 — manual positioner; 8 — CCR; 9 — piezo-scanner.

4.2.4. *Design concept of the IFS*

The opto-mechanical concept of the IFS is explained in Fig. 4.19. Unlike the majority of Fourier spectrometers, there is no reference laser in this design: the precision of scanning is fully determined by capacitive sensor in the piezo-scanner (Physik Instrumente, P625.1CD). Two identical lenses of 40 mm focal length — infinite conjugated visible achromatic triplets (Thorlabs TRH254-040-A) — are hidden in cylindrical holes milled in aluminum body and fixed with epoxy: one after the imaging optics, and the second just in front of the photodetector. In the imaging mode, a high-speed CCD camera with the sampling rate 500 frames per second (Basler 640 × 480) was used as the photodetector, while in non-imaging applications it could be replaced with the photo-multiplying tube (PMT, see below in this section). In between the lenses, in a rectangular compartment machined in a bulk aluminum body, a one-inch 50:50 beam-splitting cube (Thorlabs, BS013) is placed and fixed by a set screw in the middle of the cover. Two aluminum coated visible-domain CCR (Spectrum Scientific, HR-254-3-AL) are installed one on the X–Y adjustment stage (Thorlabs CXY1), and the second — on the piezo-scanner. For coarse manual adjustment of optical paths in two shoulders of the interferometer, the piezo-scanner is installed on a linear positioner with the ball-screw (Suruga, K101-20M-M).

4.2.5. *Data acquisition and processing*

During scanning, the CCD camera samples temporal oscillations of image intensity in each pixel, producing digital record of the interferogram — the functional dependence of light intensity on the displacement of the CCR. According to the Whittaker–Kotelnikov sampling theorem, a sinusoidal function can be totally reconstructed from its discrete values if periodicity of sampling is shorter than the period. This theorem establishes a relation between the camera sampling rate f, scanner stroke S, scanning time T, and minimum wavelength $\lambda_{\min}$ that can be reconstructed. Since a single displacement of the scanner produces twice as long variation of the optical path,

$$\frac{2S}{f \cdot T} < \lambda_{\min}.$$

The design was done for $\lambda_{\min} = 300$ nm; sampling rate $f = 500\ \mathrm{c}^{-1}$; scanning time $T = 3\,\mathrm{c}$. For these values, the above formula gives the maximum stroke $S_{\max} = 225\ \mu$m. The longer would be the stroke at the same scanning time, the shorter would be oscillations in the interferogram, and the camera samplings would miss them. Thus, from the point of view of sampling precision, the shorter the stroke is, the better. On the other hand, spectral resolution must be high enough to resolve spectral oscillations in the reflected light. From this point of view, the stroke must be as long as possible. Indeed, if the wavenumber is $k = 1/\lambda$ and spectral intensity of light $I(k)$, then the interferogram V as a function of the displacement of the CCR x is

$$V(x) = \int_{-\infty}^{+\infty} I(k)\cos(2\pi k 2x)dk.$$

Inverse Fourier transform of the interferogram reconstructs the spectrum:

$$I'(p) = \int_{-S/2}^{+S/2} V(x)\cos(2\pi p 2x)dx = \frac{S}{2}\int_{-\infty}^{+\infty} I(k)h(k-p)dk,$$

with

$$h(y) = \frac{\sin(2\pi y S)}{2\pi y S} \equiv \operatorname{sinc}(2\pi y S)$$

being the instrument function, determining spectral resolution Δk: the narrower it is the better the resolution. Numerical computations show that two narrow spectral lines located at k and $k + \Delta k$ can be resolved by a spectrometer with the instrument function $h(y)$ if $\Delta k S > 1$, thus defining spectral resolution criterion as

$$S > \frac{\lambda^2}{\Delta\lambda},$$

with $\Delta\lambda$ in this formula being the spectral resolution in the wavelength domain. The value of $\Delta\lambda$ is determined by the period of spectral oscillations Λ that may be expected in the reflected light. When the IFS works in a spectral interferometry mode, measuring thickness of films (Chapter 8),

$$\Lambda = \frac{\lambda^2}{2nt},$$

where n is real part of the refractive index of the medium and t — thickness of the structure. The thickest semiconductor structures today are 3D NAND memory chips, which may be of about 5 μm deep. With the minimum penetration wavelength $\lambda \approx 0.7$ μm, $n \approx 3$, and $t \approx 5$ μm, this formula gives Λ ~16 nm. In order to precisely sample this oscillation, spectral resolution must be an order of magnitude better, i.e. $\Delta\lambda$ =1–2 nm. For example, standard spectrometers, available on the market, are characterized by $\Delta\lambda \approx$1 nm around $\lambda = 0.6$ μm. With these values, the above formula for Λ gives the minimum stroke of the interferometer $S_{\text{min}} = 360$ μm, which is bigger than the value $S_{\text{max}} = 225$ μm obtained previously. Sacrificing slightly spectral resolution and increasing insignificantly the minimum wavelength of the device, the stroke was chosen to be 350 μm — enough to work in the visible domain, starting from around 350nm, and having spectral resolution $\approx$ 1.5 nm.

The key component of any Fourier spectrometer is the scanning module — the mechanical part that drives one of the reflectors in the Michelson interferometer. This device must be stable and accurate to a nano-meter scale in order to ensure reproducible precise spectral measurements. In visible domain, where the length of the mechanical drive does not exceed millimeter range, it is possible to rely on the precision of a piezo-scanner with capacitive sensor instead of laser interferometry, commonly used in infrared Fourier spectrometers. Very reliable results may be obtained with the Physik Instrumente (PI) P-625.1CD linear piezo stage with capacitive feedback and closed-loop control module E-665. Characteristics of this device are shown in Table 4.1.

Table 4.1. Characteristics of P-625.1 piezo stage.

Open loop travel	Closed loop travel	Closed loop resolution	Linearity error	Repeatability	Stiffness	Resonant frequency with 20g load	Material	Dimension	Mass
600 μm	500 μm	0.5 nm	0.03%	$\pm$5 nm	0.1 N/μm	180 Hz	Al	$60 \times 60 \times 15\,\mathrm{mm}^3$	240 g

Table 4.2. Linearity test of the piezo-scanner.

Input voltage (Volt)	Displacement (μm)	Nonlinearity (μm)	Nonlinearity (%)
0.0000	0.0000	0.0000	0.0000
0.6250	31.2059	−0.0451	−0.0090
1.2500	62.4180	−0.0838	−0.0168
1.8750	93.6680	−0.0848	−0.0170
2.5000	124.9343	−0.0694	−0.0139
3.1250	156.2012	−0.0536	−0.0107
3.7500	187.4671	−0.0386	−0.0077
4.3750	218.7503	−0.0063	−0.0013
5.0000	250.0426	0.0350	0.0070
5.6250	281.3176	0.0591	0.0118
6.2500	312.5083	0.0889	0.0178
6.8750	343.8636	0.1033	0.0207
7.5000	375.1258	0.1144	0.0229
8.1250	406.3692	0.1069	0.0214
8.7500	437.6034	0.0901	0.0180
9.3750	468.8083	0.0441	0.0088
10.0000	500.0151	0.0000	0.0000

Of the entire 500 μm closed-loop travel range of the P-625 scanner, only 350 μm were used in order to minimize travel errors and, at the same time, satisfy the requirements for spectral resolution.

Precision of spectral measurements depends on linearity and stability of displacements of the moving CCR in the interferometer. Without reference laser, commonly used in Fourier spectrometers, it solely relies on the capacitive sensor embedded into the piezo-scanner. Therefore, both linearity and stability measurements were made to ensure performance of the spectrometer. Linearity was measured, using Zygo ZMI 1000 laser interferometer, and results are presented in Table 4.2.

The last column of this table is presented graphically in Fig. 4.20, showing that maximum nonlinearity of the translational motion relative to control voltage is less than 0.03%.

In order to measure spectral stability of the device during multiple scans, the IFS was connected to the stabilized He–Ne Zeeman laser HP 5517B through optical fiber. Typical spectral curve, produced by IFS, is shown in Fig. 4.21. The idea of the experiment was to record spectra after every IFS scan, and to measure variations in positions of the maxima on the wavelength axis. Before this experiment, the output of the Fourier transform was calibrated against the He–Ne laser to the fourth digit precision as 632.8 nm. In experiments, it turned out that the discrete value of the wavelength, corresponding to the maximum, never changed, and remained to be 632.8(40) nm at all scans. In order to measure fine variations, the adjacent points were considered, as in Fig. 4.22. Fitting parabola in three adjacent

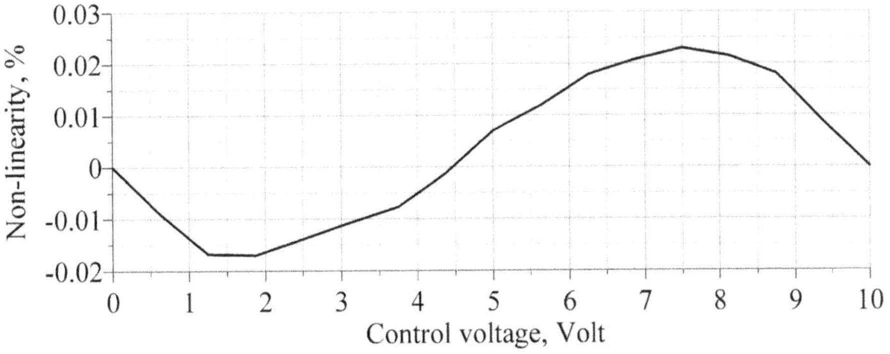

Fig. 4.20. Nonlinearity of CCR motion as a function of control voltage.

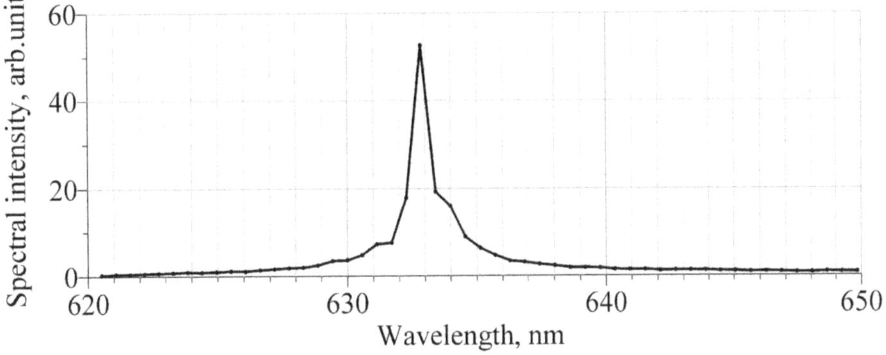

Fig. 4.21. Typical spectrum of a stabilized He–Ne laser, obtained by IFS.

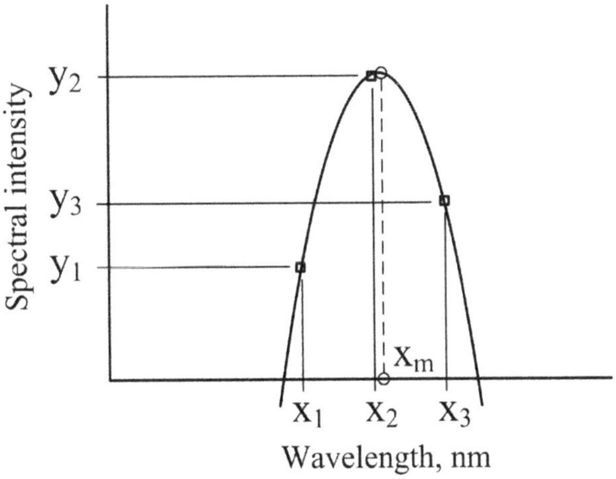

Fig. 4.22. Parabolic fit gives precise position of the maximum x_m.

points and finding its maximum, gives precise location of the peak wavelength x_m. Then, it is possible to find an average of these values $\overline{x}_m$ over N scans, compute variations in each scan, starting from the first, var $= x_m - \overline{x}_m$, and real-mean-square value for them: $\sigma = \sqrt{\text{var}^2/N}$.

Numerically, the result gave $\overline{x}_m = 632.8513$ nm and $\sigma = 3.25 \times 10^{-3}$nm. Graphically, the evolution of variations as a function of the scan number is presented in Fig. 4.23.

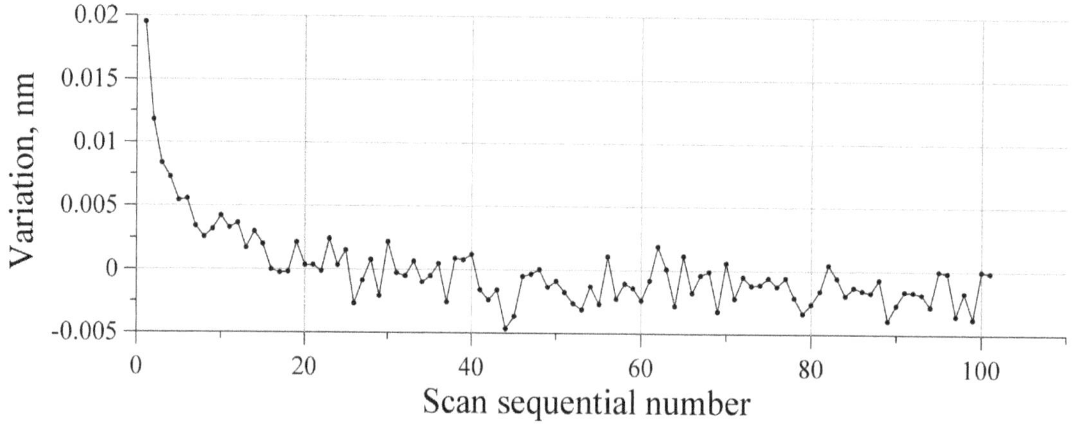

Fig. 4.23. Evolution of variations during first 100 scans of the spectrometer.

Table 4.3. H6780-20 quantum efficiency.

Wavelength (nm)	300	400	500	600	700	800
Quantum efficiency (%)	3.4	8.9	18.3	15.9	12.8	7.4

Table 4.4. H6780-20 characteristics.

Dimensions	Max. output curr.	Sensitive area	Gain	Sensitivity	Rise time	Weight
$50 \times 22 \times 22\,mm^3$	$100\mu A$	Ø8 mm	10^2–10^6	3.9×10^4 A/W	0.78 ns	80 g

The second critical component in terms of importance is the photodetector. It is known that the most sensitive high-speed photodetector, currently in existence, is the photo-multiplying tube (PMT). Hamamatsu photonics offers a variety of compact, reliable, and simple in operation PMT modules, incorporating both the PMT unit itself and miniature high-voltage power supply, working from low voltage +12 V. Various types of photo-cathodes may be ordered, spanning entire diapason of wavelengths from ultra-violet (UV) 200 nm to near-infrared (NIR) 900 nm. Since the design guidelines required ordinary optical components, mostly transparent in visible (VIS) and NIR spectral ranges, the Hamamatsu PMT module H6780-20 was chosen with maximum spectral efficiency in this range (Tables 4.3 and 4.4).

The next critical component is the beamsplitter with two essential requirements: it must be compensated, provide uniform spectral division, and be non-polarizing. The compensated beamsplitter introduces equal phase delays in both waves at its output. For instance, a front-coated semi-transparent mirror is not a compensated beamsplitter because the wave transmitted through the front semi-transparent layer acquires triple phase delay on passing the bulk glass relative to the wave reflected from the front surface. In visible domain, the best choice for the compensated beamsplitter is a standard beam-splitting cube. Another requirement is spectral uniformity of the splitting coefficient. It is practically impossible to manufacture a non-polarizing beam-splitting cube with uniform spectral characteristic of splitting:

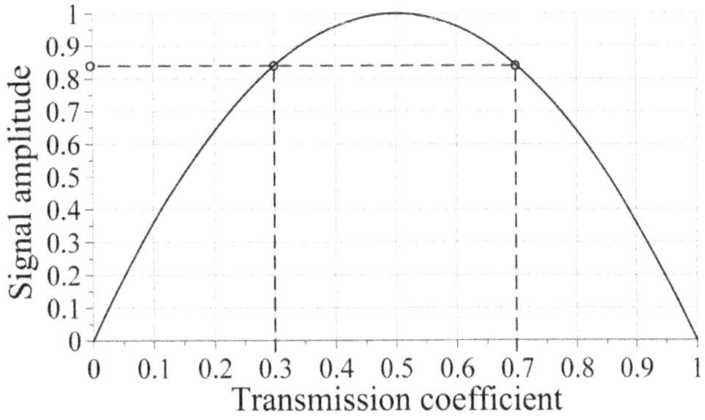

Fig. 4.24. Normalized signal amplitude as a function of the division coefficient.

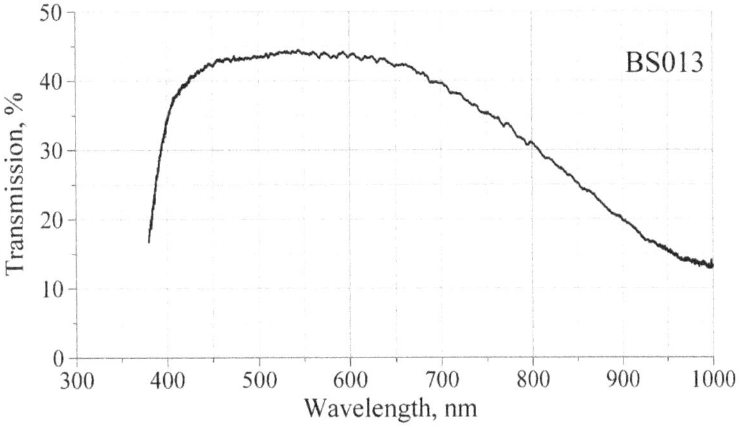

Fig. 4.25. Spectral transmission of the BS013 beamsplitter.

there is always up to 20% variation over the entire spectral range. This issue may cause fears that spectral response of the Fourier spectrometer will also be highly non-uniform, with equal scale of non-uniformity in the output signal. Luckily, this is not the case. Indeed, suppose the energy transmission coefficient is t. Then the reflection is $1-t$, and modulated component of the electrical signal, created by interference of the waves at the photodetector, is proportional to $t \cdot (1-t)$. This is a quadratic function of t, reaching its maximum at $t = 0.5$ (Fig. 4.24). Since it is a quadratic function, large deviations of t cause less significant variations of the signal. For instance, deviations of transmission up to 40% (from $t = 0.5$ to $t = 0.3$) cause only 15% variations in the signal.

With all the aforementioned considerations, the ThorLabs one inch BS013 beam-splitting cube designed for spectral interval 400–700 nm was used. Manufacturer's specifications claimed non-polarized transmission between 45% and 30% in this spectral interval. The data measured with Ocean Optics HR4000CG-UV-NIR spectrometer and SpectraView software showed similar picture presented in Fig. 4.25. With the transmission coefficient varying more than twofold from 0.45 at 550 nm to 0.2 at 900 nm, the chart in Fig. 4.24 makes it possible to expect only 35% variations of the signal in this spectral interval. However, rapid fall of beamsplitting efficiency below 400 nm, amplified by the fall of quantum efficiency of the PMT (Table 4.3), led to significant loss of sensitivity of the IFS below 400 nm — the feature that will be seen in the experimental results, reported in the subsequent sections.

Figure 4.26 presents schematically the design of the IFS in three orthogonal projections. The interferometer frame is composed of two components: the fork and the base, to which the fork is bolted.

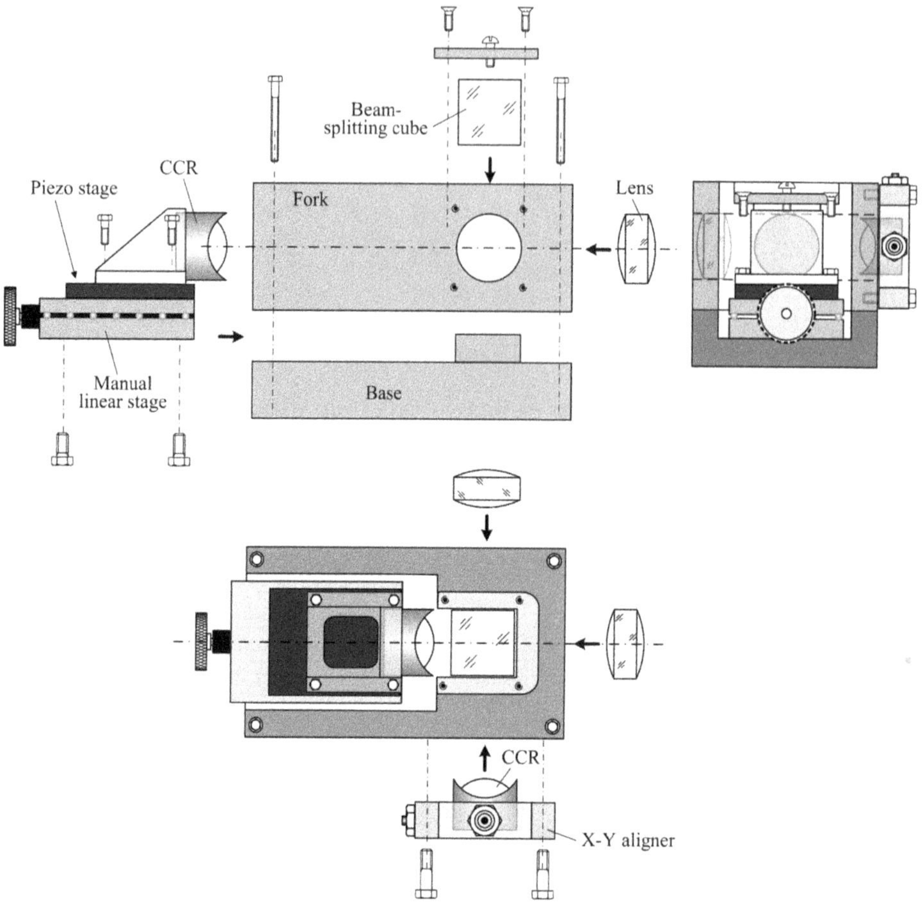

Fig. 4.26. Three projections of the IFS design scheme.

The base carries the manual linear stage, modified from the motorized Suruga K101-20MS-M stage, on top of which the piezo-stage is installed. The purpose of the manual stage is to align zero-displacement position of the moving CCR in the middle of the piezo-stage stroke. The two identical one-inch CCRs (Spectrum Scientific Inc., HR-254-3-AL) with aluminum coating, designed for visible domain, were installed, using fast-hardening two-component epoxy (AXIA EP-04, Alteco). Bolting the CCRs into their positions was considered inappropriate in this case, because possible deformations could degrade performance of the interferometer.

The fork was machined from one piece of aluminum with three identical one-inch circular holes for the two lenses (ThorLabs Hastings Triplets TRH254-040-A) of 40.0 mm focal lengths. The lenses were permanently fixed inside the holes, using the same type of epoxy.

The ramp electrical signal, necessary to drive the piezo-scanner, and synchronized data acquisition in non-imaging modes were organized, using National Instruments DAQ board PCI-6221 (37 pin) and LabVIEW software package. The algorithm of spectral reconstruction was taken from Chapter 2. The only thing that has to be mentioned is the noise-reduction technique. The principle disadvantage of the Fourier spectrometer against the grating spectrometer is the absence of accumulation during measurement. In grating spectrometers, the CCD photodetector accumulates photoelectrons in each pixel during the entire exposure time, thus reducing random fluctuations (noise) of the signal. For example, if the photoelectron rate is Poissonian, then the signal-to-noise ratio (SNR) is equal to $\sqrt{n}$, where n

is the average number of photoelectrons accumulated during exposure time. In Fourier-transform spectrometers, there is no accumulation: during scanning, the DAQ board continuously samples temporal oscillations of the signal, producing digital record of the interferogram — the functional dependence of light intensity on the displacement of the CCR. However, the sampling rate of digitization may be different, and this opens the possibility of increasing SNR. Indeed, according to the Whittaker–Kotelnikov sampling theorem, a sinusoidal function can be totally reconstructed from its discrete values if periodicity of sampling is shorter than the period. This theorem establishes a relation between the sampling rate ν, scanner stroke S, scanning time T, and minimum wavelength $\lambda_{\min}$ that can be reconstructed. Since a single displacement of the scanner produces twice as long variation of the optical path,

$$\frac{2S}{\nu \cdot T} < \lambda_{\min}.$$

In Section 3.1, it was explained that the scanning stroke S was chosen to be 350 μm, and $T = 3$ s. Thus, for the minimum wavelength $\lambda_{\min}$ =200 nm, this formula gives the minimum sampling rate $\nu = 1,167$ Hz. If the digitizer can sample the signal with a higher rate, then all the samples n obtained between ν may be used for averaging the signal, increasing SNR by a factor of $\sqrt{n}$. This feature establishes a rule: the higher the sampling rate is, the higher SNR can be achieved. The only limitation is the speed of the photodetector: if its time constant τ exceeds the inverse sampling rate ν^{-1}, then the averaging will be applied to partially coherent (not random) values, and the effect of increasing SNR will be reduced or even nulled. In our case, the time constant of the PMT is approximately 1 ns (Table 4.4) — much higher speed of response than the maximum sampling rate 250 kHz of the DAQ board PCI-6221. Therefore, in the experiments, the sampling rate was set to 250 kHz, and the running average algorithm (available in LabVIEW) with 80 points of averaging was used. With these values, the actual sampling rate was equal to 250,000/80 = 3,125 Hz — a value that guaranteed reconstruction of spectral components at 200 nm. Expectable increase in SNR is approximately $\sqrt{80} \approx 9$.

When the CCD camera is replaced by a wide-area photodetector, the device may work as a single-channel spectrometer in visible and near-infrared domains — within spectral intervals of the photodetector. Typical spectra are shown in Fig. 4.27. Quantum efficiency of the photodetector drops sharply in ultraviolet (below 400 nm) and in near-infrared (above 1.2 μm, Table 4.2). However, relatively small blue peak of excitation in the spectrum of the LED ceiling lamp is mainly not due to deficiency of sensitivity in this spectral interval but is produced by blue filters inside the lamps, installed to make the light more comfortable for eyes. The spectra of a standard white-light LED without these filters, presented in the Section 4.5, show much stronger response in the blue part of spectrum.

The mercury doublet around 570 nm is commonly used to estimate spectral resolution. Figure 4.28 shows this region magnified, proving spectral resolution of approximately 1 nm.

In the imaging mode, reflectometer must provide uniform interference picture over entire field of view, covered by the CCD matrix. Theoretically, total uniformity can be reached only for the interference picture without fringes. This becomes clear from simple geometrical considerations, outlined in Fig. 4.29.

Indeed, consider two monochromatic waves with the same wavelength λ, coming from the same point 1 in the image plane and interfering in the plane of the CCD sensor with optical paths in the shoulders L_1 and L_2, and axial scanning distance z:

$$|E(x)e^{ik(L_1+z)} + E(x)e^{ikL_2}|^2 = 2I(x)\{1 + \cos[2\pi k(L_1 - L_2 + z)]\}.$$

Here $k = 1/\lambda$ is the wavenumber and $I(x) = |E(x)|^2$ is the intensity of the image. If optical paths in the shoulders are perfectly aligned, as it must be for proper operation of the spectrometer, then $L_1 - L_2 = 0$, there are no fringes, and the signal in each point x of the image oscillates sinusoidally and in phase, as

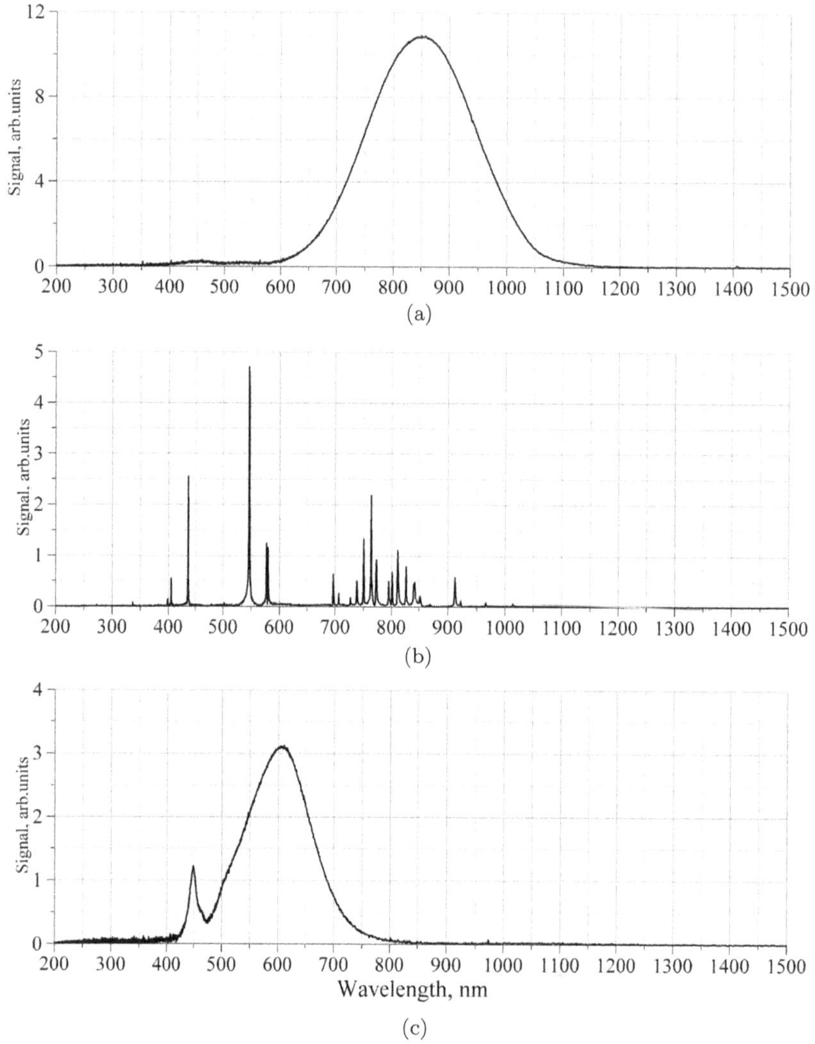

Fig. 4.27. Typical spectra: (a) white light source; (b) Hg–Ar lamp; (c) ceiling LED lamp.

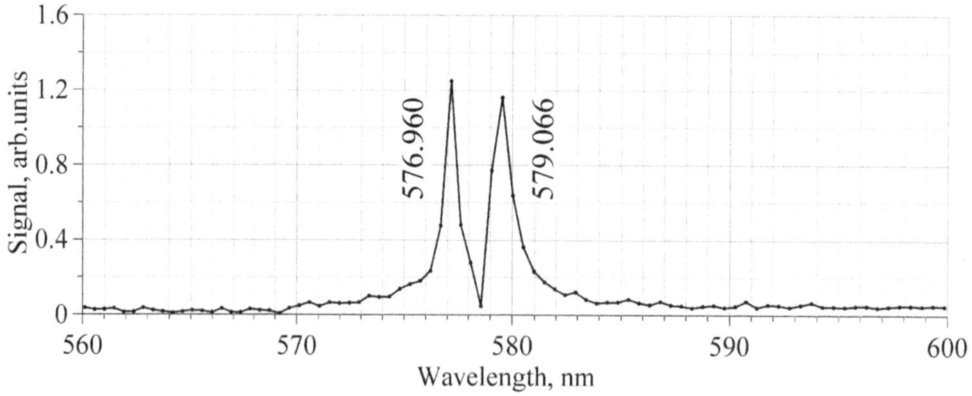

Fig. 4.28. Close view of the mercury doublet around 570 nm: the two lines separated by ≈2 nm are resolved.

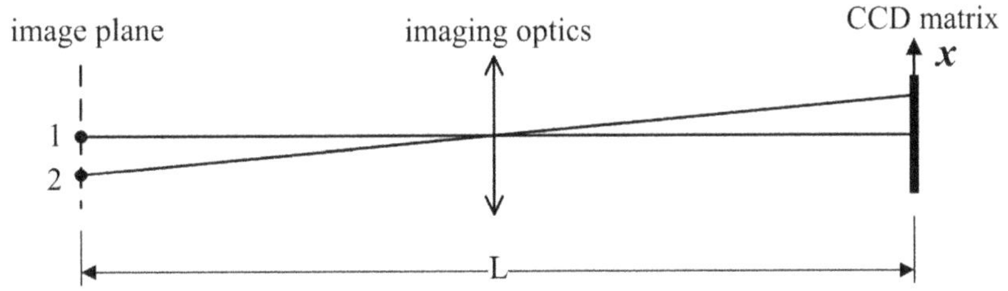

Fig. 4.29. Generalized interpretation of optical paths in imaging interferometer.

z increases:

$$|E(x)e^{ik(L_1+z)} + E(x)e^{ikL_1}|^2 = 2I(x)[1 + \cos(2\pi kz)].$$

The Fourier transform, applied to this formula with the functional coordinate z, gives the delta-function in the spectral domain, which means monochromaticity of light, in accordance with the initial assumption.

However, if the interferometer is slightly misaligned, introducing small angle α between interfering waves, then $L_1 - L_2 = L\alpha x$, fringes appear in the image, and phases of oscillations become different at different points x:

$$|E(x)e^{ik(L_1+z)} + E(x)e^{ikL_2}|^2 = 2I(x)\{1 + \cos[2\pi k(L\alpha x + z)]\}.$$

Here, the term proportional to x in the argument of the cosine describes fringes uniformly distributed over the picture. Nonetheless, even with fringes, the Fourier transform, applied to this distribution with the functional coordinate z at every single pixel $x = const$, again gives the delta function in the spectral domain. It means that spectral resolution of the imaging spectrometer is not affected by fringes. The number of fringes should be minimized, as it was explained above in Fig. 4.14, for the two reasons. Firstly, because the bigger the number of fringes the weaker their contrast is. Secondly, even with high contrast of fringes, the modulation signal may be infinitely small when a single-channel photodetector is used, and all partial signals from every pixel x are integrated.

Considering rays, coming from the point 2 in Fig. 4.29, the expression for signal should be modified, taking into account the increase in L:

$$L' \approx \frac{L}{\cos(2x/L)} \approx L\left(1 + \frac{2x^2}{L^2}\right).$$

Hence, the fringes are not uniformly distributed along the image: their density increases from the center to peripheral areas of the picture due to higher-order terms in x. Also, the fringes are colored, since their density is proportional to $k = \lambda^{-1}$. All these features are clearly seen in Fig. 4.30, portraying the interferometric image of a laboratory scene.

4.2.6. *Typical performance*

The prototype of the imaging reflectometer is shown in Fig. 4.31.

In it, the industrial microscope unit (Mitutoyo VMU) with white light illumination is connected to the IFS through four-meter imaging optical fiber bundle (FiberTech RoMack, LEONI Fiber Optics Inc; SCHOTT Americas; PN 093879). The whole bundle is composed of many thinner conduits of square cross sections, each composed of elementary round optical fibers (Fig. 4.32).

Although decreasing quality of the image and truncating spectral range, this configuration provides significant advantage by eliminating thermal and vibrational influence that working process machine may relay to the reflectometer.

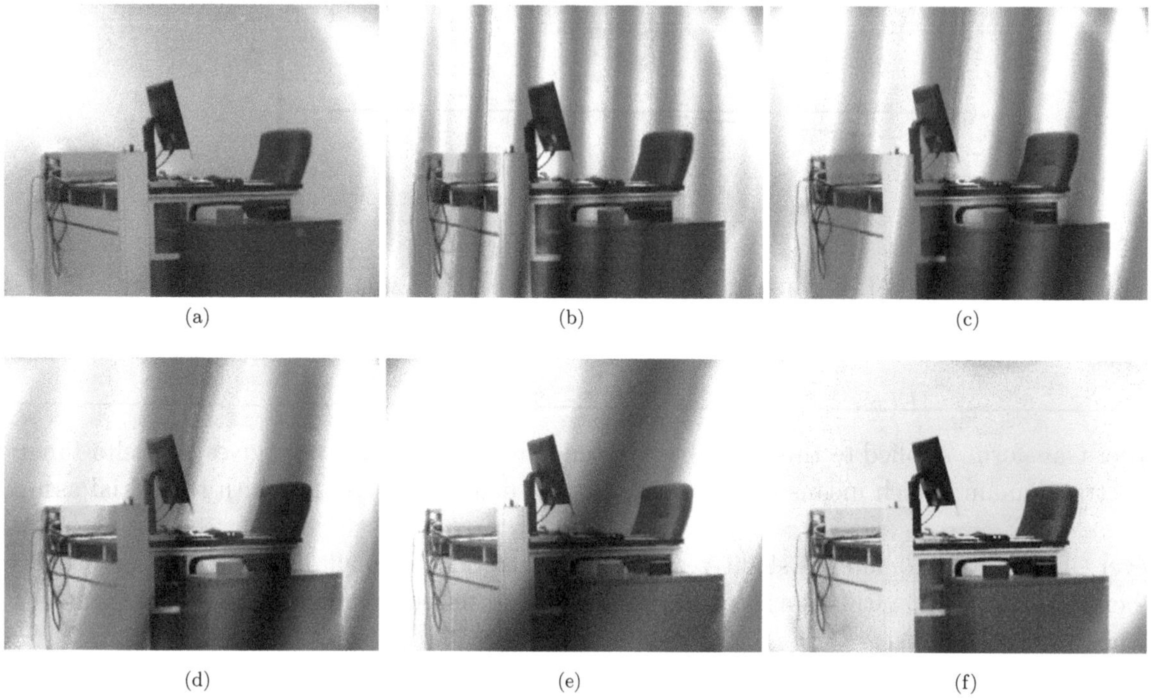

Fig. 4.30. Pictures, obtained with the imaging reflectometer equipped with a 25 mm focal length front lens. (a) Fringe contrast is zeroed by setting big z outside coherence range of light; (b)–(e) $z \approx 0$, fringe contrast is a maximum, number of fringes N decreases from 8 to 1 by aligning lateral position of the CCR; (f) nearly perfect alignment of CCRs — number of fringes $N < 1$, fringe contrast is maximized ($z \approx 0$). Field of view 25°.

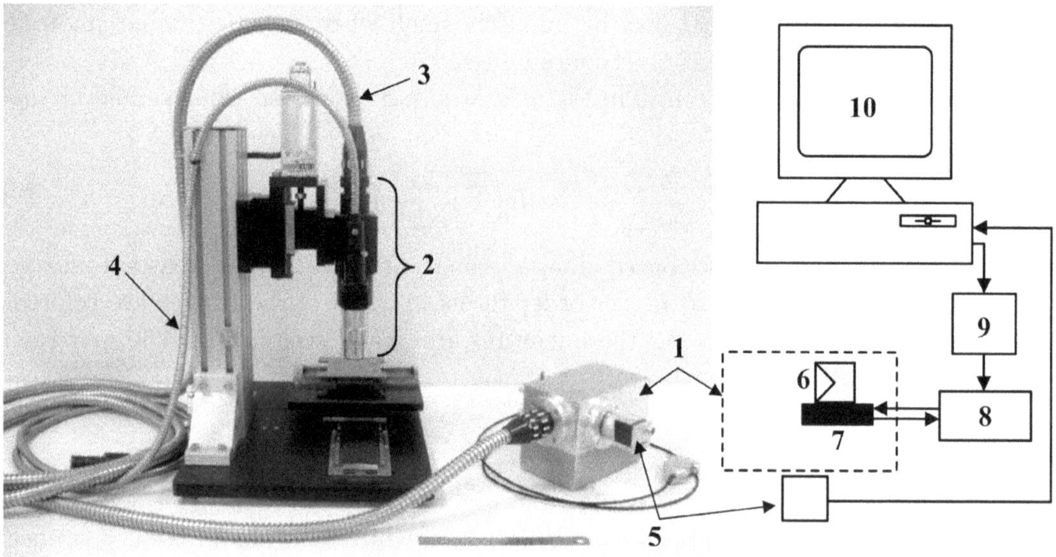

Fig. 4.31. Prototype of the imaging reflectometer: opto-mechanical parts (at left) and the control scheme (at right). (1) IFS; (2) microscope unit; (3) imaging optical fiber bundle; (4) white light fiber bundle; (5) CCD camera; (6) CCR; (7) piezo-scanner; (8) controller of the piezo-scanner; (9) DAC; (10) computer.

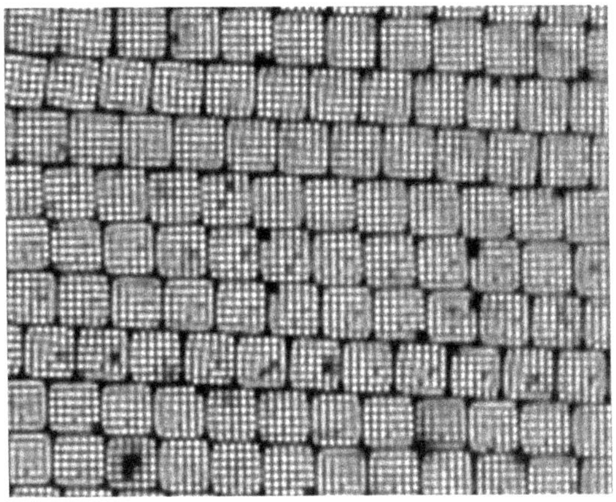

Fig. 4.32. Microscopic structure of the imaging optical fiber bundle.

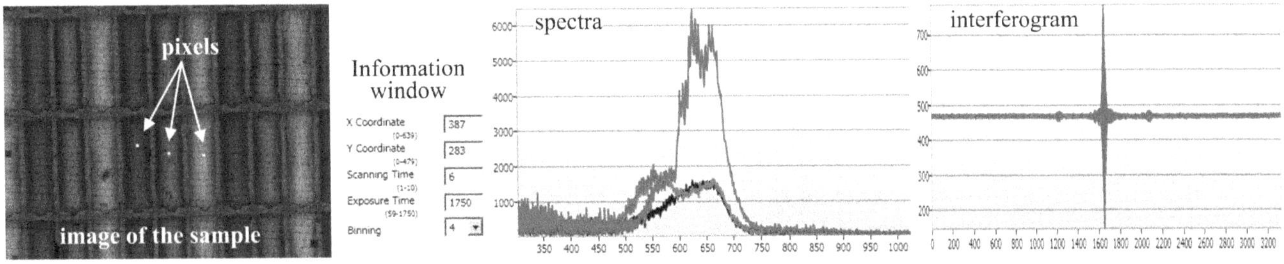

Fig. 4.33. Principal components of the graphical user interface (explained in the text). Object dimensions 0.5 mm × 0.4 mm; objective focal length 20 mm.

The principle of spectral reflectometry (Chapter 8) is implemented by illuminating the sample with white light, coming from the white light source (not shown in Fig. 4.31), and measuring the spectrum of the reflected light. For that, the piezo-scanner linearly moves the CCR in one of the shoulders of the interferometer, causing oscillations of light intensity in every pixel of the image on the sensitive element of the CCD camera. Linear motion of the CCR is guaranteed by the controller of the piezo-scanner, which compares the signals from the capacitive displacement sensor on the scanner module and the ramp voltage on the input of the controller, generated digitally in the computer and transformed into analog voltage by the digital-to-analog converter (DAC; National Instruments USB-6003). The amplitude of the ramp voltage is 7 V, which corresponds to 350 μm of total displacement of the piezo-scanner. The high-speed CCD camera samples oscillations of light and sends these data in a digital form into computer. The Fourier transform of these oscillations reconstructs the spectrum of the reflected light.

An operator can control the measurements, using graphical user interface shown in Fig. 4.33. It contains three basic windows: image of the sample as captured by CCD camera; spectra measured in chosen pixels; interferogram after each sampling. An operator chooses the pixels of interest, points the mouse cursor on them and saves the coordinates in the image coordinate frame. In particular, Fig. 4.33 portrays the pixels chosen for measurements, and information window to the right of the image indicates the coordinates: "387" and "283". Also, the scanning time and exposure time may be edited. In order to increase signal-to-noise ratio, binning is usually applied, which is indicated in the information window. After each scan, the recorded oscillation of light intensity in the chosen pixel is displayed in the interferogram window. This information is important for monitoring two characteristics: saturation of the CCD camera and how accurate the initial position of the piezo-scanner is.

Saturation causes ghost reflexes in the spectrum, because nonlinearly transformed interference signal produces artifacts in its Fourier transform. Saturation can be easily diagnosed as asymmetry in the upper and lower parts of the interferogram.

Some examples of the results, producing by the imaging Fourier spectrometer on various industrial samples, give additional insight into possibilities of this type of spectrometers. No special comments are made to the figures below, because different applications may be evaluated from different points of view. All the pictures in Fig. 4.34 are structured in 3 columns: image, spectrum of the reflected light, interferogram. White dots on the images indicate the pixel, where the measurement was made.

4.2.7. *Sensitivity*

How to increase sensitivity of spectral measurements in visible domain if quantum efficiency of the detector is already at its upper limit and the area of the input aperture cannot be increased in grating spectrometers without loss of spectral resolution? With numerical apertures of compact fiber-optic spectrometers and IFS being almost the same ($\sim$0.2), sensitivity of the IFS may be increased by scaling area of the input aperture: with the diameter of input optical aperture 6 mm and quantum efficiency 0.2 (usual values of photo-multiplying tubes (PMT)) the product of the input area by quantum efficiency could be made equal to approximately 5 mm^2, comparing to $0.16 \times 0.9 = 0.14$ mm^2 for grating spectrometers. Thus, more than an order of magnitude higher sensitivity could possibly be achieved. With equal values of exposure time and spectral resolution, the IFS shows significantly higher sensitivity than the best grating spectrometers currently available, producing spectral measurements at as low input optical intensity as 6×10^{-13} W/mm^2.

In order to compare sensitivity of the Fourier and grating spectrometers in visible domain, the competitor for the IFS among grating spectrometers should be chosen. Since the IFS described here is a compact instrument, it must be compared to its own class. Nowadays, state-of-the-art compact grating spectrometers differ insignificantly in sensitivity from one another, because they all use nearly perfect gratings and CCD matrices — the best components, available on the market. Therefore, the guiding rule of choice was to use a compact grating spectrometer that was most popular in applications, and as such, the Verity SD1024G (Verity Instruments Inc.) was chosen — a device that is currently used in hundreds in semiconductor industry. Specifically, the device with the input slit of 25 μm width by 4mm height was used, equipped with back-thinned thermo-electrically cooled two-dimensional CCD matrix of 1024×128 pixels with full vertical binning over 128 pixels. Its spectral range is 200–800 nm.

For testing sensitivity, the experimental arrangement as in Fig. 4.35 was assembled. In order to verify also spectral resolution of the two spectrometers, the Hg–Ar calibration source HG-1 (Ocean Optics Inc.) was used with well-known mercury doublet 576.960–579.066 nm. Natural separation of these two lines is approximately 2 nm, and seeing them separately in the spectra guarantees spectral resolution of about 1nm.

Comparison of sensitivity was done in three steps on medium, low, and ultra-low optical fluxes. The term "medium" was ascribed to the intensity, at which the Verity signal did not show much noise relative to standard applications. "Low" optical flux was obtained after installing additional ND filters to the level, at which the spectrum of the calibration source could barely be visible on the background of noise. Finally, additional ND filter was installed, making the Verity signal below the noise level. This level of optical flux, named "ultra-low", was then measured by digital power meter PM100D (ThorLabs). Because of extremely low optical power, this measurement was done indirectly: initially, at high optical power, the attenuation of the ND filter was accurately measured, and then the "ultra-low" power at the front end of optical fibers was estimated by dividing the measured input power by the attenuation

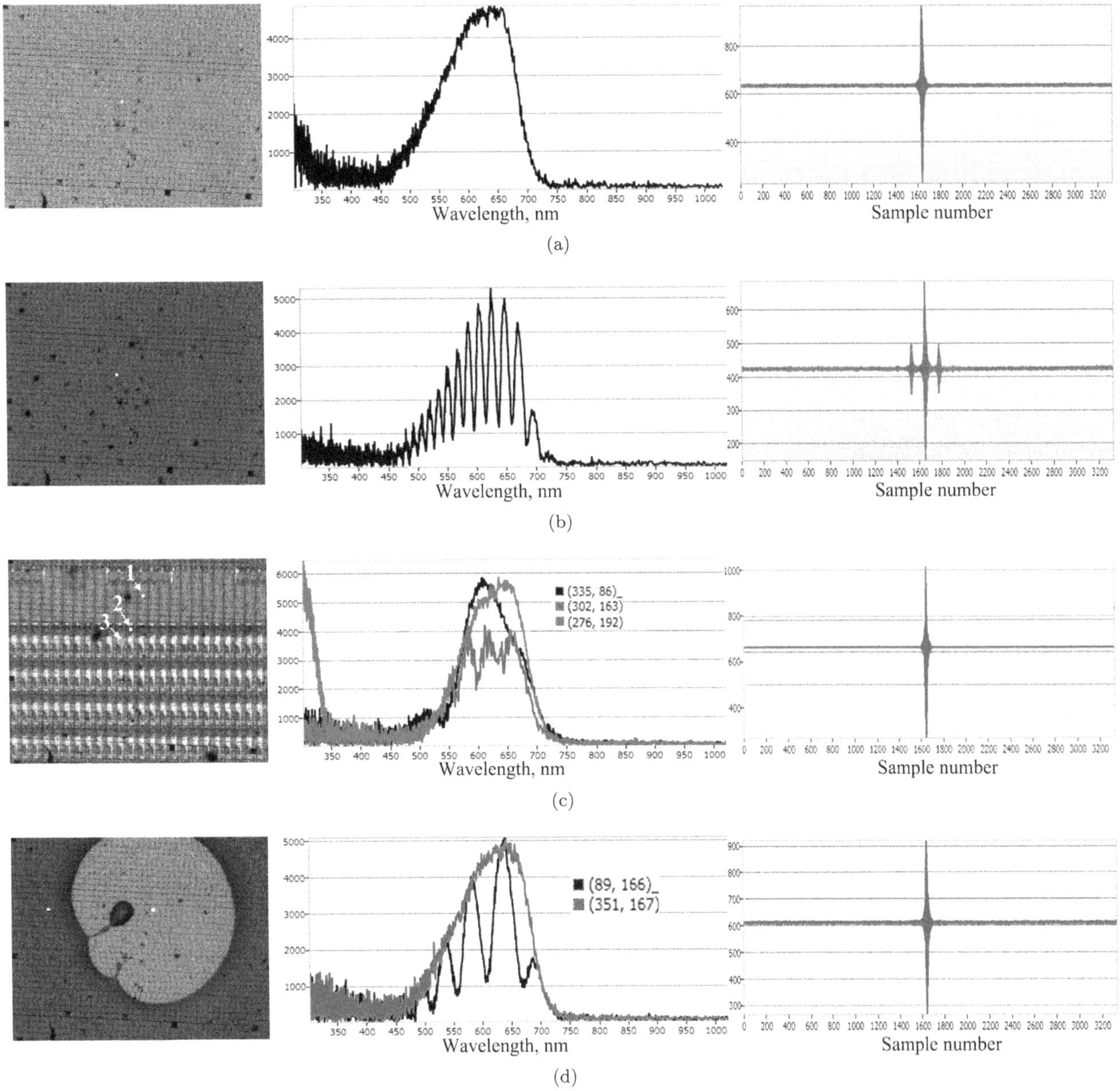

Fig. 4.34. (a) Bare silicon wafer; (b) silicon wafer coated with 6 μm photoresist; (c) mobile phone LCD display (the interferogram relates to the pixel (335,86)); (d) spinner defect on silicon wafer (rounded light area is bare silicon, not coated with photoresist; the interferogram relates to the pixel (351,167)). Numerical aperture of the objective lens

coefficient of the last ND filter. This ultimate "ultra-low" optical intensity turned out to be as low as 6×10^{-13} W/mm^2.

The results of the comparison are summarized in Figs. 4.36–4.39. They were measured at the same exposure time 3 s set for Verity spectrometer, as was the scanning time of the IFS. Claimed equality of spectral resolution of the two spectrometers is proven in Fig. 4.37, showing magnified spectral interval of the Hg doublet. Thus, if we put apart expectable lack of sensitivity of the IFS below 400 nm (Table 4.3), the IFS evidently shows superior sensitivity with respect to Verity.

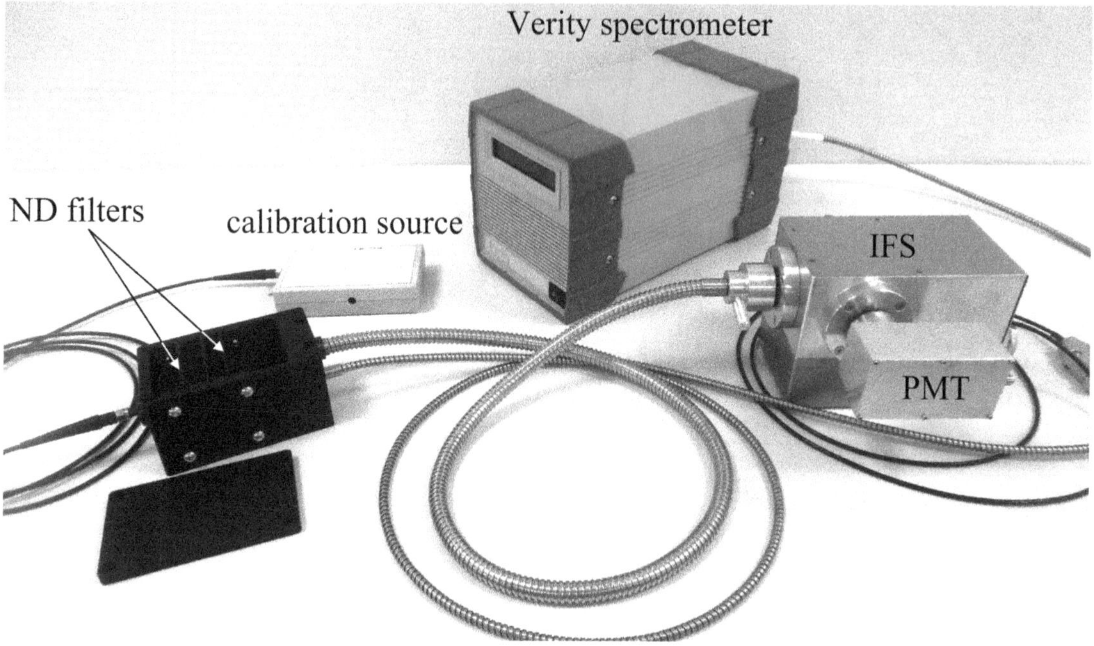

Fig. 4.35. A set of neutral-density (ND) filters was used to attenuate input optical flux from the calibration source.

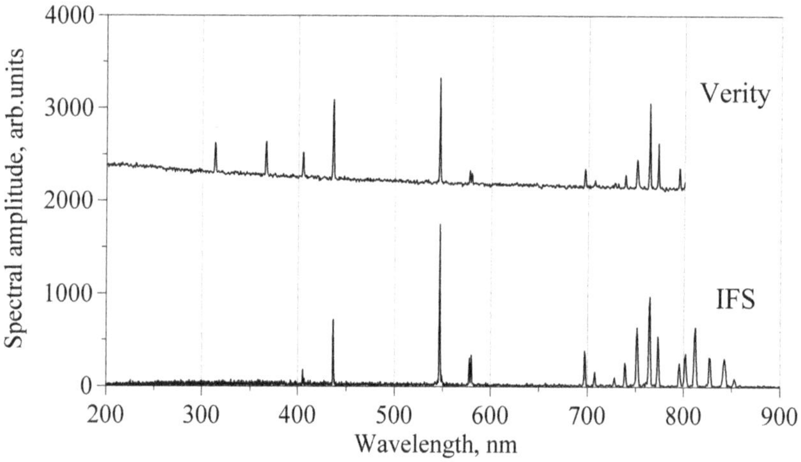

Fig. 4.36. Comparison of spectra from Verity and IFS at medium optical flux.

However, sensitivity of the IFS depends on the type of the spectrum — the phenomenon, which will be addressed below.

The IFS performs differently on smooth and line-type spectra. For the light with smooth spectra, like produced by incandescent lamps, coherence is low, and the interference modulation is concentrated only in a narrow interval near equal optical paths in the two shoulders of the interferometer. Physically it means that when one of the two CCRs moves from the beginning to the end of its travel, only small portion of modulation energy — the product of modulation power by the time, during which this modulation is observable on the background of noise — is recorded. The rest part of the scanning is filled only with noise, reducing the SNR. Alternatively, when the light is produced by narrow spectral line with high temporal coherence, the photodetector records modulation during the entire travel of the

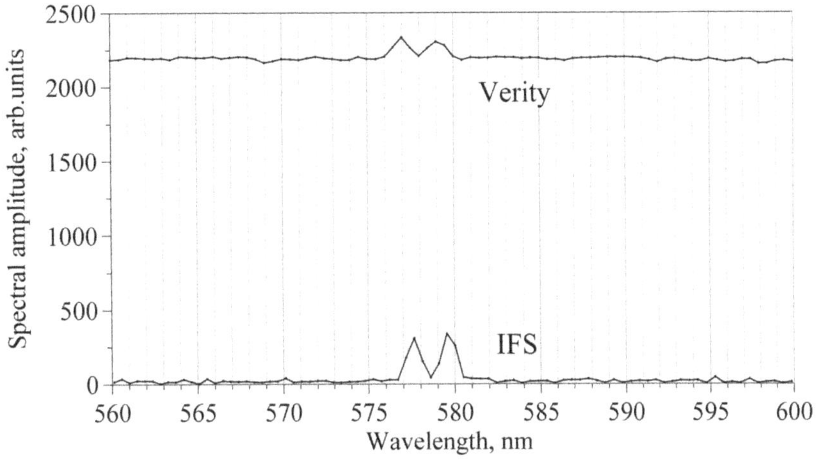

Fig. 4.37. Magnified mercury doublet 576.960–579.066 nm. Medium optical flux.

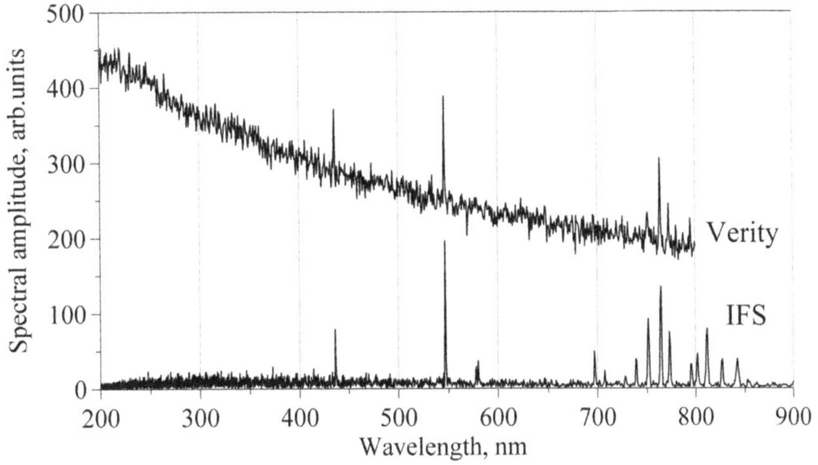

Fig. 4.38. Comparison of spectra from Verity and IFS at low optical flux. The Hg doublet is not seen in the Verity spectrum at all.

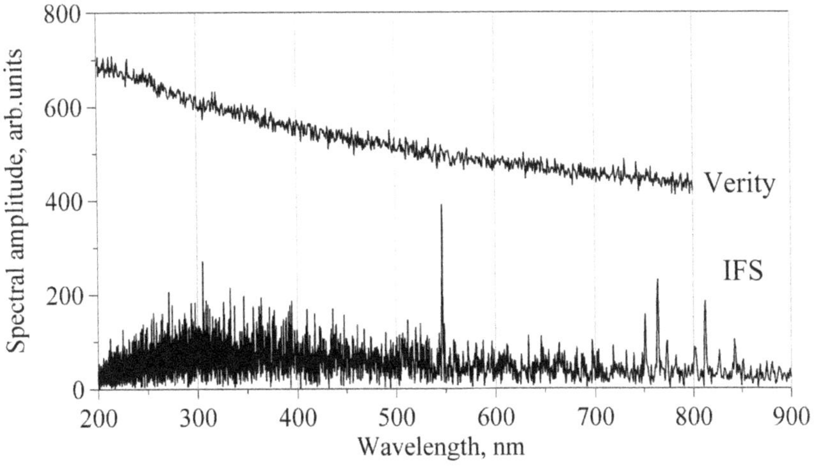

Fig. 4.39. Comparison of spectra from Verity and IFS at ultra-low optical flux: 6×10^{-13} W/mm^2.

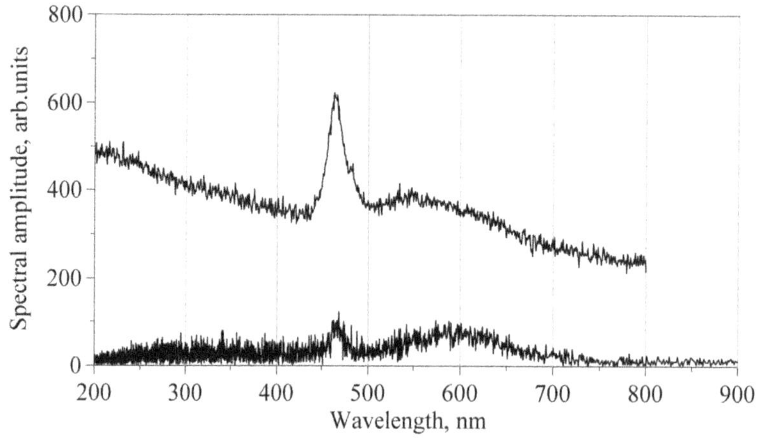

Fig. 4.40. Comparison of spectra from Verity and IFS with white-light LED as the light source.

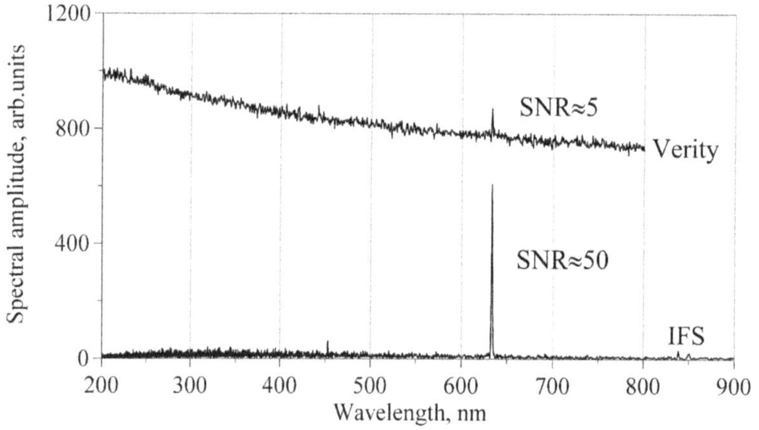

Fig. 4.41. Comparison of spectra from Verity and IFS with He–Ne laser as the light source.

CCR, increasing the total energy of modulation during scan and making the SNR high. It means that the Fourier-transform spectrometer may be expected to be especially sensitive on the line-type spectra, produced, for example, by lasers, interference filters, plasma radiation, etc.

Figure 4.40 compares spectra obtained by Verity spectrometer and the IFS with the white-light LED (light-emitting diode) as the light source. Clearly, the IFS does not show any superiority, on the contrary, it shows poorer sensitivity than Verity. However, with weak He–Ne laser radiation, the IFS is roughly 10 times more sensitive (Fig. 4.41).

Supplemental Reading

W.H. Steel, *Interferometry*, Cambridge University Press, 2nd edn., 2009.

P. Jacquinot, The luminosity of spectrometers with prisms, Grating, or Fabry–Perot etalons, *J. Opt. Soc. Am.*, **44**(10), pp. 761–765 (1954).

M. R. Descour, The throughput advantage in imaging Fourier-transform spectrometers, *Proc. SPIE*, **2819**, pp. 285–290 (1997).

R. G. Sellar, G. D. Boreman, Comparison of relative signal-to-noise ratios of different classes of imaging spectrometer, *Appl. Opt.*, **44**(9), pp. 1614–1624 (2005).

Chapter 5

Gated Intensified Spectrometers

5.1. The Principle and Characteristics of Image Intensifiers

To begin with, consider generalized concept of a gated spectrometer (Fig. 5.1).

Its key component, which brings numerous new features, is the image intensifier (Fig. 5.2). All components of it are assembled in a vacuumized enclosure and sealed in an isolating compound with four leads for the gate pulse (usually a coaxial cable), high voltage, and ground. High-voltage power supply (HV) sets variable voltage from 0 to 5 kV on the micro-channel plate (MCP), which performs amplification of the photoelectron flux from the photocathode. Close proximity between the photocathode and MCP ensures small divergence of photoelectrons needed for high image quality. Luminescent screen is covered in phosphor that radiates almost monochromatic light, usually green component around 500 nm, when excited by secondary electrons, emerging from the MCP. Diameters of the input aperture range from 18 to 100 mm.

The concept of image intensification is based on spatially-resolved amplification of photoelectrons in a micro-channel plate (MCP). The MCP is a high-technology product. It is a thin, usually about 0.5 mm, plate of a dielectric material with numerous densely packed tiny holes — micropores, piercing it from one side to another. Diameter of each micropore is typically 10 μm and their spacing, determining spatial resolution, is about 20 μm. Ratio of the length to diameter — the aspect ratio — ranges from 40 to 60. Inner surface of each micropore is covered in emissive material, producing multiple secondary electrons on each incoming photoelectron. In order to avoid direct flight of photoelectrons through micropores, the latter are slightly tilted to about 8°. For higher gain, two such plates may be stacked together in a sandwich with opposite tilts, making what is called the Chevron MCP (Fig.5.2). Photoelectrons generated by incident light on the photocathode reach the MCP in free propagation and generate

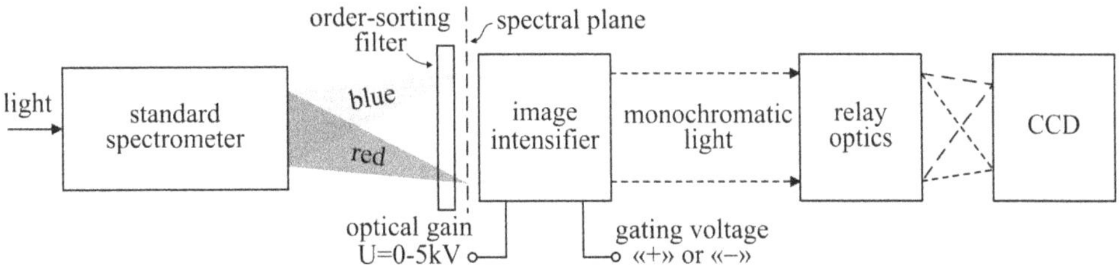

Fig. 5.1. The gated spectrometer is a combination of an ordinary spectrometer with image intensifier accompanied by relay optics that conveys the image in monochromatic light to photodetector, usually a charge-coupled device (CCD).

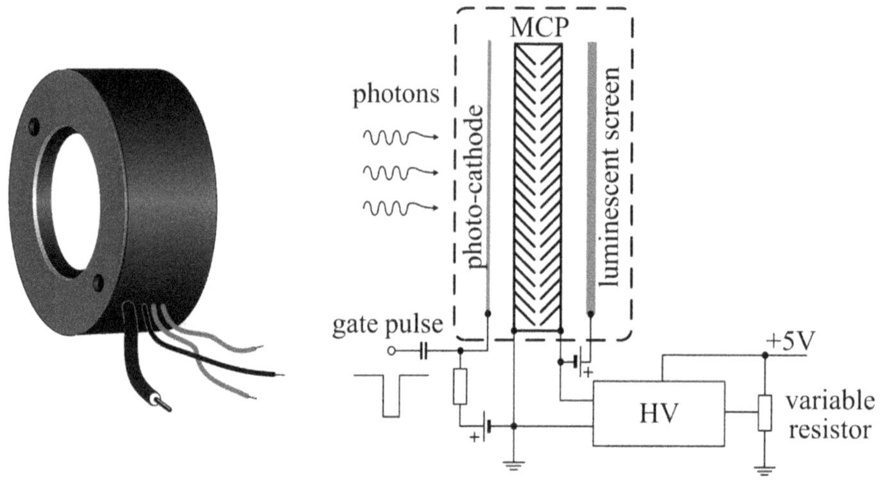

Fig. 5.2. Typical scheme of an image intensifier.

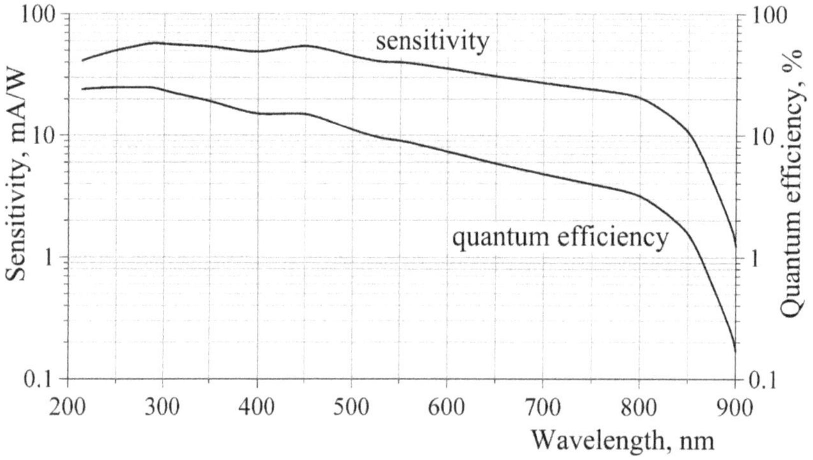

Fig. 5.3. Spectral characteristic of the "S20" photocathode.

avalanches of secondary electrons inside each micropore. These avalanches are then accelerated to the luminescent screen, producing intensified light on collision.

Is the spectral response of the photocathode sufficient to work in a wide range of wavelengths? The answer is yes, although near-infrared wing of spectrum will suffer from lower gain (Fig. 5.3).

Gating function is performed by applying permissive electrical pulses to photocathode. Polarity of these pulses depends on the polarity of the permanent voltage applied to photocathode. Negative permanent voltage across the photocathode-MCP gap accelerates photoelectrons, therefore to stop them the gate pulse must be positive and of a higher potential than the permanent voltage. In this mode, short gating pulses interrupt optical flux. In another option — the most frequently used — much lower positive permanent voltage $\sim$ 50 V locks photoelectrons near the photocathode. To unlock them and accelerate, the negative gate pulse of about 300 V must be applied. In this mode, gating pulses open spectrometer for input light. Making 300 V amplifier for handling nanosecond-scale pulses is not a simple job that can be done easily. Therefore, manufacturers of image intensifiers always supply proprietary gating modules, specifying minimum pulse width that can be achieved with them. But even with this

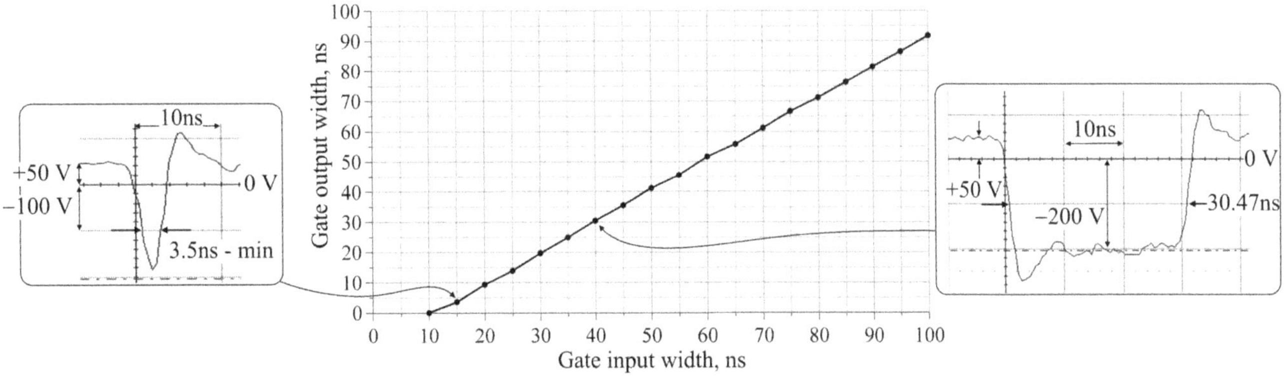

Fig. 5.4. Typical oscilloscope traces that can be expected from a gating module of image intensifier.

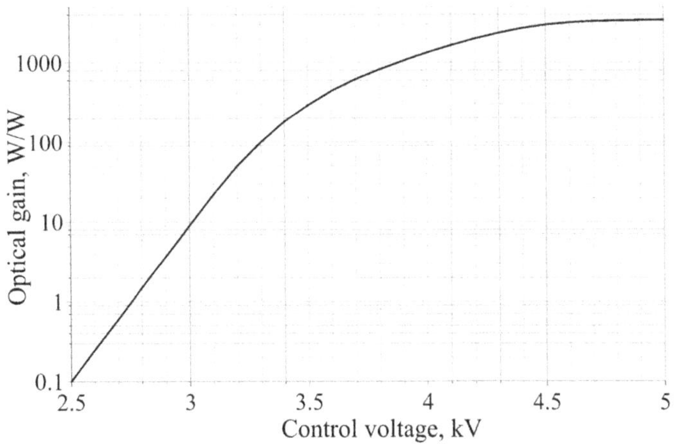

Fig. 5.5. Typical gain curve of a one-stage MCP at 400 nm monochromatic input flux.

well-designed electronics, no one guarantees exact open time of the image intensifier at the lower limit of the pulse width. Calibration curve is always necessary, like the one shown in Fig. 5.4.

Another control wire of the image intensifier is the high-voltage accelerating potential on the MCP. It controls the optical gain: the higher the accelerating potential is, the bigger the number of secondary electrons and the brighter luminosity of the phosphor screen are. Optical gain is measured in units Watt/Watt — the ratio of power of the input monochromatic light at some specific wavelength to power of the phosphor screen luminosity at its characteristic wavelength. It is a highly nonlinear function of the control voltage (Fig. 5.5).

5.2. Sensitivity and Noise

Optical gain is the feature that dramatically increases sensitivity of the spectrometer. To begin with, consider a portion of theory on this issue. The measure of the quality of the signal is the signal-to-noise ratio (SNR). The SNR of the intensified and non-intensified detectors can be compared theoretically, assuming Poisson statistics of photo-electrons. Figure 5.6 presents schematically the detection process in the non-intensified and intensified photodetectors.

Consider the case when the flux I (photons per second) of the primary photons is constant during the exposure time T. Without image intensifier, each photon creates a carrier inside a pixel with quantum efficiency η. This process is random, and the total amount m of carriers generated during the exposure

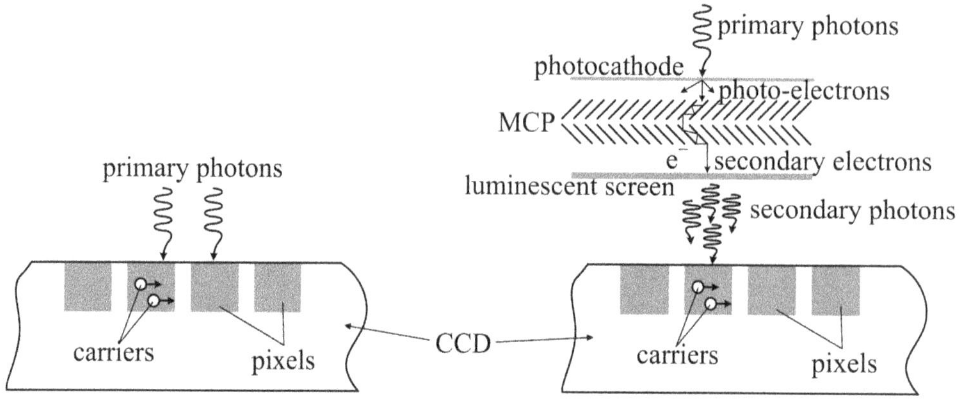

Fig. 5.6. Generalized schemes of the non-intensified (at left) and intensified (at right) photodetectors. Charge-coupled devices (CCDs) are supposed to be used in both cases.

time is described by the Poisson probability function:

$$P(m) = \frac{\nu^m e^{-\nu}}{m!}$$

with $\nu = I \cdot T \cdot \eta$. Also, there is a noise current n, generated by thermal or any other spontaneous process inside the photodetector. Then the detector signal s is a sum

$$s = n + m.$$

In spectroscopy, the dark-current signal $\overline{n}$ is always subtracted from the spectrum, so that the signal to be analyzed is

$$s = n + m - \bar{n}.$$

The SNR is commonly determined as the ratio of the average signal $\overline{s}$ to the square root of the dispersion $\overline{(s-\overline{s})^2}$ (the line above denotes averaging):

$$SNR = \frac{\overline{s}}{\sqrt{\overline{(s-\overline{s})^2}}}.$$

Having in mind that for the Poisson statistics

$$\overline{m} = \nu, \quad \overline{m^2} = \nu^2 + \nu,$$

it is easy to obtain for the non-intensified detector

$$SNR = \frac{\nu}{\sqrt{\sigma_n^2 + \nu}} = \frac{N\eta}{\sqrt{\sigma_n^2 + N\eta}},$$

where $\sigma_n^2 = \overline{n^2} - \overline{n}^2$ — the detector noise and $N = I \cdot T$ — the number of photons.

For the intensified detector not only the carrier generation process is random, but also the process of intensification, which brings additional noise. However, this additional noise may be less important than the detector noise σ_n^2, and then it is possible to expect a better SNR. Consider this in more detail. Again, let the input photon flux I be constant. Then the number M of the photons, reaching the detector, is random with the Poisson statistics

$$P(M) = \frac{\varepsilon^M e^{-\varepsilon}}{M!}$$

with $\varepsilon = I \cdot T \cdot \mu G$. Here G is the intensifier gain and μ — quantum efficiency of the photocathode. Thus, unlike the first case, the number of photons at the detector is random, therefore the probability of generating m carriers is now

$$P(m) = \sum_{M=0}^{\infty} \frac{(M\eta)^m e^{-M\eta}}{m!} \cdot \frac{\varepsilon^M e^{-\varepsilon}}{M!}.$$

With some straightforward manipulations, one can obtain the following relations:

$$\overline{m} \equiv \sum_{m=0}^{\infty} mP(m) = \eta\varepsilon,$$

$$\overline{m^2} \equiv \sum_{m=0}^{\infty} m^2 P(m) = \eta^2(\varepsilon^2 + \varepsilon) + \eta\varepsilon,$$

$$\overline{m^2} - \overline{m}^2 = (1 + \eta)\eta\varepsilon.$$

The SNR is then

$$SNR = \frac{N\eta(\mu G)}{\sqrt{\sigma_n^2 + (1 + \eta)(\mu G)N\eta}}.$$

Physically, the product μG represents the optical gain in units Watt/Watt. It shows how many photons are generated at the output of the intensifier per one photon at its input.

Figure 5.7 compares the SNR as the function of the number of input photon N.

Obviously, with the high enough μG, the intensified detector is always better than the non-intensified one.

Consider two limiting cases: very low input flux such as $N\eta(\mu G) << \sigma_n^2$, and very strong flux $N \to \infty$. At low flux, the non-intensified detector gives

$$\text{SNR} = \frac{N\eta}{\sigma_n},$$

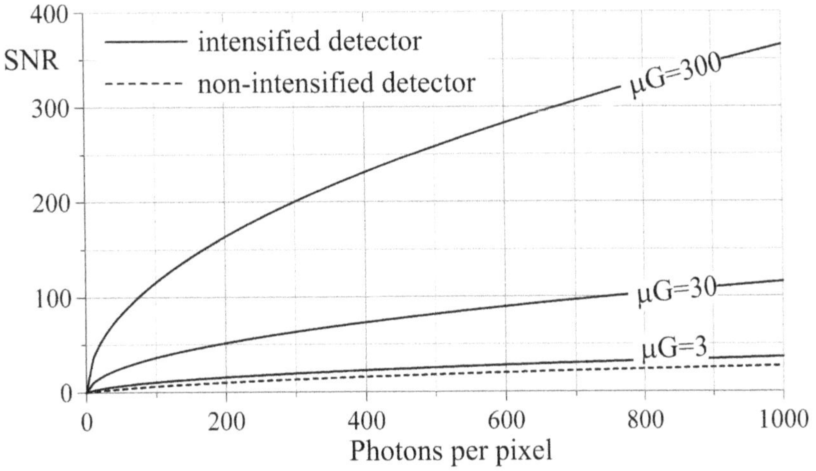

Fig. 5.7. SNR of the intensified photodetector. Quantum efficiency is taken $\eta = 0.8$ and the noise level $\sigma_n^2 = 100$. The non-intensified photodetector is shown in the dashed line.

whereas the intensified detector will have

$$\mathrm{SNR} = \frac{N\eta}{\sigma_n}(\mu G),$$

i.e. μG times bigger SNR. At very strong fluxes, this superiority is roughly $\sqrt{\mu G}$ times smaller:

$$\mathrm{SNR} = \sqrt{N\eta}$$

for the non-intensified and

$$\mathrm{SNR} = \frac{1}{\sqrt{1+\eta}}\sqrt{N\eta(\mu G)}$$

for the intensified detector. At first glance, it may look strange that we are considering the case of very strong flux with image intensifier: if the flux is strong then why do we need intensifier? However, there is one very important practical consideration why intensifier is needed even with strong input fluxes: stabilization. We shall consider this option of the intensified spectrometer below.

5.3. Additional Advantages of Image Intensifiers

The obvious advantage of combining any spectrometer with a gated image intensifier is the extension of its functionality to time-resolved operation. However, such a combination also gives several unexpected advantages that are very important from the practical point of view. The first one is the increase in sensitivity. Additional optical gain provided by an image intensifier opens the possibility to create super-sensitive spectrometers, which are in great demand for remote LIBS and micro-spectroscopy.

The second hidden advantage is associated with the dynamic range of linear detectors. Consider Table 5.1, summarizing basic parameters of the three most commonly used linear detectors.

The higher the sensitivity is, the narrower the dynamic range is. Suppose, for example, we want a spectrometer for plasma diagnostics, where optical flux may vary by orders of magnitude. Then we need the Hamamatsu S8378-1024Q detector with the widest dynamic range. But its sensitivity is poor, therefore, in order to work with weak fluxes we have to increase the time constant, which may be unacceptable. The opposite option is to choose highly sensitive Toshiba TCD 1304AP and install additional neutral optical filters every time when the optical flux is too strong. This, however, is always the last resort because any hardware modification requires much of resources.

An elegant solution comes with an image intensifier that can be used either as an optical amplifier or as a variable optical attenuator (Fig. 5.8). The first combination should be the choice for those who need super-high sensitivity, inaccessible by any other means. The second combination may be the common choice, providing wide dynamic range in a single measurement. Moreover, these two options may be

Table 5.1. Parameters of some most commonly used CCD sensors.

Parameter	Toshiba TCD 1304AP	Toshiba TCD 1205DG	Hamamatsu S8378-1024Q
Spectral range, nm	190–1100	200–1100	200–1050
Number of elements	3648	2048	1024
Pixel width, micron	8	14	25
Sensitivity, V/(lx·s)	160	80	22/4.4
Dynamic range	300	400	1200/4600
ADC resolution	12 bit; 4096 counts	12 bit; 4096 counts	14 bit; 16384 counts

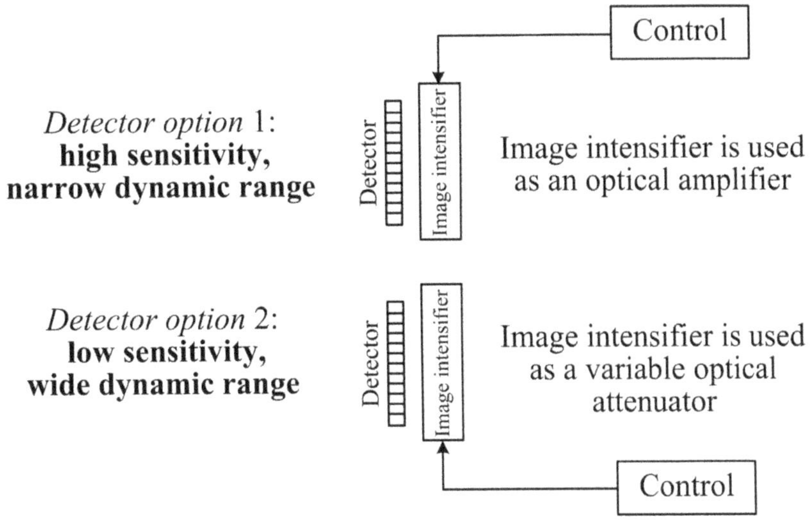

Fig. 5.8. Sensitivity and dynamic range options.

combined to achieve a completely new feature: automatic gain control, which is discussed in more detail in the last section of this chapter.

5.4. Coupling Image Intensifier to a Sensor

Commercially available gated spectrometers are always a combination of two separately sold parts: the spectrometer itself and the intensified CCD camera (ICCD) with proprietary software. The design of a typical spectrometer was already explained in Chapter 1, now it is time to explain what is inside the ICCD (Fig. 5.9).

The ICCD available on the market, commonly contains an assembly of image intensifier, fiber-optic plate, and CCD matrix. The CCD is cooled by thermo-electric element (TE). Inside are also high-voltage power supply (HV) and gating module (G) for image intensifier. Firstly, the photodetector is always a CCD matrix. Cost of the ICCD is so high that manufacturers have no motivation to make it a bit cheaper by replacing the matrix with a linear array, like in compact spectrometers. And this is

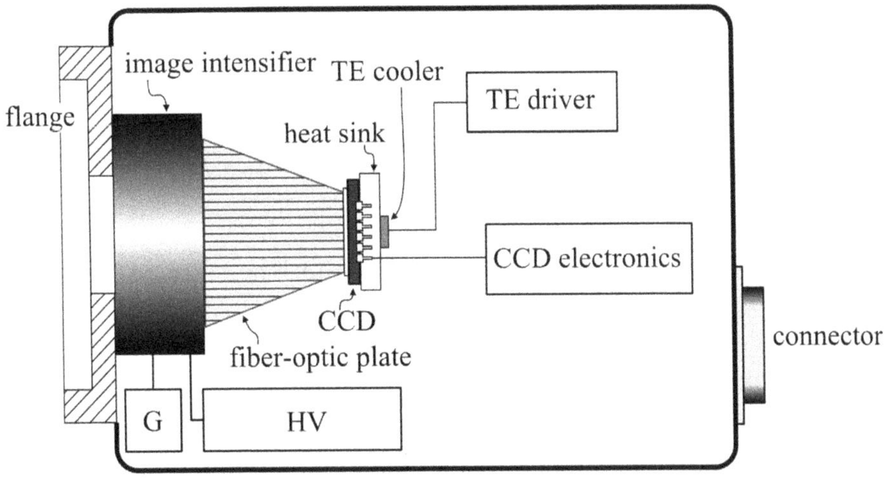

Fig. 5.9. Basic components of a typical ICCD.

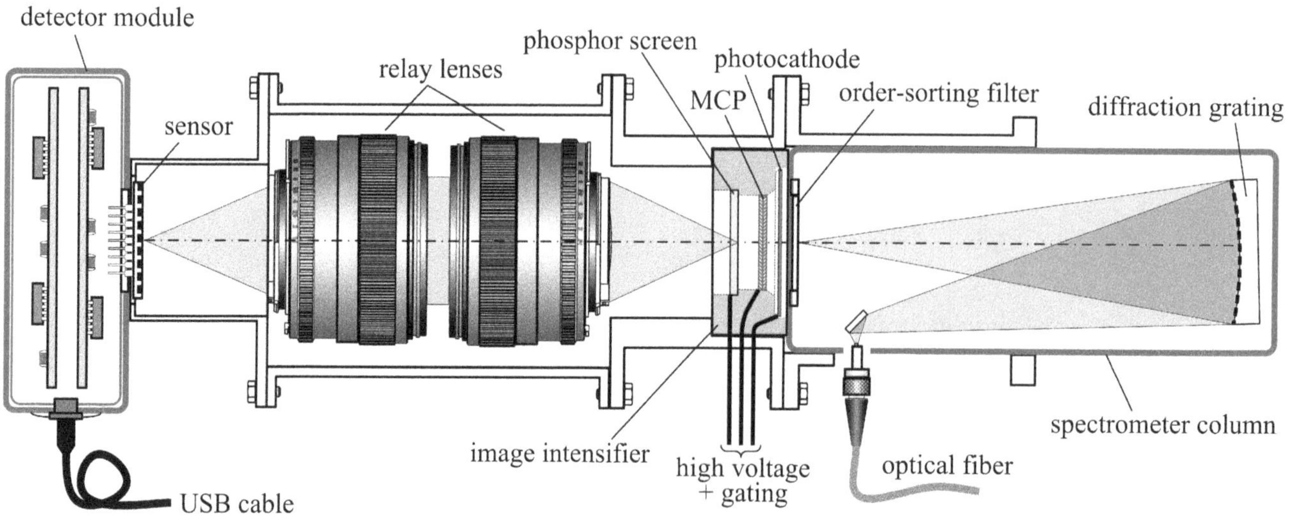

Fig. 5.10. Design concept. "MCP" stands for the micro-channel plate, namely the Photek MCP-125 image intensifier.

right decision because ICCDs separately find numerous applications in visualization of fast phenomena. Next, in order to minimize noise, the CCD matrix is always thermo-electrically cooled. This requires additional electrical circuit and fans. Finally, the image intensifier and the CCD matrix are permanently connected together in a solid block, being interfaced through imaging fiber-optic plate. This block, being proprietary technological achievement of a manufacturer, is the most expensive part of an ICCD, requiring high-grade facilities for assembling. Since the light at the output of the fiber-optic plate diverges from each single fiber, the input window of the CCD matrix must be removed and the sensor glued directly to the plate, with only thin layer of an optical bonder between them.

The scheme in Fig. 5.9 is compact and reliable. However, in order to be able to apply the spectral deconvolution technique and, therefore, to increase spectral resolution to its maximum, the optical point-spread function must have regular (not random) shape over the entire spectral range. Irregular distribution of the individual light guides in the fiber-optic plate cannot match perfect organization of the detector pixels. Therefore, the fiber-optic plate, composed of randomly distributed fibers, should be substituted with the direct optical coupling through high-quality relay optics. One possible scheme of that type is shown in Fig. 5.10. Using the prototype assembled according to this scheme, typical performance of gated intensified spectrometers is discussed in the next section.

5.5. Typical Performance of a Gated Intensified Spectrometer

5.5.1. *Spectral performance*

In the design described above in Fig.5.10, there are two main limitations to spectral range: photocathode spectral sensitivity and the input aperture of the image intensifier. The S20 photocathode of the Photek MCP is sensitive from 200nm to 900 nm (Fig. 5.3), which does not differ significantly from the sensitivity range of any solid-state detector. The only inferiority is probably in the IR wing of the spectrum, particularly beyond 1 μm. On the other hand, the photo-emissive detector shows better sensitivity in the UV wing.

Another factor is purely geometrical: *de-facto* standard for linear detectors is their 30 mm length, whereas the input aperture of the image intensifiers is also *de-facto* standardized to 25 and 40 mm diameters. Since the capacitance of the larger tablets is bigger, their gating capabilities are worse. Therefore, good gating performance can be achieved only with 25 mm apertures, which leads to 20% loss of linear space. In principle, there is no big problem to design an asymmetrical relay optics, transferring 25 mm of input aperture to 30 mm on the detector, and correspondingly to decrease linear dispersion on the grating side. However, in the prototype that is reported here, the relay optics transfer ratio is 1:1, which leads to 20% loss of the useful spectral range, provided by the grating. However, this loss mostly lies in the spectral regions out of the photocathode sensitivity.

Figure 5.11 shows typical Ar+Hg spectrum obtained with the optical gain 300. At right, magnified snapshot of the narrow spectral interval, marked with the dashed rectangle in the spectrum at left, shows that the Hg doublet 576.960 nm and 579.066 nm is resolved according to the Rayleigh 80% criterion, signifying spectral resolution 2 nm in the raw spectrum.

Many applications, especially laser-induced break-down spectroscopy (LIBS), require both good spectral resolution and wide spectral coverage. With the detector size fixed, these two parameters are competing: good spectral resolution requires high linear dispersion, which makes spectral coverage narrower. For this reason, many LIBS systems use several different spectrometers, each covering a relatively narrow spectral interval. Such a multichannel operation makes the entire system bulky, expensive, and potentially unreliable. There is also another reason to get rid of multiple channels. Below it is shown that gating spectrometers can significantly increase efficiency of the remote LIBS. However, if someone would only dream to install gated intensified detectors in each channel, then such a system would be obviously impractical for the reasons of cost and complexity. Luckily, there is a good solution of how to achieve both good spectral resolution and wide spectral range in one spectrometer: deconvolution. This mathematical procedure delivers substantial increase in resolution, if only the SNR is high, and the intensified spectrometer is just a device with very high SNR. This possibility is discussed in the next section.

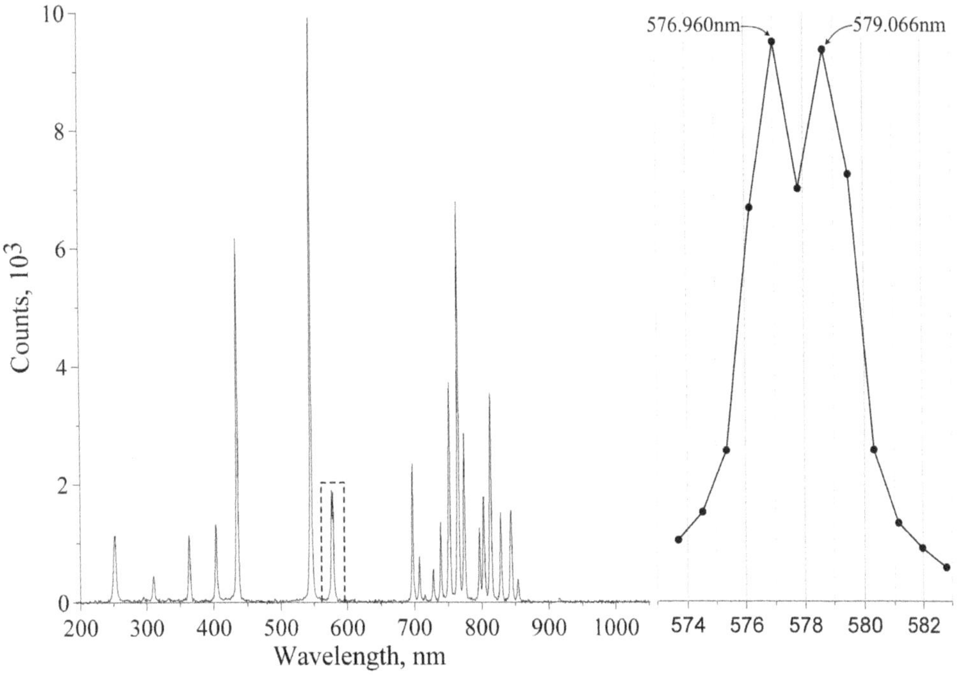

Fig. 5.11. Ar–Hg lamp spectrum.

5.5.2. *Computer-enhanced spectral resolution*

Spectrometer is an imaging system, projecting the spectrum in the plane of a photo-detector. According to the theory of linear optical systems, the resultant spectrum $s(\lambda)$ is the convolution of the original spectral function $s_0(\lambda)$ with the point-spread function $h(x)$ of the imaging system:

$$s(\lambda) = \int s_0(x)\, h(\lambda - x)dx + n.$$

Here n is the noise. The problem of deconvolution is to restore the original spectrum s_0 from the measured result $s(\lambda)$ if the point-spread function $h(x)$ is known. There are two basic approaches: filtering and fitting. Consider them separately.

Filtering relies on the fact that, in the absence of noise $n = 0$, the Fourier transforms S, S_0, and H of the three functions s, s_0, and h relate to each other by a simple product:

$$S = S_0 \cdot H.$$

Then the only thing we have to do is to divide S by H and to perform the inverse Fourier transform. However, the function H has zeroes and there is noise, which means that the result of such a simple manipulation will be wrong. Instead, the so-called Wiener filter should be used, giving not the exact function S_0 but the optimal statistical estimate of it $\hat{S}_0$:

$$\hat{S}_0 = \frac{H^*}{|H|^2 + \frac{1}{q}} \cdot S.$$

The asterisk means complex conjugate. The term $1/q$ in the denominator stabilizes against zeroes of H and the parameter q has the physical meaning of the SNR. From this formula, the general conclusion follows most easily: the higher the SNR is, the better the original spectrum can be restored. It means that deconvolution can be effectively applied only to low-noise measurements, which emphasizes again the importance of low-noise optical intensification.

The speed of the Wiener filter, if implemented with the fast Fourier transform, is sufficient for real-time operation. The only thing needed is to measure the point-spread function $h(x)$ of a spectrometer. With good spatial uniformity of the multi-channel plate and high-quality relay optics, the point-spread function $h(x)$ may be constant in the entire spectral interval 200–900 nm. In this case, $h(x)$ can be measured around any wavelength within the working spectral interval, for example at the wavelength of a He–Ne laser (632.8 nm). The result of the Wiener deconvolution applied to the same sample Ar+Hg spectrum that was shown in Fig. 5.11 is presented in Fig. 5.12(a). Now the Hg doublet 576.960–579.066 nm is 100% resolved, and other lines are narrowed.

For comparison, the original spectrum is also included in this graph, intentionally shifted up and right for better visual perception. However, further progress cannot be achieved with the Wiener filter on this sample because the number of points within each spectral line is insufficient to effectively perform the Fourier transform. Therefore, finer detectors with greater number of pixels, like Toshiba TCD 1304AP with 3648 elements, are needed for efficient implementation of the Wiener filter algorithm.

Luckily, the second approach based on fitting procedures gives impressive results even on small number of pixels per spectral line. The idea is to take some initial guess about $s_0(\lambda)$, make the convolution on it and compare the result with the measured function $s(\lambda)$. Then change slightly the guess function and try again. Find the change that delivers the most rapid convergence to the measured function, and try again. This is basically the well-known steepest-descend algorithm. A variety of much more efficient algorithms was developed during the last fifty years, among which the Levenberg–Marquardt routine is the best in terms of speed and reliability. It can be found in many software packages, for

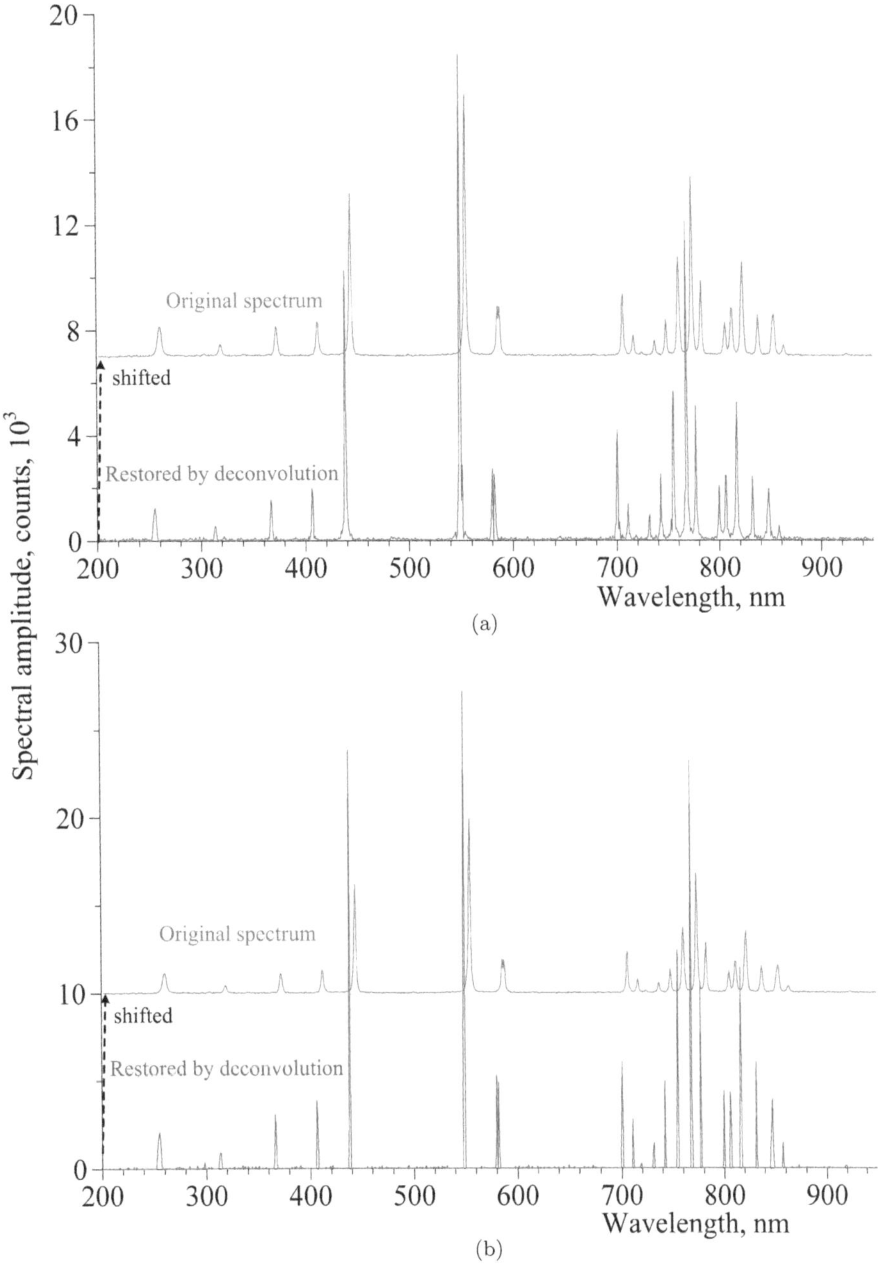

Fig. 5.12. (a) Wiener filter deconvolution. (b) Levenberg–Marquardt deconvolution.

example in the Visual Numerics IMSL Fortran library. Although such computations cannot follow real-time measurements like the Wiener filter does, they can be useful in many cases when the time is not a factor and extremely high spectral resolution is needed. Figure 5.12(b) shows how deconvolution can bring spectral resolution down to a single-pixel limit. This is the same Ar+Hg sample spectrum as in the previous figure with the original spectrum being shifted up and slightly right in order to separate it from the restored spectrum. The most impressive feature of this picture is that the strongest Hg line at 546.074 nm, being spread over a dozen of pixels in the raw spectrum, after deconvolution is entirely localized at one pixel. This is also the case for the 576.960−579.066 nm doublet (Fig. 5.13) and for the majority of other spectral lines. This signifies a mathematical limit to restoration. With 1024

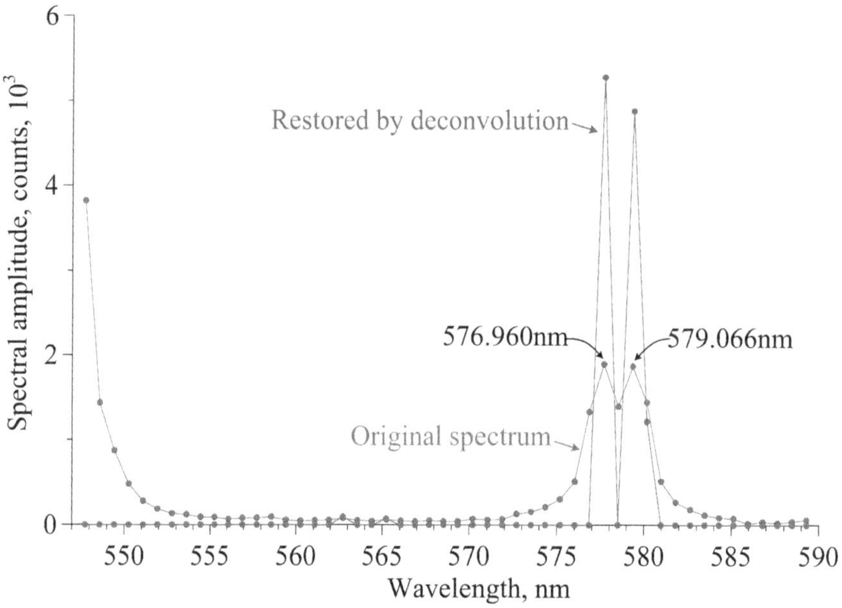

Fig. 5.13. Hg doublet 576.960–579.066 nm after the Levenberg–Marquardt deconvolution. Dots show pixels.

pixels in the detector and spectral range 900 nm, this corresponds to spectral resolution 0.9 nm. In order to increase it, a larger number of pixels should be used, for example the Toshiba TCD 1304AP detector with 3648 elements. Without any modifications to optical configuration, three-fold increase in the number of pixels will result in spectral resolution 0.3 nm.

However, careful examination of Fig. 5.12 shows that spectral peaks in the UV wing below 400 nm are restored not as good as in the visible and NIR parts. Mathematically it means that the point-spread function in the UV part of the spectrum is somewhat different from the one used for restoration. The physical reason for this is a very small defocusing in the UV region due to uncompensated tilt between the grating image plane and the plane of the detector.

5.5.3. *Gating capabilities*

The width of the optical gate cannot be measured directly, therefore, indirect experiments should be used to prove nanosecond optical sampling. Particularly, the technique of calibrated delays is very straightforward and convincing. In it, a short light pulse travels through optical fibers of different length, and the gating device must resolve the delay difference, introduced by the difference in fiber length. In the experiments, a short light pulse was generated by a high-speed laser diode powered by a nanosecond function generator. The light pulse width was set to a minimum 10 ns with 5 ns edges at the function generator end that corresponded to almost triangular pulse. The full width at half-maximum (FWHM) of such pulse should be 5 ns. The 1GHz oscilloscope trace recorded by a high-speed photo-multiplier with 80 MHz amplifier shows the FWHM about 6ns (Fig. 5.14(a)).

The gating pulse on the spectrometer was set to the minimum 3.5 ns, and the optical gain was set to 300. Initially, the laser diode was connected to the spectrometer through a 1 m long optical fiber. The delay between the function generator pulse and the spectrometer triggering pulse was adjusted to the maximum spectral amplitude (temporal coincidence of the optical pulse from the laser diode and the spectrometer gate), which, with this particular fiber, turned out to be $\tau_{\max} = 450$ ns (unreferenced

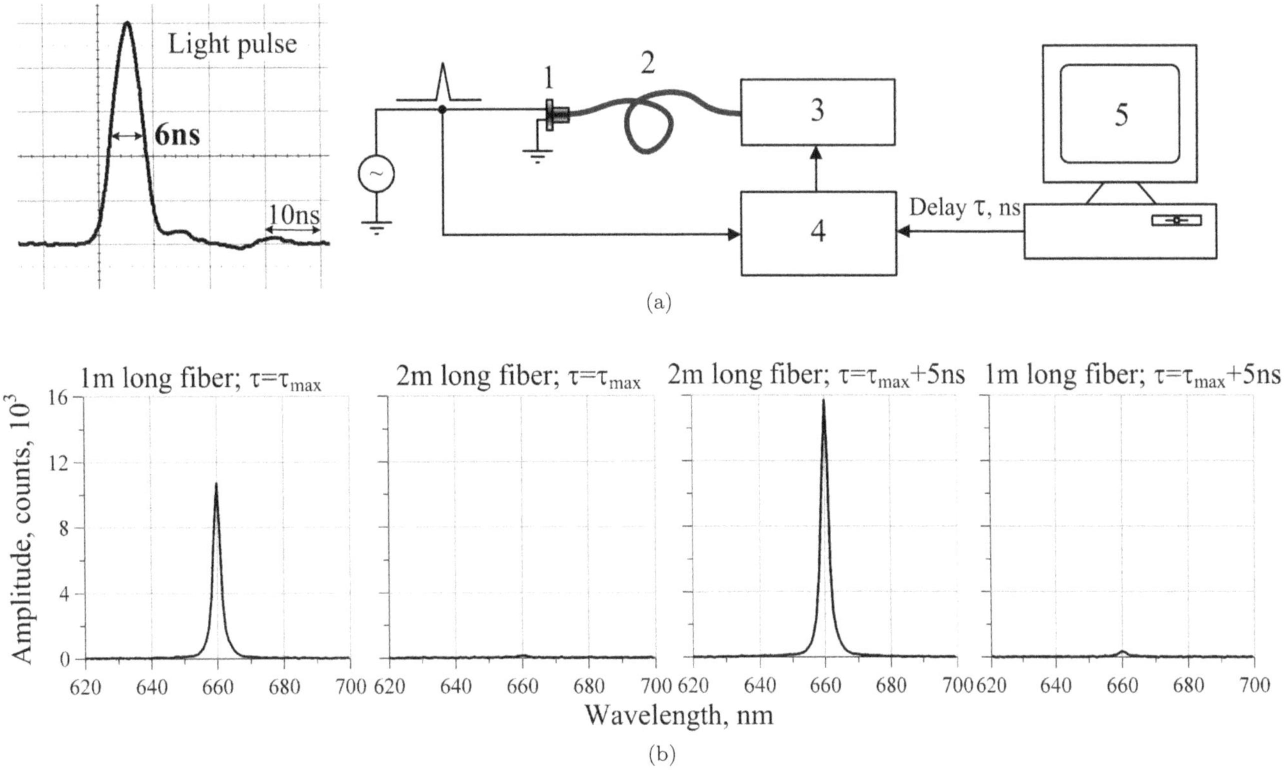

Fig. 5.14. (a) Optical sampling scheme. 1 — laser diode; 2 — optical fiber; 3 — spectrometer; 4 — digital delay generator; 5 — computer. (b) Optically gated spectra.

value). This spectrum is shown in the first picture of Fig. 5.14(b). Next, this fiber was substituted with the 2 m long one. The amplitude dropped 75 times (second picture), but after increasing the delay by only 5 ns it became even bigger than with the shorter fiber (third picture). This increase in spectral amplitude can be attributed to discrete variation of the delay time, limited to 5 ns per step. Therefore, initial positioning of the spectrometer trigger relative to the laser pulse might not coincide with its maximum. Finally, with the delay unchanged and the original configuration restored, i.e. 1 m long fiber, the amplitude of the spectrum became much smaller than it was in the very beginning (the last picture). This result clearly shows that the spectrometer readily resolves 1 m long section of optical fiber, i.e. 3 ns delay.

The next experiment shows how the shape of the optical pulse can be measured by scanning (sweeping) the delay. The idea is clear from Fig. 5.15: if the gate pulse is much shorter (narrower) than the optical pulse then the amplitude of the signal as a function of delay $a(\tau)$ presents the envelope of the optical pulse.

Settings of the experiment were as follows: gate pulse width 3.5 ns, optical gain 300, fiber 2 m long with the core diameter 0.4 mm. The delay was tuned by 5 ns increments, covering 20 ns, i.e. roughly 2 times pulse width. Fig. 5.16(a) shows how the spectrum and amplitudes evolve with the delay.

Combining the amplitude data and delays into a table, it is possible to make a curve of the optical pulse, as shown in Fig. 5.16(b). The FWHM of this curve is about 6 ns, which is in good agreement with 6ns measured directly in the chain photo-multiplier-amplifier-oscilloscope.

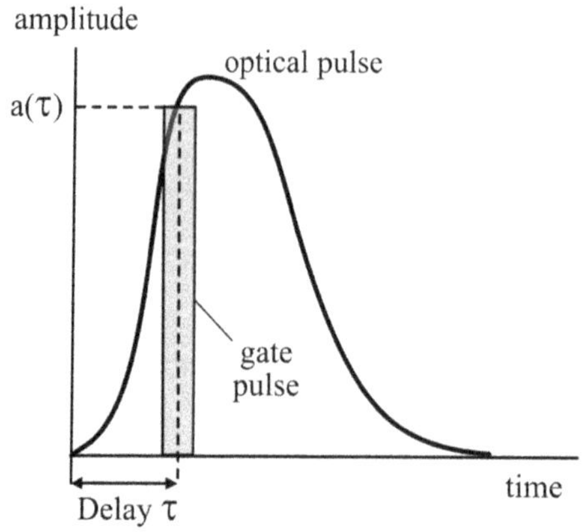

Fig. 5.15. Variable delay technique for measuring optical pulse shape.

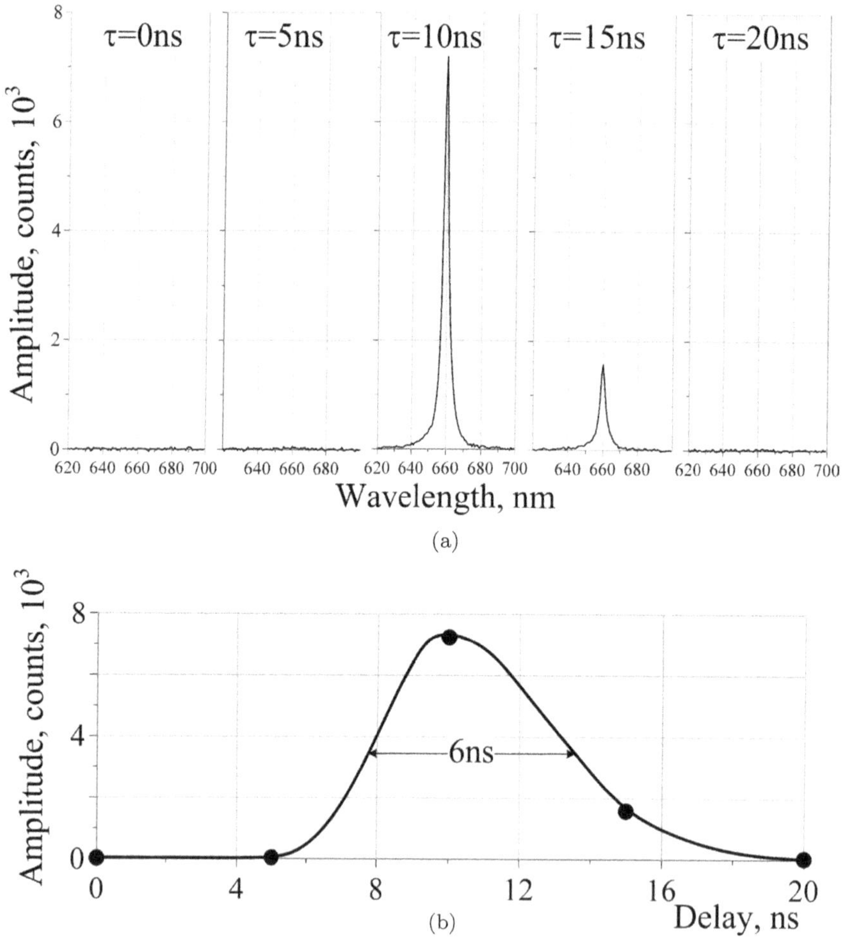

Fig. 5.16. (a) A sequence of delay-incremented spectra. (b) Restored shape of the optical pulse.

5.5.4. *Time-resolved fluorescence spectroscopy*

Temporal evolution of the fluorescence emission is a powerful tool for studying molecular kinetics. In this technique, a strong short optical excitation pulse initiates molecular transitions, producing induced fluorescence at some characteristic spectral lines. Usually, such decay process is short-lived, with the

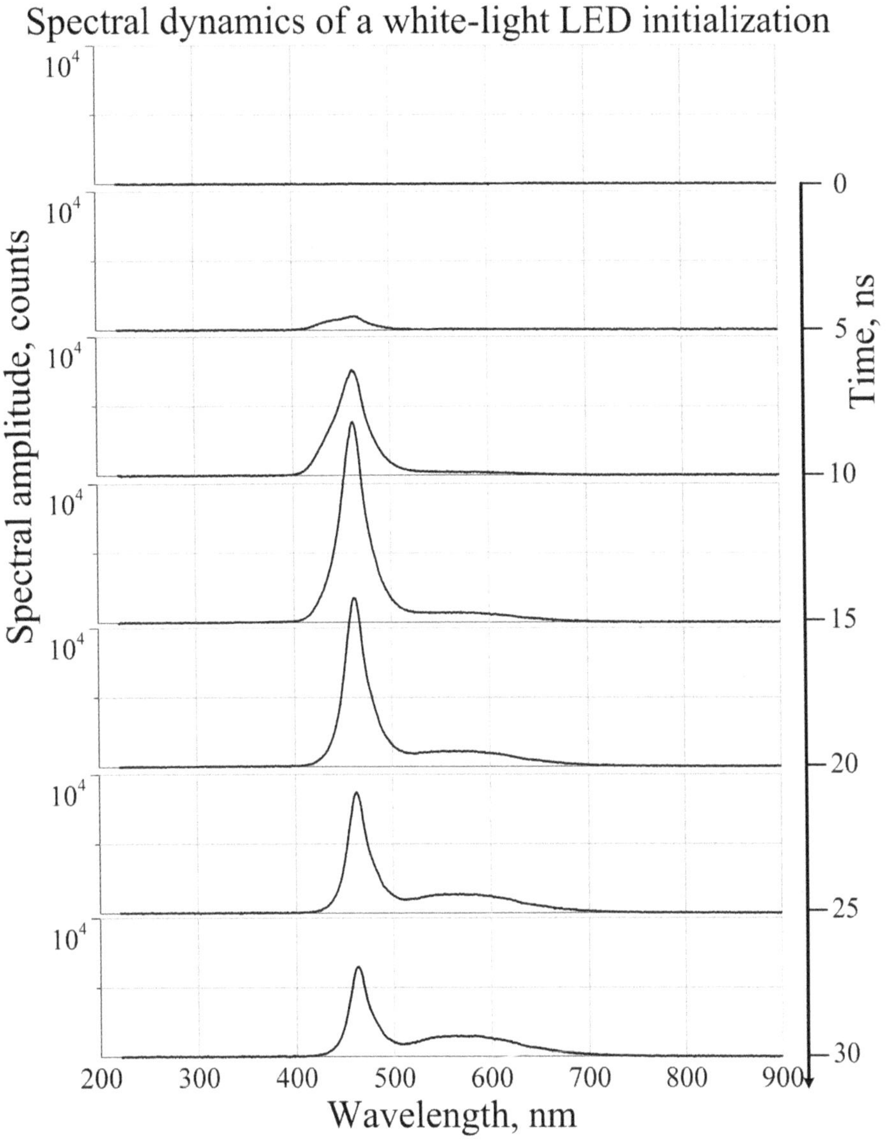

Fig. 5.17. Time-resolved spectra of the white-light LED.

timescale about nanoseconds or even picoseconds. Gated spectrometer is a common instrument for these measurements, and wide spectral coverage between 200 nm and 900 nm with one grating and embedded order sorting filter are additional benefits. The results below show how it can be used for monitoring excitation dynamics of an ordinary white-light light-emitting diode (LED).

Physics of the so-called whit-light LED is based on broadband fluorescence of a special phosphoric substance, covering the emitting surface of a blue LED. Applying a short nanosecond-scale electrical pulse to a white-light LED, one can observe spectral dynamics of its emission. The scheme of this experiment is similar to that shown in Fig. 5.14(a) except that the monochromatic laser diode is substituted with the whit-light LED. Excitation pulse width is 10 ns. Detailed spectral dynamics of a white-light LED is shown in Fig. 5.17 as a series of spectra obtained during first 30ns of excitation.

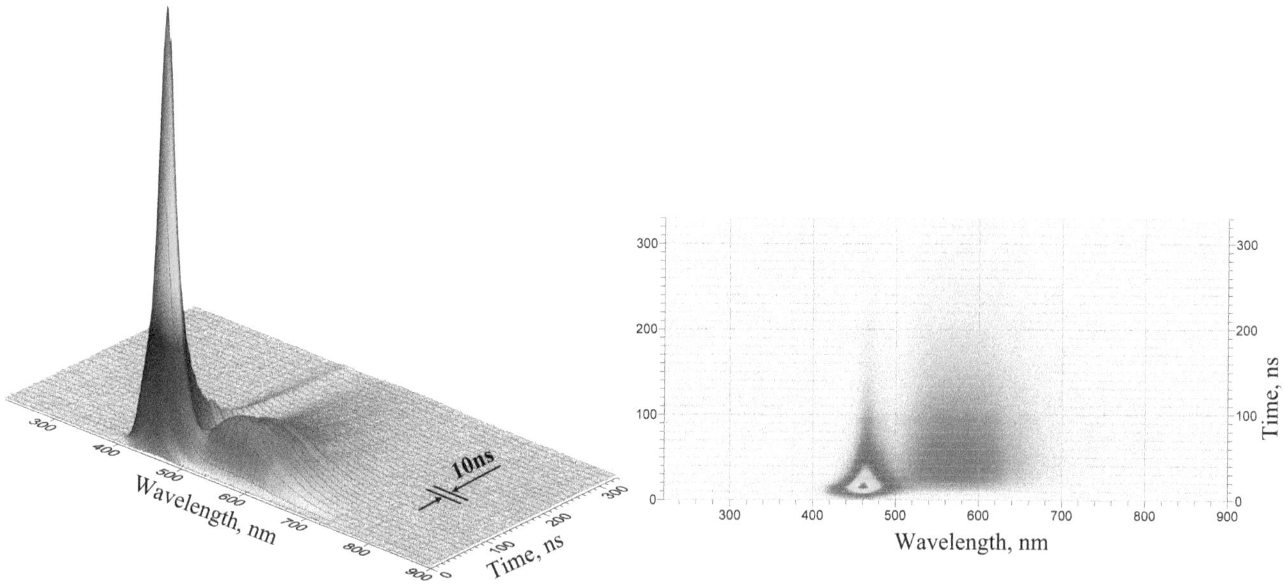

Fig. 5.18. 3D and 2D presentations of the white-light LED dynamics.

In the first 10 ns, only blue component at 460 nm emerges from the LED, which corresponds to stimulated emission from the diode junction, propagating through the phosphoric layer. No longer-wavelength fluorescence is visible during this period because some finite time is needed to populate the fluorescence energy levels. At the 15th nanosecond, blue emission reaches maximum and fluorescence starts in a wider spectral interval between 500 nm and 700 nm. It takes 15ns more for the fluorescence flux to make a significant portion of the overall radiation. It is also interesting that spectral width of the blue emission narrows with time from roughly 100 nm at the 10th nanosecond to about 50 nm at the 30th nanosecond.

The entire picture of spectral dynamics on a larger timescale can be better seen in 3D or 2D presentations, as shown in Fig. 5.18. Fluorescence emission spreads over 200 ns, which is 20 times longer than the duration of the excitation pulse. Spectral narrowing of the blue emission with time can be best seen at the 2D picture.

5.5.5. *Modulation-sensitive spectroscopy*

There are applications, such as plasma diagnostics or modulated fluorescence spectroscopy, where weak modulated optical signal must be resolved on the background of strong non-modulated optical flux. Gated intensified spectrometer is the perfect solution for these applications. Using reference signal, the phase of which is locked to the modulated source, it is possible to record two consecutive spectra with the opposite phases and to subtract them. The result of this mathematical operation is the spectrum of the modulated component only. Some variation of this technique is sometimes referred to as the phase-resolved spectroscopy.

The scheme of this experiment is explained in Fig. 5.19. A non-modulated gas-discharge lamp is combined with a high-speed LED, modulated at 2MHz. The average flux of the LED is much smaller than that of the lamp, and the ability of the gated spectrometer to select the LED spectrum from the combined flux is very impressive. For that, two consecutive spectra — even and odd — were recorded with the additional delay 250 ns added during acquisition of the odd spectrum (all this is done automatically).

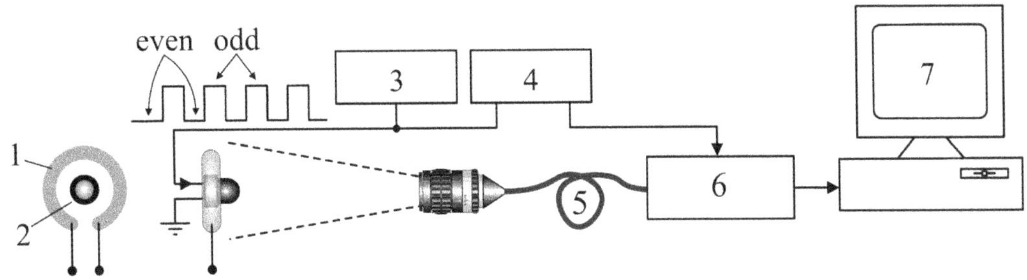

Fig. 5.19. The scheme of the experiment. 1 — gas-discharge lamp; 2 — light-emitting diode (LED); 3 — function generator; 4 — digital delay generator; 5 — optical fiber; 6 — spectrometer; 7 — computer.

The results are shown in Fig. 5.20. The difference between the two consecutively recorded spectra is not easily seen until these spectra are subtracted and the result is shown in the trace below. Although this curve is obtained as the mathematical operation, physically it represents the spectrum of the LED. It becomes obvious when compared to the spectrum recorded with the lamp turned off (the last trace in the picture). The only difference between the computed spectrum and the measured one is the subtle spectral maximum in the green light domain near 530 nm caused by ambient illumination in the laboratory. The lamp is made of ordinary glass, so that no UV radiation is detected below 400 nm.

5.5.6. *Enhanced sensitivity spectroscopy*

In many applications, like remote LIBS or biological research, researchers deal with weak optical fluxes on the verge of sensitivity of traditional spectrometers. The gated spectrometer with an image intensifier in its optical channel, makes it possible to amplify optical signal with the signal-to-noise ratio (SNR) higher than that of a traditional electronic amplifier. Therefore, such spectrometer can possibly present additional advantage of higher sensitivity with respect to traditional spectrometers with non-intensified detectors. In order to unleash this capability, the device must work in a continuous-wave (CW) or quasi-CW mode. This is clear from Fig. 5.21.

Ability to work with higher SNR is not obvious because the relay lenses, inserted together with the image intensifier, introduce additional signal loss with respect to traditional spectrometers. Also, the levels of spontaneous and Poisson noise of the image intensifier are unknown. Nonetheless, the measurements showed that image intensifier in optical chain of the spectrometer not only gives general amplification to the signal, but higher SNR as well.

Figure 5.22 shows the scheme of the experiment.

Sensitivity of the prototype, described in Section 5.4, was experimentally compared to the Spectral Products SM-240/6 non-intensified portable spectrometer, working in a CW mode. Although the Photek gating module can work in a genuine CW mode, the version of the digital delay generator did not support this mode, therefore quasi-CW mode was set for the image intensifier with 93% duty cycle at 190 kHz repetition rate. Both spectrometers worked with 300 ms integration time. Initially, the iris on the lens was set for the minimum SNR on SM-240/6, and this spectrum was recorded (first trace in Fig. 5.23). Then the fiber was switched to the intensified spectrometer with the gain $G = 300$ (second trace): spectral maxima can barely be visible. This is not a surprise because SM-240/6 uses Sony ILX-511 detector with sensitivity 200 V/(lx·s), whereas the prototype of the intensified spectrometer — the Hamamatsu S8378-1024Q chip with sensitivity only 22 V/(lx·s), i.e. almost ten times less.

After that, the gain was set to $G = 3000$ (third trace). Obviously, the SNR of the intensified spectrometer with the gain $G = 3000$ is substantially better than that of the SM-240/6. It looks like SNR

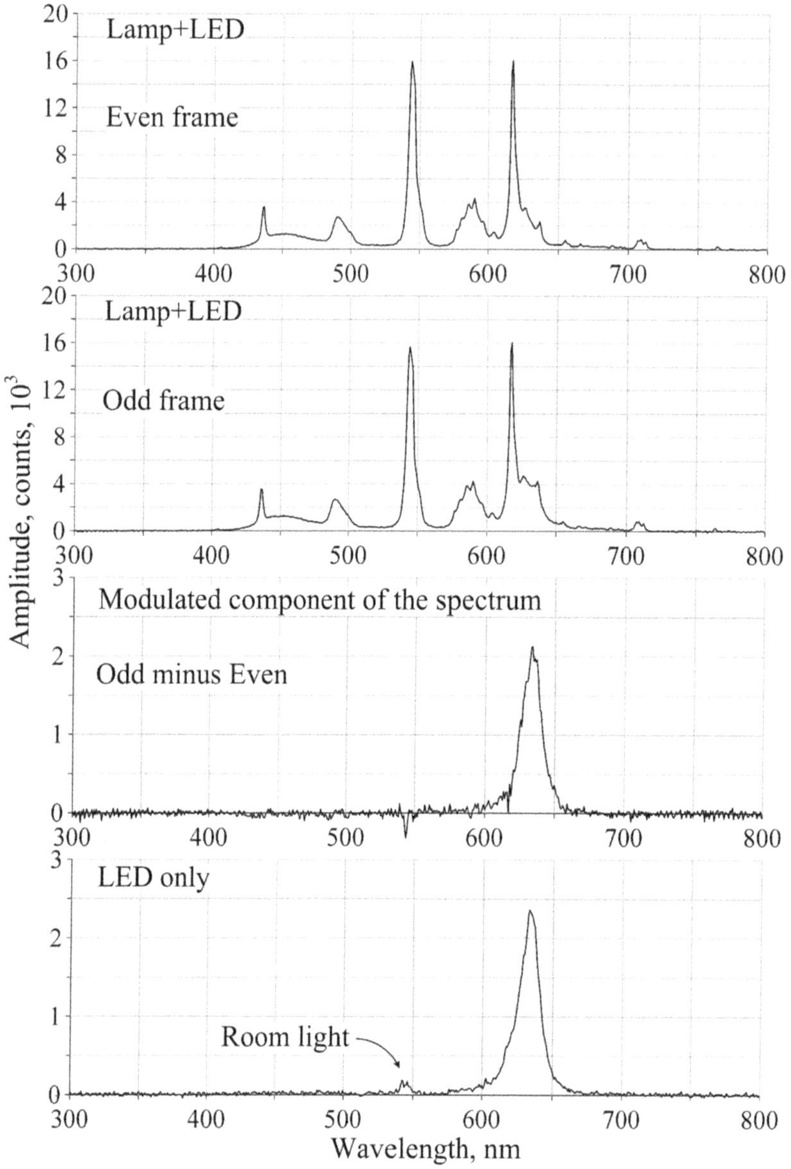

Fig. 5.20. Weak modulated component around 630 nm is recovered from strong background.

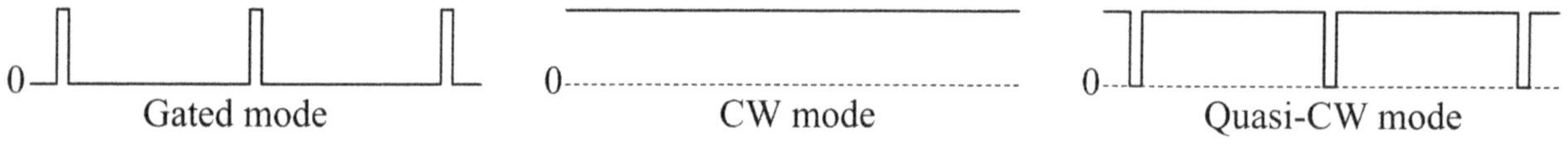

Fig. 5.21. Modes of operation.

increases with the gain up to the gain maximum value 3000, which means that Poisson noise of the photocathode is still smaller than the noise of the Hamamatsu S8378-1024Q chip. Even some features of the continuous spectrum between 400 nm and 500 nm can be recognized in this trace, comparing it to the clean spectrum of the same lamp shown in the "even" trace of the Fig. 5.20 in the previous section. Thus, the existing combination of a 22 V/(lx·s) detector with an image intensifier gives significantly better SNR than a highly sensitive 200 V/(lx·s) detector alone. It means that the combination of an

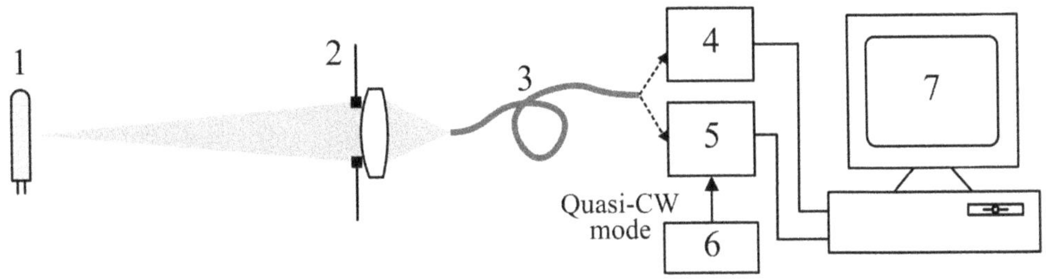

Fig. 5.22. The test of sensitivity. 1 — gas-discharge lamp; 2 — iris diaphragm; 3 — optical fiber; 4 — ordinary compact spectrometer SM-240/6; 5 — gated intensified spectrometer; 6 — digital delay generator; 7 — computer.

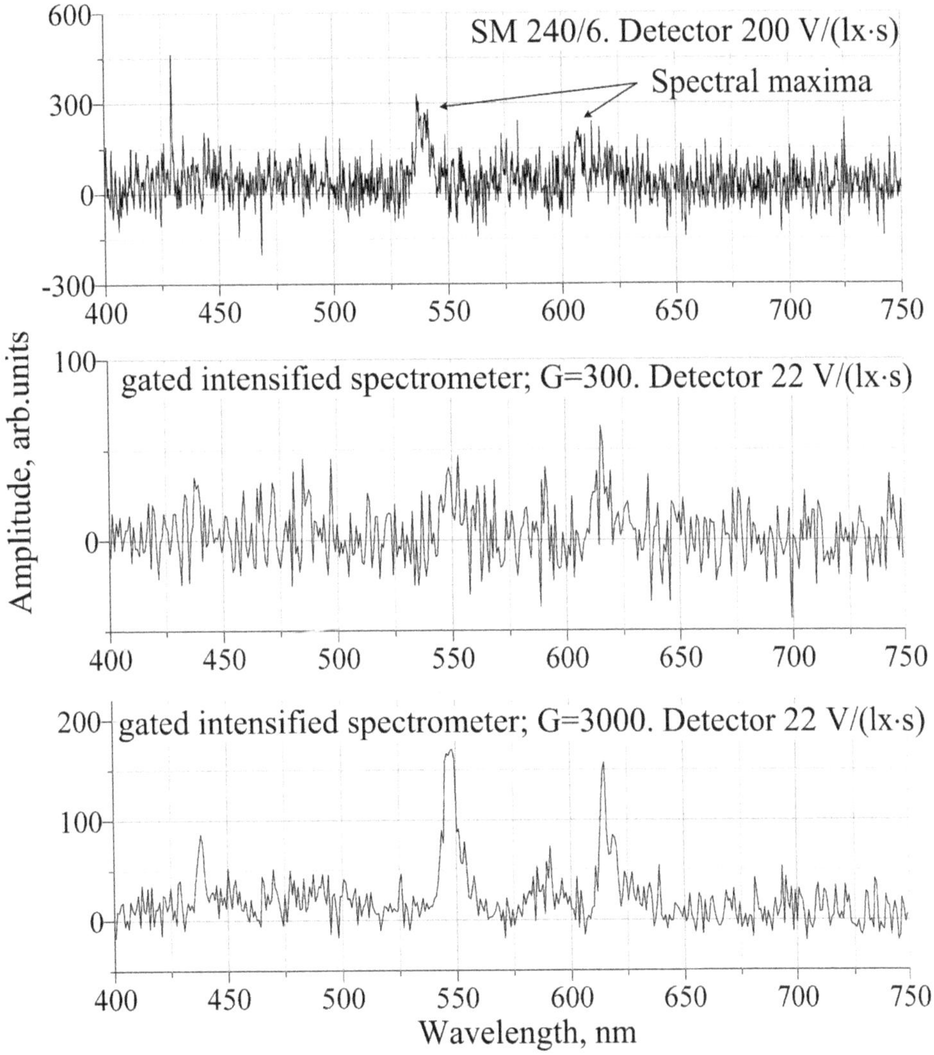

Fig. 5.23. Results of sensitivity test.

image intensifier with a more sensitive detector, like Toshiba TCD-1304 with 160 V/(lx·s), will produce even further increase in sensitivity with respect to the existing configuration of the prototype.

5.5.7. *Micro-spectroscopy*

There are many applications in semiconductor manufacturing, where spectral information of microscopic objects is of importance. Technically, micro-spectroscopy presents the combination of microscopy and spectroscopy. In microscopy, it is well known that the higher the magnification is, the stronger the illumination must be. But the illumination cannot be made too strong for the two reasons: the image sensor (usually the CCD chip) must not be saturated, and the sample must not be overheated or damaged. The dynamic range of a CCD matrix is defined as the ratio of the maximum number of electrons that can be collected in one pixel to the pixel noise, and commonly lies between 10^3 and 10^4. Suppose now we want to decompose the optical flux, coming to a single pixel, into a spectrum. Then this optical flux will be spread over at least 10^3 spectral elements, so that the signal-to-noise ratio (SNR) in each one will be at best $(10^3 \div 10^4)/10^3 \sim 1 \div 10$. However, for a quantitative analysis, the SNR 50 is needed at least, and the dynamic range must be more than 10^2. Thus, the sensitivity and dynamic range of a spectrometer are the key factors for micro-spectroscopy. Therefore, gated intensified spectrometers are the first candidates for these applications even if gating is not needed.

A simplest micro-spectrometer can be assembled around any standard microscope, providing that a fixed (known) area of its field of view or a single pixel is optically connected to a spectrometer. However, it is advantageous to create a visible mark within the field of view, explicitly showing what particular area of the image is connected to the spectrometer. Optical scheme of this experiment is shown in Fig. 5.24.

The green laser diode sets a green spot in the microscope field of view through a semi-transparent mirror, which marks the position to which the spectrometer is optically connected. During spectral measurements, the green laser diode is turned off, so that the spectrometer — either one or another — receives all light from the sample. The idea of the experiment was to compare quality of spectra obtained

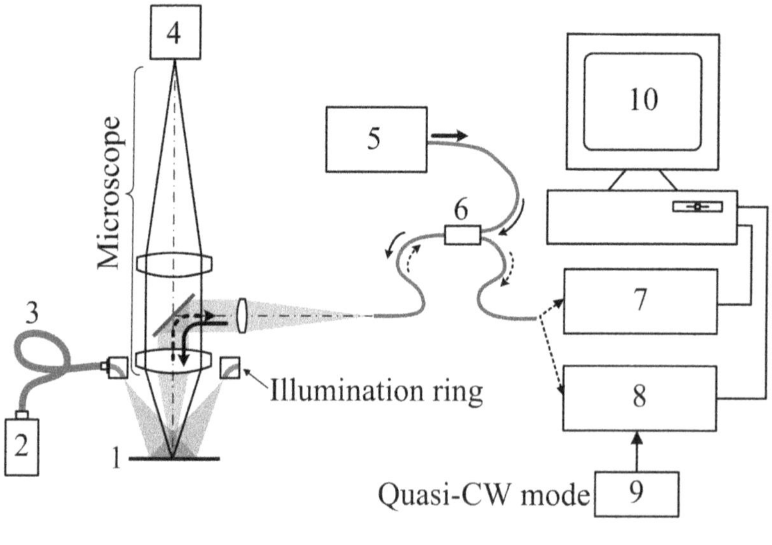

Fig. 5.24. Experimental arrangement of micro-spectroscopy. 1 — sample; 2 — white-light source; 3 — illumination fiber bundle; 4 — CCD camera; 5 — green laser diode; 6 — bifurcated optical fiber; 7 — grating spectrometer SM 240/6; 8 — gated intensified spectrometer; 9 — digital delay generator; 10 — computer.

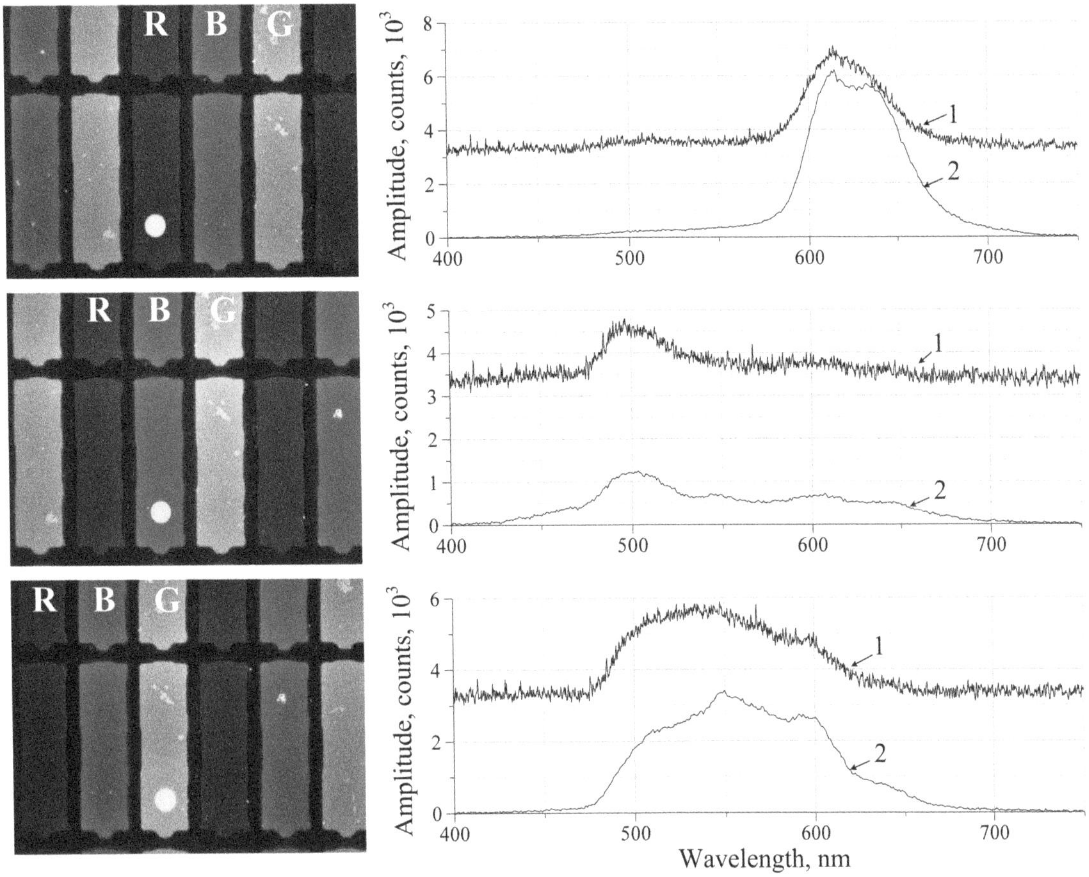

Fig. 5.25. Micro-spectra of a color filter matrix of a computer display. Green bright disk marks the measurement position. 1 — ordinary spectrometer; 2 — intensified spectrometer.

by a standard Spectral Products spectrometer SM 240/6 with the Sony ILX-511 detector and the gated intensified spectrometer. In the previous section, it was already noted that sensitivity of the Sony ILX-511 detector is an order of magnitude higher than that of the Hamamatsu S8378-1024Q detector used in the prototype of the gated spectrometer. Figure 5.25 presents the results obtained on a colour filter matrix of a computer display as a sample.

Each red, green, and blue segment is 250 μm long and 50 μm wide. Each row of images shows the microscope field of view with the green LED spot marking position of the area, to which the spectrometer is optically connected, and the spectrum recorded by GIS with the level of illumination, producing sufficiently bright image. These are the reflection-type spectra, which differ from the transmission-type ones. In these measurements, the optical gain was set to 3000, demonstrating excellent SNR with the time constant only 300 ms. To facilitate comparison with the SM 240/6, the spectra recorded with this spectrometer are added to the graphs with some vertical displacement. Clearly, intensified spectrometer provides a better SNR, despite an order of magnitude lower sensitivity of its detector.

5.5.8. *Laser-induced breakdown spectroscopy*

High-power laser pulse, being focused onto a target, causes evaporation of a thin layer of the material in the focal spot, thus creating a radiating source. Its spectral emission is a characteristic information about the material, which may be used for chemical analysis. This technique is called laser-induced breakdown spectroscopy (LIBS). LIBS is inherently a remote-sensing technique, but when the target is located several meters and more away from the sensor, such a technique is called remote LIBS in order to emphasize the scale of range. The two most important properties of gated intensified spectrometers — gating and enhanced sensitivity — are vital for the remote LIBS.

If the target is located at the distance R from the detector then the signal, coming to the detector input aperture of the area S, is proportional to its solid angle as viewed from the target, i.e. S/R^2. Thus, signal decreases quadratically with range, which makes sensitivity the most important factor. The majority of commercially available LIBS systems use ordinary non-gated and non-intensified spectrometers. For them, the only way to increase sensitivity is to increase exposure time T on the detector. Then the number of photons N increases as

$$N = \nu T + \eta T,$$

where ν is the constant signal flux and η — noise rate. Neglecting all other sources of noise and assuming Poisson statistics, the signal-to-noise ratio

$$\mathrm{SNR} = \frac{\nu T}{\sqrt{\nu T + \eta T}} = \frac{\nu}{\sqrt{\nu + \eta}}\sqrt{T}$$

increases as square root of the exposure time. This remains true also for pulsed operation, if only one considers ν as the average signal flux. However, for remote LIBS it is necessary to take into account background radiation, like daylight. Then the number of photons will be

$$N = \nu T + \eta T + BT$$

where B is the background. Although SNR does decrease to somewhat smaller value

$$\mathrm{SNR} = \frac{\nu}{\sqrt{\nu + \eta + B}}\sqrt{T},$$

it is not the main problem. The problem is saturation of the detector by the background. When T increases, the term BT drives the signal to a saturation level that blinds the detector at some areas of the spectrum where the background is intense. Although the plasma on the target may be significantly brighter than the background, it is short-lived, few nanoseconds at best, while the background is constant. Therefore, the integral contribution of the background may be bigger than that of the plasma.

The only efficient solution to this problem is gating. If gated image intensifier blocks all optical flux from reaching the detector between consecutive laser shots (Fig. 5.26) then no excessive background will be accumulated.

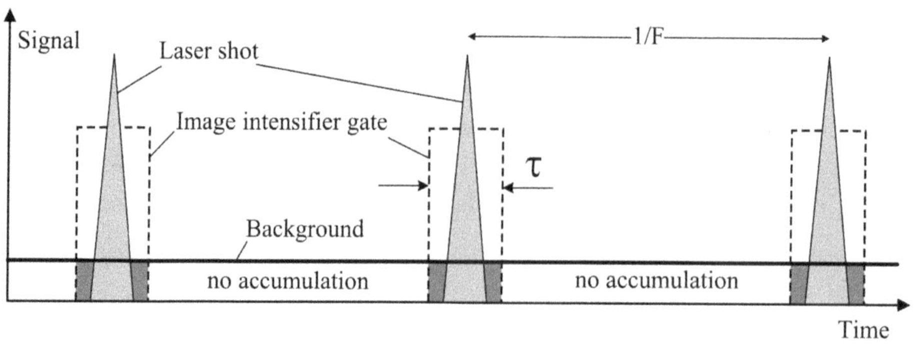

Fig. 5.26. Optical gate blocks the background in between laser shots.

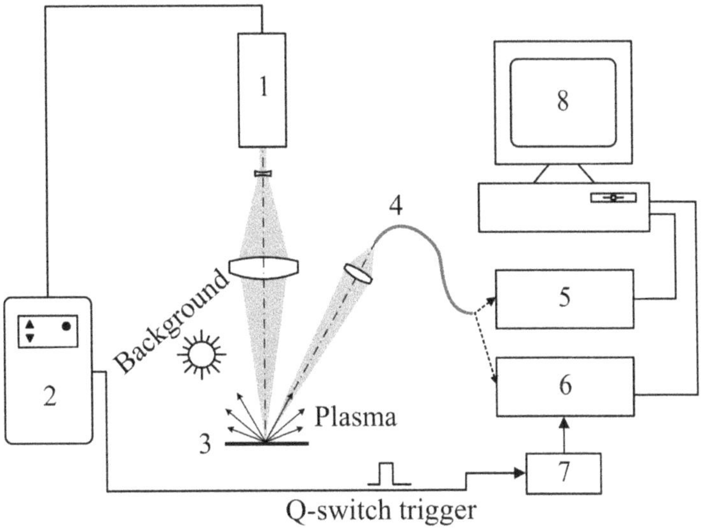

Fig. 5.27. LIBS experiment. 1 — Nd:YAG laser; 2 — laser control unit; 3 — sample; 4 — optical fiber; 5 — ordinary spectrometer SM 240/6; 6 — gated spectrometer; 7 — digital delay generator; 8 — computer.

With F being the pulse repetition rate and τ — the gate duration, the number of photons

$$N = \nu T + \eta T + B\tau(T \cdot F) = \nu T + \eta T + BT(\tau F)$$

contains $(\tau F)^{-1}$ times less background than in the non-gated operation. For example, with $\tau = 10$ ns and $F = 10$ Hz this number is 10^7. It means that integration time can be made 10^7 times longer than in the non-gated mode. Moreover, during the gate, optical signal may be additionally amplified, thus not only preventing saturation of the detector but also increasing SNR.

The Quantel Nd:YAG laser model Ultra was the core of the prototype LIBS system assembled as shown in Fig. 5.27 in order to verify the concept above.

Its output 8 ns 75 mJ (maximum) pulse at 1.06 μm created plasma source on a copper target, radiating in a wide spectral interval between the UV and IR. Therefore, quartz pick-up optics and a UV-grade fiber were used at the detector end to deliver the UV portion of the spectrum to the spectrometers. Figure 5.28(a) helps to realize how big the difference between the background signals obtained by the gated and non-gated spectrometers is. In all the experiments, the gate pulse width for the gated spectrometer was set to 500 ns, which is much longer than the laser pulse width in order to ensure accumulation of even long-decay spectral components of plasma. Note, that in order for the signal of the gated spectrometer to be visible in the figure, it has to be multiplied by 10^2.

After the laser is turned on, the plasma emission spectrum appears above the background. Figure 5.28(b) presents this result for the non-gated spectrometer, and Fig. 5.28(c) — for the gated one on the same wavelength scale. These two figures look very different mainly because of the dominating background in the non-gated spectrometer. Although the background is disturbing in case of the non-gated spectrometer, it is possible to eliminate it by subtraction, if only the detector is not saturated. This is the case of the exposure time T = 0.5 s. However, if we would like to increase the useful signal by increasing the exposure time to T = 1 s, it could not be possible because the background signal saturates the detector, making subtraction useless. Figure 5.28(c) clearly shows that the gated spectrometer does not have any of these problems. Also, the advantage of a single wide-range spectrometer over a system of

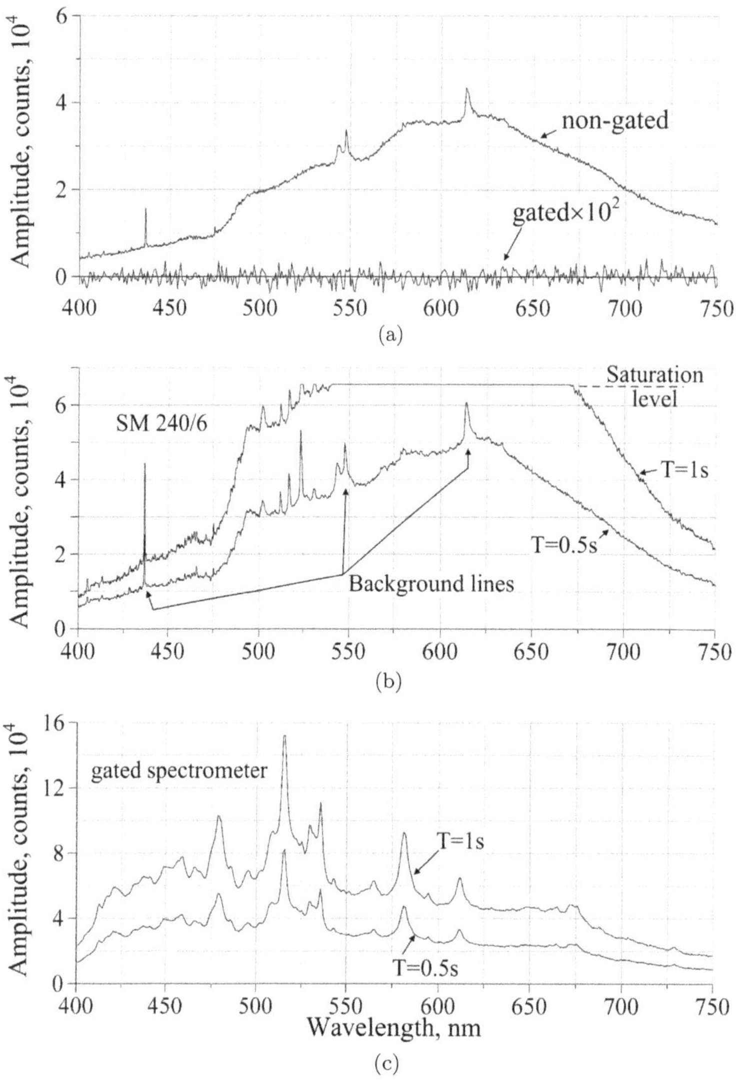

Fig. 5.28. (a) Laser off. Exposure time $T = 0.5$ s for both the spectrometers. Gated spectrometer pulse width 500 ns; optical gain 3×10^3. (b) Laser on. Non-gated spectrometer. (b) Laser on. Gated spectrometer. Optical gain 3×10^3.

narrow-range ones is obvious: no stitching of separately recorded spectra is needed in order to compare the amplitudes of the UV, visible and infrared lines.

5.6. Automatic Gain Control in Gated Intensified Spectrometers

Image intensifier installed in optical path of a compact spectrometer may act not only as a fast gating unit, which is widely used for time-resolved measurements, but also as a variable attenuator-amplifier in a continuous-wave mode. This opens a possibility of an automatic gain control — a new feature in spectroscopy. With it, the user is relieved from the necessity to manually adjust signal level at a certain value — it is done automatically by means of an electronic feedback loop. It is even more important that automatic gain control is done without changing exposure time — an additional benefit in time-resolved experiments. The concept, algorithm, design considerations, and experimental results are presented.

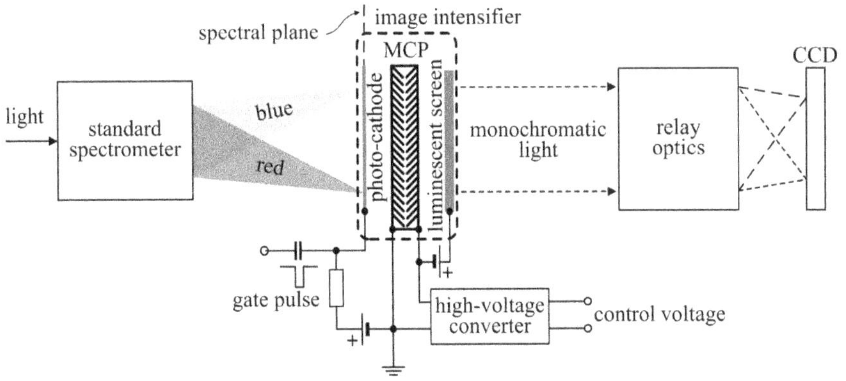

Fig. 5.29. Image intensifier installed in the spectral plane of a spectrometer may either attenuate or amplify optical flux relayed to the image sensor: CCD — charged coupled device.

Automatic gain control (AGC) is a technical feature of a spectrometer that maintains certain, preliminary chosen, spectral amplitude in a processed spectrum at a constant level, whatever low or high the actual input optical flux is. If $S(\lambda, t)$ is the input optical spectrum as a function of wavelength λ and time t, then the AGC function multiplies the entire spectrum $S(\lambda, t)$ by variable gain G

$$S'(\lambda, t) = G \cdot S(\lambda, t)$$

so that to maintain

$$S'(\lambda_0, t) = const$$

at the desired wavelength λ_0.

There are many applications where AGC may be useful. To begin with, consider a common situation when spectrometer is connected to an unknown source and is turned on for the first time on this source. The result is usually either saturation or too low signal. The user then starts to adjust exposure time to make spectrum friendly for analysis. This takes some time and efforts, during which the source may change or even disappear. AGC makes it simple: the user turns the spectrometer on, and, in a second or two, the spectrum on the display of a computer sets to a stabilized level without even changing the exposure time. Speaking about exposure time, it is nowadays inevitable but bad practice to change it in order to maintain the level of the processed spectrum. Working with dynamic sources, exposure time should be chosen not by considerations of gain but by speed of the process.

Neutral density optical filters may be used to avoid saturation, but they change the spectrum itself. And of course, neutral filters cannot increase the signal if it is too low for analysis.

In other cases, optical flux may vary so significantly during measurement that dynamic range of the spectrometer cannot follow it. This is a typical situation in plasma diagnostics. Here, AGC is an indispensable technique.

The principle of an image intensifier is outline in Section 5.1. Basically, image intensifier contains a photocathode, a micro-channel plate (MCP), and a luminescent screen (Fig. 5.29).

Being placed in the spectral plane of an ordinary spectrometer, it acts as either attenuator or amplifier of optical flux, depending on the control voltage. Without any voltage applied to control electrode, image intensifier totally blocks optical flux. Electrical pulses applied to the gating electrode enable pulsed operation. For the AGC feature, pulsed operation is irrelevant, and we shall not discuss it in this section.

The concept of AGC is explained in Fig. 5.30.

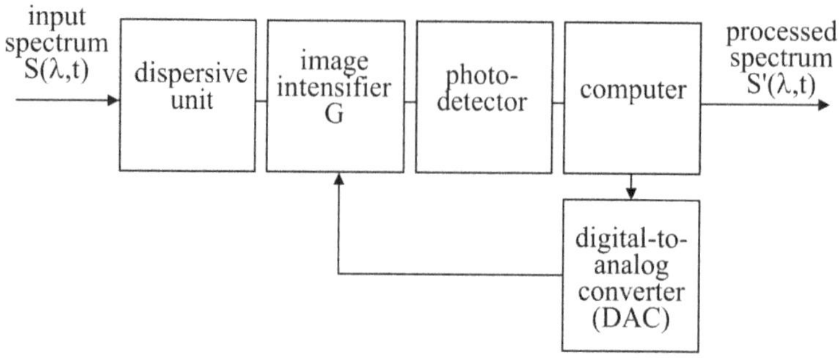

Fig. 5.30. Block-diagram of AGC.

The image intensifier, photo-detector of the spectrometer, and computer are connected in a negative feedback loop, whose purpose is to maintain $S'(\lambda_0, t) = const$. If, for any reason, spectral amplitude $S'(\lambda_0, t)$ in a processed spectrum exceeds the desired value s, set by the user, computer generates differential signal $S'(\lambda_0, t) - s$, which drives the gain of image intensifier G in such a way that to decrease this difference. How quick and accurate the response is depends on the algorithm that computes the difference.

We shall evaluate the AGC concept, using the simplest possible algorithm. The CCD, usually used in compact spectrometers, cannot generate continuous signal: it generates discrete readings $S'_n(\lambda)$ in an infinite sequence of measurements $n = 1, 2, \ldots$. Each consecutive action is separated from the previous one by the time interval needed to acquire the signal from the detector in one single measurement. The user chooses the wavelength of interest λ_0 and determines the reference spectral amplitude s for this wavelength. Then the computer generates control voltage u_{n+1} for the image intensifier as follows:

$$u_{n+1} = u_n + c \cdot [s - S'_n(\lambda_0)],$$

where c is the constant. Starting value u_0 may be any voltage within the input range of the image intensifier. In such an algorithm, as it will be shown below, both precision and speed are determined entirely by the value of parameter c.

Design of the compact spectrometer with image intensifier was reported in every detail in Section 5.4. The only modification made for implementation of AGC was the National Instruments data acquisition board PCI-6221 installed in the computer (Fig. 5.31). This board was used only as a digital-to-analog converter in order to produce analog control voltages u_n for the image intensifier. The voltage range for these signals was set from zero to +5V which matched the input voltage range of the high-voltage converter. The control algorithm was written in LabVIEW.

It is not easy to devise an experiment, convincingly showing performance of AGC. How to make it obvious to everyone that a series of processed spectra with almost equal amplitudes were created by orders of magnitude different optical fluxes? One of the solutions is to use variable tungsten halogen lamp: the bigger the current, the stronger the flux; and as the temperature increases, spectral maximum shifts to shorter wavelengths. Additionally, it is possible to compare spectra obtained with AGC spectrometer to those obtained with an ordinary spectrometer. The scheme of such an experiment is shown in Fig. 5.32. Note that optical power values, given in the figure, relate to fiber bundles composed of seven individual fibers with the core diameter 0.6 mm each, arranged in a hexagonal geometry with a central fiber. Only one central fiber relays light to spectrometers, and how much of this flux is transmitted through slits is not known. Thus, power values in Fig. 5.32 may be considered only as a variation scale, and not as the exact values, received by spectrometers.

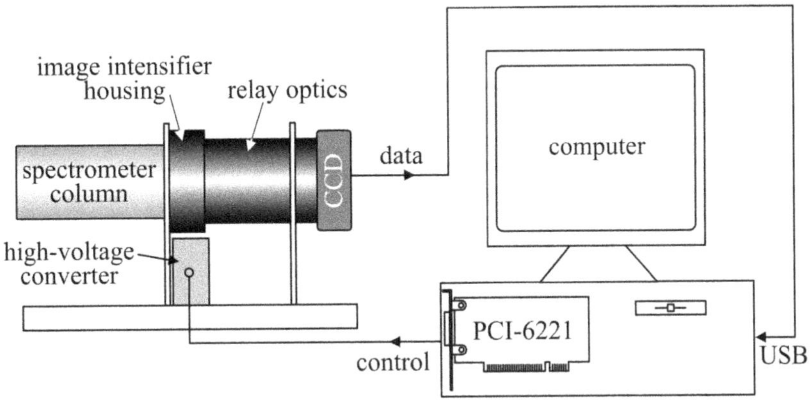

Fig. 5.31. Schematic of the system.

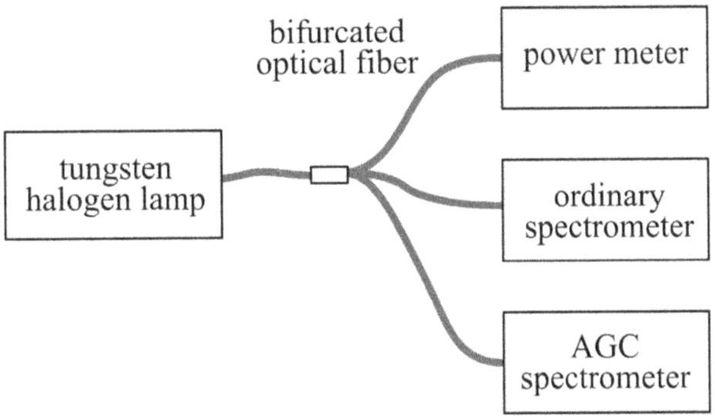

Fig. 5.32. Experimental arrangement. The power meter: ThorLabs PM100D; the ordinary spectrometer: Ocean Optics HR4000CG-UV-NIR.

The results are presented in Fig. 5.33 and are almost self-explanatory.

The lamp source was equipped with the reflector, blocking ultraviolet and infrared wings of spectrum. Therefore, the spectrum is mostly localized in visible domain between 400 nm and 700 nm. The AGC spectrometer was programmed for $\lambda_0 = 565$ nm and the reference amplitude $s = 10^4$. Therefore, it presents all the spectra drawn through this single point, no mater what the optical flux is (Fig. 5.33(a)). The ordinary spectrometer performs as expected: only two levels of optical flux at 10 μW and 100 μW can be analyzed without changing exposure time — the only way to adjust the level of the signal in the ordinary compact spectrometer.

Dynamics of AGC, controlled by the simplest algorithm described above, is fully determined by one parameter c. Transfer characteristics are presented in Fig. 5.34. In this figure, the vertical axis shows control voltage u_n that drives high-voltage converter of the image intensifier. The horizontal axis shows time elapsed since the start of the program. The starting voltage u_0 was programmed to be +5 V, giving maximum optical gain at the very first moment. As the program starts, it tends to decrease control voltage to the level, at which spectral amplitude at $\lambda_0 = 565$ nm is equal to $s = 10^4$. From Fig. 5.34 it follows that this voltage is about 1.5 V.

Great variety of more efficient algorithms is possible. However, algorithmic topic is beyond the scope of the present book.

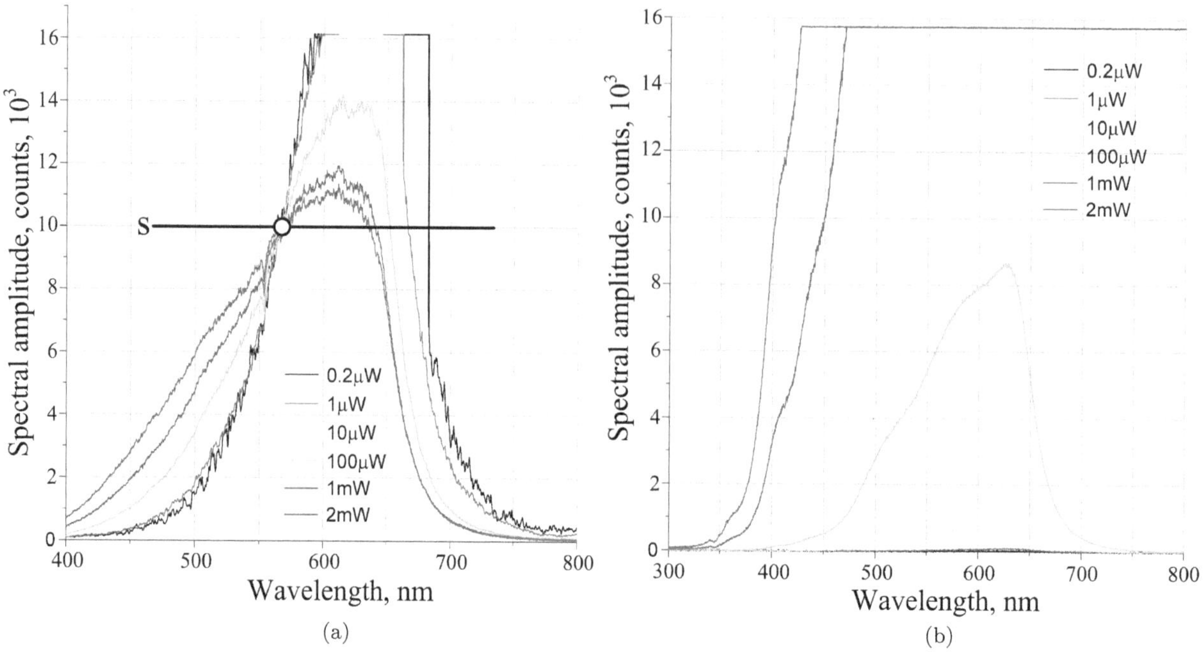

Fig. 5.33. Series of spectra from tungsten-halogen lamp obtained with (a) AGC spectrometer and (b) ordinary spectrometer. Exposure time in both the spectrometers 5 ms; stabilization wavelength $\lambda_0 = 565$ nm; reference amplitude $s = 10^4$. Optical power values, as measured by the power meter, are showed in the legends.

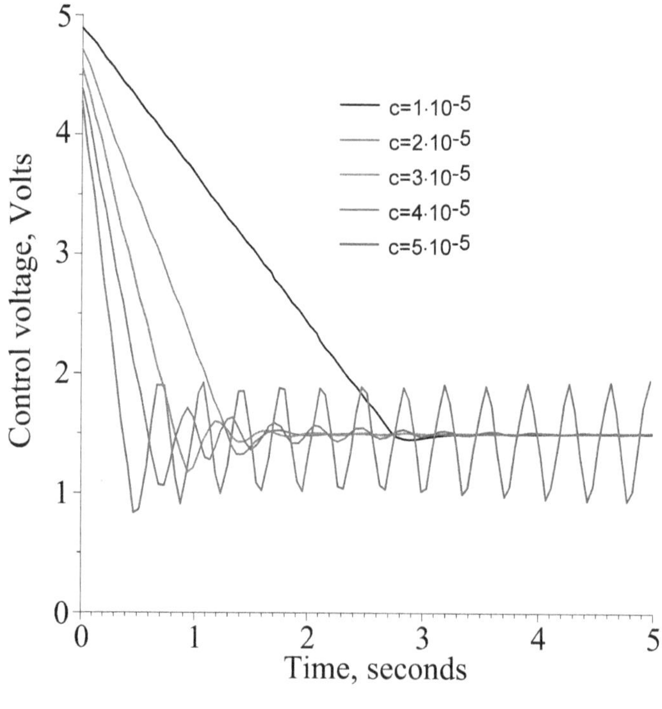

Fig. 5.34. Stabilization dynamics of AGC.

Supplemental Reading

S. Musazzi, U. Perini (eds.), *Laser-Induced Breakdown Spectroscopy: Theory and Applications*, Springer Series in Optical Sciences, vol. 182, Springer, 2014.

C.B. Johnson (ed.), *Image Intensifiers and Applications II*, Proceedings of SPIE, vol. 4128 SPIE, 2000.

C.B. Johnson, B.N. Laprade (eds.), *Electron Tubes and Image Intensifiers*, Proceedings of SPIE, vol. 1655, 1992.

Chapter 6

Modulation-Sensitive and Frequency-Selective Spectroscopy

6.1. Modulation-Sensitive Fourier-Spectroscopy

Traditional Fourier-spectroscopy deals with the spectral intensity $I(\lambda)$ only. Then, as it was explained in Chapter 2, taking into account spectral dispersion of optics $\psi(\lambda)$ and non-zero initial position of the moving mirror x_0, and summing over the entire spectrum, we may write down the photodetector signal as proportional to

$$S(x) = \int I(\lambda) \cos\left[\psi(\lambda) + \frac{4\pi}{\lambda}(x - x_0)\right] d\lambda.$$

This formula looks similar to Fourier transform of the input spectrum $I(\lambda)$ but it is not the Fourier transform because the wavelength stands in denominator.

In the current chapter, we are considering modulated optical signals. When input optical flux is modulated at frequency Ω, it can be presented without constant component as

$$I(\lambda) = I(\lambda)\cos[\Omega t + \phi(\lambda)],$$

where $\phi(\lambda)$ is the phase shift of a particular spectral component at the wavelength λ. Thus, spectral intensity $I(\lambda)$ and the phase shift $\phi(\lambda)$ are mixed together under the integral. Modulation-sensitive Fourier-spectroscopy not only can extract the modulated part $I(\lambda)$ of the optical spectrum, but also offers additional possibility of measuring phase shift $\phi(\lambda)$, which is of great importance for plasma diagnostics and biology. In biology, for instance, phase shift determines the lifetime τ of an optical transition:

$$\tau = \frac{\tan\phi}{\Omega},$$

and it can be measured, using synchronous detector. Indeed, if optical transition with the lifetime τ is excited by oscillating function $a\cos(\Omega t)$, then population of the excited state n satisfies differential equation

$$\frac{dn}{dt} = -\frac{n}{\tau} + a\cos(\Omega t).$$

The solution may be found in the general form

$$n = x\cos(\Omega t) + y\sin(\Omega t).$$

Substitution of this form into the differential equation leads to a system of algebraic equations, equalizing terms with oscillating components $\cos(\Omega t)$ and $\sin(\Omega t)$:

$$\begin{cases} x\,\Omega = y\tau^{-1}; \\ y\,\Omega = -x\tau^{-1} + a. \end{cases}$$

From here

$$x = \frac{a\tau}{\Omega^2\tau^2 + 1}; \quad y = \frac{a\Omega\tau^2}{\Omega^2\tau^2 + 1},$$

and

$$n = \frac{a\tau}{\sqrt{1 + \Omega^2\tau^2}} \cdot \cos(\Omega t - \phi); \quad \tan\phi = \Omega\tau.$$

Some mathematics below shows how $I(\lambda)$ and $\phi(\lambda)$ can be separated in the measurements.

Substitute $I(\lambda)$ into formula for $S(x)$, decompose cosine of the sum into the sum of products, group the result into sine and cosine components to obtain the following:

$$S(x) = A(x) \cdot \cos[\Omega t + \alpha(x)]$$

$$A(x) = \sqrt{\left[\int I(\lambda)\cos\phi(\lambda) \cdot \cos\Phi(x,\lambda)d\lambda\right]^2 + \left[\int I(\lambda)\sin\phi(\lambda) \cdot \cos\Phi(x,\lambda)d\lambda\right]^2},$$

$$\cos\alpha(x) = \frac{\int I(\lambda)\cos\phi(\lambda) \cdot \cos\Phi(x,\lambda)d\lambda}{A(x)},$$

$$\sin\alpha(x) = \frac{\int I(\lambda)\sin\phi(\lambda) \cdot \cos\Phi(x,\lambda)d\lambda}{A(x)},$$

$$\Phi(x,\lambda) = \psi(\lambda) + \frac{4\pi}{\lambda}(x - x_0).$$

Here $A(x)$ and $\alpha(x)$ are the amplitude and phase of high-frequency modulation of the photodetector signal. The following products are important for reconstruction:

$$A(x) \cdot \cos\alpha(x) = \int I(\lambda)\cos\phi(\lambda) \cdot \cos\left[\frac{4\pi}{\lambda}x + \theta(\lambda)\right] d\lambda,$$

$$A(x) \cdot \sin\alpha(x) = \int I(\lambda)\sin\phi(\lambda) \cdot \cos\left[\frac{4\pi}{\lambda}x + \theta(\lambda)\right] d\lambda,$$

where $\theta(\lambda) = \psi(\lambda) - \frac{4\pi}{\lambda}x_0$. Now, we can apply the complex Fourier transforms to these products, implying that physical reality requires $I(-\lambda) = 0$:

$$F_c(\omega) \equiv \int A(x) \cdot \cos\alpha(x) \cdot e^{-i\omega x}dx = \pi I\left(\frac{4\pi}{\omega}\right)\cos\phi\left(\frac{4\pi}{\omega}\right)e^{i\theta(4\pi/\omega)},$$

$$F_s(\omega) \equiv \int A(x) \cdot \sin\alpha(x) \cdot e^{-i\omega x}dx = \pi I\left(\frac{4\pi}{\omega}\right)\sin\phi\left(\frac{4\pi}{\omega}\right)e^{i\theta(4\pi/\omega)}.$$

Denote $\frac{4\pi}{\omega} \equiv \lambda$ to obtain

$$F_c\left(\frac{4\pi}{\lambda}\right) = \pi I(\lambda)\cos\phi(\lambda)e^{i\theta(\lambda)}, \quad F_s\left(\frac{4\pi}{\lambda}\right) = \pi I(\lambda)\sin\phi(\lambda)e^{i\theta(\lambda)}.$$

Although F_c and F_s are the complex variables, their ratio is real:

$$\tan\phi(\lambda) = \frac{F_s\left(\frac{4\pi}{\lambda}\right)}{F_c\left(\frac{4\pi}{\lambda}\right)}.$$

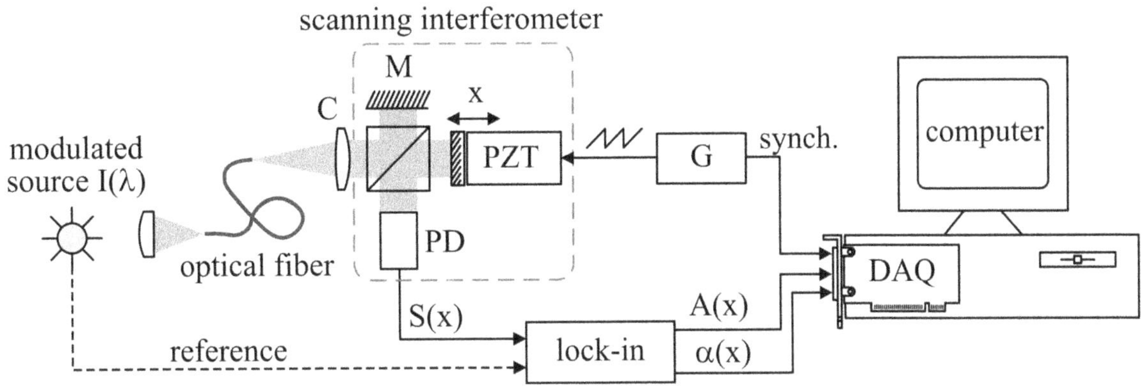

Fig. 6.1. Conceptual scheme of the modulation-sensitive Fourier spectrometer.

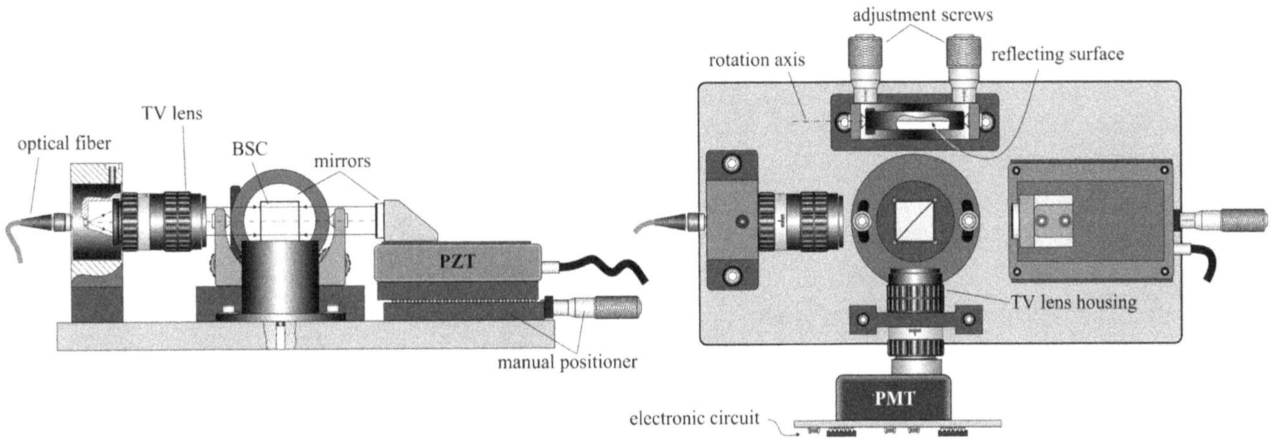

Fig. 6.2. Practical implementation of the modulation-sensitive spectrometer.

From here, the final result follows:

$$I(\lambda) = \frac{1}{\pi}\sqrt{|F_c|^2 + |F_s|^2}, \quad \phi(\lambda) = \arctan \frac{F_s\left(\frac{4\pi}{\lambda}\right)}{F_c\left(\frac{4\pi}{\lambda}\right)}.$$

Now consider how this theory works in practice. There are no commercial devices on the market, specifically designed for modulation-sensitive applications, but it is quite possible to make a laboratory prototype, and we are going to show how to do that, using readily available blocks and modules. Its concept is sketched in Fig. 6.1.

The key component is the scanning interferometer, combined of the fixed mirror (M), beam-splitting cube, and the moving mirror installed on the piezo-transducer (PZT). Collimator (C) makes parallel beam at the input of the interferometer. Saw-tooth voltage from generator (G) drives the scanning mirror linearly in one direction of x-axis and returns it quickly back. Oscillating electrical signal $S(x)$ from the photodetector (PD) comes to a synchronous detector (lock-in amplifier), which returns both the amplitude $A(x)$ and the phase $\alpha(x)$ of the electrical signal. Analog signals are digitized by data acquisition board (DAQ) installed in a computer. DAQ starts acquisition when synchronization impulse comes from the generator, marking initial position of the moving mirror. The reference signal must be available in order to implement synchronous detection.

The actual design of the modulation-sensitive prototype is clarified in Fig. 6.2.

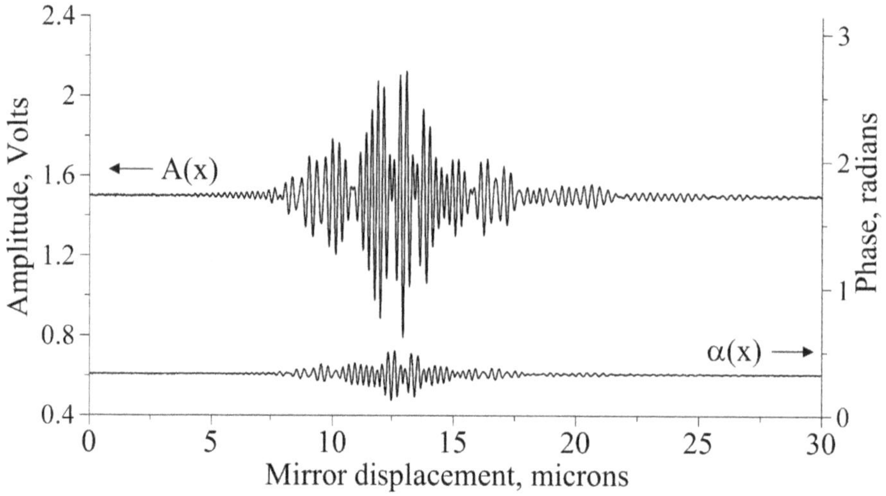

Fig. 6.3. White-light LED amplitude and phase traces at 2 MHz modulation.

Standard TV lens coupled to an optical fiber forms a collimator. TV lens always has focus adjustment ring, which makes collimation easy. Also, the iris ring on the TV lens is very useful to truncate beam diameter. Beam-splitting cube (BSC) is mounted on the rotation pedestal, enabling coarse adjustment. Final adjustment is done by the fixed mirror. For the convenience of adjustment, reflecting surface of this mirror coincides with the rotation axes: the vertical and horizontal. Fine micrometer screws are needed for precise alignment. Piezo-transducer (PZT) carries the second (non-adjustable) mirror. Its zero position must be adjusted manually by a micrometer screw. Output beams go through TV lens housing with the second iris diaphragm in it. The compact photomultiplier tube (PMT) module was used in this design. Standard TV lens with an optical fiber in its focal plane is a very comfortable tool to create collimated beam. Angular divergence α of the beam is the second parameter that determines spectral resolution $\Delta\lambda$ (see Chapter 2). For example, for $\Delta\lambda \approx 2$ nm in visible domain divergence of the beam must be smaller than 60 mrad or 3°. For comparison, an optical fiber ∅0.6 mm in the focal plane of a TV lens with focal distance 17 mm (C-mount standard) produces three times smaller divergence. The second TV lens is used at the photodetector side. In it, the lens system is completely removed, leaving only the iris with its adjustable ring and the C-mount flange. This iris serves to confine sensitive area of the photodetector, increasing contrast of fringes. The beam-splitting cube must be of non-polarizing type and meet stringent requirements on polarization. Any priorities in polarization contribute to lower contrast of fringes. Another requirement, which is never specified by manufacturers of beamsplitters, is equality of optical paths. The importance of this topic was explained in Chapter 2.

The idea of the experiment is to measure simultaneously the spectrum $I(\lambda)$ and spectral dependence of the phase $\phi(\lambda)$ of light, emitted by modulated white-light diode. Since the red fluorescence of a white-light diode presumably lags behind its blue excitation component at 460 nm, the non-trivial result may be expected. Modulation frequency was chosen to be equal to 2 MHz — not very high to preserve good modulation contrast, and not too low to obtain reasonable values of the phase. Figure 6.3 shows the amplitude $A(x)$ and phase $\alpha(x)$ signals at the two outputs of the synchronous detector (Fig. 6.1).

Mirror scans linearly $x = V \cdot t$ with constant speed $V = 30\ \mu\text{m}/2s = 15\mu\,\text{m/s}$. The left axis shows $A(x)$, the right one — $\alpha(x)$. The $A(x)$ trace also shows minor asymmetry, which means that the beamsplitter is not completely compensated.

Computed phase delay $\phi(\lambda)$ along with the restored spectral amplitude $I(\lambda)$ are shown in Fig. 6.4.

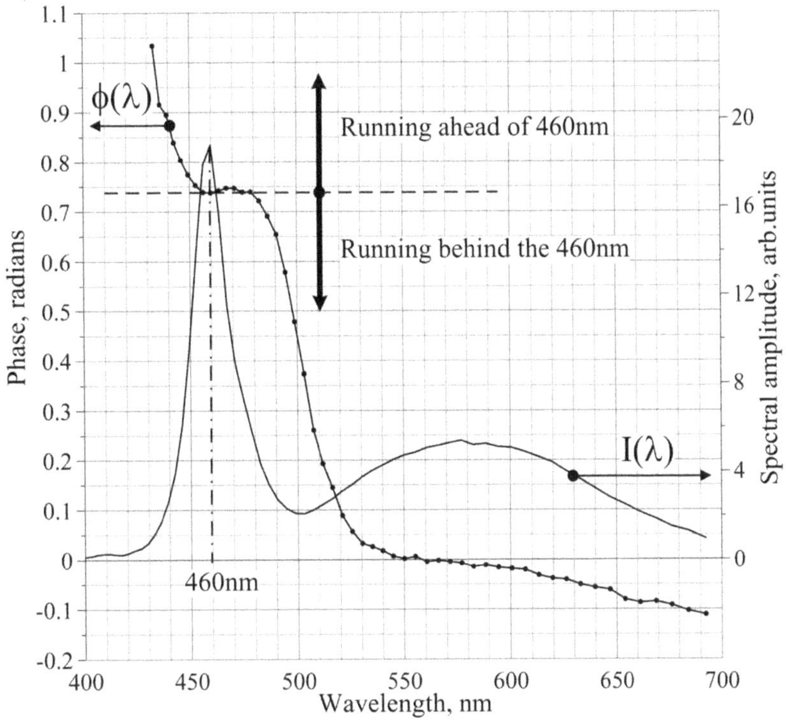

Fig. 6.4. Restored phase delay $\phi(\lambda)$ is shown in solid line with dots, representing discrete Fourier-transform nods; spectral amplitude $I(\lambda)$ is shown in a solid line.

The absolute value of the phase is of no avail for us because it is measured relative to reference electrical signal that is composed of many uncertain delays in electronic circuits and cables. We are interested in the phase relative to some fixed spectral component, for example, the maximum of blue luminescence at 460 nm because it may be considered as the pumping optical source for the red fluorescence. Figure 6.4 shows that phases of all the red spectral components are smaller relative to the phase of the 460 nm excitation, meaning that red fluorescence of the phosphor falls behind the excitation. Time delay can be computed accordingly as

$$\tau(\lambda) = \frac{\tan\phi}{\Omega} \approx \frac{\phi_\lambda - \phi_{460\,\text{nm}}}{2\pi \cdot 2 \times 10^6}[s].$$

Negative sign means lagging, positive — running ahead of the 460 nm pumping component. For example, for $\lambda = 500$ nm $\tau = -24$ ns, which means that this spectral component runs 24 ns after excitation. At $\lambda = 500$ nm, blue excitation wing overlaps with red fluorescence, which physically means that this spectral component does not represent a single damped oscillator and the decay constant phenomenology is inapplicable here. Alternatively, all the spectral components of red fluorescence beyond 550 nm are free from blue excitation contribution, which means that they may be considered as damped oscillators with definite decay constants. Figure 6.4 also shows roughly the same phase delay for all of them and the decay constant may be computed as $\tau \sim 60$ ns. Violet wing phase below 460 nm excites even before the central line at 460 nm, reacting faster to electrical signal.

6.2. Frequency-Selective Spectroscopy

6.2.1. *Principle of operation*

When no phase retrieval is needed, the technique described in the previous section significantly simplifies, and may be called the frequency-selective spectroscopy. Its conceptual scheme is presented in Fig. 6.5. The basic simplification is the absence of the synchronous detector, and, therefore, the reference signal is not needed. Instead of the synchronous detector, a simple band-pass filter tuned to the frequency of the modulated light source may be used. However, if the reference signal is available, synchronous detector provides better signal-to-noise ratio, and therefore would be a preferable solution.

Usually, intensity of the modulated component of light is low, bringing into consideration another important requirement — efficiency of collecting light. Therefore, the imaging scheme of the scanning interferometer should be the choice, as it was explained in Chapter 4. In it, a wide bundle of rays can be simultaneously processed, substantially increasing light-collecting efficiency. Such a device, equipped with a CCD camera for obtaining images, is known as the Imaging Fourier Spectrometer (IFS). When the CCD camera is replaced with a single-channel high-speed photodetector, it is possible to record only modulated component of light, filtering out the modulated electrical signal from the output of the photodetector (Fig. 6.5). The results presented below were obtained with the IFS described in Chapter 4, using the Hamamatsu photo-multiplying tube (PMT) module H6780-20 and Stanford Research Systems SR844 200 MHz synchronous detector (lock-in amplifier).

Experimental arrangement is explained in Fig. 6.6. White-light light-emitting diode (LED), powered by direct-current power supply, created constant, non-modulated optical signal. Laser diode (LD, Quantum Semiconductor International, QL65F6SC, 650 nm) was pumped by rectangular pulse train at 100 kHz, generated by a function generator (Agilent 33250A). These two optical signals were combined by a bifurcated optical fiber and alternatively (manually) connected either to standard grating spectrometer (Ocean Optics HR4000CG-UV-NIR), or to the IFS.

Thus, switching on or off any of the two light sources, it was possible to observe a variety of spectral combinations: LED alone, LD alone, LED together with LD. It should be emphasized again, that LED always produced constant and LD — modulated light. The results are summarized in a series of pictures below.

6.2.2. *Selectivity of modulated spectra*

Figure 6.7 compares spectra of combined LED plus LD light obtained by two spectrometers: the grating one and the IFS without filtering. In order to present both signals in an approximately equal scale, the two vertical axes are introduced: the left one for the grating spectrometer, and the right one for the IFS.

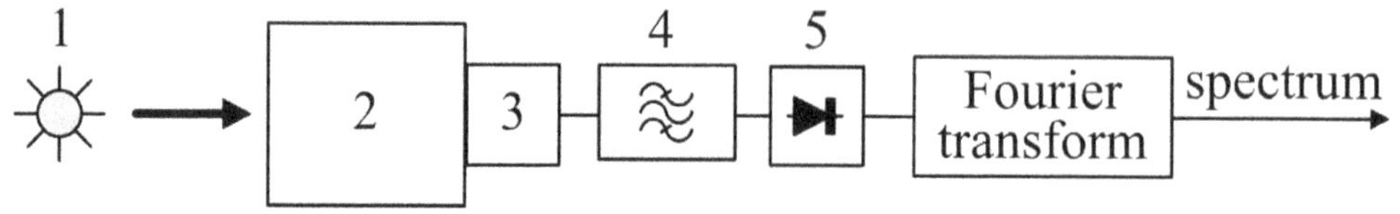

Fig. 6.5. The principle of frequency-selective spectroscopy. 1 — modulated light source; 2 — scanning interferometer; 3 — photodetector; 4 — band-pass filter; 5 — detector.

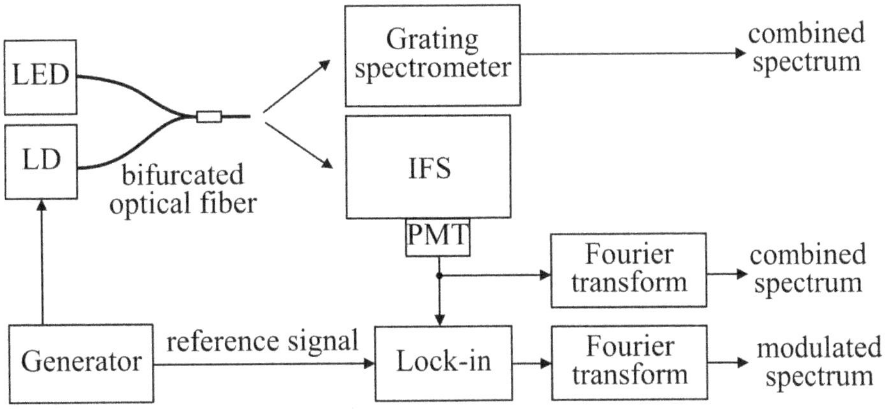

Fig. 6.6. Block-diagram of the experiment.

Although the two spectra are basically the same, the IFS shows relatively lower response in the blue part of the spectrum, which is the consolidated effect of lower spectral sensitivity of the PMT and beamsplitter transmission in this spectral interval (Table 4.3 and Fig. 4.25 in Chapter 4).

When the modulated component (LD) is switched off, the result shows a well-known spectrum of a white-light diode (Fig. 6.8).

Figure 6.9 compares shapes of the LD spectra obtained by the grating spectrometer and the IFS without filtering — they are almost identical. The only difference is much higher signal-to-noise ratio in the IFS.

Now, compare filtered spectra of the IFS with the spectra of a standard spectrometer when both non-modulated (LED) and modulated (LD) components are present (Fig. 6.10). For the grating spectrometer, the spectrum is exactly the same as in Fig. 6.7, while the IFS completely removes non-modulated component created by the LED.

The first question that comes after seeing this result is how big distortion has been introduced into the spectrum by the process of filtering? The answer is presented in Fig. 6.11: no visible distortion. The non-filtered spectrum, shown in black line, was measured when only modulated LD was turned on and the LED was turned off. In this mode, the IFS worked without lock-in amplifier (no filtering). The filtered spectrum was taken at the output of the lock-in amplifier when both non-modulated LED and modulated LD were turned on.

Finally, Fig. 6.12 compares the spectrum of the modulated LD, filtered by the IFS from the mixture of the non-modulated LED and modulated LD, and the spectrum of the modulated LD measured by the grating spectrometer with the LED being turned off. The only noticeable difference (apart of the obvious difference in signal-to-noise ratio), maybe, is a slight spectral shift. There was no such effect on the non-filtered IFS spectrum presented in Fig. 6.9.

6.2.3. *Measurement of decay times*

With PMT as a photodetector, the IFS was tested at frequencies up to 100 MHz. Such high frequency response may be used for measuring decay times of optical transitions of chemicals, which commonly lie in nanosecond region. A widely accepted technique for measuring decay times of optical transitions is pulsed excitation with subsequent measurement of spectra by means of time-resolved gated spectrometers (Chapter 5). However, there are applications where pulsed excitation is inapplicable, for instance, radio-frequency excited plasma in semiconductor manufacturing. In those cases, the method of high

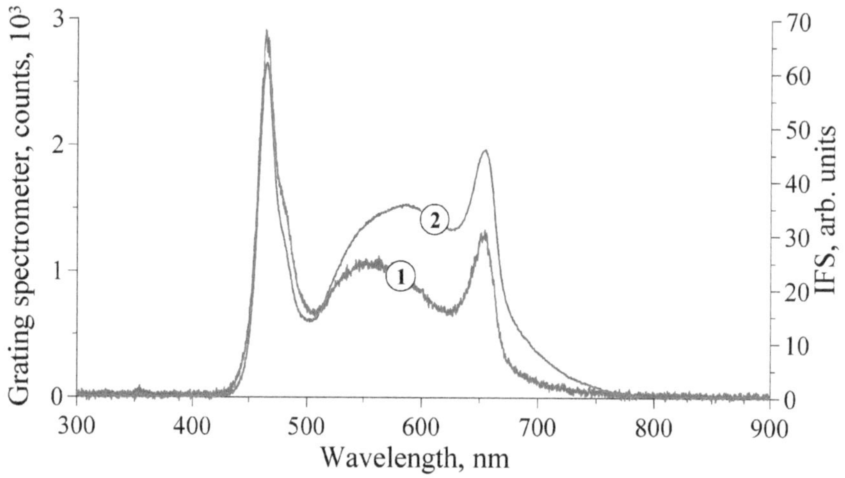

Fig. 6.7. Combined spectra of the LED and LD measured by the grating spectrometer (1) and IFS without filtering (2).

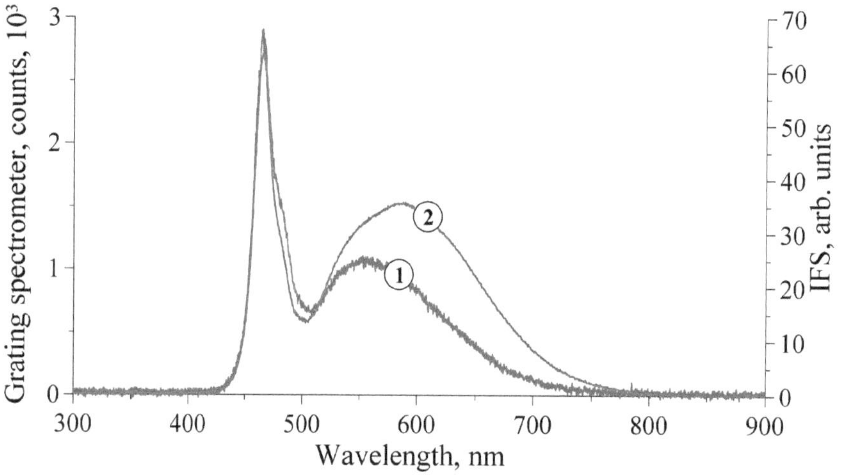

Fig. 6.8. Only constant spectra from LED. Modulated signal is switched off. (1) — grating spectrometer; (2) — IFS.

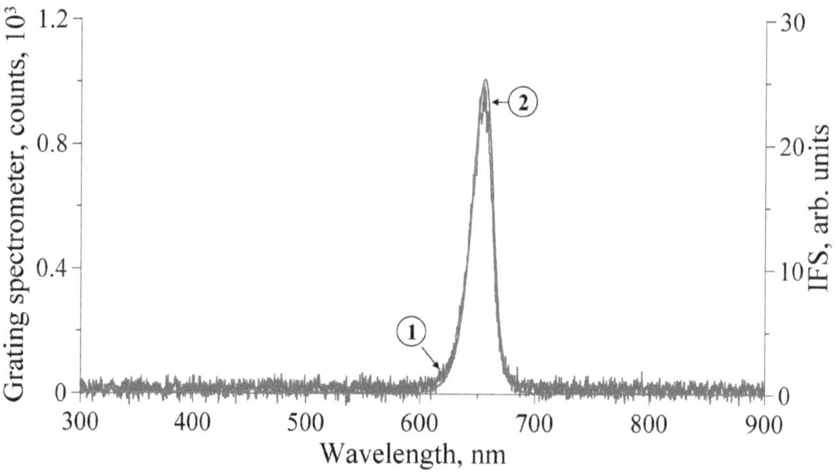

Fig. 6.9. Comparison of the LD spectra in grating spectrometer (1) and the IFS (non-filtered, (2)).

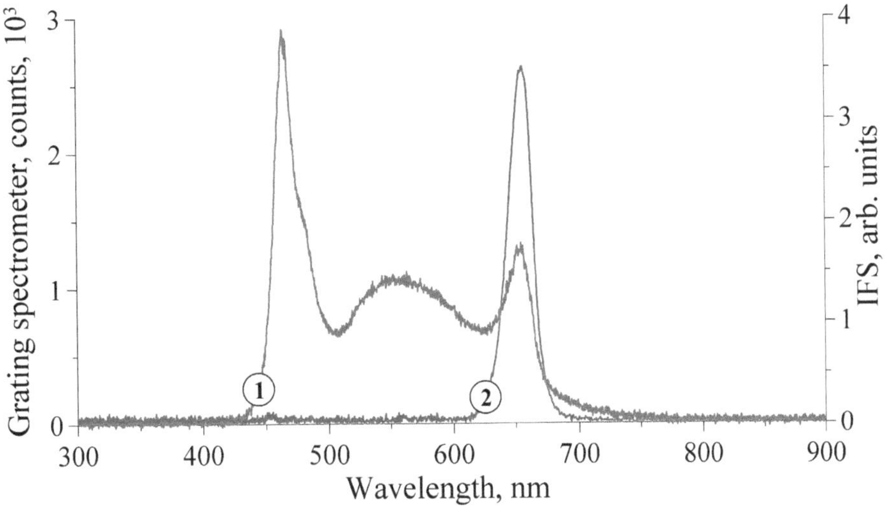

Fig. 6.10. In comparison with the grating spectrometer (1), modulated spectrum obtained from the IFS (2) shows complete removal of the non-modulated background.

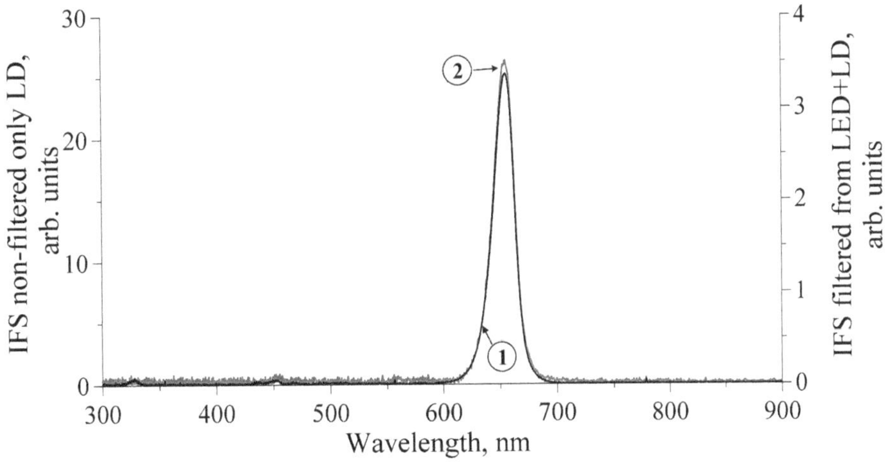

Fig. 6.11. Comparison of the non-filtered ((1), left axis) and filtered ((2), right axis) spectra of LD produced by the IFS.

frequency excitation may be used. For demonstration, decay time of a phosphor component of a white-light light-emitting diode (LED) was measured. In a white-light LED, the pumping source is a blue laser diode, emitting at approximately 450 nm. Its radiation excites the phosphor, surrounding the diode, which creates spectrally wide red component of emitting light. While frequency response of the pumping diode is fast, the phosphor decays much slower. This decay process of the phosphor may be expressed in a form of a differential equation, written for the number of excited phosphor atoms n as a function of time t:

$$\frac{dn}{dt} = -\frac{n}{T} + f(t),$$

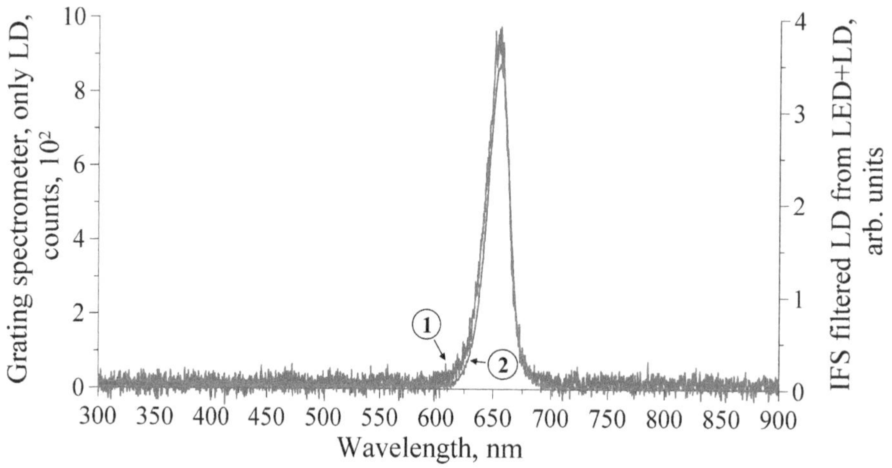

Fig. 6.12. Comparison of the filtered IFS spectrum of the modulated LD (2) with the spectrum measured by the grating spectrometer (1).

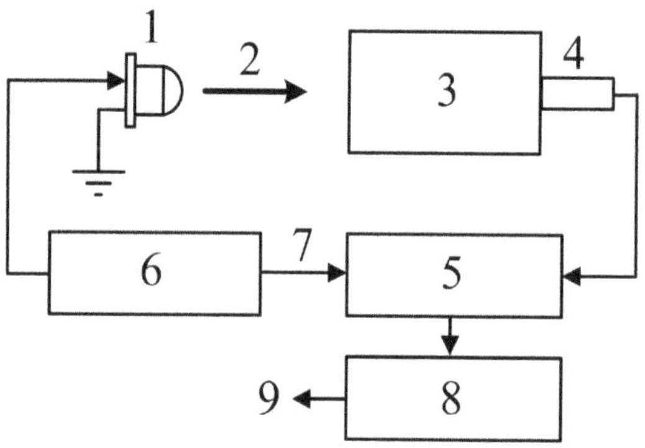

Fig. 6.13. The scheme of measuring decay time of a phosphor component of an LED. 1 — LED; 2 — light; 3 — IFS; 4 — PMT; 5 — lock-in amplifier Stanford Research Systems SR844; 6 — radio-frequency generator Agilent 33250A; 7 — reference signal; 8 — Fourier transform; 9 — spectrum.

where T is the decay time and $f(t) = w + a\cos(\omega t)$ — high-frequency excitation with the time origin chosen at zero phase. The stationary form of the general solution to this equation, i.e. when $t \to \infty$, is

$$n(t) = \frac{aT}{\sqrt{1+\omega^2T^2}}\cos(\omega t - \phi) + wT, \quad \tan\phi = \omega T.$$

From here it follows that, increasing gradually modulation frequency ω and observing the moment when the amplitude of modulation falls two times $\omega_{0.5}$, it is possible to measure decay time:

$$T = 1/\omega_{0.5}.$$

The scheme of such an experiment is shown in Fig. 6.13 and its results are shown in Fig. 6.14.

Spectral intensities of both the blue and red parts of spectra decrease with frequency in a different way: while the blue component decreases only by roughly 40% from 0.1 MHz to 6 MHz, the red wing drops sharply by a factor of 5. The physical reasons for that are also different: lower modulation efficiency of the pumping blue laser may be ascribed to radio-frequency mismatch between the generator and the semiconductor chip, while the drop of modulation in the red wing is mostly described by finite decay

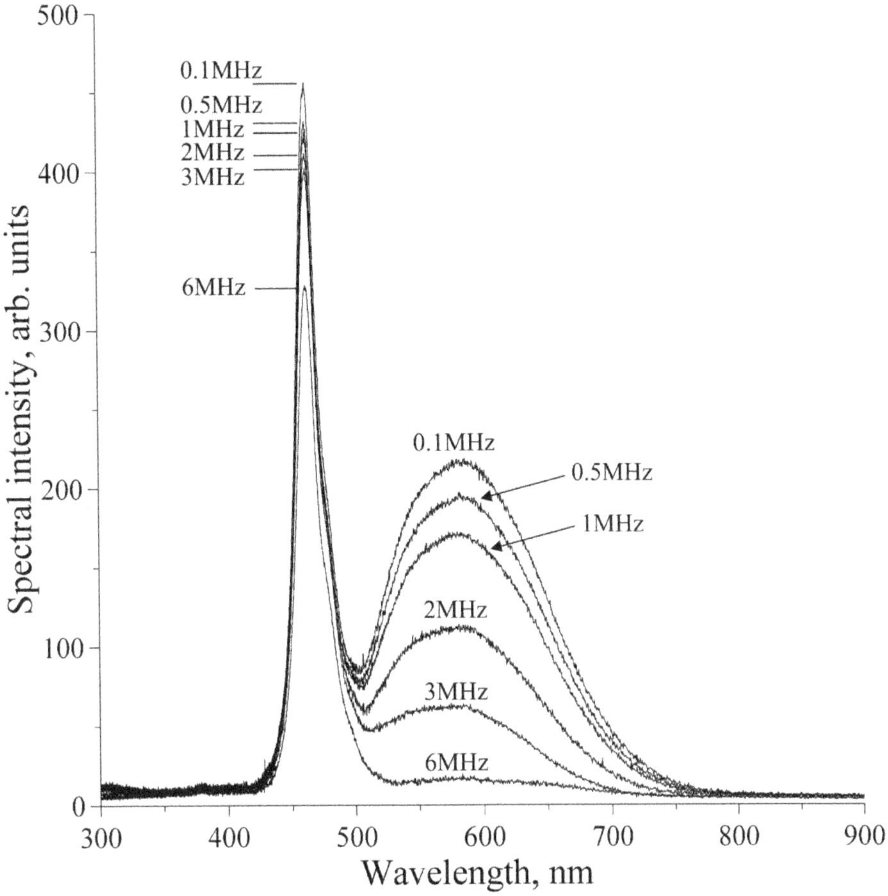

Fig. 6.14. Modulated parts of LED spectra at various frequencies.

time of the phosphor. From Fig. 6.14 it is possible to conclude that the decay time of the phosphor is roughly $T \sim (2\pi \cdot 3MHz)^{-1} \approx$ 50 ns.

The lowest frequency that is shown in Fig. 6.14 is 100kHz — the same value as in the previous experiments. However, in this experiment spectra are more noisy. The reason for that is the precaution made to avoid generation of multiple radio-frequency harmonics on the LED. Indeed, the aim of the experiment was to compare amplitudes of modulated spectra, which means that these amplitudes must not be affected by other factors, like electrical nonlinearity of the LED. Generation of higher harmonics would decrease the fundamental harmonic, thus making comparison inaccurate. Therefore, in order to obtain pure sinusoidal modulation of light on electrically nonlinear LED, the latter was excited not by rectangular pulses, like in the previous experiment, but by sinusoidal modulation on a substantial constant voltage pedestal. This pedestal, many times stronger than the modulated component of light, creates additional shot noise, visible in Fig. 6.14.

6.3. Spectroscopy of Harmonics

6.3.1. *Introduction*

Rapid advance of semiconductor industry is a driving force for research of inductively excited plasma — a tool for manufacturing electronic components by means of plasma-chemical etching. Among numerous fundamental questions of plasma physics, which were addressed and studied on plasma etching machines

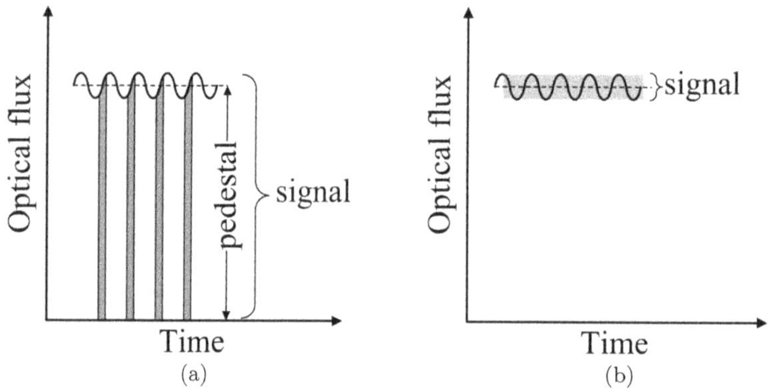

Fig. 6.15. In gated spectrometers, the signal contains both the pedestal and the modulated component (a). In modulation-sensitive spectrometers, the signal is formed only by modulated optical component of light (b).

in recent years, there is the one that was not fully resolved: whether or not the optical emission of plasma is modulated at excitation frequency, and if it is then what the modulation depth is and how the spectra of modulated components differ from the static spectrum. A series of research was done on the gases with low ionization energies, like hydrogen, oxygen, and krypton. It was shown that modulation depth of optical emission of these gases may reach 20% at various excitation frequencies, ranging from 2 MHz to 13.56 MHz, thus making measurements possible with traditional time-resolved spectroscopy. In this technique, optical emission is sensed during short pulses, locked in phase with excitation or modulation frequency of plasma. Changing this phase, it is possible to study optical spectra of plasma emission as a function of phase, obtaining versatile information about plasma dynamics. Referencing optical spectra to the phase of plasma modulation gave birth to another name of the same technique — the phase-resolved spectroscopy.

Based on gated spectrometers (Chapter 5), accurate, and versatile tool for plasma diagnostics, the phase-resolved spectroscopy cannot be applied to weakly modulated plasma phenomena because strong constant background (pedestal) suppresses weak modulated component (Fig. 6.15(a)).

The modulation-sensitive spectroscopy — the subject of this chapter — excludes constant background from measurements (Fig. 6.15(b)). In this technique, the Fourier-transform spectrometer with narrow-pass radio-frequency filter at its output is used to select solely the modulated component of the spectrum. The modulation-sensitive spectroscopy is capable of measuring modulated component of spectra with coefficients of modulation below 10^{-4} — the values completely unattainable for ordinary phase-resolved spectrometers. The current section describes how to measure modulation of optical emission spectra of noble gases in inductively-coupled plasma at excitation frequency 13.56 MHz and its harmonics.

6.3.2. *Qualitative theory*

In inductively coupled plasma, a circular coil creates magnetic field **H**, directed normally to the coil and oscillating with radio frequency, which creates the orthogonal electric field **E** (Fig. 6.16).

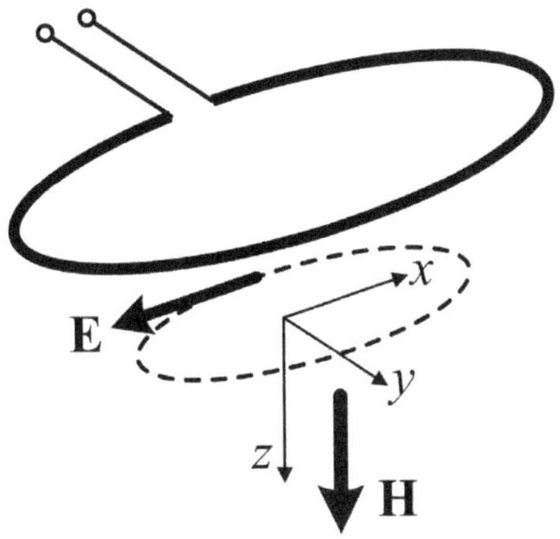

Fig. 6.16. System of coordinates.

In simplified qualitative theory, z-component of the electric field may be assumed zero. Between collisions, the equations for the components v_x, v_y, v_z of the velocity of an electron are

$$\begin{cases} \dfrac{dv_x}{dt} = \dfrac{q}{m}E(t), \\ \dfrac{dv_y}{dt} = 0, \\ \dfrac{dv_z}{dt} = 0, \end{cases}$$

where q and m are the charge and mass of the electron. The energy of the electron is

$$\varepsilon = \frac{m}{2}(v_x^2 + v_x^2 + v_x^2).$$

The electric field oscillates with angular frequency ω independently of the motion of the electron:

$$E(t) = A\cos(\omega t).$$

Here the time origin is chosen at zero phase. If the electron collided the last time at $t = t_0$, then the system of equations for the velocity components can be easily solved:

$$\begin{cases} v_x(t) = v_{x0} + \dfrac{qA}{m}\displaystyle\int_{t_0}^{t}\cos(\omega t)dt = v_{x0} + \dfrac{qA}{m\omega}(\sin\omega t - \sin\omega t_0), \\ v_y = v_{y0}, \\ v_z = v_{z0}. \end{cases}$$

This solution is valid until the next collision of the electron.

Evolution of the electron energy $\varepsilon(t)$ as a function of time t can be found, substituting formulas for velocities. We do not know, how long time will pass until the next collision, therefore, it is necessary to average $\varepsilon(t)$ over all possible inter-collision intervals $\tau = t - t_0$:

$$\langle\varepsilon(t)\rangle_\tau = \frac{m}{2}\left(\left\langle v_x^2(t)\right\rangle_\tau + v_{y0}^2 + v_{z0}^2\right).$$

Here the angle brackets denote averaging over this parameter. Since the real probability density $p_\tau(\tau)$ of this parameter is never known, it is logical to estimate it from the probability density $p_v(v)$ of the electron velocity v, assuming it to be Maxwellian:

$$p_v(v) = \frac{4\alpha^{3/2}}{\sqrt{\pi}}v^2e^{-\alpha v^2}.$$

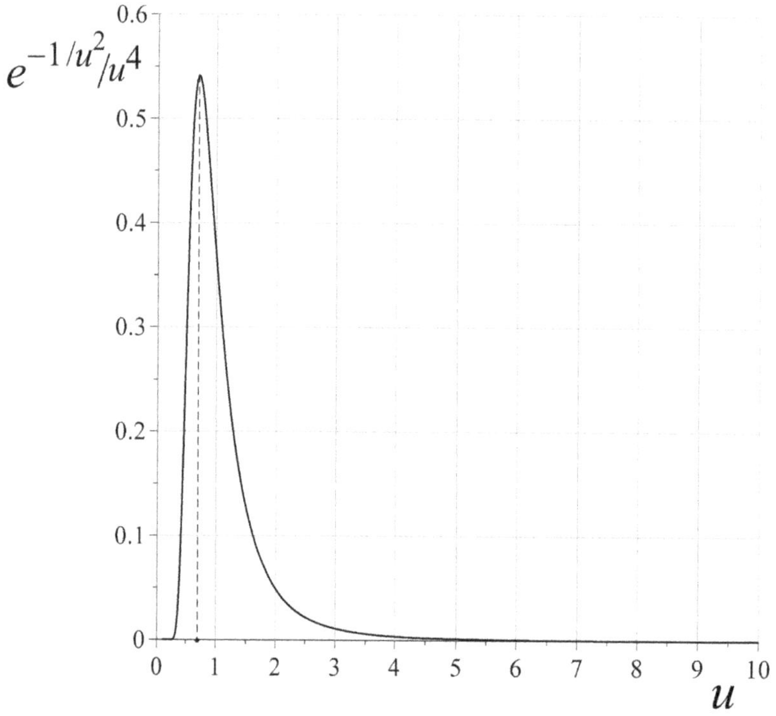

Fig. 6.17. Probability density of the inter-collision interval τ as a function of the parameter u. The maximum probability is reached at $u_m = 2^{-1/2}$.

Physically, there must be inverse proportionality between τ and v

$$\tau = \frac{l}{v}$$

with the proportionality coefficient l having the meaning of a free electron path. On this basis, it is possible to compose an equation, connecting the unknown $p_\tau(\tau)$ with the known $p_v(v)$, using equality of the probability function $P(\tau)$ computed over the two variables τ and v:

$$P(\tau) = \int_0^\tau p_\tau(\tau)d\tau = \int_{+\infty}^{v(\tau)} p_v(v)dv.$$

From here

$$p_\tau(\tau) = \frac{dP(\tau)}{d\tau} = \frac{dv}{d\tau} \cdot p_v[v(\tau)] = \frac{4}{l\sqrt{\pi\alpha}} \cdot \frac{1}{u^4}e^{-1/u^2}, \quad u \equiv \frac{\tau}{l\sqrt{\alpha}}.$$

This function is shown in Fig. 6.17. Its narrowness greatly simplifies calculations because the averaging over τ may be approximated as setting τ equal to its most probable value

$$\tau_m = l\sqrt{\frac{\alpha}{2}},$$

and dropping the index "m" for conciseness.

Additional simplification comes from the fact, that in practice, at the standard excitation frequency 13.56 MHz and pressures 10 mTorr or higher, $\omega\tau < 1$, so that $\sin\omega\tau \approx \omega\tau$. As a result,

$$\langle\varepsilon(t)\rangle_\tau \approx \varepsilon_0 + \frac{(qA\tau)^2}{4m} + \sqrt{\frac{2\varepsilon_0}{3m}} \cdot qA\tau \cdot \cos(\omega t - \phi) + \frac{(qA\tau)^2}{4m} \cdot \cos(2\omega t - \gamma).$$

with ϕ and γ being unimportant phases. Here we used

$$\frac{mv_{x0}^2}{2} = \frac{1}{3}\varepsilon_0, \quad v_{x0} = \sqrt{\frac{2\varepsilon_0}{3m}}.$$

The formula for $\langle\varepsilon(t)\rangle_\tau$ has clear physical meaning: between two consecutive collisions, an electron acquires additional energy (heating — the second term) and its energy becomes modulated with the first (the third term) and second harmonics of excitation frequency (the last term). These three additional components of the electron density tend to zero when the average inter-collision time τ tends to zero.

When long time passes after plasma ignition, the heating term becomes compensated by dissipative processes, and the stationary form of electron energy may be written as

$$\varepsilon(t) = a + b\cos(\omega t - \phi) + c\cos(2\omega t - \gamma),$$

where coefficients a, b, and c depend on the electron charge and mass, the strength of electric field, and oscillating frequency.

Electrons in plasma excite optical emission of atoms with optical flux being proportional to the number of excited transitions n. Consider a single spectral line, corresponding to the transition with the decay time T. This spectral line may be excited by electrons with various energies ε, and the efficiency of excitation depends on ε. Denoting the excitation rate $f(\varepsilon)$, the differential equation for n takes the form

$$\frac{dn}{dt} = \int f(\varepsilon)d\varepsilon - \frac{n}{T}.$$

Accordingly, the energy of electrons ε oscillates with time. Therefore, the integral in the right-hand side of this formula may be written as a time-dependent function

$$F(t) = \int f[\varepsilon(t)]d\varepsilon.$$

Such an equation may be solved analytically:

$$n(t) = e^{-t/T} \cdot \int e^{t/T} F(t)dt.$$

Here parameters b and c are small, comparing to a. Therefore, $F(t)$ may be expanded in McLaurin series over powers of the sum $b\cos(\omega t - \phi) + c\cos(2\omega t - \gamma)$. These powers will create cross-products that, eventually, create the entire spectrum of harmonics $\cos(m\omega t - \phi_m)$, $m = 1, 2, 3, \ldots$. Thus,

$$F(t) = \sum_{m=0}^{\infty} C_m \cos(m\omega t - \phi_m).$$

Substitution of $F(t)$ into $n(t)$ gives time dependence of the optical flux:

$$n(t) = n_0 + T\sum_{m=1}^{\infty} \frac{C_m}{\sqrt{1 + m^2\omega^2T^2}} \cos(m\omega t - \psi_m), \quad \psi_m = \phi_m + \arctan(m\omega T).$$

The first conclusion that follows from this formula is that all harmonics of the excitation frequency ω are expectable in the optical flux of plasma emission. The second conclusion is that it is almost impossible to predict which particular harmonic will dominate: coefficients C_m depend on numerous physical parameters, such as average electron energy ε_0, inter-collision time interval τ, excitation efficiency and decay time T of a particular spectral line, and others. Such uncertainty of the above qualitative theory and the absence of the quantitative theory, makes experimental study of the phenomenon even more interesting and challenging.

6.3.3. *Experimental installation and measurement*

The experiments were made at a standard etching chamber that is routinely used in production lines for manufacturing electronic devices on 300 mm silicon wafers. Its schematic diagram with basic dimensions is shown in Fig. 6.18.

Electric field under the antenna decreases exponentially towards the wafer. Therefore, in order to measure modulation of optical emission at different distances from the antenna, a periscope, consisting of two 45° parallel elliptical glass mirrors (ThorLabs BBE1-E02) fixed inside the dielectric body, was installed inside the chamber. With it, the observation axis passed 160 mm above the wafer, sensing the areas of plasma with strong electric field. The mirrors have dielectric coating. With the periscope removed, the viewing axis passed 15 mm above the wafer, sensing the areas of plasma with weaker electric field.

Three types of gases were available for the experiments: argon, neon, and nitrogen. The chamber was pumped out by a turbomolecular pump to a base pressure 0.2 mTorr, after which the gases could be added in a range of pressures between 1 and 100 mTorr. A typical range of electric powers that could sustain plasma discharge at these pressures was 50–500 W.

The scheme of measurements is outline in Fig. 6.19. Its key component is the modulation-sensitive spectrometer 6, which is portrayed in Fig. 6.20. Basically, it is a Fourier spectrometer, designed for visible domain and described in detail in Chapter 4. Its dispersive element is the Michelson interferometer with two corner cube reflectors.

When light is modulated at high frequency, the photodetector signal is combined of two independent components: the so-called static part, containing information only about non-modulated component, and high-frequency part with information about modulated component. These two components can be effectively separated, using the high-frequency lock-in amplifier 7 (Stanford Research Systems, SR844, 200 MHz upper limit). The lock-in amplifier is actually a synchronous detector, requiring reference signal that is strictly in phase with the modulated signal. This reference signal is created by the function generator 8 (Keysight 81150A, 120 MHz upper limit) that has additional capability of generating phase-locked harmonics of the fundamental harmonic 13.56 MHz at its second output, which is connected to the lock-in amplifier 7. The fundamental harmonic from its first output is used to synchronize the power generator 1 that delivers high-power electrical signal to the antenna on the chamber 2. Thus, modulation of optical emission of plasma and reference electrical signal at the input of the lock-in amplifier 6 are

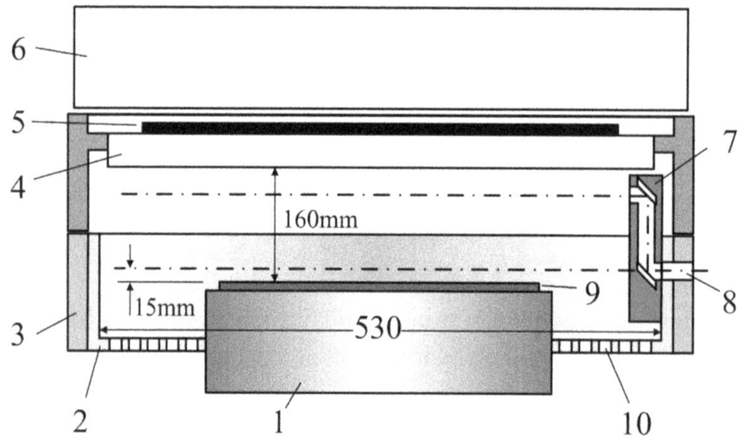

Fig. 6.18. Schematic diagram of 13.56 MHz inductively coupled plasma chamber. 1 — pedestal; 2 — liner; 3 — lower wall; 4 — quartz plate; 5 — antenna; 6 — matcher; 7 — periscope; 8 — viewport; 9 — wafer; 10 — pumping canals.

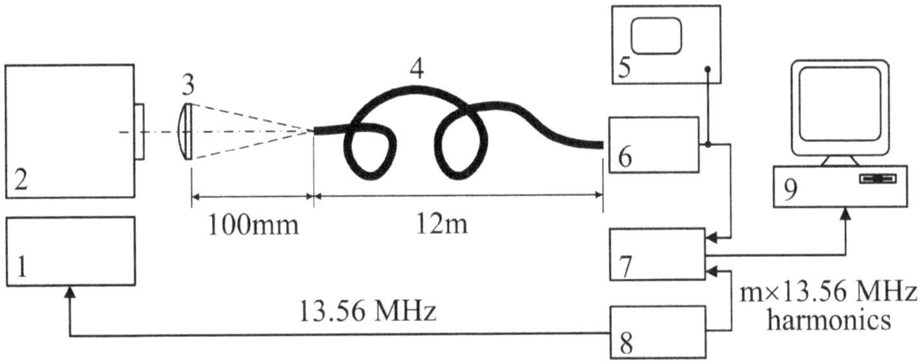

Fig. 6.19. Generalized scheme of measurements. 1 — power generator; 2 — the chamber; 3 — collecting lens; 4 — optical fiber bundle; 5 — oscilloscope; 6 — modulation-sensitive spectrometer; 7 — lock-in amplifier; 8 — function generator; 9 — computer.

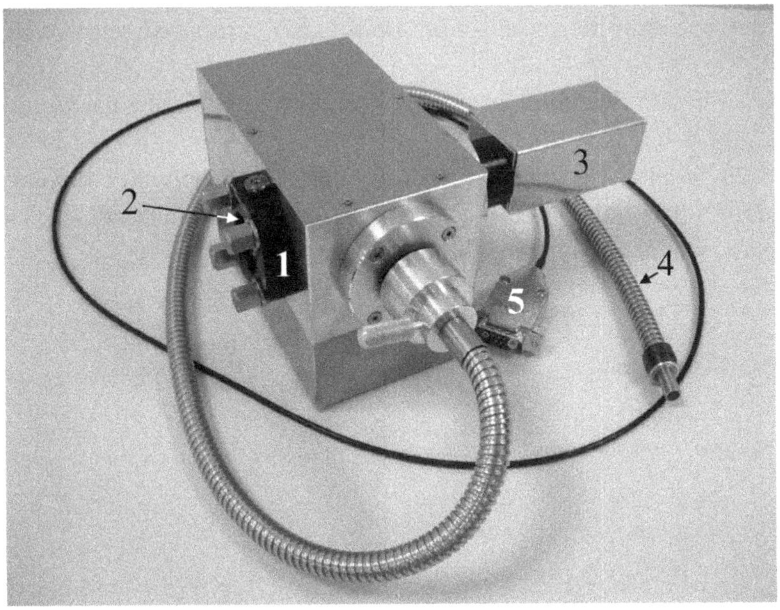

Fig. 6.20. Modulation-sensitive spectrometer. 1 — lateral aligner; 2 — corner-cube reflector; 3 — photodetector; 4 — optical fiber; 5 — connector for the piezo-stage.

always in phase, making efficient filtering of electrical signal from the spectrometer 6. The output signal of the lock-in amplifier 7 is a slowly varying analog electrical signal, containing information about spectrum of only one modulated optical component, namely the one that is referenced by the signal from the function generator 8. This slowly varying analog electrical signal is converted into digital form and reconstructed digitally in the computer 9 to obtain the spectrum. Thus, choosing the harmonic number at the second output of the function generator 8, it is possible to measure optical spectra of any harmonic of the light, emitted from the chamber 2.

In order to ensure good collection of light, coming out of the chamber, the one-inch plano-convex lens 3 (ThorLabs LA4380) was coupled to the optical fiber bundle 4. Optical diameter of the fiber bundle was 8 mm (customized product from ThorLabs).

High-quality spectral measurements could be obtained only with averaging of many consecutive spectra. Each individual spectral measurement required 3 seconds to complete. Although the total number of averaged spectra was not limited by technical reasons, not more than 100 accumulations were usually made. Thus, the total measurement time was 5 minutes — the interval, during which plasma did not change noticeably, owing to excellent stability of the chamber.

6.3.4. *Experimental results*

The entire amount of experimental data obtained on Ar and He at various pressures from 1 mT to 500 mT, and various powers from 20 W to 500 W is summarized below as typical spectra of the four first harmonics within spectral interval 400–900 nm, and the effect of chamber conditions.

Typical spectra are presented in Figs. 6.21–6.29 for Ar and He for the same experimental conditions: 10 mT pressure and 200 W power. The term "static" used in captions below refers to an ordinary function of the spectrometer, when it measures spectrum of the sum of non-modulated and modulated components. Values of the signal, plotted along the vertical axes as "arbitrary units", have different scales for the static and harmonic spectra: gains of amplifiers in these two modes of operation are different. Nonetheless, amplitudes of harmonics may be roughly compared relative to the noise level in each measurement.

The second harmonic of helium was its last visible harmonic. All higher harmonics were not visible on the level of noise, as shown in Fig. 6.29.

From Figs. 6.21 to 6.29, it may be seen that spectra of some harmonics display more noise than the others. The reason for that is not understood yet, and may be associated with several factors like high-frequency interference within the lock-in amplifier or plasma noise.

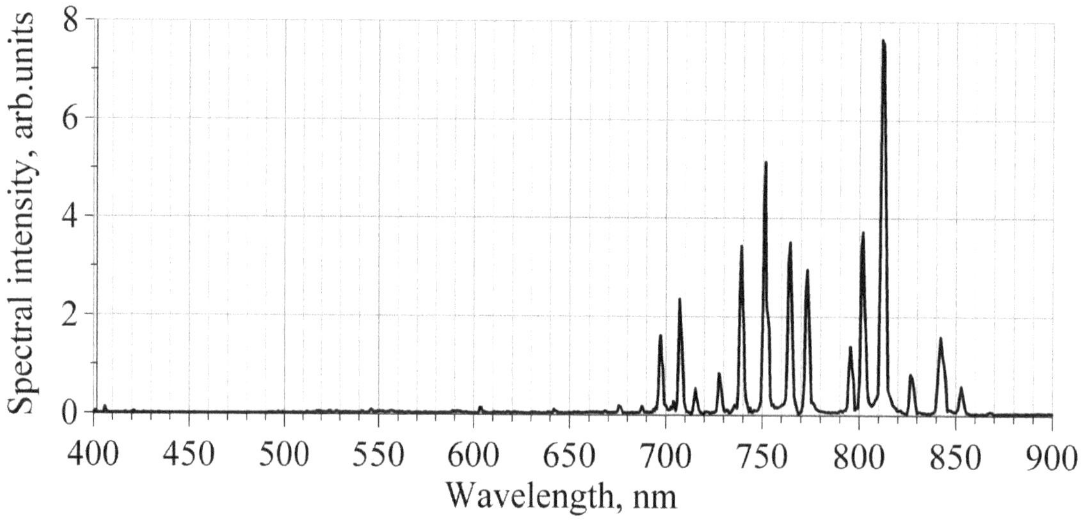

Fig. 6.21. Ar static spectrum.

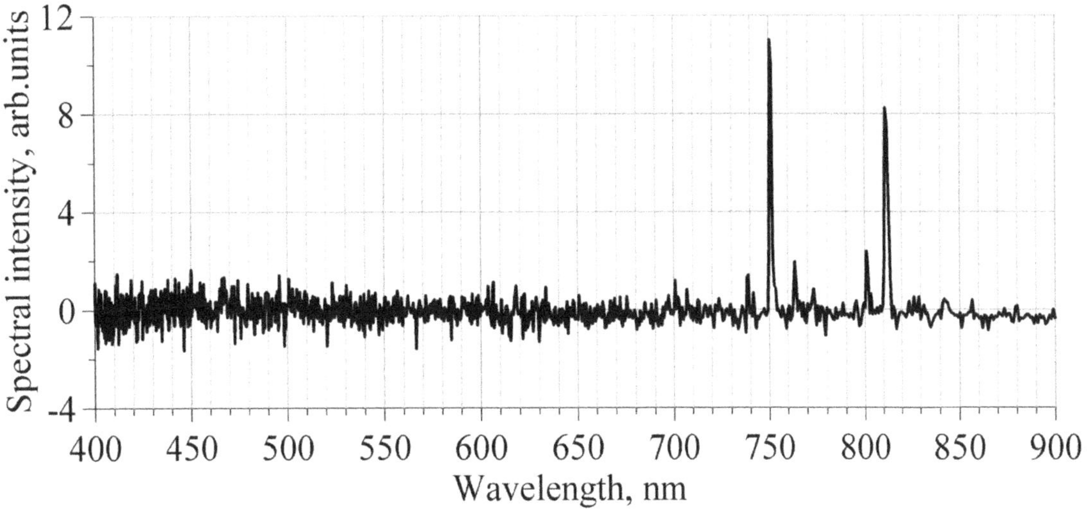

Fig. 6.22. Ar 1st harmonic 13.56 MHz.

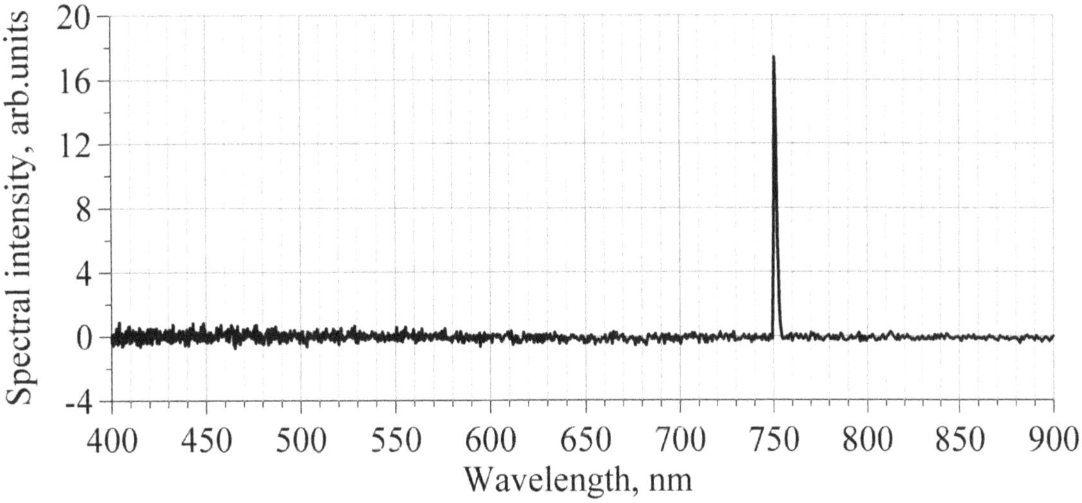

Fig. 6.23. Ar 2nd harmonic 27.12 MHz.

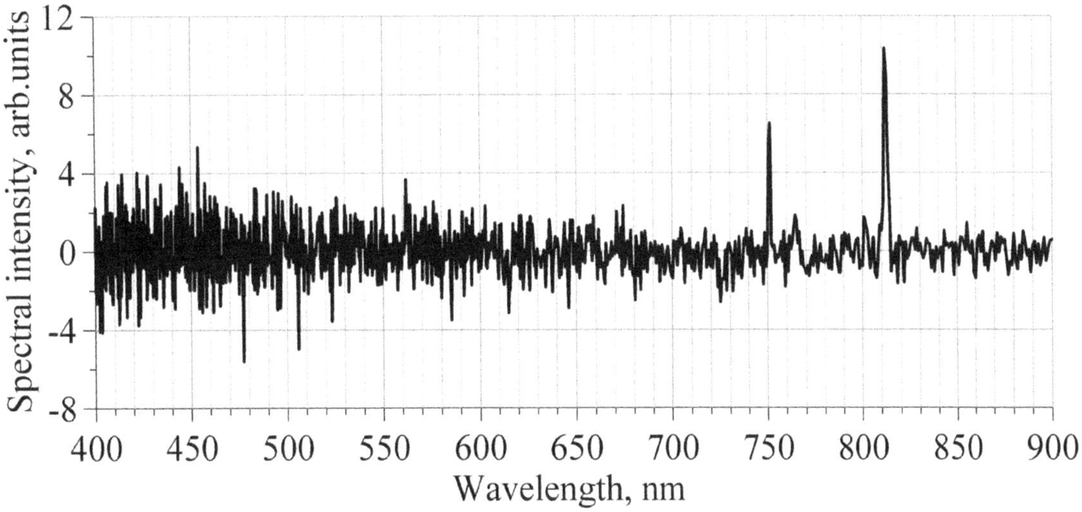

Fig. 6.24. Ar 3rd harmonic 40.68 MHz.

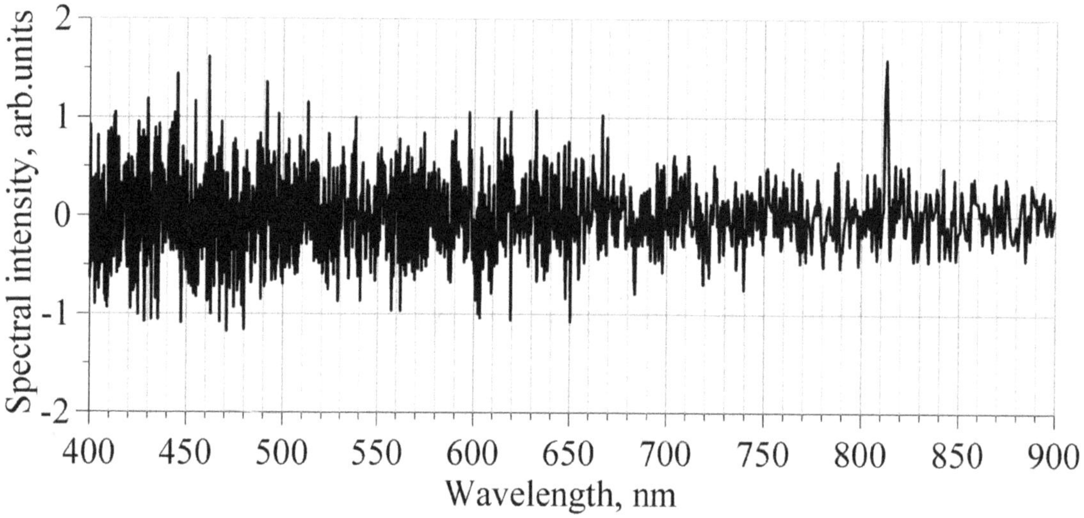

Fig. 6.25. Ar 4th harmonic 54.24 MHz.

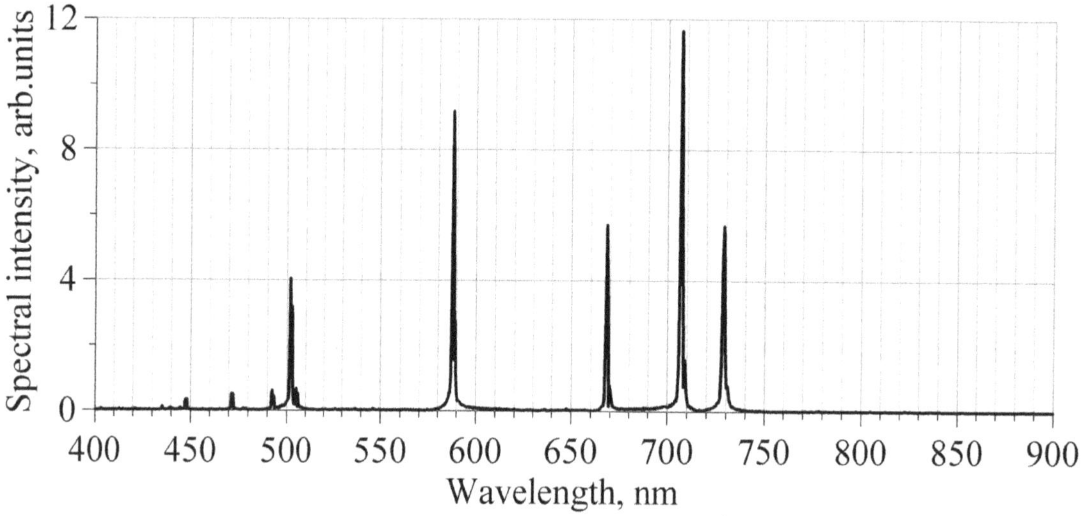

Fig. 6.26. He static spectrum.

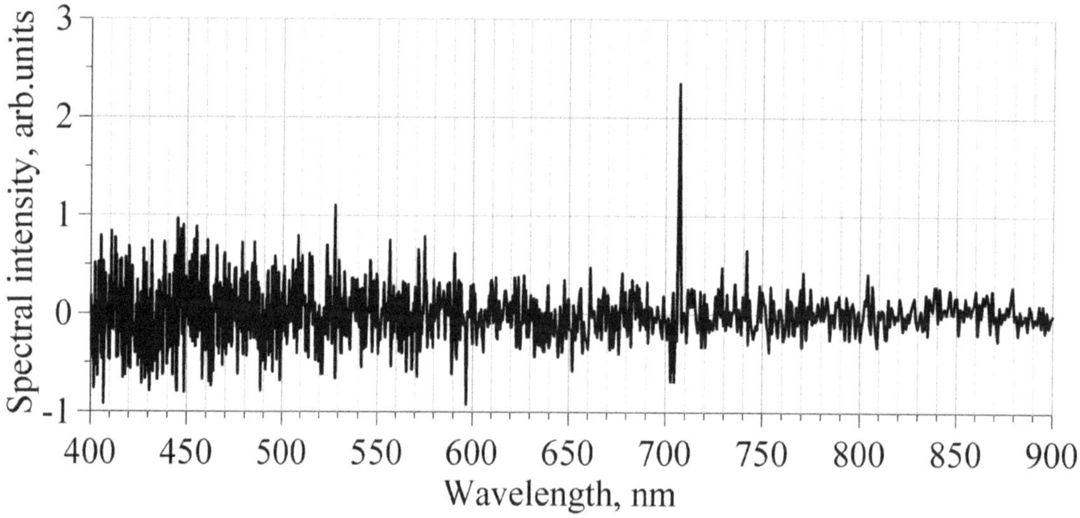

Fig. 6.27. He 1st harmonic 13.56 MHz.

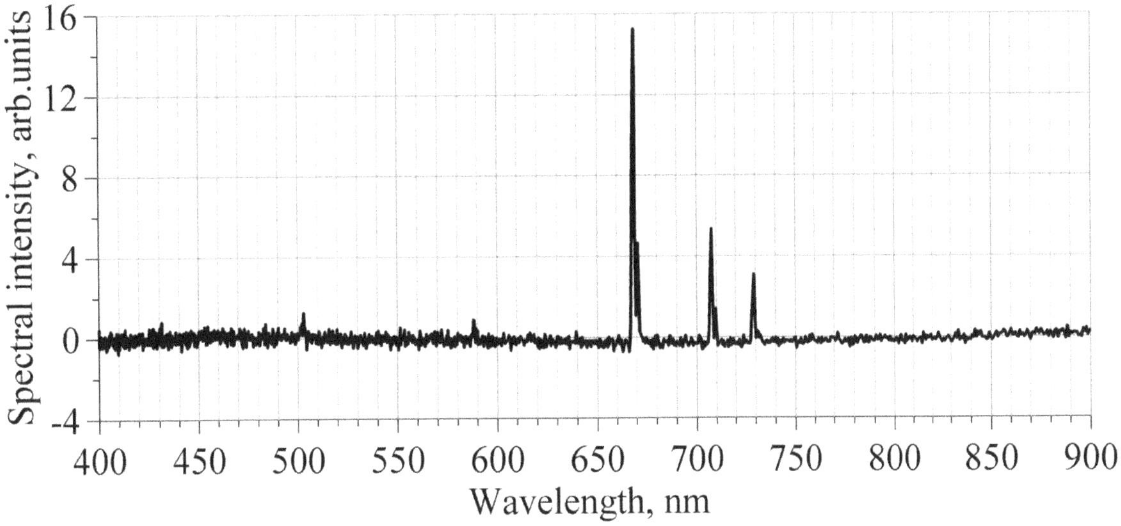

Fig. 6.28. He 2nd harmonic 27.12 MHz.

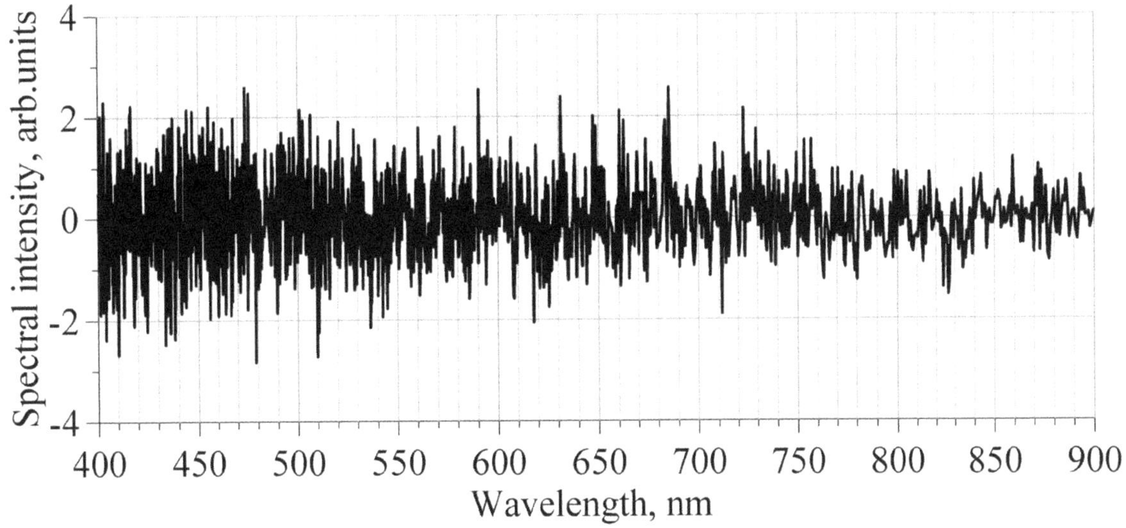

Fig. 6.29. He 3rd harmonic 40.68 MHz.

6.3.5. *Effect of chamber condition*

Spectra of harmonics may provide useful information about status of the plasma chamber — an overall electrodynamic condition that determines quality of etching. In semiconductor manufacturing, progressive deterioration of electrodynamic conditions of chambers is known as chamber ageing. Monitoring spectra of harmonics may be used to detect ageing on early stages because condition of the chamber strongly affects spectra of harmonics. For example, Figs. 6.30 and 6.31 show the spectra of the first and the second harmonics of argon and helium before and after opening the chamber.

6.3.6. *Measurement of modulation depth*

The question that was not answered in the previous sections is what is the modulation depth of optical harmonics? The modulation depth is the ratio of oscillating intensity to the average intensity of the

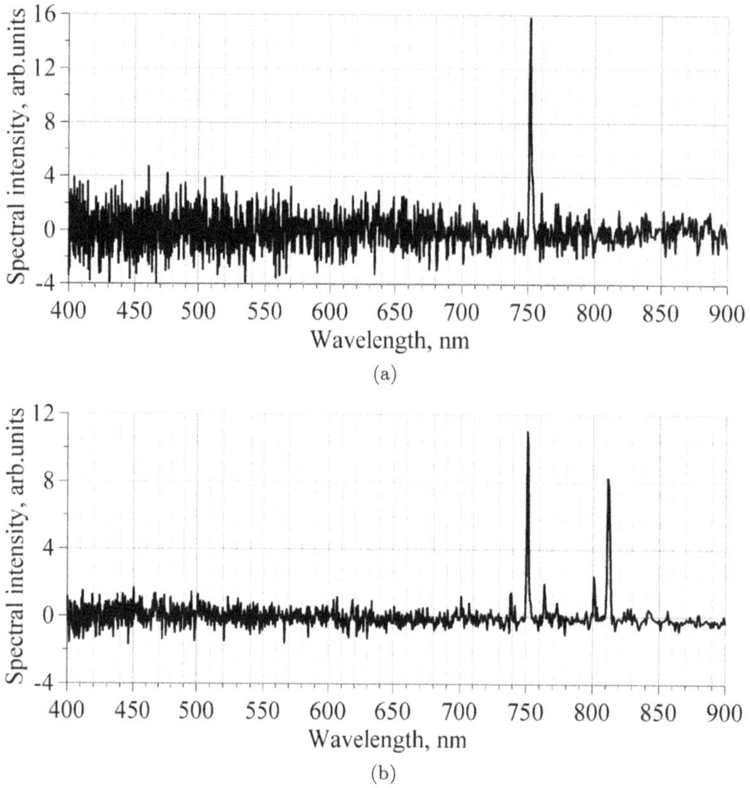

Fig. 6.30. Two Ar spectra of the 1st harmonics before (a) and after (b) opening the chamber.

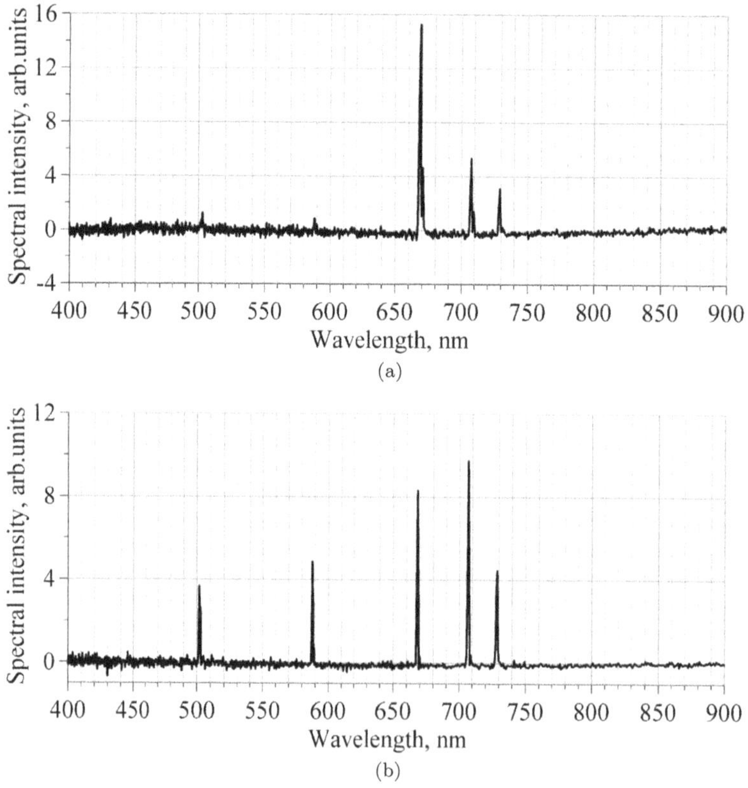

Fig. 6.31. Two He spectra of the 2nd harmonics before (a) and after (b) opening the chamber.

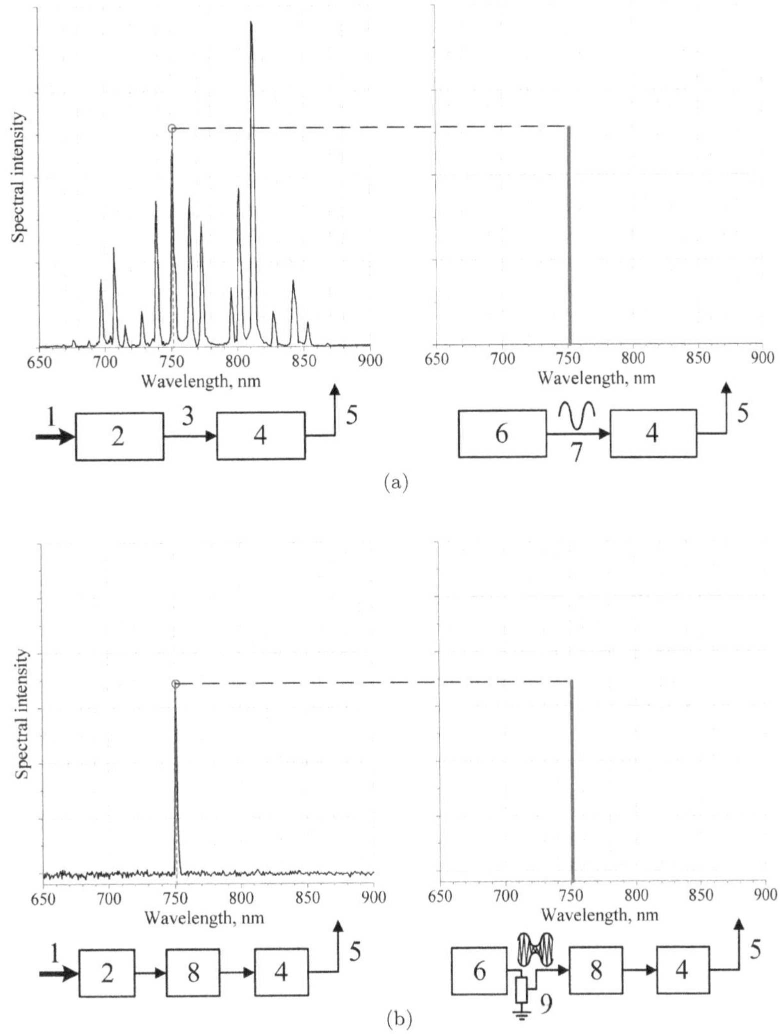

Fig. 6.32. 1 — optical signal; 2 — spectrometer; 3 — electrical signal; 4 — Fourier transform; 5 — spectrum; 6 — function generator; 7 — electrical signal at frequency ω_0; 8 — lock-in amplifier; 9 — variable attenuator. The left panels of this figure show optical measurements, and the right ones — their electrical simulation.

specific optical emission line. For measuring the values of modulation depth below 10^{-4}, a very sensitive two-stage technique may be implemented (Fig. 6.32).

In order to understand it, it must be explained that every single spectral line develops sinusoidal oscillations of light at the output of the scanning Michelson interferometer, depending on the speed of scanning. The frequency of oscillations is a fingerprint of a particular spectral line and scanning speed. Specifically, for the spectrometer used in this work, the spectral line at 750 nm developed oscillations at 355.9 Hz. Therefore, low-frequency electrical signal from a function generator at exactly $\omega_0 = 355.9$ Hz simulates the spectral line 750 nm, measured by the Fourier spectrometer. At the first stage of measurements, the amplitude of this low-frequency electrical signal is adjusted to equalize the amplitude of any chosen spectral line in the static (not modulation-sensitive) mode of the spectrometer (Fig. 6.32(a)). This adjustment can be done by varying the amplitude of electrical signal, using controls on the front panel of the function generator. At the second stage (Fig. 6.32(b)), the function generator is switched to high-frequency modulation mode, in which the 13.56 MHz signal is modulated at $\omega_0 = 355.9$ Hz. This modulated electrical signal, being connected to the lock-in amplifier, simulates the modulated spectrum at the same spectral line 750 nm. However, the minimum resolvable amplitude that the function generator can usually provide (on the scale of millivolts) is orders of magnitude higher than real signal from the spectrometer. Therefore, in order to equalized them, additional variable attenuator must be used. Thus, adjusting the amplitude of electrical modulation at the output of the function generator, it is possible to equalize the amplitudes of both the real and simulated modulated spectra. Comparing these electrical amplitudes — adjusted at the first and second stages — it is possible to estimate the modulation depth of a particular spectral line. With this technique, as small modulations as 10^{-6} can be measured. Particularly, in all cases, described above, the depth of modulation was in the interval 10^{-2}–10^{-6}.

Supplemental Reading

J.R. Lakowicz, *Principles of Fluorescence Spectroscopy*, 2nd edn., Springer Science, 2013.

S. Shulda, R.M. Richards, Modulation excitation spectroscopy with phase-sensitive detection for surface analysis, Chapter 5 In: *New Materials for Catalytic Applications*, V.I. Parvulescu, E. Kemnitz (eds.), Elsevier, 2016.

R.Grosskloss, P. Kersten, W. Demröder, Sensitive amplitude- and phase-modulated absorption spectroscopy with a continuously tunable diode laser, *Appl. Phys. B, Photophys. Laser Chem.*, **58**(2), pp. 137–142 (1994).

Chapter 7

Optical Diagnostics in Plasma Etching Machines

7.1. Observation Conditions on Plasma Chambers

Optical emission from plasma reaches photo-receivers through vacuum-tight optical windows as shown in Fig. 7.1.

The main purpose of these windows in the viewports is to provide technicians with the possibility to visually control situation inside the chamber and, particularly, to see position of the wafer after it is loaded. For this reason, the optical axis of the viewport always passes approximately in the plane of the wafer, just several millimeters above its surface. The active plasma created here initiates plasma-chemical reactions on the wafer. It is namely the active plasma that has to be monitored. However, the electrons drift within the entire chamber and create plasma between the chamber walls and the chuck — areas 14 as shown in Fig. 7.1. Through these areas, the products of plasma-chemical reaction are removed by turbo-molecular pump, and parameters of plasma here may differ significantly from those of the active plasma 13 above the wafer. Optical emission from plasma contains strong ultra-violet radiation (see the table below), which is harmful for human eyes. Therefore, as the safety requirement, the observation viewports always contain glass windows that block ultra-violet portion of optical emission. For measurement purposes, on the contrary, ultra-violet radiation is important — the reason why every plasma chamber has the second technical viewport with the quartz window that is transparent for

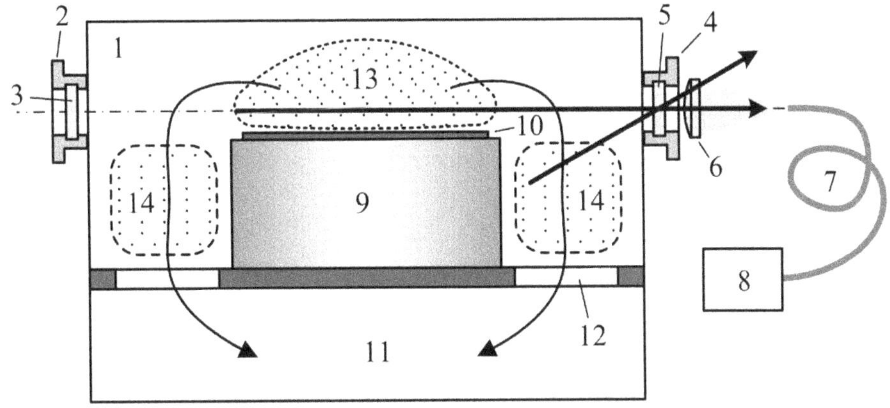

Fig. 7.1. Optical configuration of plasma chamber: 1 — plasma chamber; 2 — observation viewport; 3 — glass window; 4 — technical viewport; 5 — quartz window; 6 — lens; 7 — optical fiber; 8 — spectrometer; 9 — electrostatic chuck; 10 — wafer; 11 — turbo-molecular pump; 12 — openings; 13 — active plasma area; 14 — inactive plasma.

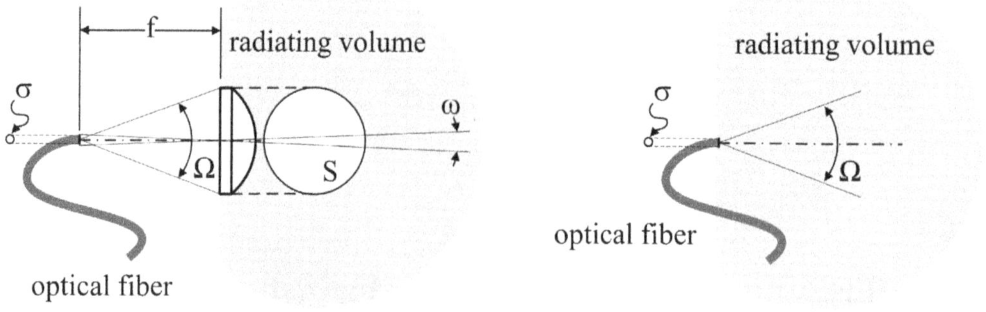

Fig. 7.2. Comparison of energy efficiency of two optical schemes: with and without the focusing lens.

ultra-violet radiation. Spectrometers and other measurement equipment are always connected to the technical viewport.

In order to select optical emission of active plasma from radiation of inactive areas, the lens is usually installed in front of the optical fiber, focusing light from designated areas onto its front tip. It is shown in Fig. 7.1 by straight black arrows that, in this configuration, emission from inactive plasma misses the front end of the fiber and, therefore, is excluded from analysis. It is a very common mistake to assume that the bigger the front diameter of the lens is, the bigger the received optical flux is. Consider Fig. 7.2.

Let the brightness of radiating volume be B [Watt/(Srad $\cdot$ cm^2)]. Then, with the lens, the optical power transmitted through optical fiber is

$$W = B \cdot S \cdot \omega = B \cdot S \cdot \left(\frac{\sigma}{f^2}\right) = B \cdot \sigma \cdot \left(\frac{S}{f^2}\right) = B \cdot \sigma \cdot \Omega,$$

where S and ω are the front area and the solid angle of the field of view of the lens, σ is the area of the front tip of the fiber, and Ω is the solid angle of the focused light cone created by the lens. The maximum solid angle of focusing Ω cannot be made arbitrary big because the optical fiber can transmit waves only within its numerical aperture, which is typically about 0.2. It means that the angle of focusing Ω must equal the maximum solid angle of rays that can be transmitted by the optical fiber. Next, consider the single optical fiber, receiving optical flux from the radiating volume without the lens (the right-hand part of Fig. 7.2). Since its solid angle is equal to Ω, the received optical power is equal to

$$W = B \cdot \sigma \cdot \Omega,$$

exactly the value received, using the lens.

Tables 7.1–7.3 summarize most commonly used spectral lines in plasma diagnostics.

7.2. Basic Applications

7.2.1. *Endpoint detection*

During plasma-chemical etch, material of a certain layer of the device structure is being removed through openings in the photoresist layer deposited on top of the structure. As the etching starts, the plasma volume fills with byproducts of this reaction. These species produce characteristic optical emission that is being monitored by a spectrometer. When the material within the photoresist openings is removed, the flow of byproducts into plasma volume abruptly stops, thus marking the end of the etch process. At this very moment, plasma must be switched off in order to prevent undesirable etching of the material below. The command for this is generated when the spectrometer shows abrupt step in optical emission

Table 7.1. Representative spectral lines for endpoint detection (in alphabetic order).

Monitored species	Wavelength (nm)
Al	308.2, 309.3, 396.1
AlCl	261.4
As	235.0
C_2	516.5
CF_2	251.9
Cl	741.4
CN	289.8, 304.2, 387.0
CO	292.5, 302.8, 313.8, 325.3, 482.5, 483.5, 519.8
F	703.7, 712.8
Ga	417.2
H	486.1, 656.5
In	325.6
N	674.0
N_2	315.9, 337.1
NO	247.9, 288.5, 289.3, 303.5, 304.3, 319.8, 320.7, 337.7, 338.6
O	777.2, 844.7
OH	281.1, 306.4, 308.9
S	469.5
Si	288.2
SiCl	287.1
SiF	440.1, 777.0

Table 7.2. Common spectral lines in etching applications (by wavelength).

200 nm	201.2 – Au	240.0 – CF	254.4 – Cl_2	264.8 – AlCl	271.1 – CF_2	281.0 – SiCl	288.2 – Si
	202.1 – Au	241.4 – SiO	255.1 – W	264.7 – Ta	271.5 – Ta	281.1 – OH	288.5 – NO
	205.5 – Cr	242.8 – Au	255.3 – P	265.1 – Ge	272.2 – BCl	281.4 – N_2	289.3 – NO
	206.8 – Ge	247.4 – CF	255.8 – CF	265.9 – Pt	272.2 – NO	282.0 – N_2	289.8 – CN
	221.6 – Ni	247.9 – NO	256.1 – Cl_2	266.5 – BCl	272.4 – W	282.4 – SiCl	290.0 – CO_2
	228.8 – Cd	248.3 – Fe	257.8 – He	266.9 – SiO	274.8 – Au	283.3 – CO	292.1 – CF_2
	229.9 – SiO	248.7 – SiO	258.0– CCl	267.6 – Au	275.0 – CF_2	283.3 – Pb	292.5 – CO
	232.0 – Ni	248.8 – CF_2	259.5 – CF_2	268.0 – NO	277.8 – CCl	283.7 – C	293.0 – Pt
	234.4 – SiO	249.8 – B	259.6 – NO	268.3 – AlCl	278.8 – CCl	286.0 – BCl	295.3 – N_2
	235.0 – As	251.9 – CF_2	261.4 – AlCl	269.4 – SiO	280.0 – CF_2	286.0 – NO	296.2 – N_2
	237.0 – NO	253.6 – P	262.8 – Pt	269.8 – CO	280.2 – Pb	287.1 – SiCl	297.7 – N_2
	238.9 – CO	253.7 – Hg	262.9 – CF_2	270.2 – Pt	280.7 – SiCl	288.0 – CO_2	
300 nm	301.2 – Ni	311.7 – N_2	327.4 – Cu	338.3 – Ag	358.1 – Fe	371.1 – N_2	389.0 – Zr
	301.2 – Ta	312.3 – Au	328.1 – Ag	338.6 – NO	358.4 – NO	372.0 – Fe	390.2 – SiCl
	302.1 – OH	313.2 – Mo	328.5 – N_2	339.2– Zr	358.6 – CN	373.7 – Fe	390.3 – Mo
	302.8 – CO	313.4 – CO	330.6 – CO	340.5 – Pd	359.0 – CN	375.5 – N_2	391.2 – O
	303.5 – NO	313.6 – N_2	330.9 – N_2	341.5 – Ni	359.3 – Cr	376.8 – Gd	394.3 – N_2
	304.2 – CN	313.8 – CO	331.1 – Ta	343.8 – Zr	360.1 – Zr	379.8 – Mo	394.4 – Al
	304.3 – NO	315.9 – N_2	333.9 – N_2	346.2 – Ni	360.5 – Cr	380.5 – N_2	395.5 – SiF_2
	306.4 – OH	317.0 – Mo	334.6 – SiF	346.6 – Cd	361.0 – Pd	382.0 – He	396.0 – Al
	306.5 – Pt	319.3 – Mo	335.0 – Gd	349.3 – CO	361.1 – Cd	385.8 – N_2	396.5 – He
	306.7 – OH	319.8 – NO	335.0 – N_2	349.6 – Zr	363.5 – Pd	386.2 – CN	397.3 – O
	307.0 – CCl	320.7 – NO	335.0 – Ti	350.0 – N_2	363.5 – Ti	386.4 – Mo	399.8 – N_2
	307.8 – OH	321.4 – CF_2	336.0 – NH	352.5 – Ni	364.2 – N_2	387.0 – CN	399.9 – Ti

(*Continued*)

Table 7.2. (*Continued*).

	308.2 – Al	324.7 – Cu	336.3 – SiF	353.7 – N_2	365.0 – Hg	387.1 – CN	
	308.9 – OH	325.3 – CO	337.0 – CO_2	357.2 – NO	365.3 – Ti	388.3 – CN	
	309.3 – Al	325.6 – In	337.1 – N_2	357.7 – N_2	367.2 – N_2	388.9 – He	
	310.4 – N_2	326.8 – N_2	337.7 – NO	357.9 – Cr	368.3 – Pb	389.5 – N_2	
400 nm	400.0 – Ti	407.7 – TiF	418.3 – TiCl	425.9 – AlH	436.8 – SiF	464.9 – O	483.5 – CO
	400.9 – W	408.1 – Zr	419.0 – O	426.7 – C	438.8 – He	468.8 – Zr	486.1 – H
	402.6 – He	408.7 – SiN	419.3 – TiCl	427.0 – N_2	440.1 – SiF	469.0 – Zr	488.0 – Ar
	404.7 – Hg	409.5 – N_2	419.7 – CN	427.5 – Cr	440.7 – SiN	469.5 – S	489.7 – Cl
	405.1 – SiN	410.2 – In	420.0 – N_2	429.5 – W	444.3 – SiN	471.3 – He	492.2 – He
	405.8 – Pb	411.6 – SiN	420.4 – SiN	431.4 – CH	447.1 – He	473.7 – C_2	
	405.9 – N_2	412.7 – SiN	421.6 – CN	434.0 – H	451.1 – In	474.2 – Ge	
	407.2 – W	414.2 – N_2	423.9 – SiN	434.5 – N_2	451.1 – CO	476.5 – Ar	
	407.4 – W	417.2 – Ga	424.1 – AlH	434.8 – Ar	460.0 – CO	479.4 – Cl	
	407.6 – O	418.1 – CN	425.4 – Cr	435.8 – Hg	460.2 – P	482.5 – CO	
500 nm	501.6 – He	508.6 – Cd	520.9 – Ag	546.0 – Hg	585.8 – Cn	590.6 – N_2	
	504.1 – Si	516.5 – C_2	521.8 – Cl	561.0 – CO	585.4 – N_2	595.9 – N_2	
	504.8 – He	519.8 – CO	542.3 – Cl	575.5 – N_2	587.6 – He		
	505.5 – Si	520.8 – Cr	546.6 – Ag	580.4 – N_2	589.3 – Ge		
600 nm	601.4 – N_2	623.9 – F	639.7 – Ga	646.9 – N_2	662.4 – N_2	683.4 – F	696.5 – Ar
	607.0 – N_2	632.3 – N_2	641.3 – Ga	647.8 – CN	667.8 – He	685.4 – F	696.6 – F
	608.0 – CO	634.7 – Si	641.4 – F	654.5 – N_2	670.5 – N_2	685.6 – F	
	615.6 – O	634.8 – F	643.8 – Cd	656.3 – H	674.0 – N	687.0 – F	
	615.7 – O	637.1 – Si	645.6 – O	656.5 – H	677.4 – F	690.2 – F	
	615.8 – O	639.5 – N_2	646.0 – P	662.0 – CO	678.9 – N_2	691.0 – F	
700 nm	703.7 – F	720.2 – F	732.6 – C	741.4 – Cl	751.5 – F	763.5 – Ar	777.2 – O
	706.5 – He	725.4 – O	733.2 – F	742.6 – F	755.2 – F	772.4 – Ar	780.0 – F
	706.7 – Ar	725.6 – Cl	738.4 – Ar	750.4 – Ar	757.3 – F	775.3 – N_2	787.3 – CN
	712.8 – F	727.3 – N_2	738.7 – N_2	750.4 – N_2	760.7 – F	775.5 – F	789.6 – N_2
	716.5 – N_2	728.1 – He	739.9 – F	751.5 – Ar	762.6 – N_2	777.0 – SiF	794.8 – Ar
800 nm	844.7 – O						

Table 7.3. Common spectral lines in etching applications (by species).

Letter	Species	Wavelength (nm)
A	Ag	328.1, 338.3, 520.9, 546.6
	Al	308.2, 309.3, 394.4, 396.1
	AlCl	261.4, 264.8, 268.3
	AlH	424.1, 425.9
	Ar	434.8, 476.5, 488.0, 696.5, 706.7, 738.4, 750.4, 751.5, 763.5, 772.4, 794.8
	As	235.0
	Au	201.2, 202.1, 242.8, 267.6, 274.8, 312.3
B	B	249.8
	BCl	266.5, 272.2, 286.0
C	C	283.7, 426.7, 732.6
	C_2	473.7, 516.5
	CCl	258.0, 277.8, 278.8, 307.0, 460.0
	Cd	228.8, 346.6, 361.1, 508.6, 643.8
	CF	240.0, 247.4, 255.8
	CF_2	248.8, 251.9, 259.5, 262.9, 271.1, 275.0, 280.0, 292.1, 321.4
	CH	431.4
	Cl	479.4, 489.7, 521.8, 542.3, 725.6, 741.4
	Cl_2	254.4, 256.1
	CN	289.8, 304.2, 358.6, 359.0, 386.2, 387.0, 387.1, 388.3, 418.1, 419.7, 421.6, 585.8, 647.8, 787.3
	CO	238.9, 269.8, 283.3, 292.5, 302.8, 313.4, 313.8, 325.3, 330.6, 349.3, 451.1, 482.5, 483.5, 519.8, 561.0, 608.0, 662.0
	CO_2+	337.0

(*Continued*)

Table 7.3. (*Continued*)

Letter	Species	Wavelength (nm)
	Cr	205.5, 357.9, 359.3, 360.5, 425.4, 427.5, 520.8
	Cu	324.7, 327.4
F	F	623.9, 634.8, 641.4, 677.4, 683.4, 685.4, 685.6, 687.0, 690.2, 691.0, 696.6, 703.7, 712.8, 720.2, 733.2, 739.9, 742.6, 755.2, 757.3, 760.7, 775.5, 780.0
	Fe	248.3, 358.1, 372.0, 373.7
G	Ga	417.2, 639.7, 641.3
	Gd	335.0, 376.8
	Ge	206.8, 265.1, 474.2, 589.3
H	H	434.0, 486.1, 656.3, 656.5
	He	257.8, 382.0, 388.9, 396.5, 402.6, 438.8, 447.1, 471.3, 492.2, 501.6, 504.8, 587.6, 667.8, 706.5, 728.1
	Hg	253.7, 365.0, 404.7, 435.8, 546.1
I	In	325.6, 410.2, 451.1
M	Mo	313.2, 317.0, 319.3, 379.8, 386.4, 390.3
N	N	674.0
	N_2	281.4, 282.0, 295.3, 296.2, 297.7, 310.4, 311.7, 313.6, 315.9, 326.8, 328.5, 330.9, 333.9, 335.0, 337.1, 350.0, 353.7, 357.7, 364.2, 367.2, 371.1, 375.5, 380.5, 385.8, 389.5, 394.3, 399.8, 405.9, 409.5, 414.2, 420.0, 427.0, 434.5, 575.5, 580.4, 585.4, 590.6, 595.9, 601.4, 607.0, 632.3, 639.5, 646.9, 654.5, 662.4, 670.5, 678.9, 716.5, 727.3, 738.7, 750.4, 762.6, 775.3, 789.6
	NH	336.0
	Ni	221.6, 232.0, 301.2, 341.5, 346.2, 352.5
	NO	237.0, 247.9, 259.6, 268.0, 272.2, 286.0, 288.5, 289.3, 303.5, 304.3, 319.8, 320.7, 337.7, 338.6, 357.2, 358.4
O	O	391.2, 397.3, 407.6, 419.0, 464.9, 615.6, 615.7, 615.8, 645.6, 725.4, 777.2, 844.7
	OH	281.1, 302.1, 306.4, 306.7, 307.8, 308.9
P	P	253.6, 255.3, 460.2, 646.0
	Pb	280.2, 283.3, 368.3, 405.8
	Pd	340.5, 361.0, 363.5
	Pt	262.8, 265.9, 270.2, 293.0, 306.5
S	S	469.5
	Si	288.2, 504.1, 505.5, 634.7, 637.1
	SiCl	280.7, 281.0, 282.4, 287.1
	SiF	334.6, 336.3, 436.8, 440.1, 777.0
	SiF_2	390.2, 395.5
	SiN	405.1, 408.7, 411.6, 412.7, 420.4, 423.9, 440.7, 444.3
	SiO	229.9, 234.4, 241.4, 248.7, 266.9, 269.4
T	Ta	264.7, 271.5, 301.2, 331.1
	Ti	335.0, 363.5, 365.3, 399.9
	TiCl	418.3, 419.3
	TiF	407.7
W	W	255.1, 272.4, 400.9, 407.4, 429.5
Z	Zn	339.2, 343.8, 349.6, 360.1, 389.0, 407.3, 408.1, 468.8

of the byproduct. Intensity of the optical emission at a designated wavelength as a function of time is called the trend, which is schematically shown in Fig. 7.3.

The noise in electrical signal is an important feature because, in modern semiconductor technology, the ratio of the open area to the entire area of the wafer may be less than 1%. Consequently, the relative value of the endpoint step on the trend is of the same scale and must be reliably detected on the background of noise. The noise in the trend is mostly composed of electrical noise of the photodetector, noise of plasma oscillations, and random character of the etch process. Moreover, due to variation of the etch rate over the surface of the wafer, the real endpoint step on the trend is not as sharp as it is drawn in the figure above. Indeed, consider spatially non-uniform radial distribution of plasma over the wafer (Fig. 7.4).

For analytical calculations, it is possible to approximate smooth radial variation of the etch rate $v(r)$ quadratically, i.e.

$$v(r) = v_C(1 - \alpha r^2),$$

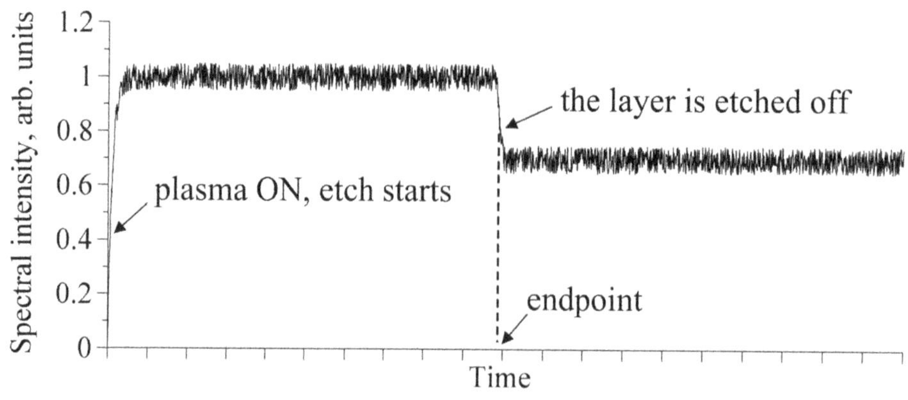

Fig. 7.3. Idealized presentation of a trend with endpoint.

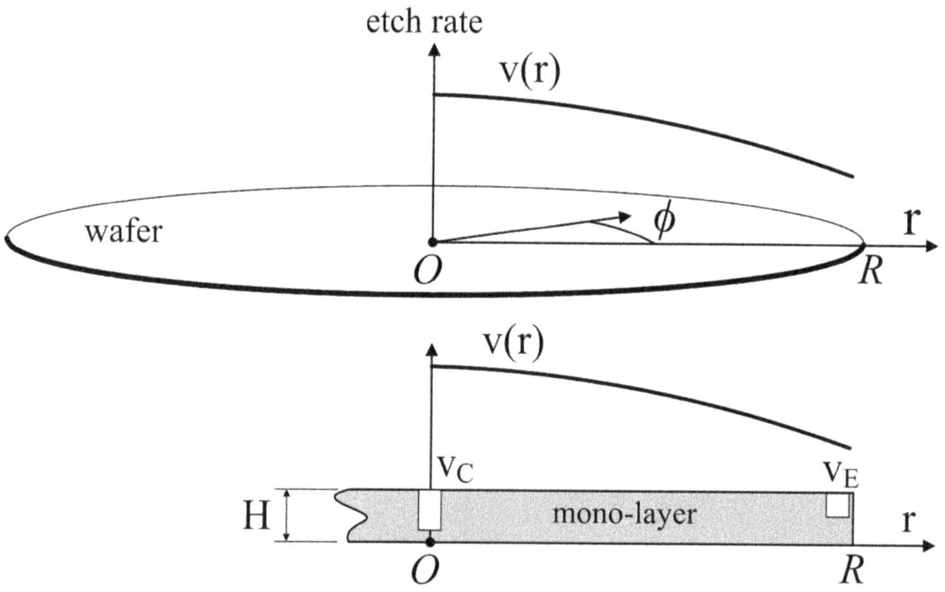

Fig. 7.4. Etching of a monolayer. Radial system of coordinates and notations.

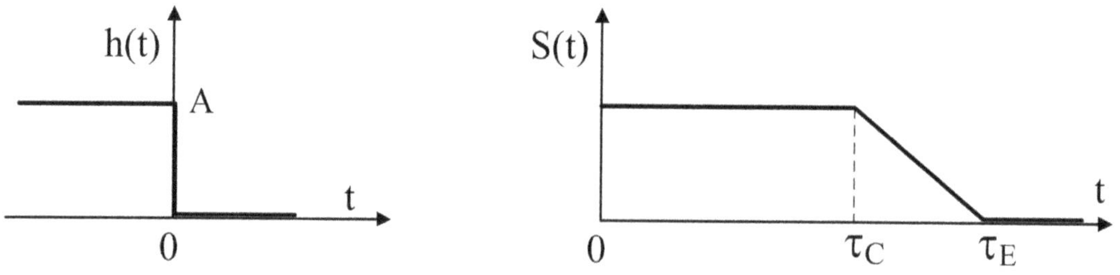

Fig. 7.5. The local (at left) and integral (at right) endpoint trends.

where v_C is the etch rate at the center and $v_E = v_C(1 - \alpha R^2)$ — the etch rate at the edge of the wafer. In each individual point of the wafer, the endpoint trend is a sharp function of time $h(t)$ (the left-hand part of Fig. 7.5).

Contribution to the endpoint trend from such a small area $d\sigma = d\phi r dr$ on the wafer is equal to

$$dS(t) = h[t - \tau_C(1 + \alpha r^2)]d\phi r dr,$$

where

$$\tau_C = \frac{H}{v_C}$$

is the endpoint time at the center. Integrating over the entire surface of the wafer, we obtain the entire trend signal as in the right-hand part of Fig. 7.5:

$$S(t) = 2\pi \int_0^R h[t - \tau_C(1 + \alpha r^2)]rdr = \begin{cases} 0, & t \geq \tau_E, \\ \pi R^2 A \cdot \dfrac{\tau_E - t}{\tau_E - \tau_C}, & \tau_C \leq t \leq \tau_E, \\ \pi R^2 A, & t \leq \tau_C \end{cases}$$

with

$$\tau_E = \frac{H}{v_E}$$

being the endpoint time at the edge of the wafer. The radial non-uniformity is an important parameter of any process machine, and finding τ_C and τ_E from the endpoint curve, it is possible to measure its relative value:

$$\frac{v_C - v_E}{v_C} = \frac{\tau_E - \tau_C}{\tau_E}$$

or

$$\frac{v_C - v_E}{\frac{1}{2}(v_C + v_E)} = 2\frac{\tau_E - \tau_C}{\tau_C + \tau_E}.$$

For not very low open area processes, the endpoint can be reliably detected, using standard spectrometers (Chapter 1), programmed to monitor characteristic spectral lines of byproducts listed in Tables 7.1–7.3. However, for low open area less than 1% and for some peculiar etch processes, the endpoint is not reliably seen on the background of noise. The "brute force" method of improving endpoint contrast in such problems is to implement principal component analysis to identify spectral lines that strongly contribute to endpoint variation, and then to include these lines in a combination with proportionality coefficients equal to factors with which these lines were identified. Apart of this method, there is another approach, having very clear physical meaning. The best way to explain it is to summarize the main steps in a table (Table 7.4), and then to present its mathematical proof.

Now, consider the mathematical proof for this technique. In the spectrometer, each channel $i = 1, 2, \ldots, M$ contains regular signal and noise with zero mean. Accordingly, variations ΔS_i are also noisy with the mean values s_i and noise n_i:

$$\Delta S_i = s_i + n_i.$$

The term "noise" is defined as real-mean-square (rms) variation of ΔS_i:

$$\langle(\Delta S_i - \langle\Delta S_i\rangle)^2\rangle = \langle n_i^2\rangle.$$

Assuming equal noise in all channels,

$$\langle n_i^2\rangle \equiv \sigma^2.$$

Signal-to-noise ratio (SNR) in a single spectral channel is defined as follows:

$$\text{SNR} = \frac{\langle\Delta S_i\rangle}{\sqrt{\langle(\Delta S_i - \langle\Delta S_i\rangle)^2\rangle}} = \frac{s_i}{\sqrt{\langle(\Delta S_i - \langle\Delta S_i\rangle)^2\rangle}} = \frac{s_i}{\sigma}.$$

Table 7.4. Adaptive endpoint algorithm.

Phase	Step	Description
Preparation on large open area	1	Measure full spectrum trend $S(\lambda, t)$ on large open area. This will be an array of $M \times N$ points, where M — number of spectral points and N — number of time points
	2	Reliably identify time t_{EPD} of the endpoint
	3	At t_{EPD}, compute variations $\Delta S_i = S(\lambda_i, t_{\mathrm{EPD}} + \delta) - S(\lambda_i, t_{\mathrm{EPD}} - \delta)$ for each wavelength λ_i. It will be an array ΔS_i of M points for wavelengths, typically covering the interval from 200 nm to 800 nm
	4	Compute a scaling vector $$a_i = \frac{\Delta S_i}{\sqrt{\sum_{i=1}^{M} \Delta S_i^2}}.$$ ***This is a fixed process-specific vector, same for any etching machine on the same process.*** It defines optimal endpoint detection
Regular application to small open area	1	Measure full spectrum trend $S(\lambda, t)$ and compute in real time t variations $\Delta S = S(\lambda, t + \delta) - S(\lambda, t - \delta)$. This is an array ΔS_{ij}, with i — wavelength index, j — real time index.
	2	At each moment of time j compute $p_j = \sum_{i=1}^{M} a_i \Delta S_{ij}$.
	3	Follow the trend of p_j. At the endpoint, variation $\Delta p_j = p_{j+1} - p_j$ will be the maximum, and signal-to-noise ratio will be the maximum possible. All this is proved mathematically in the next section

In each channel, noise is uncorrelated:

$$\langle n_i \cdot n_j \rangle = 0, \quad i \neq j.$$

Consider the sum over all spectral channels $\sum_i \Delta S_i$. The rms variation of it is

$$\left\langle \left(\sum_i \Delta S_i - \left\langle \sum_i \Delta S_i \right\rangle \right)^2 \right\rangle = \left\langle \left(\sum_i s_i + \sum_i n_i - \sum_i s_i \right)^2 \right\rangle = N \cdot \sigma^2.$$

If all the signals are equal $s_i = s$, the SNR for the sum is

$$\mathrm{SNR} = \frac{\langle \sum_i \Delta S_i \rangle}{\sqrt{\langle (\sum_i \Delta S_i - \langle \sum_i \Delta S_i \rangle)^2 \rangle}} = \frac{s}{\sigma}\sqrt{N},$$

i.e. $\sqrt{N}$ times better than in a single channel.

However, if the signals s_i are not equal, the advantage in SNR will not be that big. The problem, which we attempt to solve, is as follows: find the algorithm, maximizing SNR for non-equal signals. For that, consider a set of multiples a_i, $\quad i = 1, 2, \ldots, M$ and the sum

$$u = \sum_i a_i \cdot \Delta S_i.$$

The SNR for this variable is

$$\mathrm{SNR} = \frac{\langle \sum_i a_i \Delta S_i \rangle}{\sqrt{\langle (\sum_i a_i \Delta S_i - \langle \sum_i a_i \Delta S_i \rangle)^2 \rangle}} = \frac{\sum_i a_i s_i}{\sigma \sqrt{\sum_i a_i^2}}.$$

From this formula, it follows that all the a_i may be scaled without change of SNR. Therefore, it is our choice how to scale them. Choose

$$\sum_i a_i^2 = 1.$$

It means that the vector $\vec{a}$ has unity length. Formula for SNR can be then rewritten in a form of a scalar product:

$$\mathrm{SNR} = \frac{1}{\sigma}\vec{a}\vec{s}.$$

Since vector $\vec{a}$ has unity length, the scalar product reaches its maximum when $\vec{a}$ is collinear to $\vec{s}$:

$$\vec{a} = \frac{1}{\sqrt{\sum_i s_i^2}}\begin{pmatrix} s_1 \\ s_2 \\ \cdots \\ s_M \end{pmatrix}.$$

This is the basic formula for the adaptive endpoint algorithm described in Table 7.4. The only thing that is left to do is to explain the physical meaning of this algorithm. What is the signal? For the endpoint detection, signal means variation of some spectral component. If spectral trend is $S(\lambda, t)$, then

$$s_i \equiv s(\lambda_i) = S(\lambda_i, t_{\mathrm{EPD}} + \delta) - S(\lambda_i, t_{\mathrm{EPD}} - \delta) \equiv \Delta S(\lambda_i, t).$$

Therefore, the formula for the scaling vector $\vec{a}$ means that this vector must be directed along the gradient of the spectral trend at the moment of the endpoint. The more accurately this gradient is measured, the better. That is why the first phase should be done on large open area wafers in order to ensure accurate measurement.

As an example consider the etch process on the 4×4 cm^2 coupon that is monitored by the trend of the CO characteristic line at 483 nm. Use of small test coupons instead of the full-size 300 mm wafer is a standard technique to simulate etching on low open area wafers. Particularly, the 4×4 cm^2 coupon corresponds to about 1% open area — a challenging value for endpoint detection. The variation map of this process is shown in Fig. 7.6. The endpoint is somewhere around 150 s, and the scaling vector a_i derived here is shown in Fig. 7.7 as a function of wavelength.

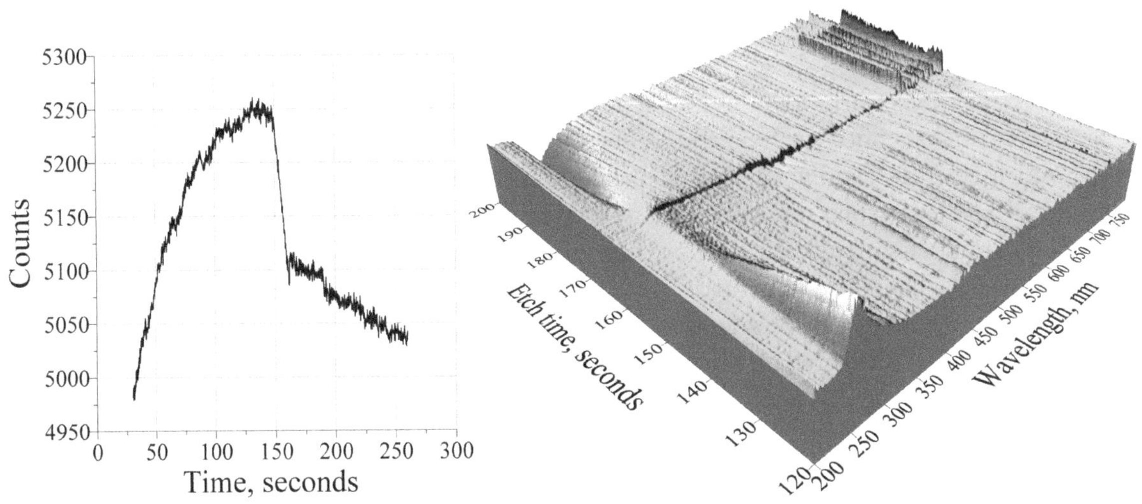

(a) (b)

Fig. 7.6. Trend at 483 nm (a) and variation map $\Delta S(\lambda, t)$ (b) for the etch process.

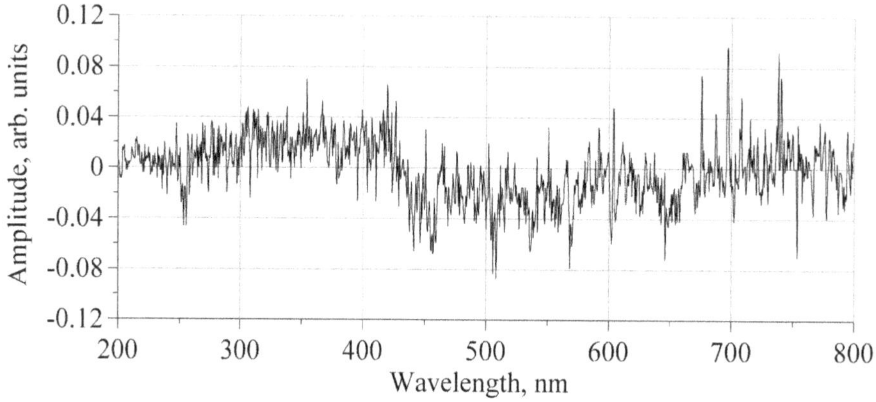

Fig. 7.7. Scaling vector a_i.

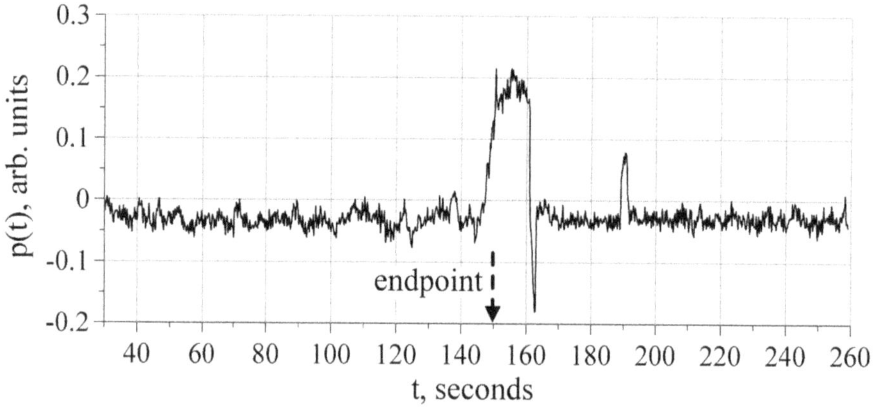

Fig. 7.8. Endpoint function.

Application of the adaptive algorithm, explained above, means real-time computation of the endpoint function

$$p(t) = \sum_{i=1}^{M} a_i \Delta S_i(t),$$

which gives the result that clearly identifies the endpoint (Fig. 7.8).

When manufacturing technology requires higher etch rates, the power delivered to plasma has to be increased, which leads to significant rise of plasma fluctuations. These fluctuations are faster than sampling speed of ordinary spectrometers can follow, acting as random noise in the trend. Plasma fluctuations, i.e. variations of electron density, act similarly on both the byproducts of etching and other species in plasma. Therefore, measuring optical emission of plasma at some inactive species, it is possible to decrease the influence of plasma fluctuations and reduce the noise by dividing or subtracting the two signals — the trend measured on the active byproduct and the trend on inactive species. For that, the sampling speed of the spectral device must be significantly increased to the scale of megahertz. The straightforward solution is to use photomultipliers with interference filters in two independent separate spectral channels. Such combination is inferior to spectrometers in terms of convenience of changing the working wavelength of the trend, because adjustment to every new process requires exchange of interference filters. Nevertheless, the result may worth it. Before addressing this technique in detail, some light should be shed on photomultipliers and interference filters.

7.2.2. *Photomultiplier tubes*

A photomultiplier tube (PMT) is the most sensitive high-speed photodetector available. In the laboratory, its time comes when optical flux becomes too low for photodiodes or phototransistors. Amazingly, at low signals, PMTs with quantum efficiency of only 15% completely outperform photodiodes with their quantum efficiency up to 90%. The reason is low-noise inner amplification by secondary electron emission. Therefore, PMT requires high voltage of a kilovolt scale, and voltage-dividing system of resistors connected in series. All this sounds like too much for easy use in the laboratory, and it is true. Understanding that, manufacturers offer completely finalized ready-to-use devices called PMT modules, looking like small, the size of a matchbox, metal boxes with only few wires/connectors for low-voltage power, gain adjustment, and the output (Fig. 7.9). All other components, like low-to-high voltage converter, resistive divider, and the PMT itself, are hidden inside. The only thing the user has to do is to interface the PMT module with an amplifier. But to do that correctly, it is necessary to understand some basic rules and limitations, discussed in this section.

PMT is a vacuum tube, containing the photocathode followed by focusing electrodes, electron multiplier with many accelerating electrodes (dynodes), and an electron collector (anode). Specific design of the electrodes may vary from one type to another, but the basic concept is the same (Fig. 7.10). The only classifying difference is the type of the photocathode: the reflection or the transmission type. They are also called front- and side-illumination types. The dynodes form an electron multiplier with typical gain about millions. Capacitors maintain constant voltage on the last dynodes during pulsed operation, when current increases in avalanche from photocathode to anode. Anode is not connected inside the vacuum tube. The transmission type requires less space for the electrode system, being therefore the choice for ultra-miniature designs used in PMT modules.

The key parameter of a PMT is its quantum efficiency that includes quantum efficiency of the photocathode itself and optical efficiency of collecting the photons. The most advanced photocathode material, commonly referred to as S20 or multialkali photocathode, is a Na–K–Cs–Sb thin-film compound, manufactured by depositing consecutive layers of sodium, potassium, and cesium onto the thick antimony layer in high vacuum. Neither the exact chemical composition or structure of this sandwich, nor true functions of individual components are known. The manufacturing process is controlled by measuring light transmission coefficient and photoemission current during deposition. Refractive index of the final film varies, depending on the manufacturing process, and is known to range from 2.5 to 4.5.

There are three reflecting interfaces in the optical path: air-glass, glass-photocathode, and photocathode-vacuum (Fig. 7.11).

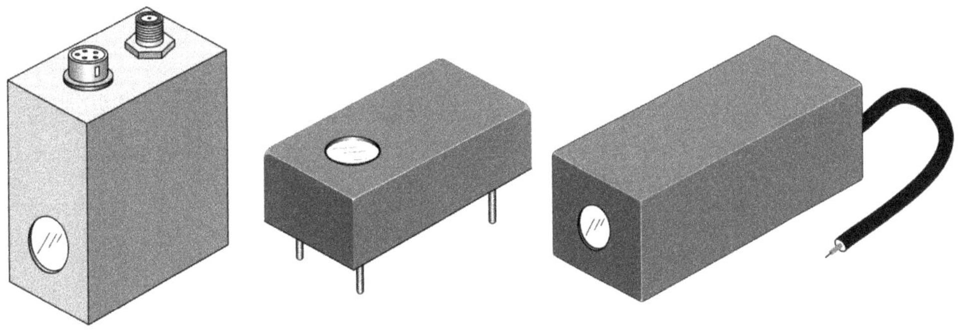

Fig. 7.9. PMT modules come in a variety of form-factors. They require low-voltage power, usually 5V or 12V direct current, and can be easily interfaced with simple electronic components to adjust the PMT gain.

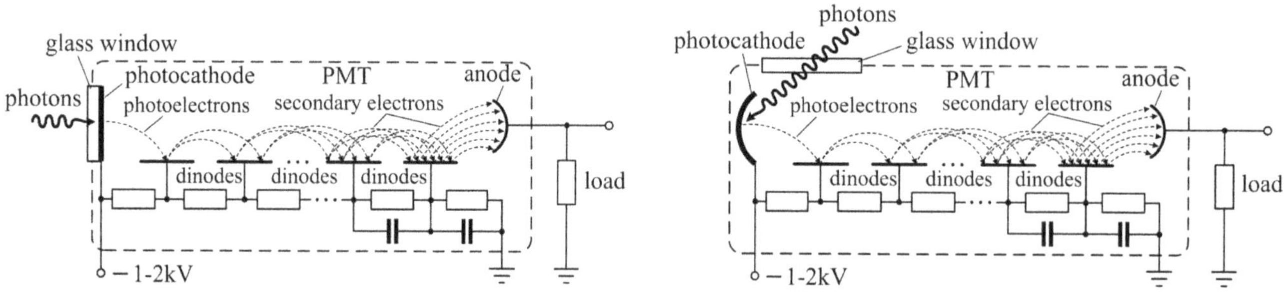

Fig. 7.10. Schematic diagram of a PMT: transmission (at left) and reflection (at right) types.

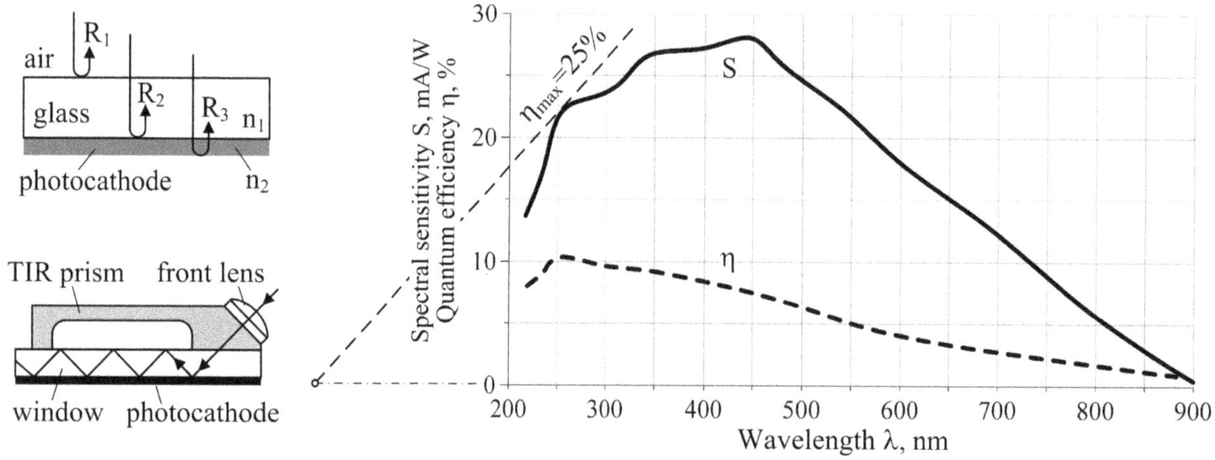

Fig. 7.11. Reflection budget of a PMT window and typical sensitivity curves.

The first two contribute negatively, attenuating the optical flux that reaches the photocathode. Some additional improvement can be made by a total internal reflection (TIR) prism glued to the window. The aim of the prism is to direct the rays at approximately 45° to the window surface without considerable reflection at the front interface. The second leg of the prism serves only for stronger hold on the window. Manufacturers of PMTs do not offer this option. Refractive index of glass or fused silica (for UV applications) is typically $n_1 = 1.5$. The Fresnel formula then gives reflection coefficient at the first interface

$$R_1 = \left|\frac{n_1 - 1}{n_1 + 1}\right|^2 \approx 4\%,$$

which can be reduced to less than 1% by additional antireflection coating onto the glass. Relatively high refractive index of the photocathode film $n_2 \sim 3$–4 adds much stronger reflection from the second interface

$$R_2 = \left|\frac{n_2 - n_1}{n_2 + n_1}\right|^2 \sim 10\text{–}20\%,$$

which cannot be compensated for by any coating. Finally, reflection from the bottom interface with the reflection coefficient

$$R_3 = \left|\frac{n_2 - 1}{n_2 + 1}\right|^2 \sim 20\text{–}30\%$$

acts positively, returning transmitted light back to the photocathode, thus contributing to better usage of light. Total quantum efficiency of an S20 photocathode may peak above 20% in the blue part of visible spectrum (Fig. 7.11).

Quantum efficiency η is always recalculated from experimentally measured spectral sensitivity S, using the formula:

$$S[\mathrm{mA/W}] = \eta \cdot 0.806 \cdot \lambda[\mathrm{nm}].$$

Typical PMT gain delivered by electron multiplier may be as high as $G \sim 10^6$–10^7, which often poses a question about noise: does large gain introduce additional noise? The answer is yes, but it is not the full answer. The full answer is that the electron multiplier is the least noisy amplifier that can be made, and the electrical signal-to-noise ratio at the output of the PMT remains practically the same as the optical one at its input. This conclusion requires some mathematics. With optical intensity F, exposure time T, and photocathode quantum efficiency η, the average number of photoelectrons is

$$\nu = F \cdot T \cdot \eta.$$

The actual number of photoelectrons m is a random variable with Poisson probability distribution

$$p(m) = \frac{\nu^m e^{-\nu}}{m!}.$$

Each photoelectron creates on the average G secondary electrons at the output of the multiplier, so that the average number of electrons after multiplication is $G \cdot \nu$. Probability of creating M secondary electrons from exactly m photoelectrons may again be assumed Poissonian with the average value $G \cdot m$:

$$p(M|\,m) = \frac{(m\,G)^M}{M!} e^{-m\,G}.$$

Probability of generating M secondary electrons at the PMT output with the input optical flux, creating ν photoelectrons on the average, is the sum of partial probability functions:

$$P(M) = \sum_{m=0}^{\infty} p(M|\,m) \cdot p(m) = \sum_{m=0}^{\infty} \frac{(mG)^M e^{-mG}}{M!} \cdot \frac{\nu^m e^{-\nu}}{m!},$$

understanding that $0! = 1$. The average number of secondary electrons $\overline{M}$ must be $G \cdot \nu$, and it is indeed so, which can be proved by rearrangement of the sum

$$\overline{M} = \sum_{M=0}^{\infty} \sum_{m=0}^{\infty} M \frac{(mG)^M e^{-mG}}{M!} \cdot \frac{\nu^m e^{-\nu}}{m!} = \sum_{m=0}^{\infty} \frac{\nu^m e^{-\nu} e^{-mG}}{m!} \left[\sum_{M=0}^{\infty} M \frac{(mG)^M}{M!} \cdot \right] = G \cdot \nu.$$

We call the noise the real-mean-square variation σ of photo- or secondary-electrons:

$$\sigma = \sqrt{\overline{m^2} - \overline{m}^2},$$

and the signal-to-noise ratio (SNR)

$$\mathrm{SNR} = \frac{\overline{m}}{\sigma}.$$

For Poisson distribution of photoelectrons, $\overline{m} = \nu$, $\sigma = \sqrt{\overline{m}} = \sqrt{\nu}$, and

$$\mathrm{SNR_{input}} = \sqrt{\nu}.$$

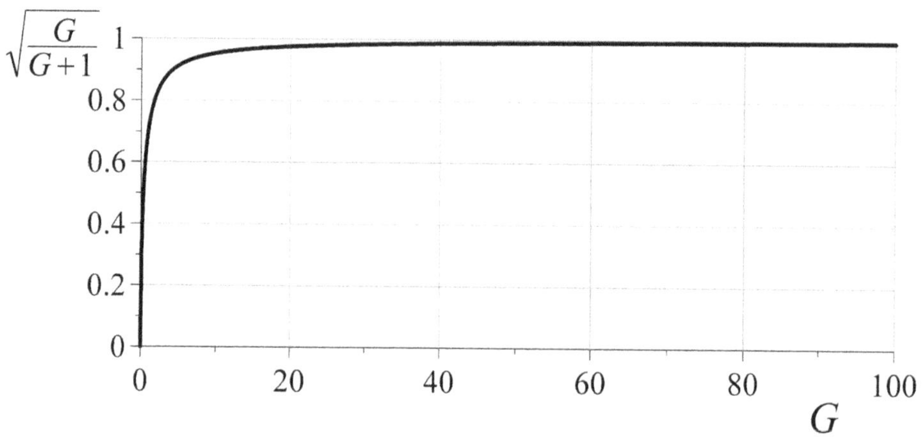

Fig. 7.12. Function $\sqrt{G/(G+1)}$ quickly reaches its maximum value.

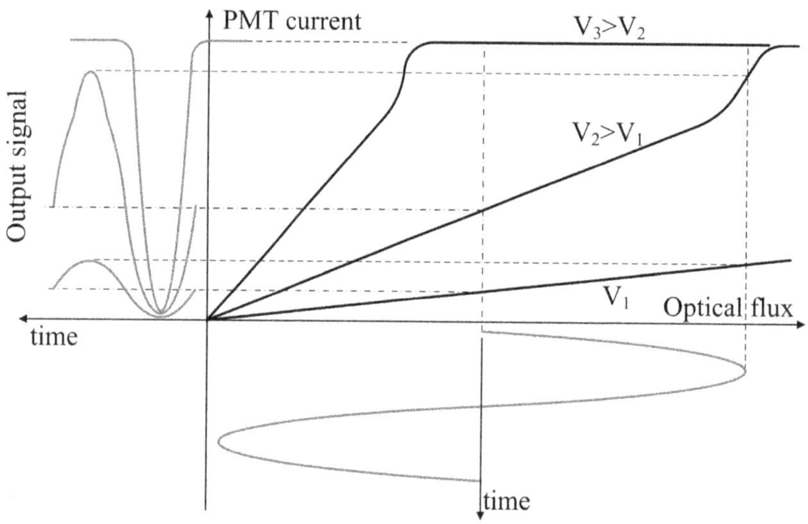

Fig. 7.13. Electro-optical characteristics of PMTs.

For the secondary electrons, some mathematical manipulation gives their average square number

$$\overline{M^2} = \sum_{M=0}^{\infty} M^2 P(M) = G^2(\nu^2 + \nu) + G\,\nu$$

that leads to $\sigma = \sqrt{G\nu(G+1)}$ and

$$\mathrm{SNR}_{\mathrm{output}} = \frac{G\nu}{\sqrt{G\nu(G+1)}} = \sqrt{\frac{G}{G+1}} \cdot \mathrm{SNR}_{\mathrm{input}}.$$

Theoretically, since always $\sqrt{G/(G+1)} < 1$, the signal-to-noise ratio at the PMT output is always smaller than at its input (Fig. 7.12). However, practically, they are almost the same already at as small gains as 10^2.

PMTs are designed for low-light applications, therefore they saturate quickly as the optical flux increases, especially the PMT modules that use miniature photomultipliers with small photocathode area (Fig. 7.13). Nonetheless, with moderate optical fluxes, PMT shows exceptional linearity.

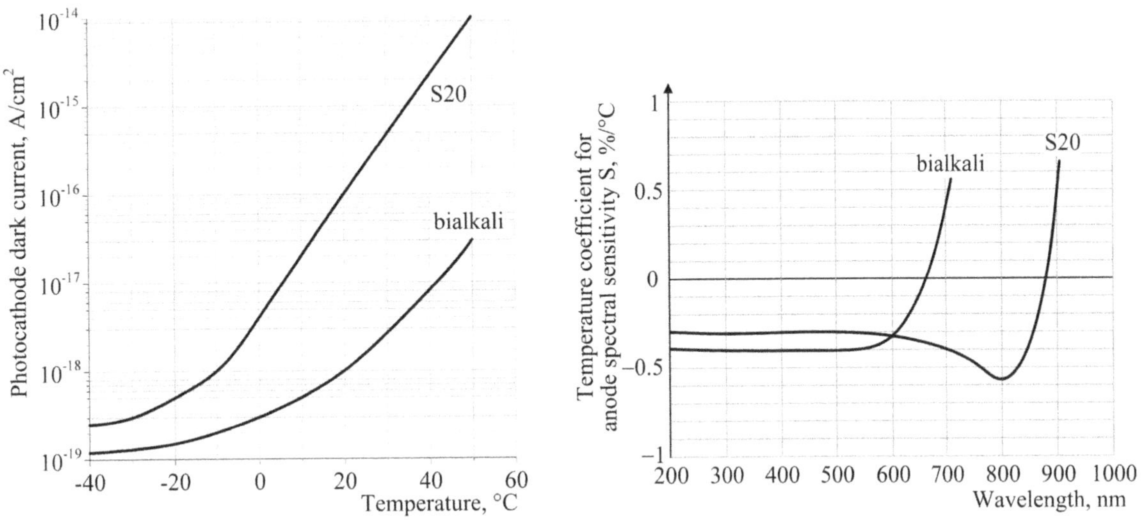

Fig. 7.14. Cooling reduces dark current by orders of magnitude.

Saturation is preceded by sharp increase of gain. Higher cathode voltage V makes higher gain and leads to earlier saturation. Saturation current may be roughly considered as a constant for a particular PMT. The smaller the photocathode and dynodes area is, the smaller is the maximum current delivered to the anode. With the input optical flux increasing, the photocurrent at the last sections of the dynode system increases to values comparable to those in the resistive divider circuit. This causes voltage redistribution, resulting in visible increase of the PMT gain. As it happens, further increase of optical flux quickly leads to saturation, when output current cannot follow variations of the input optical flux. The gain of the PMT, i.e. the slope of the current-flux curve, can be tuned by changing cathode voltage V. Higher V results in earlier saturation, producing nonlinear response. Sometimes, it may be difficult to realize what has happened, when you clearly see the signal at the oscilloscope, but the amplitude does not react to light variations. The solution is simple: set lower PMT gain.

As PMTs are mostly used in low-light applications, the dark current is a very important factor. Dark current originates from thermal electron (thermionic) emission from the photocathode. As such, it can be reduced by cooling the photocathode or the entire PMT. There are two processes, associated with the cooling: reduction of the dark current and change of spectral sensitivity (Fig. 7.14). Luckily, the second process acts positively in UV-visible domain, making cooling an efficient technique, when utmost sensitivity is required. Spectral sensitivity increases with cooling in UV-visible domain (temperature coefficient is negative), but decreases in near infrared around 900 nm. Note that the left figure shows the dark current of the photocathode alone, whereas the right figure shows anode sensitivity, which includes temperature changes of the electron multiplier.

Cooling is not a simple procedure, especially on large-area PMTs. In order to avoid condensation on the input window, the PMT should be placed inside a sealed housing filled with dried gas. Thermoelectrically cooled PMT modules are available in the market. The most efficient way to create cooled PMT modules is to use miniature metal-package photomultipliers: small size eases requirements to thermoelectric cooler, and metal case significantly improves thermal conductivity. Not only this type of PMTs is most suitable for cooling, but it is used in the majority of compact PMT modules as well (Fig. 7.15).

PMT is a very fast photodetector, with time resolution about several nanosecond. Metal-package PMTs and PMT modules are among the best on this scale because small size of the electrodes makes transient time minimal.

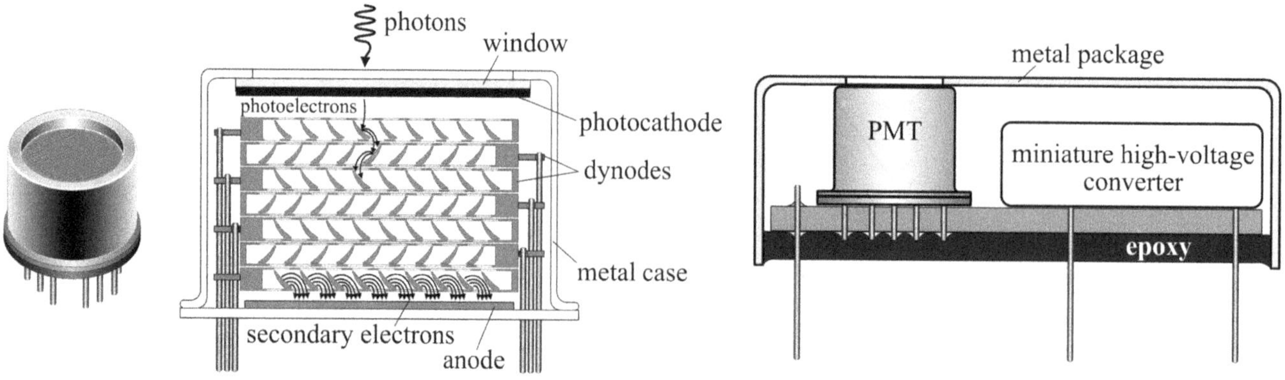

Fig. 7.15. Metal-package PMTs (at left) are used in compact PMT modules (at right). Dynode system is made of thin finely perforated metal foil (in the middle).

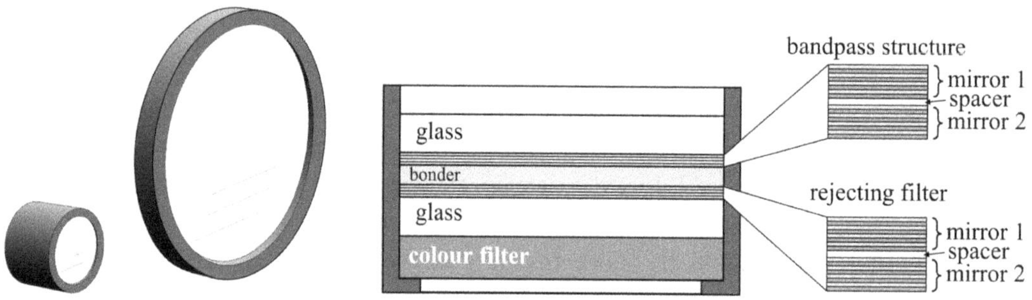

Fig. 7.16. Interference filters (IFs) are essentially closely spaced interferometers Fabry-Perot with multilayer mirrors.

7.2.3. *Interference filters*

The indispensable feature of grating spectrometers in the endpoint detection is the ability to arbitrarily select the wavelength with potentially high spectral resolution. On the other hand, small area of a slit, not exceeding 5×0.3 mm^2, sets the limit to energy efficiency of grating spectrometers. It is almost hundred of times less than a standard photodetector like photodiode or photomultiplier with 10×10 mm^2 sensitive aperture can accept. Interference filters (IFs), with clear apertures up to 50 mm and transmission-bandwidth from 20%–1 nm to 70%–40 nm at the central wavelength, leave no chance to grating spectrometers to compete in delivering optical flux from wide-area sources to photodetectors (Fig. 7.16).

Basically, IFs are thin-film interferometers Fabry-Perot (chapter 3). In them, mirrors are made of many altering layers of transparent dielectric films with different refractive indices. When optical thickness of each layer is exactly a quarter of the wavelength, such a structure creates constructive interference and behaves as a mirror. According to the Airy formula (attention: not the Airy function), transparency of the Fabry-Perot structure is proportional to

$$\frac{1}{1 + f\sin^2(\pi\frac{2nL}{\lambda}\cos\theta)}, \quad f = \frac{4R}{(1-R)^2}$$

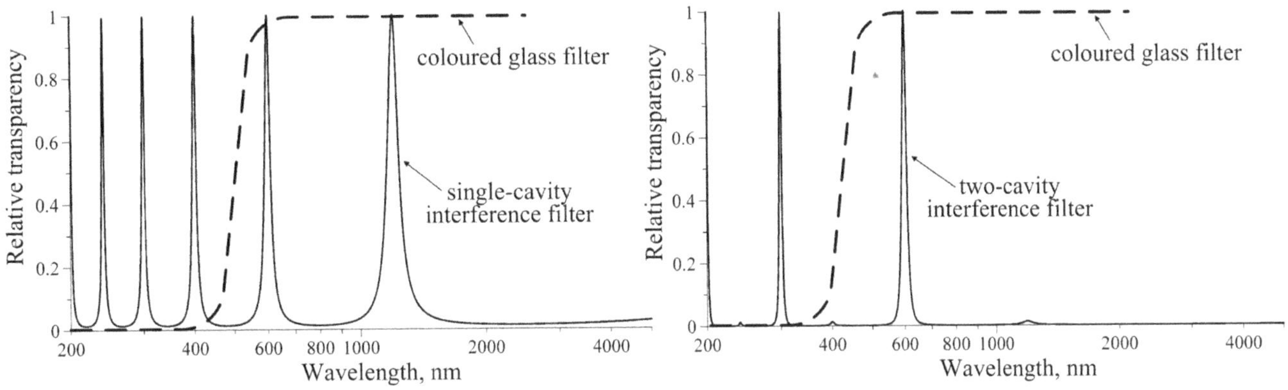

Fig. 7.17. Theoretical spectral transparency of a single Fabry-Perot cavity with $R = 0.8$ and $2nL = 1200$ nm (at left). Spectral peaks at 400 nm and below can be stopped by a colored glass filter as shown in the dashed line. The second IF with $2nL = 600$ nm not only suppresses parasitic maximum at 1200 nm but also sharpens the design peak at 600 nm (at right).

with λ being the wavelength, n and L — refractive index and thickness of the spacer, θ — angle of incidence, and R — energy reflection coefficient of the mirrors. It reaches maxima when

$$2nL\cos\theta = m\lambda, \quad m = \pm 1, \pm 2, \pm 3, \ldots .$$

What is the width $\Delta\lambda$ of the interference maxima in the wavelength domain? Recalling that full width at half-maximum of each interference peak in the domain of phases $\delta \equiv 4\pi nL\cos\theta/\lambda$ is $\varepsilon = 4/\sqrt{f}$, and assuming $\theta = 0$, we can easily derive from

$$\varepsilon = \left|\frac{d\delta}{d\lambda}\right| \Delta\lambda$$

that

$$\frac{\Delta\lambda}{\lambda} = \frac{\lambda}{\sqrt{f}\pi nL} = \frac{1}{mF},$$

where

$$F \approx \frac{\pi}{1 - R}$$

is the finesse (Chapter 3). Therefore, if we want narrow spectral transmission then both the interference order m and the finesse F must be high. Big finesse means that reflectivity of mirrors must be close to unity, which is quite achievable with dielectric multilayer coatings. As to the interference order m, it cannot be made high because then consecutive orders become inseparable. Therefore, in IFs, $m = 1-3$. Suppose, for example, we want to make an IF for 600 nm at normal incidence with reflecting stack with $R = 0.8$. This reflectivity gives finesse only 15. In order to make relative spectral width narrower we have to choose bigger interference order, say $m = 2$. For these parameters, theoretical transmission curve for a single cavity given by the Airy formula is shown in Fig. 7.17.

Now, the problem is how to block spectral maxima other than 600 nm. The shorter wavelengths can be blocked by a red filter, but the longer wavelengths cannot. This is usually done by inserting the second interference filter with $m = 1$. Transmission of such a sandwich is the product of the two individual curves, which gives already acceptable result (Fig. 7.17). Not only transmission at 1200 nm is blocked but also the peak at 600 nm is narrower.

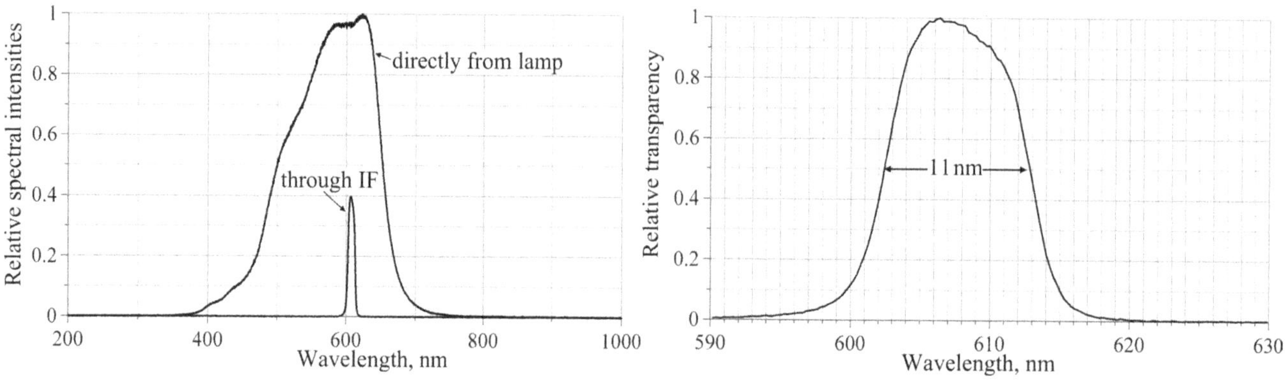

Fig. 7.18. Spectral transmission of an IF designed for 610 nm. Design width 10 nm. Broad angular illumination.

An example of real transmission curve of a most popular IF with 10 nm bandwidth is shown in Fig. 7.18. Comparison with direct flux of tungsten-halogen lamp shows transmission efficiency better than 40% (at left). Spectral width is approximately 11 nm at half maximum (at right).

Some attention should be paid to angular properties of interference filters (IFs). From the aforementioned, it follows that the resonant wavelength decreases with the angle of incidence as $\cos\theta$. Therefore, if very narrow spectral line is needed then it is better to place the IF in a parallel beam. In this configuration, some spectral tuning is possible by tilting the IF relative to the beam. On the contrary, if the beam is diverging then some very minor widening of spectral transmission may be expected.

7.2.4. *Subtraction of plasma fluctuations*

As was explained in Section 7.2.1, plasma fluctuations can be subtracted from the endpoint signal using two photomultipliers with interference filters adjusted to two independent spectral lines. The first spectral line must correspond to characteristic radiation of the endpoint species, and the second one — to some chemically neutral agent that does not participate in chemical reaction, like argon, for instance. However, the best choice for the second spectral line is the one that changes oppositely to the first spectral line during endpoint. If electrical amplitudes of these two signals are equalized, then their subtraction not only removes plasma fluctuations, but additionally doubles the endpoint variation. Example of such two-channel photoreceiver is shown in Fig. 7.19. Diameter of the front-end windows on the photomultipliers (Hamamatsu H6780-20) is 8 mm, which makes it possible to use wide optical fiber bundles of the same optical diameter — the feature that dramatically increases sensitivity of the receiver, comparing to grating spectrometers. For easy exchange of interference filters, removable metal cartridges keep together the half-inch filter capsules and the ferrules of the optical fibers, which have exactly the same diameter as the filters.

For subtraction of plasma fluctuations, it is important to equalize electrical signals A and B in two channels before any mathematical computations. This is supposed to be done manually by adjusting gains of the photomultipliers from the front panel of the device. When this is done, the signal A itself and the reference signal B undergo equal plasma fluctuations F:

$$A + F \quad \text{and} \quad B + F.$$

Thus, subtraction completely removes fluctuations F:

$$A + F - (B + F) = A - B.$$

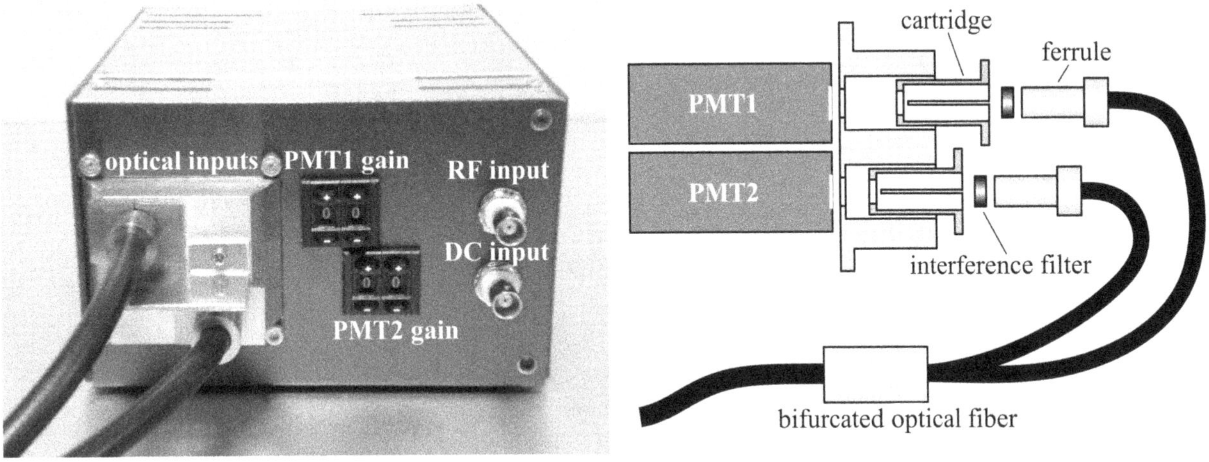

Fig. 7.19. Possible design of the two-channel photoreceiver.

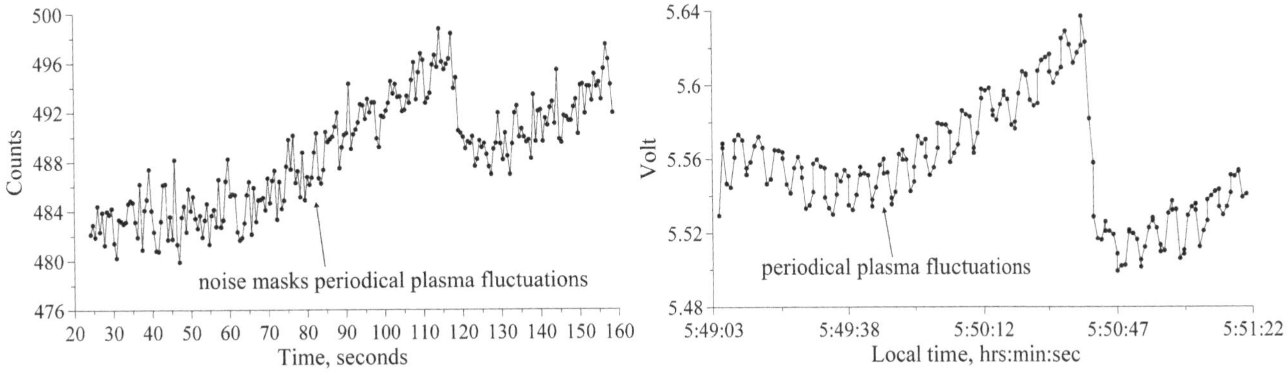

Fig. 7.20. Endpoint trends at the wavelength 387 nm (CN spectral line) measured simultaneously by the grating spectrometer (at left) and by two-channel photoreceiver (at right).

Grating spectrometers do not have equalization option: signal channel with the amplitude a and added plasma fluctuations f differ in amplitude from the reference channel $B + F$. Therefore, subtraction

$$a + f - (B + F) = a - B + f - F,$$

or normalization

$$\frac{a+f}{B+F} \approx \frac{a}{B}\left(1 + \frac{f}{a} - \frac{F}{B}\right),$$

which is commonly used in algorithms supplied with the spectrometer, both contain f and F, meaning that fluctuations f and F are not removed.

Performance of this technique was tested on etching the nitride layer on 1×1 cm^2 coupon of silicon wafer. Use of small test coupons instead of the full-size 300 mm wafer is a standard technique to simulate etching on low open area wafers. Particularly, the 1×1 cm^2 coupon corresponds to about 0.1% open area — a very challenging value for endpoint detection. Figure 7.20 compares the fragments of two endpoint trends of the same coupon obtained simultaneously, using the highly sensitive grating spectrometer of the type shown in Fig. 7.3 and the two-channel photoreceiver shown in Fig. 7.19. In both devices, trends were spectrally adjusted to CN characteristic lines around 387 nm.

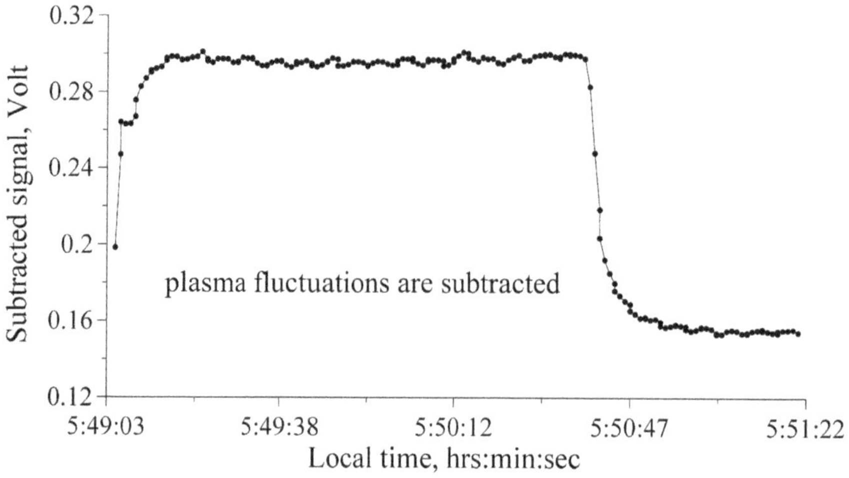

Fig. 7.21. Subtraction of plasma fluctuations in the trend signal.

It must be kept in mind that spatial non-uniformity of plasma does not influence the endpoint trend on the coupon, because its size is much smaller than the possible area, at which noticeable non-uniformity of plasma may reveal. Therefore, the endpoint drop of the trend curve is sharp.

Figure 7.20 clearly shows regular periodical plasma fluctuations caused by numerous control systems of the plasma chamber, including gas filling system, radio-frequency matchers, etc. In order to subtract them, the second channel of the two-channel photoreceiver was spectrally adjusted to the CO line 483 nm — one of the two oppositely varying spectral lines of CO, commonly used for endpoint detection. The result of direct subtraction of the two channels is presented in Fig. 7.21. Clearly, the decision on the endpoint can be made much more reliably.

7.3. Diagnostics of Pulsed Plasma

In recent decades, pulsed plasma became the basic technology for etching nanoscale patterns on silicon wafers. It demonstrated several advantages over continuous plasma: higher etching rate, better uniformity, less structural damage to the pattern. Three basic types of pulsing are used in semiconductor manufacturing: the source pulsing, i.e. periodical turning on and off the radio-frequency generator on the chamber; bias pulsing, i.e. periodical on-off switching of the electrical potential on the chuck; and the combination of the two aforementioned. In the source pulsing, plasma completely interrupts during turning off the radio-frequency generator, and then ignites again as the radio-frequency power returns. In this mode of operation, optical emission spectroscopy is a valuable tool to monitor plasma ignition status and radicals lifetimes. To begin with, consider plasma ignition sensors.

7.3.1. *Plasma ignition sensors*

In the source pulsing mode, the primary parameter is delay of plasma ignition relative to the electrical synchronization pulse that turns the radio-frequency generator on. For reliable measurements, temporal resolution of the sensor should be less than one microsecond. An ordinary photomultiplier (PMT) alone easily satisfies this requirement. However, in order to preserve high speed after connecting to read-out electronic circuits, some precautions should be made and common mistakes avoided. Consider these rules in more detail.

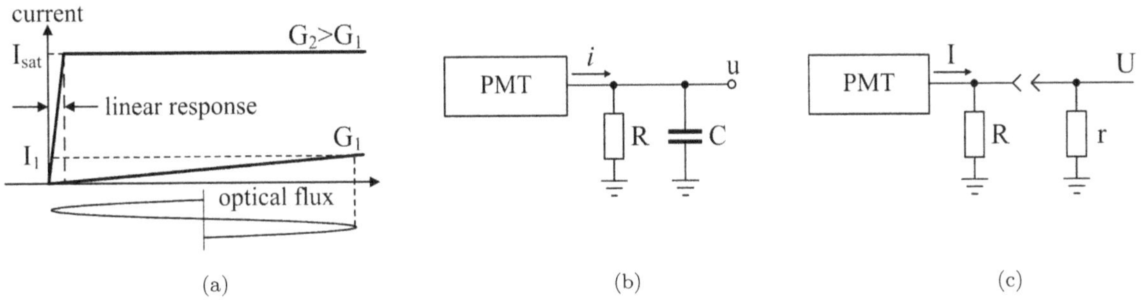

Fig. 7.22. Three reason why a photomultiplier requires additional amplifier.

Applications, dealing with continuous signals, definitely require additional amplifier after the PMT in order to increase dynamic range, improve frequency response and load capacity. These three reasons are summarized graphically in Fig. 7.22 and briefly commented below.

With high PMT gain, the interval of linear response, also called dynamic range, may be too narrow for the input optical swing (Fig. 7.22(a)). High output voltage obtained with large load resistor R may be compromised by low cutoff frequency due to finite load capacitance C (Fig. 7.22(b)). High load resistor R may be shunted by input resistance r of the recording device, which is typically 50 Ohm, thus lowering the voltage to the noise level (Fig. 7.22(c)). Gain G_2 gives large current swing up to I_{sat} but narrow range of linear response. Gain $G_1 < G_2$ may give necessary linear response range but smaller current swing I_1. The solution is to set PMT gain at G_1 and use additional amplifier with the gain I_{sat}/I_1. Thus, optimal performance requires adjusting of two parameters: the PMT gain and the gain of the amplifier.

At high frequency, variable component of the output voltage

$$|u| = \frac{iR}{\sqrt{1 + (RC\omega)^2}}.$$

is proportional to the load resistance R, but cutoff frequency $\omega = (RC)^{-1}$ is inversely proportional to it. The voltage amplitude can be increased by additional amplifier, but the cutoff frequency cannot. There are two solutions to this problem, both using amplifiers. In the simplest one, the amplifier does not provide negative feedback to PMT. Therefore, first priority is to guarantee cutoff frequency by lowering R, and then to compensate for the amplitude by setting proper amplifier gain. In the more efficient scheme discussed in the next paragraph, the amplifier sets negative feedback to PMT, not only amplifying the signal but also improving cutoff frequency.

Without the load, PMT output voltage $U = I \cdot R$ may be high, if R is chosen large. Connection of PMT to outer circuits with lower input resistance $r < R$ lowers the signal to $U = I \cdot Rr/(R + r)$. Therefore, it is always recommended to connect PMT to an operational amplifier, which always has high load capacity. The basic circuit that works for all the applications below 10 MHz is shown in Fig. 7.23. Due to high open-loop gain of the operational amplifier, only negligible current flows into the inverting input. Therefore, almost entire PMT current I flows through the feedback resistor R, making output voltage $U = I \cdot R$ regardless of the load. Trimming resistor $R_{\text{t}} \sim 10$ kΩ sets zero output voltage. If small output bias is not important, the non-inverting input may be grounded (dashed line). Variable feedback resistor $R < 1$ MΩ sets the amplifier gain.

For high-speed applications, the operational amplifiers with gain-bandwidth product more than 100 MHz should be used. Second amplification stage may be applied as in Fig. 7.24 when the highest bandwidth is needed. Fast laser diode generates the pulse train. Horizontal dotted lines show the trigger level.

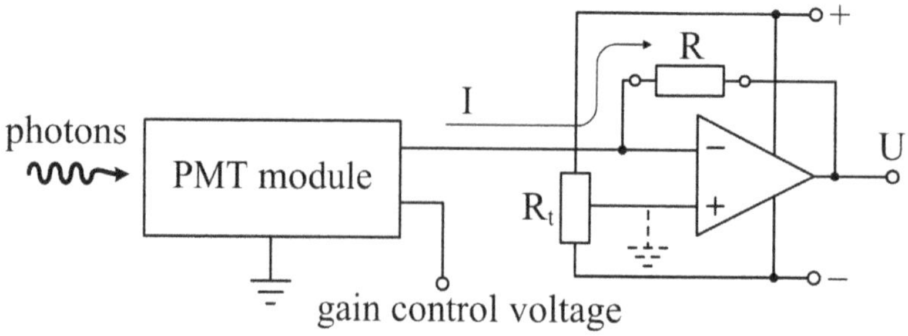

Fig. 7.23. Typical electronic scheme of amplified photomultiplier.

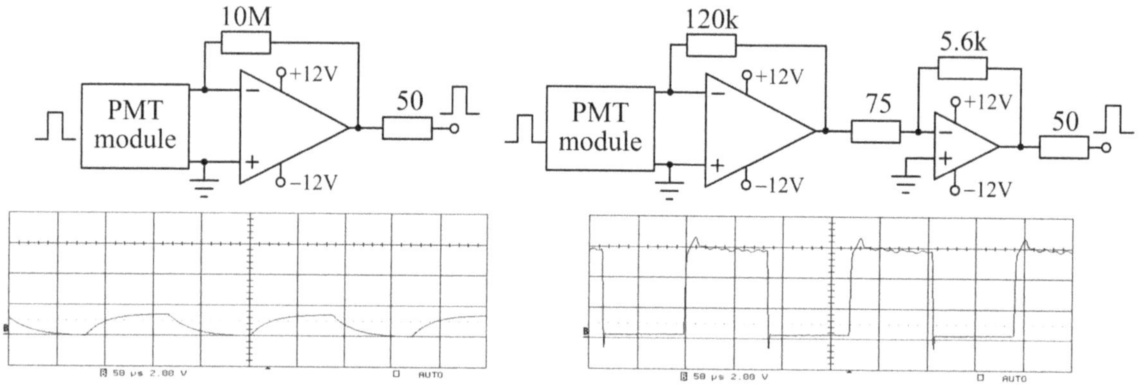

Fig. 7.24. Two consecutive operational amplifiers may improve frequency response. Scale in the left lower corner of each oscilloscope trace reads: 50 μs; 2.00 V.

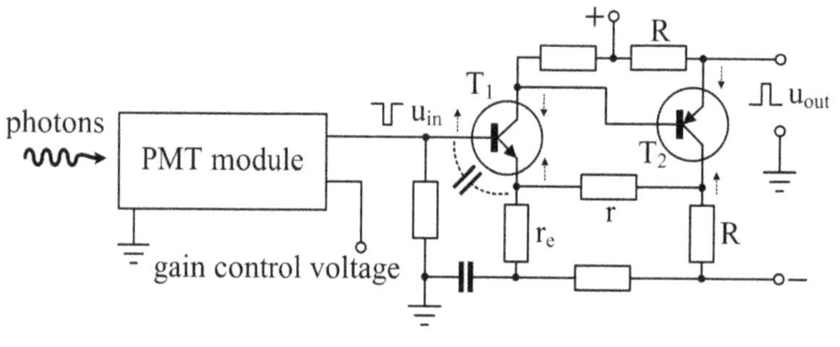

Fig. 7.25. High-speed PMT amplifier with discrete transistors.

Operational amplifiers are designed with the priority for high open-loop gain $\sim 10^6$, which inevitably sacrifices speed. This high gain plays its role with photodiodes, when it is necessary to shunt relatively large junction capacitance about hundreds of picofarad. Output capacitance of PMTs is much smaller, only few picofarad. Therefore, large negative gain is not needed, and even faster response can be obtained with discrete transistors (Fig. 7.25). Opposite polarity transistors T_1 and T_2 form an efficient amplifier with an open-loop gain proportional to the product of h_{21} of the transistors. The resistor r establishes an in-phase feedback, which is actually a negative feedback because it minimizes voltage drop on the emitter-base junction of T_1. This feedback virtually disconnects the emitter-base capacitance from the circuit, improving frequency response. Voltage gain is set by the product r/r_e, which may be $\sim 10-30$. Rise time of the order of 1 ns is achievable. Dashed arrows show phase of voltage variation.

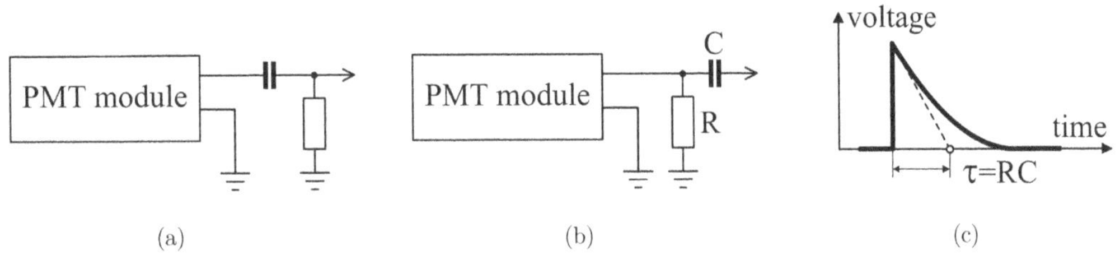

Fig. 7.26. Wrong use of capacitors in the PMT circuits.

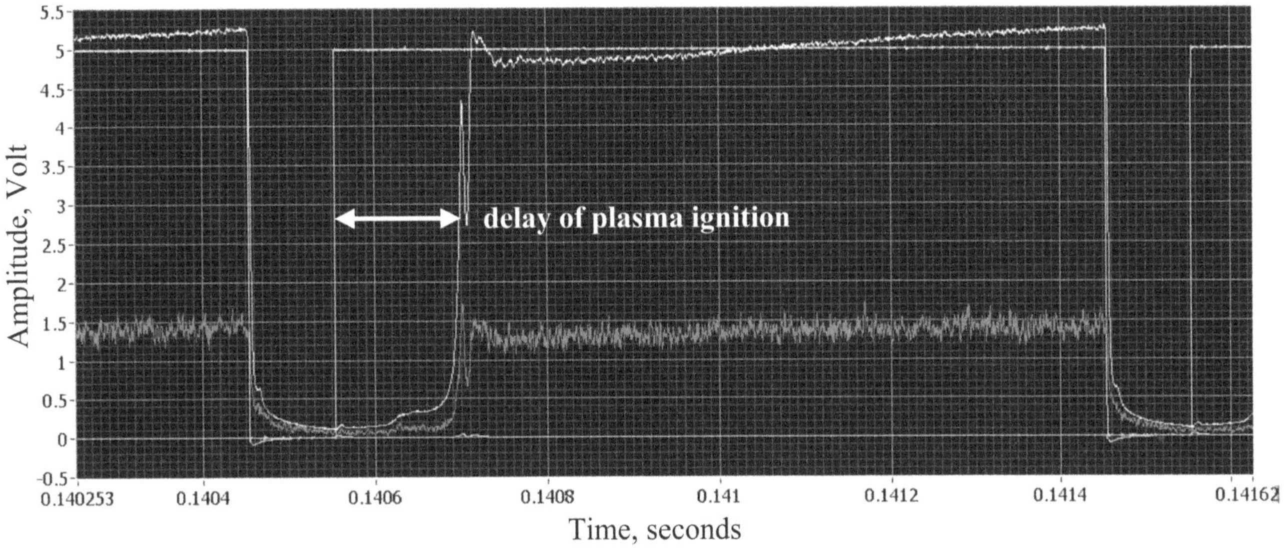

Fig. 7.27. Two optical and one electrical trigger signal recorded by two-channel photoreceiver. White line: white-light emission; red line: oxygen emission; blue line: electrical trigger signal; green line: not used. The timescale is in seconds.

Common mistakes that are known with PMTs are mostly associated with the use of capacitors (Fig. 7.26). When strong background is present together with modulated component, it is always better to filter it out somehow. The straightforward solution is to insert a capacitor somewhere, and the simplest mistake that could be made is to place it right after the PMT anode (Fig. 7.26(a)). Since the current flows in one direction, the capacitor gets charged immediately to the voltage that blocks electrons from reaching the anode. The current stops.

The scheme in Fig. 7.26(b) also works poorly, creating inexplicably long tails on short pulses (Fig. 7.26(c)). The reason is simple: when optical pulse stops, the anode becomes disconnected from any source of electrons, and the capacitor can discharge only through the resistor R. The time constant is at best RC (if the impedance of the right part of the circuit is much less than R). On sinusoidal signals, however, the circuit will work without noticeable errors, if not to mention phase delay.

After the above introduction, it is time to look at what accurately design photoreceiver may give for monitoring pulsed plasma. The two-channel photoreceiver shown in Fig. 7.19 repeatedly measures real time traces of two optical and two electrical signals on a time interval of about 1 second with 0.4 μs resolution, displays this sequence, performs statistical analysis, and stores statistical data in a file. An example of recorded traces obtained on source pulsed argon–oxygen plasma are presented in Fig. 7.27. Conditions: pulsing repetition rate 100 Hz; argon flow 100 sccm/s; oxygen flow 20 sccm/s; pressure 10 mTorr.

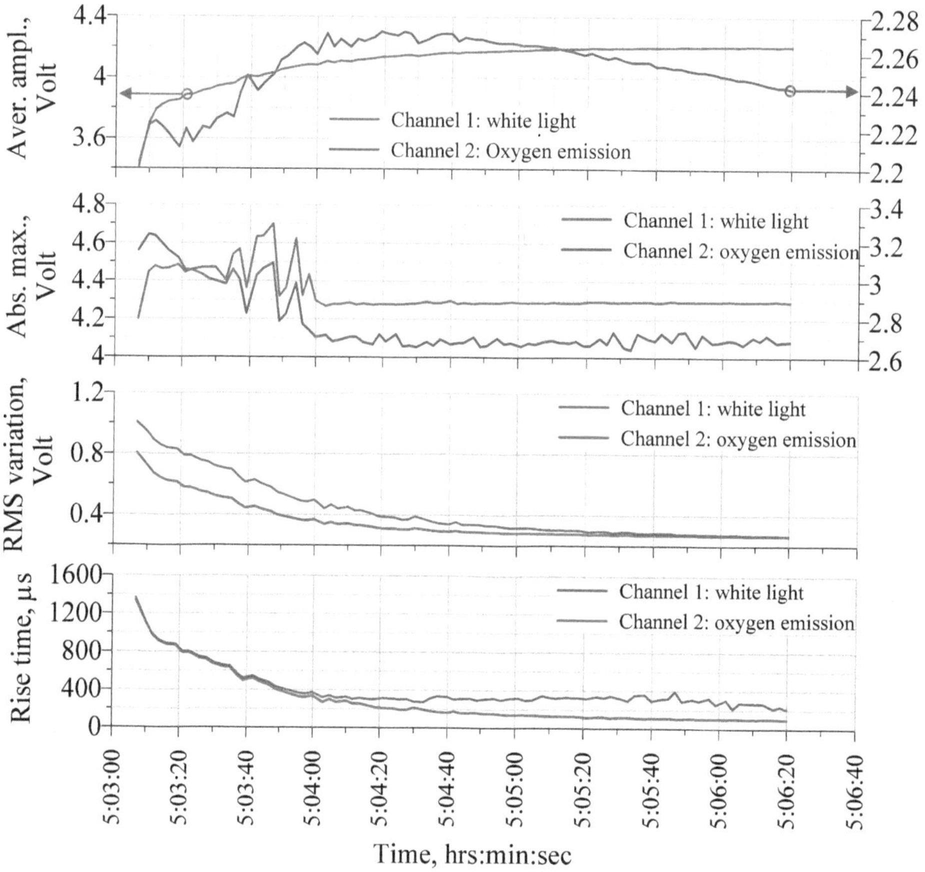

Fig. 7.28. Examples of computed trends: average amplitude, maximum amplitude, variation of amplitudes, and rise time of plasma ignition.

With evident manipulations of the recorded data, it is possible to compute and present as temporal evolution the following parameters: average intensity of optical emission; absolute maximum amplitude during pulse; real-mean-square variation of optical emission; rise time, and many others. All these parameters, being analyzed as trends on long term scale, give valuable information about stability of the process. Examples of such trends, computed for the same process explained above, are shown in Fig. 7.28.

Apart of computerized opto-electronic sensors of the type described above, technicians in manufacturing lines require simpler instruments, which may be called "pocket type tools". Such devices are supposed to be of a hand-held type and used without computers, requiring only oscilloscopes — equipment that is available in any manufacturing facilities. An example of such pocket tool is shown in Fig. 7.29. Its design is basically the same as in Fig. 7.19, with only two modifications: the photomultipliers are powered from +5 V, and their electrical outputs are not digitized but directly connected to two BNC connectors. The device is powered from USB connector, which may be found on the oscilloscope, and the outputs may be connected to the inputs of the same oscilloscope by means of standard coaxial cables.

Although only two spectral channels in plasma ignition sensors may suffice basic demands of plasma diagnostics, there are many special cases when more spectral channels must be monitored in pulsed plasma. Besides, it is inconvenient to swap manually interference filters every time when it is necessary

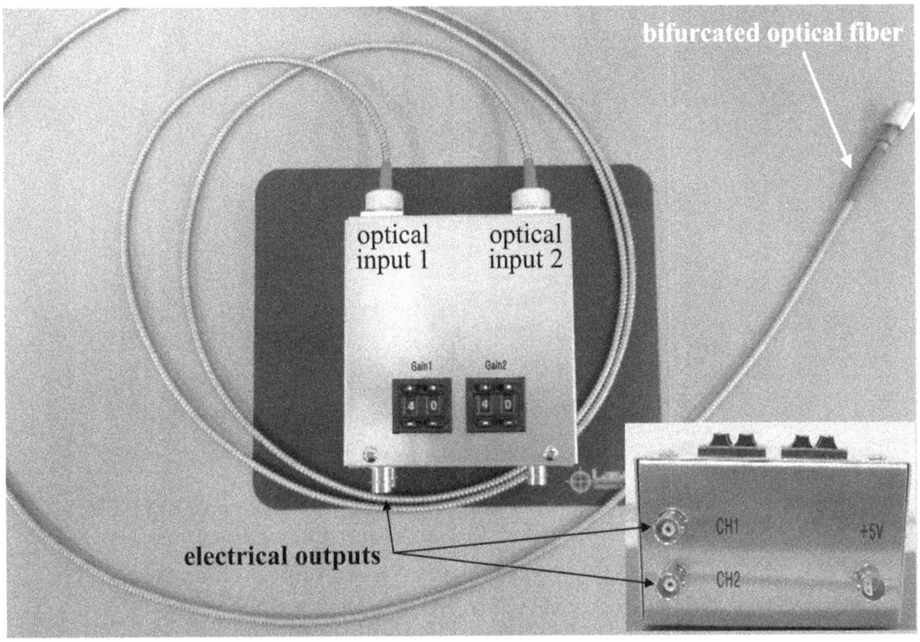

Fig. 7.29. Pocket-type opto-electronic sensor.

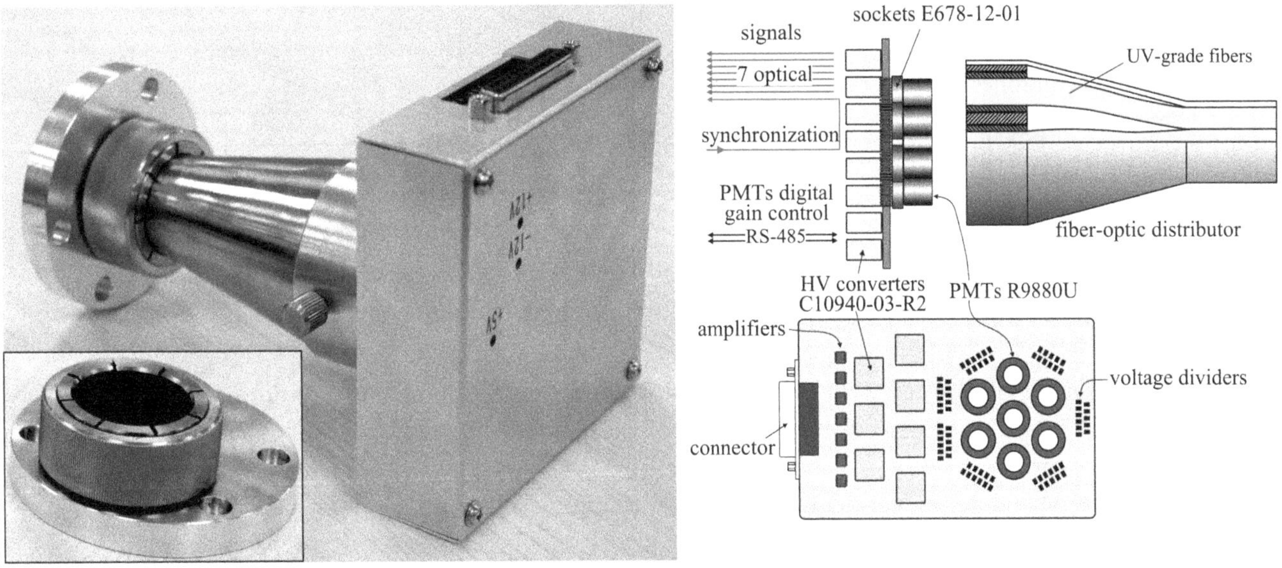

Fig. 7.30. The multi-channel sensor.

to change spectral line of interest. In order to satisfy these requests, a more complicated multi-channel sensor may be used, as the one shown in Fig. 7.30.

It is a computerized seven-channel sensor with seven pre-installed interference filters for most commonly monitored spectral lines. With that big number of filters, it is rarely necessary to change them in real applications. However, when new spectral lines are to be monitored, the cover of the sensor may be removed and new filters installed. Seven photomultipliers Hamamatsu R9880 are installed hexagonally and connected optically to the viewport of plasma chamber by means of optical fiber distributor. The entire sensor installs on the viewport by means of an imploding segmented flange, where the stainless

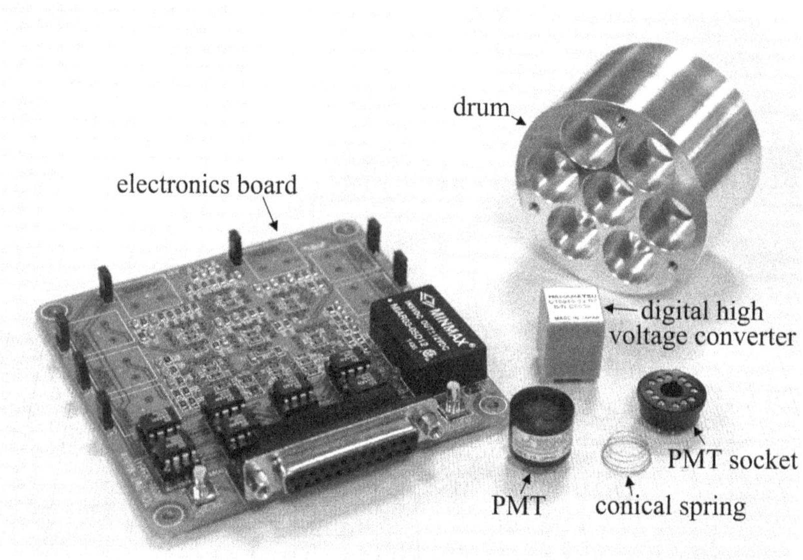

Fig. 7.31. Parts of the multi-channel sensor.

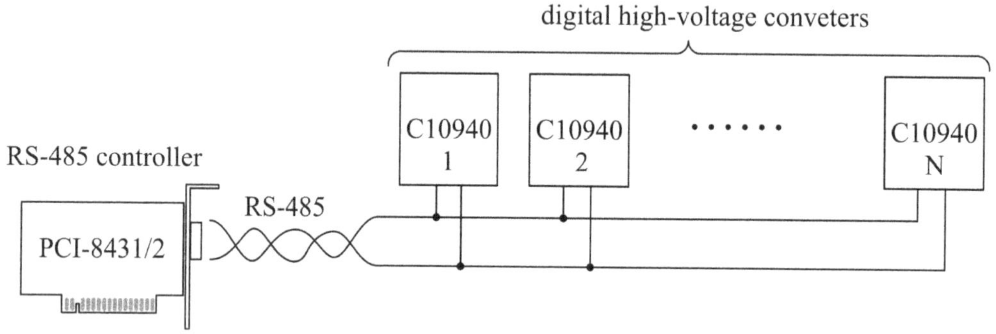

Fig. 7.32. Principle of digital control of high-voltage converters.

steel body of the sensor is clamped with the threaded ring (red in the photo). Other details of the design are clear from Fig. 7.31. Interference filters are positioned inside cylindrical holes of the drum and fixed in there by conical springs.

The multi-channel sensor implements the feature of digital photomultipliers, i.e. the gain of each photomultiplier can be independently digitally controlled from the computer. It reduces the number of manual controls on the sensor and, consequently, the size of sensor, and enables distant control when the sensor is installed on the process machine. The concept of digital control is outlined in Fig. 7.32. Seven Hamamatsu miniature digital high-voltage converters C10940-03-R2 receive the RS-485 digital code to change the high voltage applied to each photomultiplier.

Every command consists of 12 bytes. For every command, PCI-8431/2 generates succession of bytes, sends them to all the C10940 modules, and receives the reply. Only one module responds to the command whose address is included in the command. Figure 7.33 shows how it happens.

Calibration shows almost perfect linearity of the output voltage as a function of the hexadecimal code (Fig. 7.34). Uncertainty of voltage is below ±0.5 V in all the range from 0 V to 1024 V.

The fiber-optic distributor transfers light almost uniformly from its front surface to the photomultipliers (Fig. 7.35).

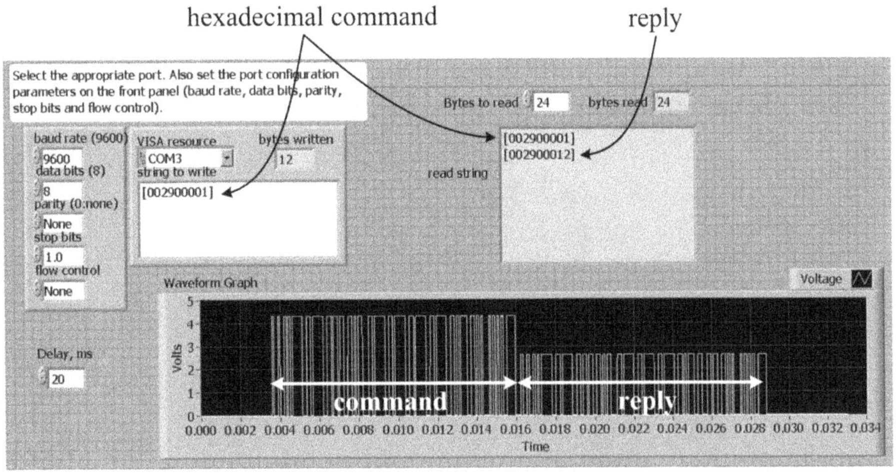

Fig. 7.33. Series of digital commands that is sent to high-voltage converters to change the output voltage.

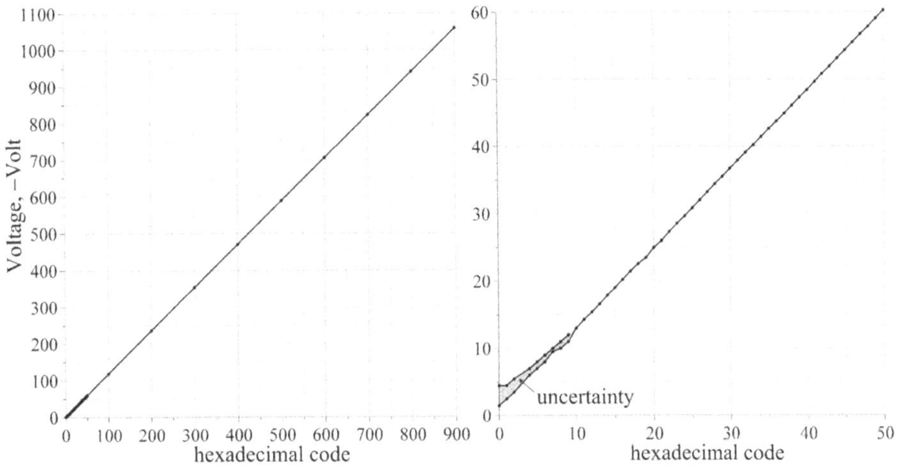

Fig. 7.34. Calibration curves.

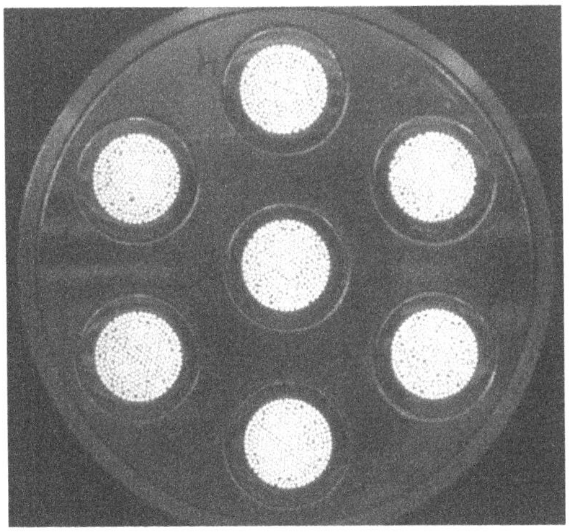

Fig. 7.35. Uniformity of light transferred from the front section to photomultipliers.

7.3.2. *Life time measurements*

Plasma pulsing enables measurement of lifetimes of radicals τ — a very important parameter for plasma-chemical reactions. When plasma ignites, electrons start excitation of species with the rate

$$P(t) = \begin{cases} 0, & t \leq 0, \\ P, & t > 0, \end{cases}$$

where t is time. The optical flux, generated by decaying species, is proportional to the number of excited species n, which is governed by the equation

$$\frac{dn}{dt} = P(t) - \frac{n}{\tau},$$

where τ is the decay time (the lifetime). Solution to this equation for $t \geq 0$ is the exponentially rising function

$$n(t) = \frac{P}{\tau}(1 - e^{-t/\tau}).$$

When the discharge stops, concentration of excited species falls according to the equation

$$\frac{dn}{dt} = -\frac{n}{\tau},$$

giving the exponentially falling solution

$$n(t) = \frac{P}{\tau}e^{-t/\tau}.$$

The entire shape of the optical flux is the combination of the rising and falling parts, as shown in Fig. 7.36. Thus, measuring the front and rear inclinations of the optical pulse, it is possible to compute the decay time (lifetime) of the species.

Two options exist for measuring lifetimes of radicals in plasma: the PMT-based high-speed photoreceivers described in the preceding section, or the gated spectrometers described in Chapter 5. The first option does not necessitate synchronization to electrical ignition pulse, making clear measurements at any delays between electrical ignition and optical emission. The second option, as it will be explained

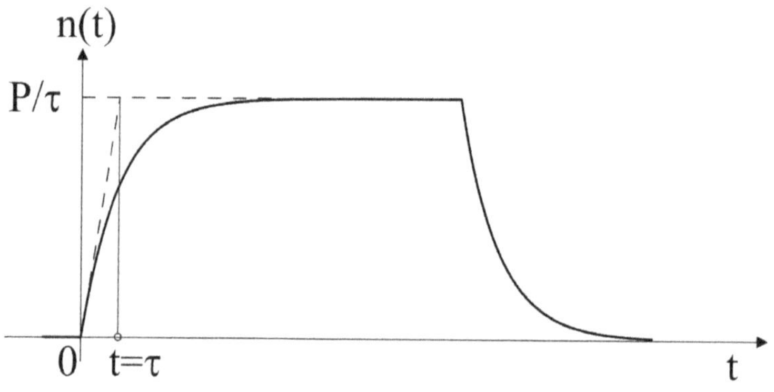

Fig. 7.36. Exponential pulse.

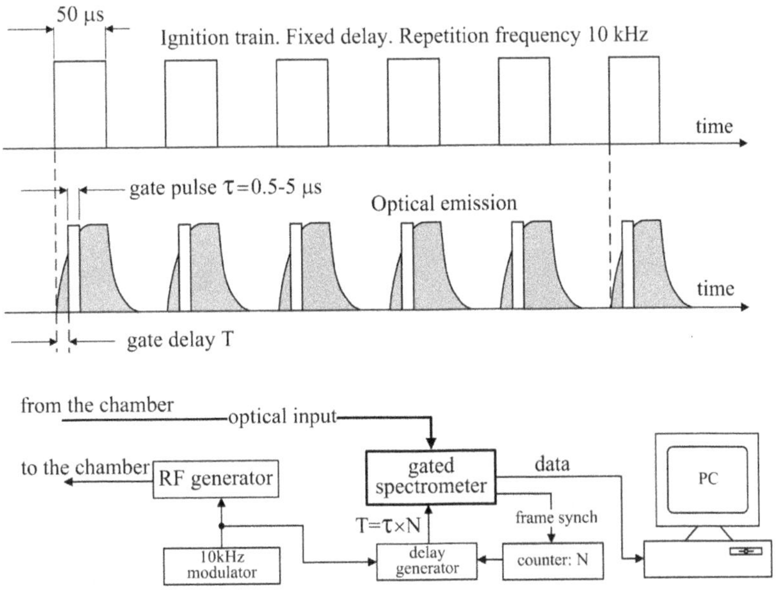

Fig. 7.37. The concept of measuring temporal evolution of optical emission in source pulsing.

below, does require such synchronization, and quality of measurements heavily depends on the stability of this delay. The delay between electrical ignition and optical emission, being quite stable in capacitively coupled plasma, is relatively unstable in inductively coupled plasma. Therefore, PMT-based photoreceivers are simpler in use and may be recommended for any plasma discharge, whereas the gated spectrometers work reliably only on capacitively coupled plasmas. However, unlike PMT-based photoreceivers with interference filters adjusted to only few fixed spectral lines, gated spectrometers provide two very useful advantages — visualization of the entire spectrum of optical emission during measurement, and programmable change of monitored spectral lines, the number of which is practically unlimited. Therefore, this technique deserves more elaborated discussion.

Temporal response of optical emission to pulsed ignition can be measured by means of synchronous gating and accumulation of optical spectra as outlined in Fig. 7.37.

In the source pulsing, the radio-frequency generator, igniting the plasma, is modulated, thus forming the high-voltage ignition train. The optical emission responds to it, producing exponential pulses like in Fig. 7.36. The gated spectrometer, connected optically to the chamber, records entire spectrum during short gating pulse of about microsecond duration. The width of the gate pulse must be chosen so narrow that to reasonably resolve the temporal evolution of optical emission during ignition. For instance, with the ignition pulse 50 μs wide, the gate pulse should be around 0.5 μs in order to obtain 100 resolved points on the exponential curve.

Once the relative delay T of the gate pulse is set, the photo-sensor in the spectrometer receives the sequence of short optical pulses, each of the duration τ. Due to physics of the photo-sensor (typically the Charge Coupled Device — the CCD), its output electrical signal is accumulated over all the optical pulses until the accumulation time expires. The accumulation time is set programmably on the spectrometer. Thus, the signal of the photo-sensor is proportional to spectral intensity of optical emission at exactly the time T from the beginning of ignition pulse. The larger the number of accumulated pulses is, the better the signal-to-noise ratio. The upper limit of the accumulated number of pulses is set by the entire time of observation, which cannot exceed several minutes — the operational time of the plasma chamber.

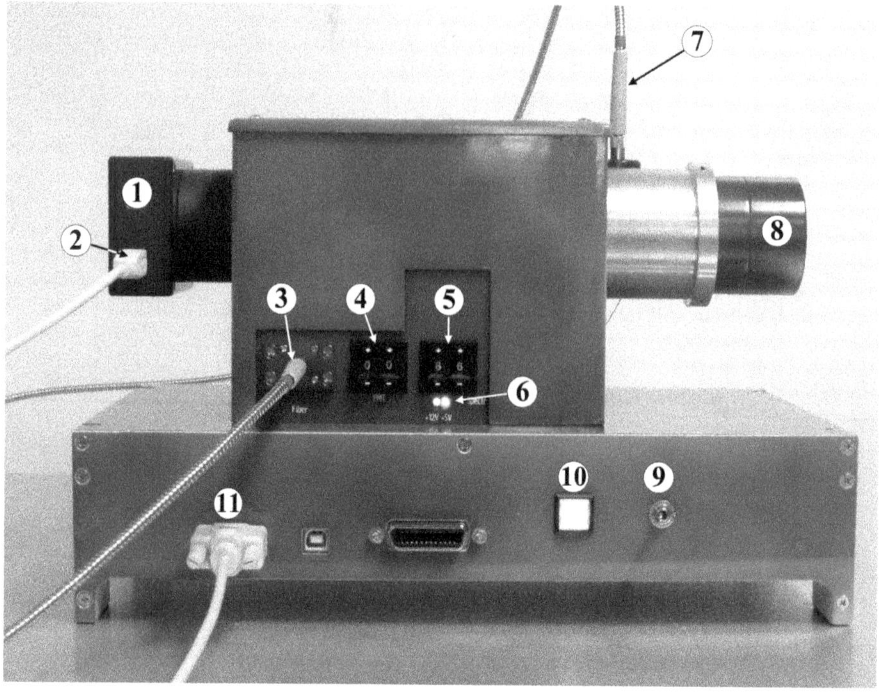

Fig. 7.38. Gated spectrometer for measuring lifetimes of radicals. 1, 2 — photo-sensor and its USB connector; 3 — optical synchronization; 4 — gain of optical synchronization; 5 — spectrometer gain; 6 — power LED; 7 — optical fiber; 8 — spectrometer column; 9 — CW - pulsing switch; 10 — high voltage LED; 11 — delay generator RS-232 connector.

Once the accumulation time of the photo-sensor expires, the spectrometer stores spectral intensities of the monitored spectral lines in the computer, sends a request to the delay generator to increase the delay by one gate interval τ: $T_{k+1} = T_k + \tau$, and the entire sequence repeats itself. Eventually, the entire optical emission pulse is being sensed by short measurements of the duration τ. Practically, since the gating pulse τ is short, the optical energy that reaches the photo-sensor during each accumulation interval is low, which requires long accumulation about several minutes. The gated spectrometer, implementing this algorithm, is portrayed in Fig. 7.38. Its optical scheme is explained in Fig. 5.10 (Chapter 5).

In order to minimize effects of random delays between electrical ignition and optical emission, the optical synchronization is introduced in the form of a fast photodetector coupled to the optical fiber looking into plasma. At the moment when plasma ignites and its optical emission starts, the photodetector generates electrical signal that can be used as synchronization for the measurement sequence. With this function activated, the spectrometer can work on inductively coupled plasma as well.

Typical results that can be obtained on capacitively coupled plasma and electrical synchronization are presented in Fig. 7.39. In this experiment, integration time was set to 2 s, so that with the total 100 samples recorded the entire measurement time was more than 3 minutes. Also, the sum of all spectral components within sensitivity domain of the spectrometer gives the white light response. It is shown in proper scale by the violet line. The dashed lines marked "Ar", "H", "F" and colored accordingly to the legend, indicate the slopes of the exponential rise for these species. Even without detailed computations, it is clear that argon lifetime is the shortest one, followed by hydrogen and fluorine.

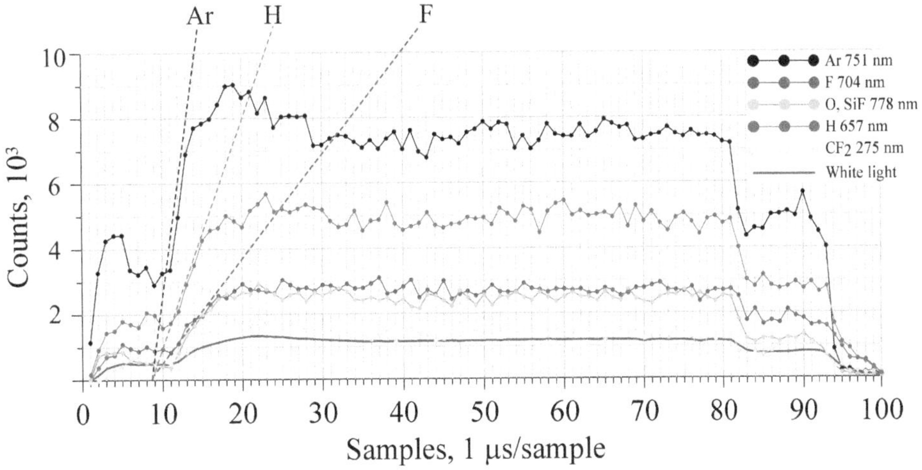

Fig. 7.39. Transition curves of optical emission for several species in capacitively coupled plasma discharge. Amplitude of the 275 nm spectral line is magnified 10 times for better comparison. 10 kHz source pulsing.

7.4. Vertical Distribution of Optical Spectra

One-dimensional imaging spectrometers (Section 4.1) may give useful information about vertical distribution of plasma above the chuck (Fig. 7.1). This is especially important in bias pulsing mode, when narrow, on the scale of 1 cm, layer above the chuck — the so-called sheath — oscillates in vertical direction, accelerating charged particles towards the wafer and enhancing etching effect. For understanding of plasma chemistry on the wafer, it is important to know what exactly species are excited in the sheath. The answer to this question may be given, analyzing modulated component of spectra emitted from the sheath. Thus, the task may be formulated as follows: measurement of modulated component of spectra emitted from thin layer of plasma above the wafer.

7.4.1. *Optical system*

From the optical point of view, observation of such relatively thin layer on the area of 300 mm is not a simple task because the imaging system must have narrow field of view in order to intercept only rays, coming parallel to the wafer surface. The solution is to use the so-called telecentric lens. Consider first the basic optical principle that determines fundamental properties of this type of a lens.

Any ray, coming to an ideal lens with the focal length f parallel to its optical axis, intercepts the optical axis in the focal plane (Fig. 7.40).

A paraxial parallel beam, coming to the lens at an angle α, focuses in the focal plane in a point r, making the same angle with the lens center:

$$\alpha = \arctan\frac{r}{f} \approx \frac{r}{f}.$$

If a diaphragm of radius r is placed in the focal plane, then the bundle of rays, originating at any point in front of the lens and passing through the diaphragm, form a cone with the generatrix angle α. This may be rephrased in terms of spatial frequencies filtering: since the focal plane of a lens is the Fourier-transform plane, only waves with spatial frequencies within small part $\pm r/f$ of the wavenumber can pass through. It is actually the basic concept of a telecentric lens.

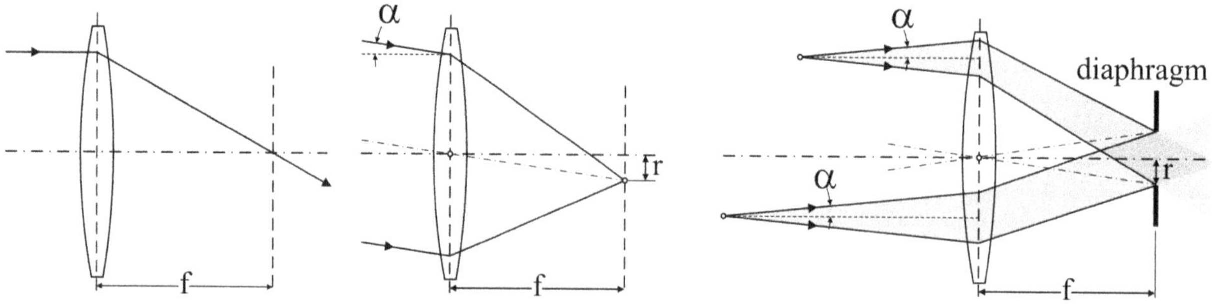

Fig. 7.40. Telecentric configuration relays only those rays that are inside a narrow fixed-angle cone around horizontal axis.

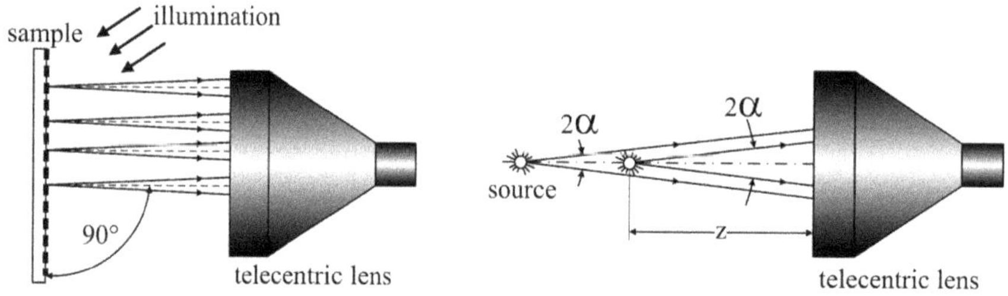

Fig. 7.41. Two basic telecentric concepts: ray cone geometry does not depend on radial position (left) and cone angle is independent of the source axial position (right).

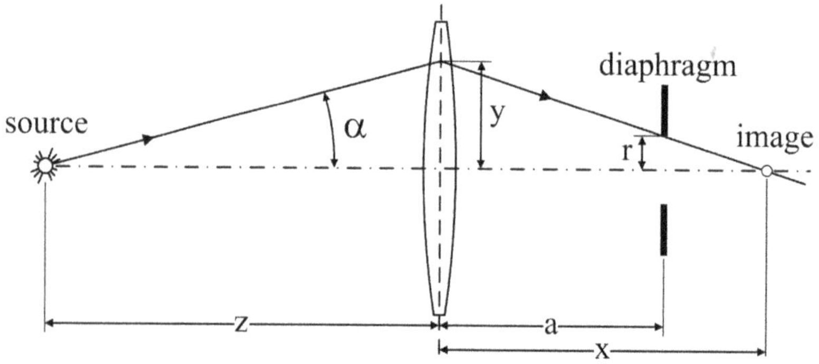

Fig. 7.42. The highest ray, passing through the system, is limited by a diaphragm.

For physical applications, like spectro-photometry, scatterometry, and spectral interferometry, telecentric configuration gurantees two most important concepts, which are portrayed in Fig. 7.41: the axis of the receiving cone of rays is always parallel to optical axis, and the cone angle 2α is constant independently of axial position z of the source. The second one is a very important photometric property, and should be proved with formulas.

Consider a source positioned at z in front of the lens (Fig. 7.42). We are going to analyze how the maximum angle α of rays, passing through the diaphragm, changes with z, and what is the condition for the case when α does not depend on z. Since

$$\alpha = \arctan \frac{y}{z},$$

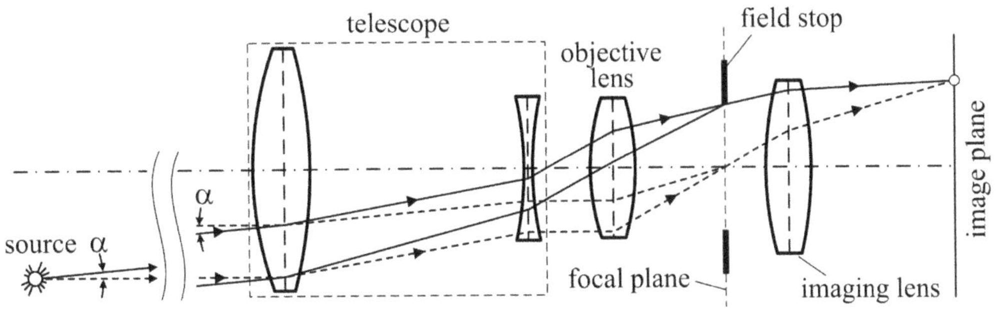

Fig. 7.43. Four steps of telecentric lens design: Galilean telescope, objective lens, field stop, and imaging lens.

it is reasonable to analyze the ratio y/z. The image of the source is formed at x behind the lens, and the lens formula

$$\frac{1}{z}+\frac{1}{x}=\frac{1}{f}$$

gives

$$x=\frac{fz}{z-f}.$$

The highest ray that can pass through the system is determined by similarity relation:

$$\frac{r}{x-a}=\frac{y}{x}.$$

From here

$$\frac{y}{z}=\frac{rf}{fz-a(z-f)},$$

showing that, in general, y/z depends on z, except for one case $a=f$, i.e. when the diaphragm is placed in the focal plane. Then the solid angle, at which the system receives radiation from the source, is independent of the position of the source, which means constant optical flux through the system. If there is a photodetector behind the diaphragm, then its signal does not depend on the position of the source. This is the photometric property of a telecentric lens.

In order to obtain an image of the source, an imaging lens must be added to the output of the basic telecentric scheme in Fig. 7.42. Thus, conceptually, a telecentric lens may be composed of only two lenses. Real optical schemes of telecentric lenses are much more complicated, and the idea of their design should be explained to avoid confusion. First, consider how the spatial filtering concept is implemented in real design, tracing only parallel rays, coming from infinity (Fig. 7.43). Step number 1: a Galilean telescope. Step number two: an short-focus objective lens, focusing the infinity image into its focal plane. Step number three: a diaphragm in this plane, which is better to call a field stop, since it is not a narrow hole any more. This is the end of spatial filtering system. Next, add the final imaging lens that forms an image of the source. The telecentric lens is ready.

There are two types of telecentric lenses basically used in practice: the object-side telecentric and the double-telecentric lenses. The difference is clear from Fig. 7.44. Telecentric lenses always end with the C-mount.

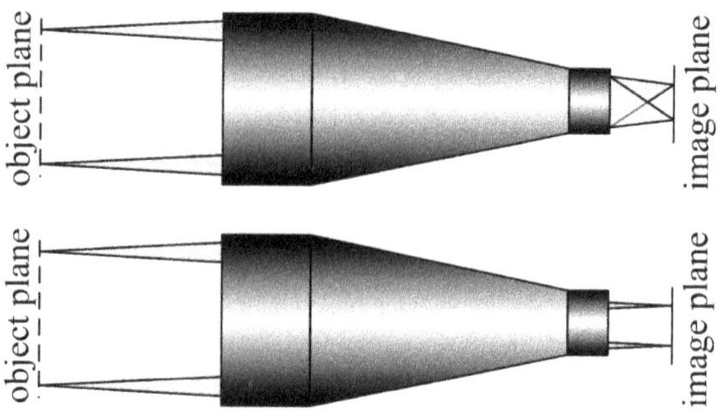

Fig. 7.44. The object-side (above) and double (below) telecentric lenses.

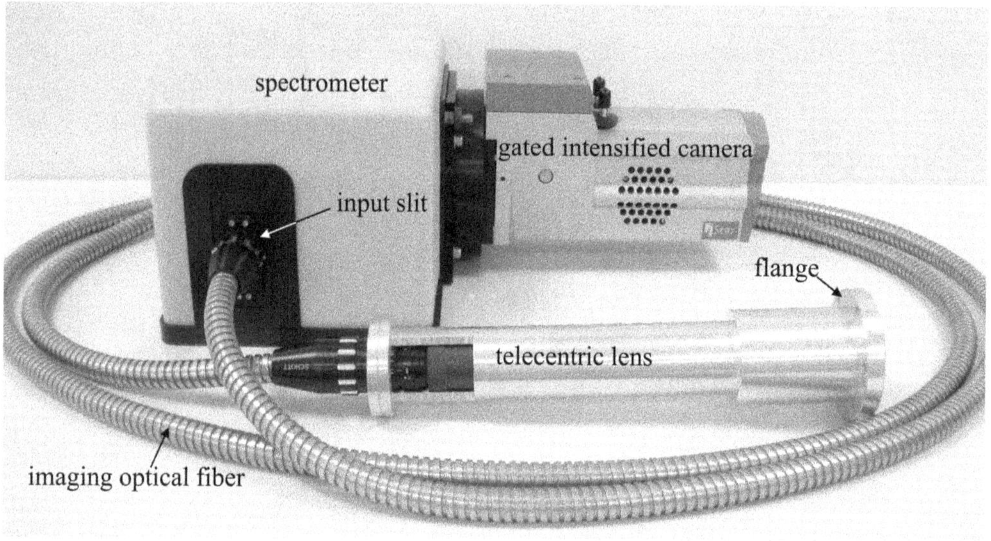

Fig. 7.45. The spectroscopic system for measuring modulated component of plasma emission: HoloSpec imaging spectrometer; Andor iStar gated intensified camera; SCHOTT imaging optical fiber; Edmund Optics telecentric lens.

7.4.2. *Spectroscopic system*

Modulated component of optical emission from plasma, spatially resolved in vertical direction, can be measured, using one-dimensional gating spectrometers (Section 4.1). The size and weight of a gated spectrometer, together with vibrations of the plasma chamber, make it impractical to install all optical components directly on the viewport of the chamber. Therefore, an imaging optical fiber should be used to relay optical image from the telecentric lens to the spectrometer. A possible configuration, suitable for experiments on plasma chambers in manufacturing lines, is shown in Fig. 7.45, and the entire experimental arrangement is outlined in Fig. 7.46.

In order to extract only modulated component of optical emission, the algorithm similar to the one shown in Fig. 7.37 may be used. It is explained in Fig. 7.47. Two sequences of gating pulses are applied consecutively to the spectrometer: with gate pulses in phase and in counter-phase to the bias pulsing.

The two spectra, obtained on two opposite gating sequences, are recorded and saved in the computer. After that, these two spectra are subtracted to obtain the spectrum of the modulated component of optical emission. An example of measurements on 600 W bias power in capacitively coupled $Ar/O_2/CHF_3$ plasma are summarized in Fig. 7.48.

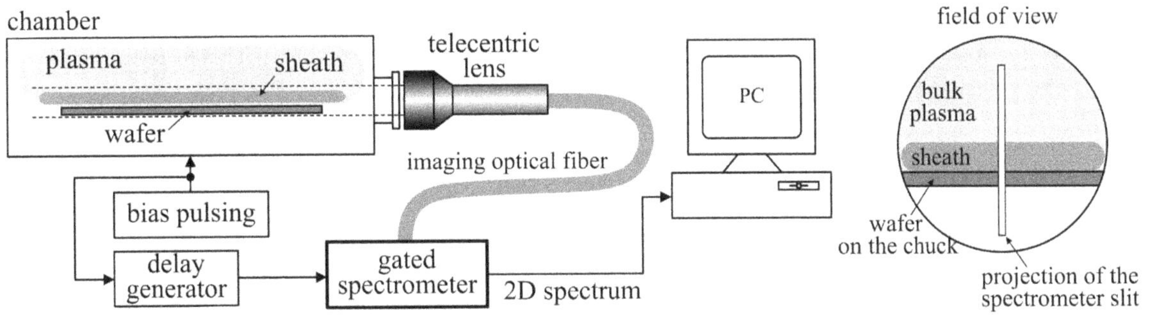

Fig. 7.46. Conceptual scheme of measuring vertical distribution of spectrum from plasma sheath.

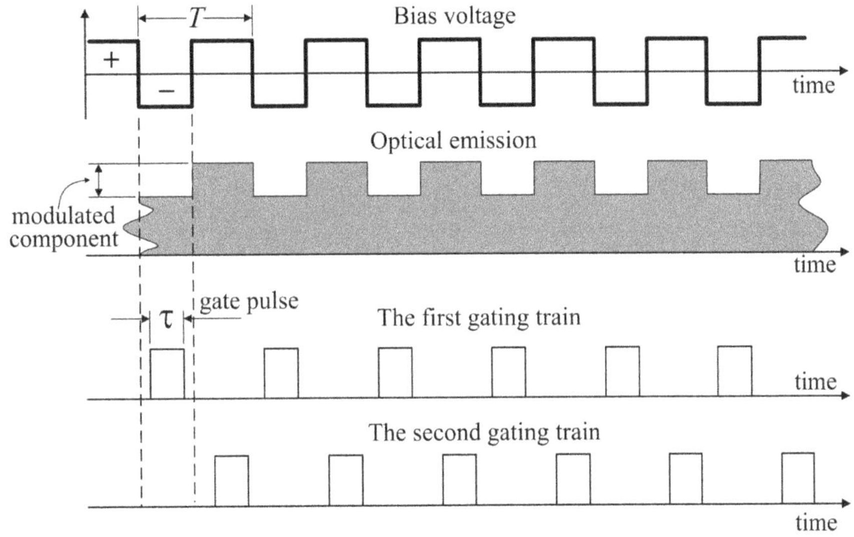

Fig. 7.47. Sequence of pulses for subtracting modulated component of the spectrum.

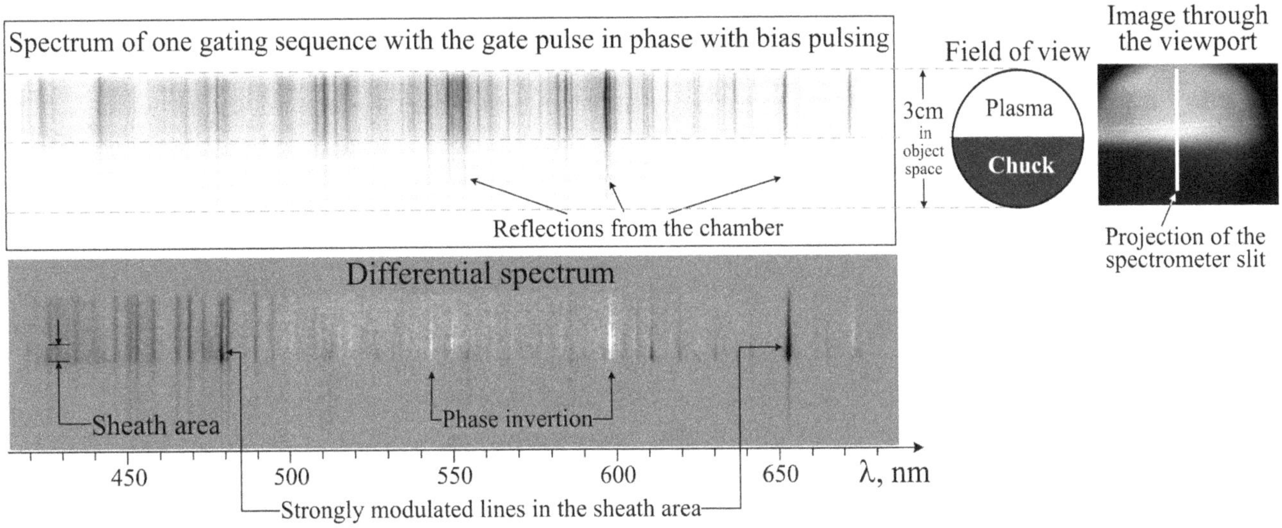

Fig. 7.48. An example of measuring modulated component of optical emission from the sheath area.

It is clearly seen that not all spectral lines are equally modulated and, moreover, some of them are inversely modulated relative to the bias modulation. Discussion of the physics of this phenomenon goes beyond the scope of the present book.

7.4.3. *Spectral analysis of the plasma sheath*

The sheath signal varies periodically in time since the sheath volume is periodically filled with electrons. There are two special time points within the modulation period — t_{max} and t_{min}, corresponding to the maximal and minimal emission intensity. Then t_{max} corresponds to the sheath volume filled with the electrons (bright sheath), and t_{min} corresponds to the sheath volume free from electrons (dark sheath). Subtraction of total optical signal, containing emission from the bulk plasma S_b and the bright sheath $S_s(t_{max})$, and the total signal, corresponding to the dark sheath $S_s(t_{min})$, gives the differential signal that is actually the signal from the volume of the sheath:

$$S_{diff} = [S_b + S_s(t_{max})] - [S_b + S_s(t_{min})] = S_s(t_{max}) - S_s(t_{min}).$$

Note that the signal from the plasma bulk is canceled. Thus, in order to obtain the signal from the sheath, we have to measure total signals, corresponding to the bright and dark sheath, respectively, and then subtract the latter from the former. This can be done by using the gating technique described above (Fig. 7.47). Gating implies that optical signal is measured only during the gating pulse. The pulse width τ should be sufficiently smaller than the modulation period T. For $T = 500$ ns (2 MHz frequency), we can set $\tau < 50$ ns. The gating pulse can be delayed with respect to the reference signal from the 2 MHz generator. In the system shown in Fig. 7.46, the two gating pulses within the period T are delayed by $T/2 = 250$ ns. The global delay relative to radio-frequency generator must be adjusted so that these two pulses correspond to the special time points t_{max} and t_{min}.

Experimental data presented below prove that the spectroscopic system works just as described above. These data were obtained in the process gas mixture $Ar/O_2/CHF_3$. Figure 7.49(a) shows measured total spectra, corresponding to the special time points (t_{max} and t_{min}), and Fig. 7.49(b) shows the differential spectrum. Note that the total spectra contain strong continuum background that does not provide any useful information, while in the differential spectra the background is removed, making analysis more comfortable. Since the system measures the whole spectrum simultaneously, one can use any suitable spectral line for process monitoring.

In the technique explained above, the measurement depends on the common delay between the reference signal from bias pulsing and the two gating trains (see Fig. 7.46). In the experiments presented in Fig. 7.52, the value of the common delay was chosen so that to make measurements at the first gating train at times exactly equal to t_{max} and at the second train — at times t_{min}. What happens when the common delay is chosen arbitrarily is shown in Fig. 7.50, where total intensities of hydrogen atom emission at the wavelength 656 nm are presented as a function of the common delay. The experimental curves demonstrate periodic behavior. These curves, when measured in advance, allow to determine the proper common delay, which should be set in such a way as to maximize the differential signal. From Fig. 7.50 it follows that the proper common delay should be in the range 220–240 ns.

Using one-dimensional imaging capabilities of the spectrometer, it is possible to measure vertical profiles of optical emission at any wavelength within the spectral range of the spectrometer. For example, Fig. 7.51 presents vertical profile of modulated component of hydrogen emission at the wavelength 656 nm at various values of bias power, providing direct experimental proof that this spectral line originates mostly from the sheath rather than from bulk plasma. These data correspond to the spectrum shown in Fig. 7.49(b).

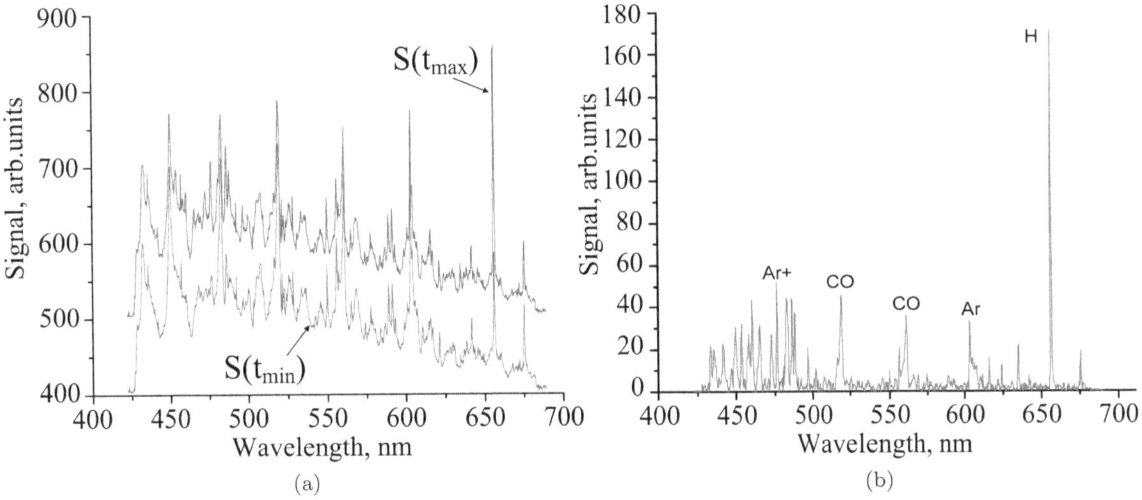

Fig. 7.49. Spectra of plasma emission: (a) as measured at the moments $t_{\max}$ and $t_{\min}$; (b) differential S_{diff}. The red and blue lines in figure (a) are intentionally separated vertically for better visual perception.

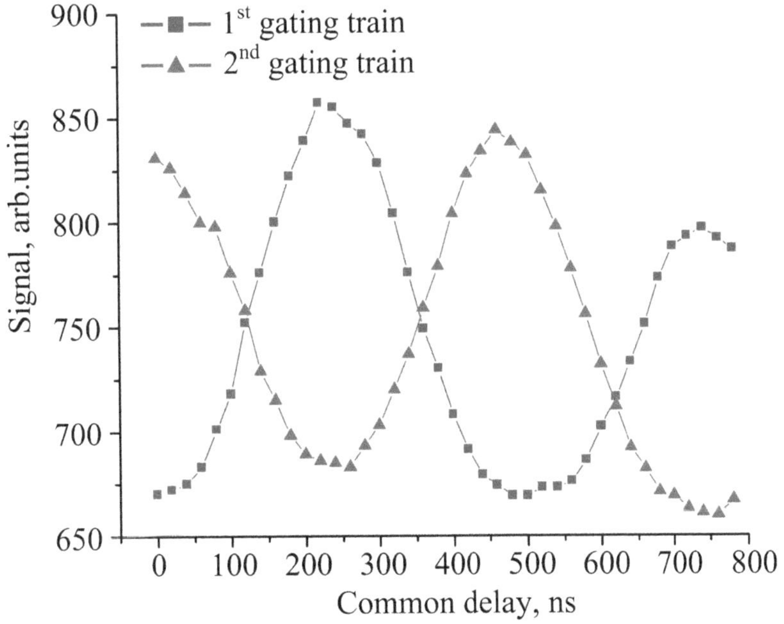

Fig. 7.50. Hydrogen optical emission at 656 nm as a function of common delay for two gating sequences.

From this figure, it follows that the hydrogen emission at 656 nm is localized within the sheath with its maximum at the level about 2 mm above the wafer. The behavior is same for different bias powers (i.e. sheath thicknesses). Note that according to Fig. 7.48, different lines in the spectrum may show quite different spatial behavior: some lines may have maximal intensity in the bulk plasma. At the same time, it is clear that the lines excited close to the wafer are the most useful for monitoring the wafer processes.

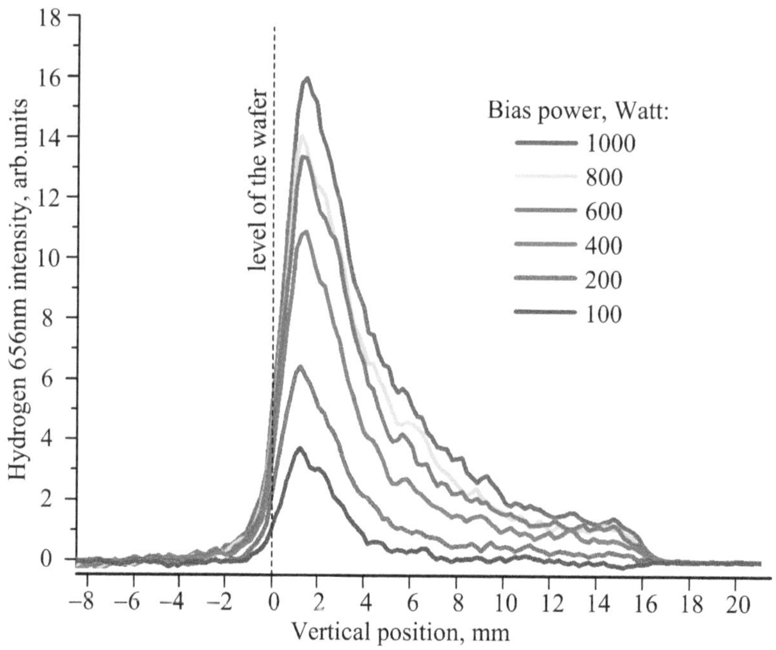

Fig. 7.51. Vertical profile of hydrogen emission at 656 nm.

7.5. Radial Distribution of Optical Emission

Spatial uniformity of plasma during etch is the key factor for high production yield. The inner geometry of plasma chambers in production lines is mostly axially symmetrical, except for minor asymmetry caused by wafer loader and optical viewports. Therefore, the dominant type of non-uniformity of plasma distribution inside the chamber is radial non-uniformity. In laboratory environment, radial distribution of plasma can be readily measured by means of intrusive optical and electrical probes, being inserted into plasma and scanned across its volume. In production lines, only non-intrusive solutions may be used, among which the so-called Abel sensor is the most appropriate variant — an optical sensor that traverses optical emission inside the chamber along numerous chords.

7.5.1. *Theoretical concept*

The idea of the Abel sensor is outlined in Fig. 7.52. A linear photodetector (Chapter 1) receives light, created by optical emission of plasma and accumulated along the chords, connecting the entry pinhole and the pixel of the photodetector.

Axially symmetrical plasma with radial distribution of optical emission $U(r)$ is thus sensed through a series of sections to produce electrical signals $F(y)$ (Fig. 7.53):

$$F(y) = A[U(r)] = 2\int_y^{\infty} \frac{U(r)}{\sqrt{r^2 - y^2}} r\,dr.$$

This formula is known as the Abel transform between functions $U(r)$ and $F(y)$. Remarkably, the inverse Abel transform exists that reconstructs analytically $U(r)$ if $F(y)$ is known:

$$U(r) = A^{-1}[F(y)] = -\frac{1}{\pi}\int_r^{\infty} \frac{dF}{dy} \cdot \frac{dy}{\sqrt{y^2 - r^2}}.$$

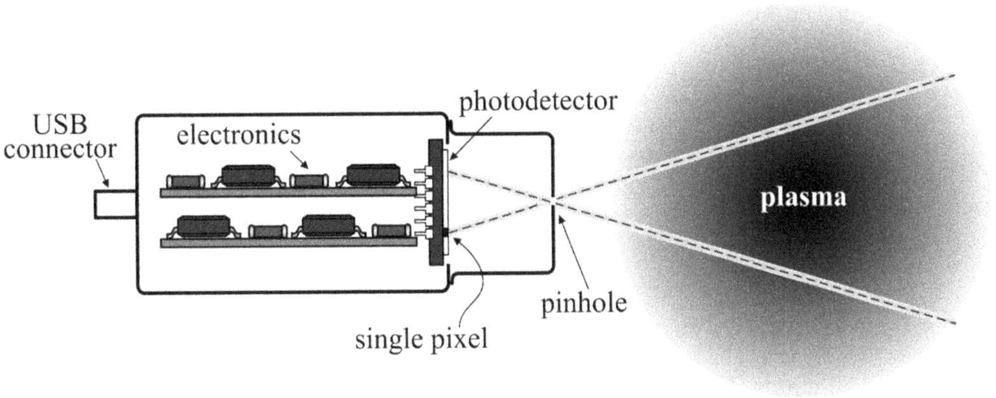

Fig. 7.52. Abel sensor.

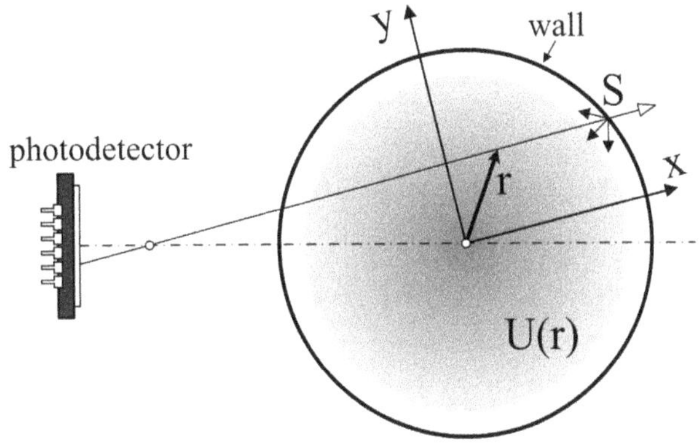

Fig. 7.53. System of coordinates.

Thus, it might seem that if only $F(y)$ is measured, then radial distribution $U(r)$ can be easily reconstructed according to the above formula. Regretfully, this is a delusion: the derivative and singularity under the integral together with inevitable noise of the photodetector, make reconstruction unstable. Therefore, practical implementation of this technique requires special mathematical methods to overcome this instability. Various algorithms were suggested for this purpose without definite winner, and the quest for the best solution continues. In the particular case of plasma non-uniformity, the reconstructed distribution $U(r)$ is expected to be smooth — a requirement that makes the algorithm described below quite acceptable.

The algorithm uses Abel binomial eigenfunctions:

$$f_n(r) = (a^2 - r^2)^{n/2}, \quad r \leq a.$$

Here a is the radius at which plasma turns to zero. Physically, it is the radius R of the circular inner volume of the plasma chamber because no plasma exists outside the wall. All the binomial eigenfunctions with $n > 0$ satisfy the condition of physical reality:

$$f_n(a) = 0.$$

The Abel transform of a binomial eigenfunction is again the binomial eigenfunction of the higher order:

$$A[f_{n-1}(r)] = 2\int_y^\infty \frac{f_{n-1}(r)}{\sqrt{r^2-y^2}} r dr = C_n \cdot f_n(y), \quad C_n = 2\int_0^{\pi/2} \cos^n x \, dx.$$

Binomial eigenfunctions are not orthogonal. However, it is always possible to construct a sum of them that approximates measured photodetector signal $F(y)$:

$$F(y) \approx \sum_{n=1}^{N} X_n \cdot C_n \cdot f_n(y),$$

where X_n are unknown coefficients. Here $F(y)$, C_n, and $f_n(y)$ are known. Coefficients X_n can be found by any fitting routine, for instance the Levenberg–Marquardt routine. With Levenberg–Marquardt routine, fitting into 500 points of y and 20 functions f_n takes only 100 ms.

Since all the eigenfunctions f_n satisfy the condition of physical reality $f_n(a) = 0$, their sum also satisfies it, so that the condition of physical reality satisfies automatically for $F(y)$ in this expansion. Moreover, all the f_n are the solutions of the Abel transform. Therefore, the inverse Abel transform of the approximation for $F(y)$ gives the result in a form of a sum:

$$U(r) = A^{-1}[F(y)] \approx \sum_{n=1}^{N} X_n \cdot f_{n-1}(r),$$

and the requirement of physical reality $U(a) = 0$ is satisfied automatically because all the f_n satisfy it. Thus, the algorithm may be summarized in two steps:

(1) Using the photodetector signal $F(y)$ and the Levenberg-Marquardt routine, find expansion coefficients X_n:

$$F(y) \approx \sum_{n=1}^{N} X_n \cdot C_n \cdot f_n(y).$$

(2) Construct the radial distribution of plasma in the form of a sum

$$U(r) = \sum_{n=1}^{N} X_n \cdot f_{n-1}(r).$$

The solution does not use derivatives $\frac{dF}{dy}$ and does not compute any integral. With 500 points of y and 20 functions f_n, computation takes only about 500 ms on the average.

Another conceptual problem, emerging in practical implementation of the Abel inversion, is reflection from walls. In order to protect walls of aluminum chambers from etching, they are covered with chemically resistant yttrium layer — a white, diffusely reflecting material with almost 100% reflectivity in visible domain. Although this reflection is diffuse, spreading over the entire 4π solid angle, some portion of it does reach the photodetector, introducing noticeable error into the result of reconstruction. In order to better realize this problem and how to solve it, consider geometrical explanation in Fig. 7.54.

Denoting brightness of reflection as w and the pinhole area σ, the optical power, coming from the element ds of the wall through the pinhole, is equal to

$$dP = \frac{wds}{4\pi} \cdot \frac{\sigma \cos\theta}{(2R\cos\theta)^2}.$$

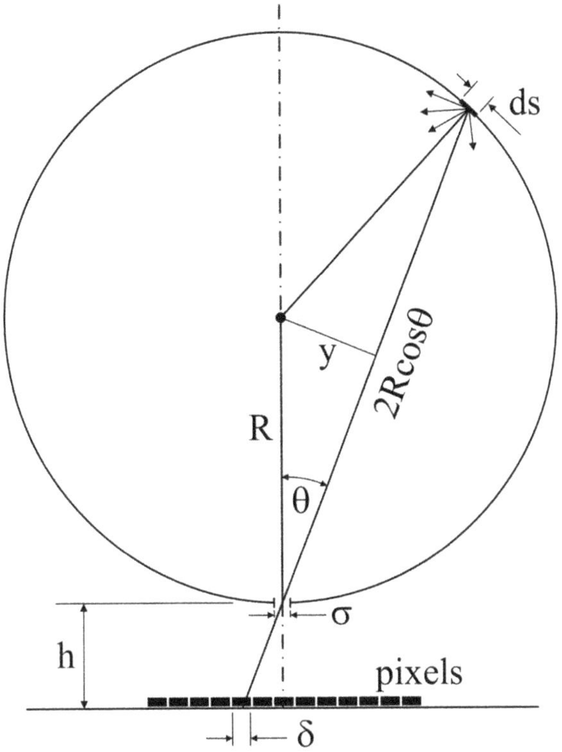

Fig. 7.54. Geometrical parameters of the task.

Reaching the photodetector, the intensity of light in the plane of the photodetector decreases by a factor of $\cos\theta$, and the power absorbed by a single pixel becomes equal to

$$dP = \frac{wds}{4\pi} \cdot \frac{\sigma \cos\theta}{(2R\cos\theta)^2} \cdot \delta\cos\theta = \frac{w\sigma\delta}{16\pi R^2} ds.$$

The reflected optical power from entire perimeter S of the wall is then equal to

$$P = \frac{w\sigma\delta}{16\pi R^2} S,$$

does not depend on θ and, consequently, on the pixel ordinal number. Thus, the problem of diffuse reflectance from walls may be accounted for by simple subtraction of some constant value from the signal of the photodetector. All plasma chambers have practically the same inner protective cover and dimensions; therefore this value, once evaluated, may be a good approximation for all the chambers.

To finalize theoretical topics, consider specific optical corrections that should be taken into account. Assume uniform distribution of plasma. Although the length of each section of plasma traversed by the line, connecting the pinhole and any single pixel of the photodetector, is proportional to $\cos\theta$, the electrical signal from this pixel is not proportional to $\cos\theta$ (Fig. 7.54). Here are two optical factors: first – projection of the pinhole area on the traversing direction is proportional to $\cos\theta$; and second — the same projection of the pixel area is also proportional to $\cos\theta$. Thus, the signal delivered by uniform plasma is proportional to $\cos^3\theta$, and prior to using the above algorithm, the photodetector signal must be divided by $\cos^2\theta$. This is the straightforward conceptual solution. A better result, compensating for mechanical defects and mismatches as well, can be obtained by dividing the raw signal by preliminary recorded and stored calibration signal, as explained in Section 7.5.3.

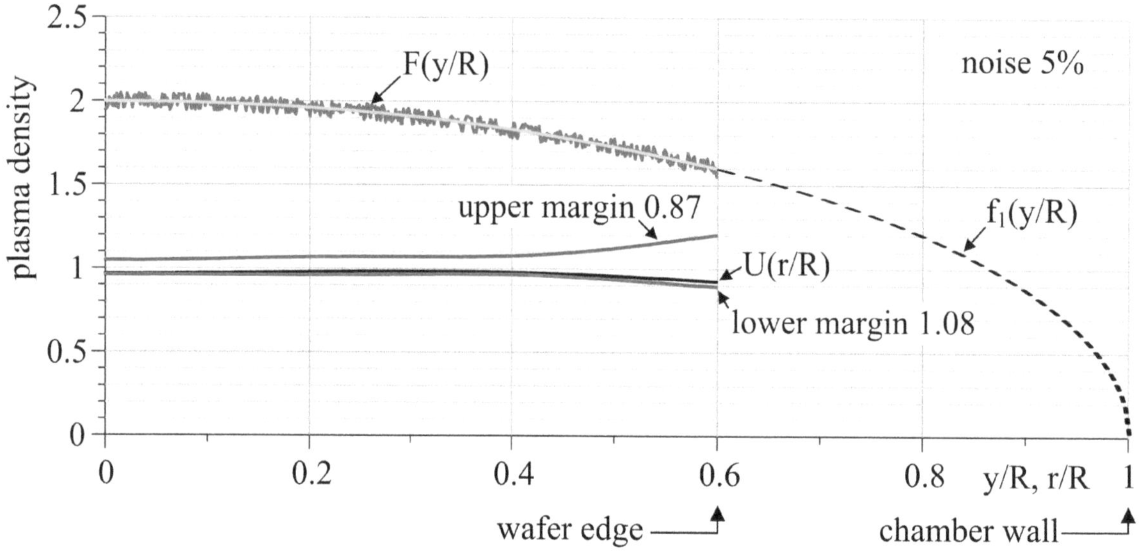

Fig. 7.55. Uniform distribution with noise level 5%.

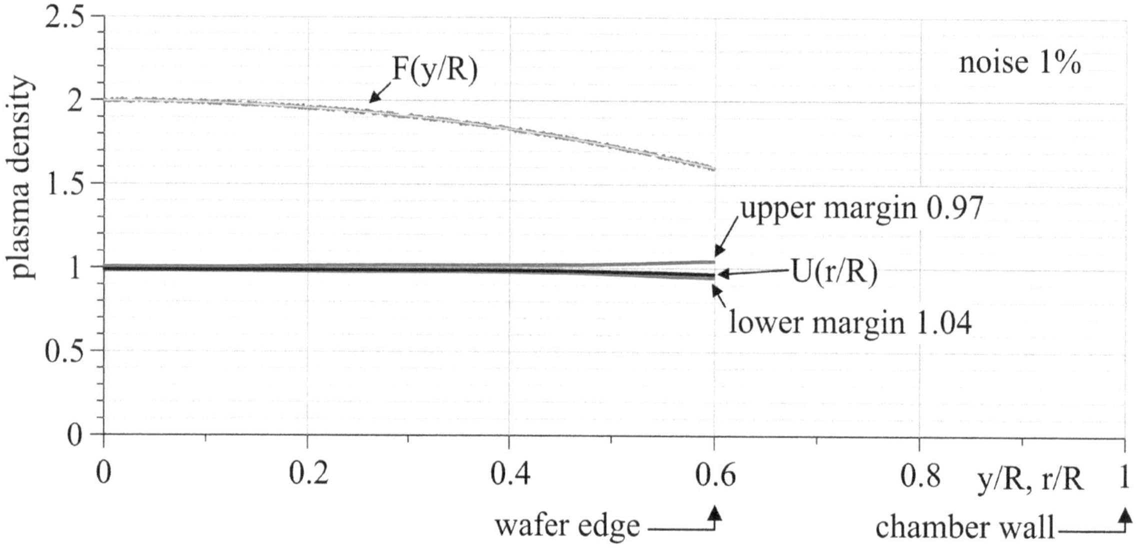

Fig. 7.56. Uniform distribution with noise level 1%.

Another optical correction to the above algorithm is related to computation of parameter y: since the plane of the photodetector is not perpendicular to the traversing line,

$$y = R \sin\left(\arctan\frac{k\delta}{h}\right),$$

where k is the ordinal number of the pixel.

7.5.2. *Simulation*

Below are some simulated results, showing how critical the noise may be for correct reconstruction. Figure 7.55 shows the result simulated for uniform plasma distribution in between $0 < y < R$. When

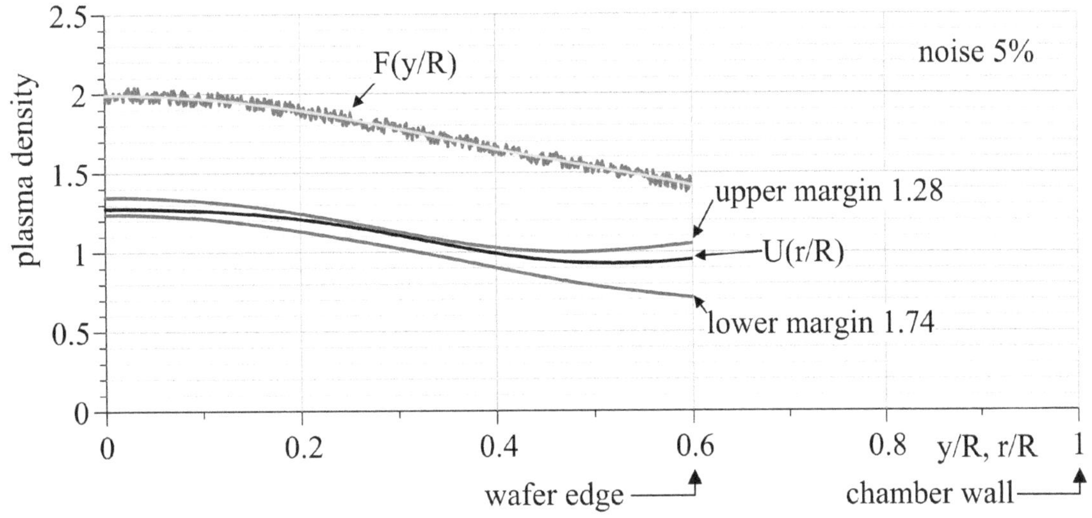

Fig. 7.57. Non-uniform distribution with noise level 5%.

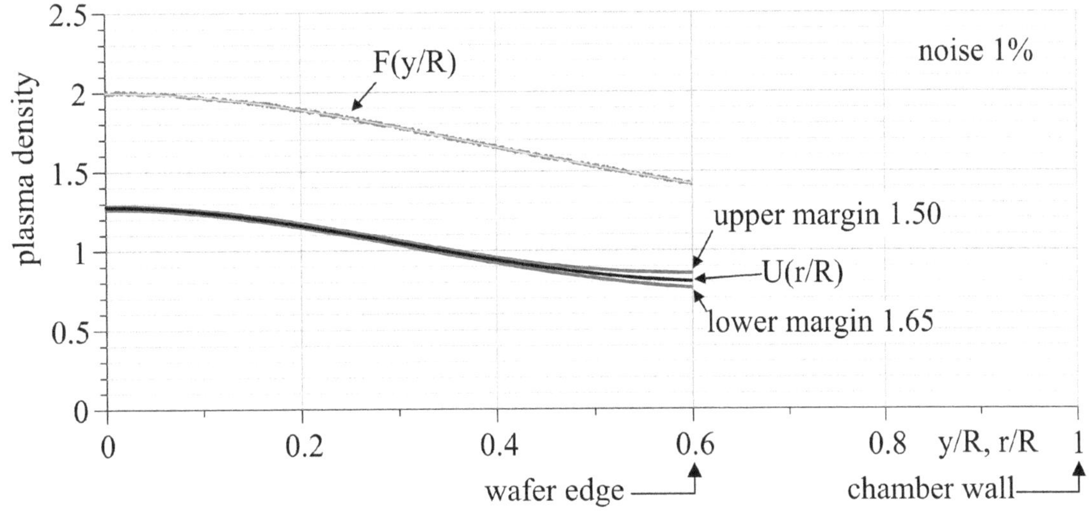

Fig. 7.58. Non-uniform distribution as in Fig. 7.60 with noise level 1%.

plasma is uniformly distributed inside the chamber, $F(y) = const \cdot f_1(y)$, or merely $\cos\theta$. The reconstructed function $U(r)$ then must be a straight horizontal line. The field of view of the sensor covers only the wafer itself, and can never see tangential rays from the chamber walls. Therefore, reconstruction will always be truncated to some maximum radius $r_{\max} < R$. In simulation, we assumed $r_{\max} = 0.6R$. Figure 7.56 shows noisy measured photodetector signal $F(y)$ (blue line), fitted series of the Abel eigenfunctions (green line on the blue one), reconstructed radial distribution of plasma $U(r)$ (black line), and the error margins for it (red lines above and below). Error margins show how uncertain the reconstruction may be, depending on noise. Also, the figure shows the ratios of plasma density in the middle to that at the edge of the wafer, computed for the upper and lower margins. The closer to unity these values are, the better the reconstruction is.

Figures 7.57 and 7.58 show an example of non-uniform plasma distribution, when $F(y)$ differs from $f_1(y)$.

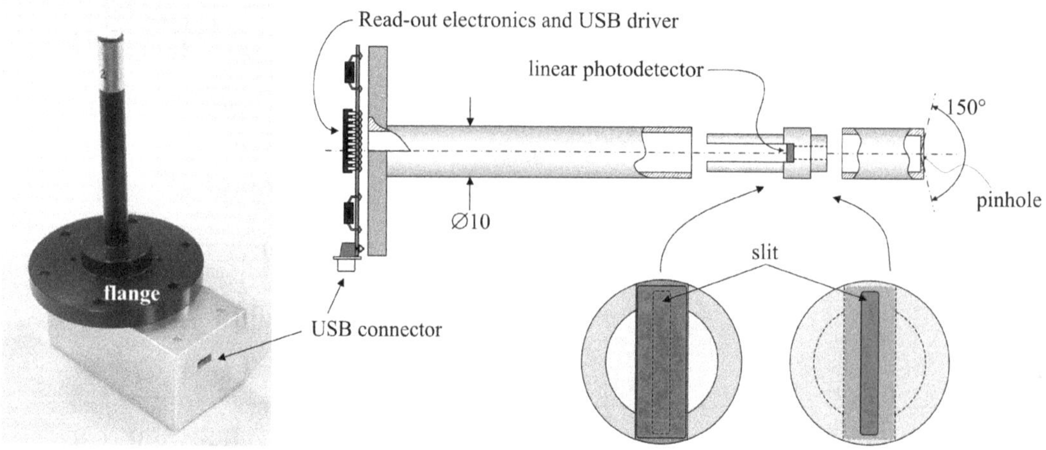

Fig. 7.59. Possible view of the Abel sensor with miniature linear photodetector.

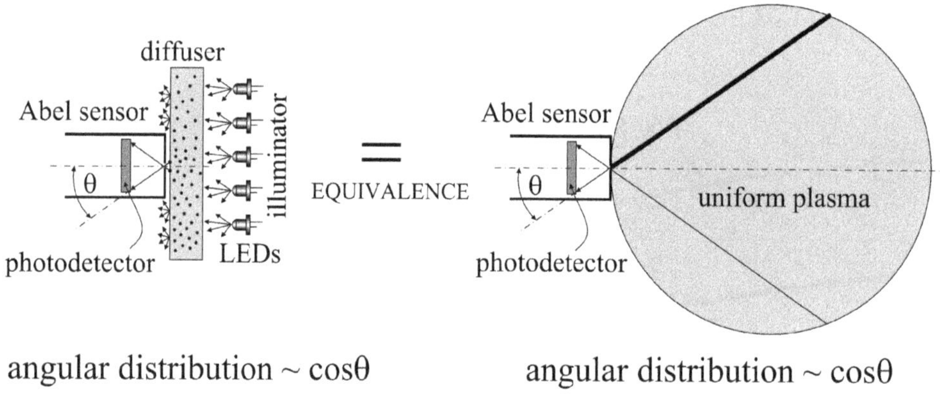

Fig. 7.60. Optical diffuser is almost ideal calibrator for the Abel sensor. LEDs: light-emitting diodes.

7.5.3. *Design concept and experimental results*

In order to guarantee accurate reconstruction, field of view of the sensor must be wide enough. For that, the tip of the sensor with the pinhole in it must be placed as close to the inner wall of the chamber as possible. On the other hand, manufacturers of plasma chambers always try to minimize the size of viewports in order to achieve highest uniformity of plasma. Therefore, the Abel sensor must be designed in a form of a thin and long tubular body that can be inserted deeply into narrow viewport. An example of such a design is shown in Fig. 7.59. It uses miniature linear photodetector TSL 1401CL with 128 pixels.

Reliable results can be obtained only on calibrated sensors, i.e. sensors tested on the media with known angular distribution of light. Remarkably, an optical diffuser is an excellent media for calibration: the Lambertian law of light scattering produces exactly the same angular distribution as uniform plasma does (Fig. 7.60).

This conclusion is corroborated by experimental results presented in Fig. 7.61, showing the sensor signal as the function of the angle

$$\theta = \arctan \frac{(k - k_0)\delta}{h},$$

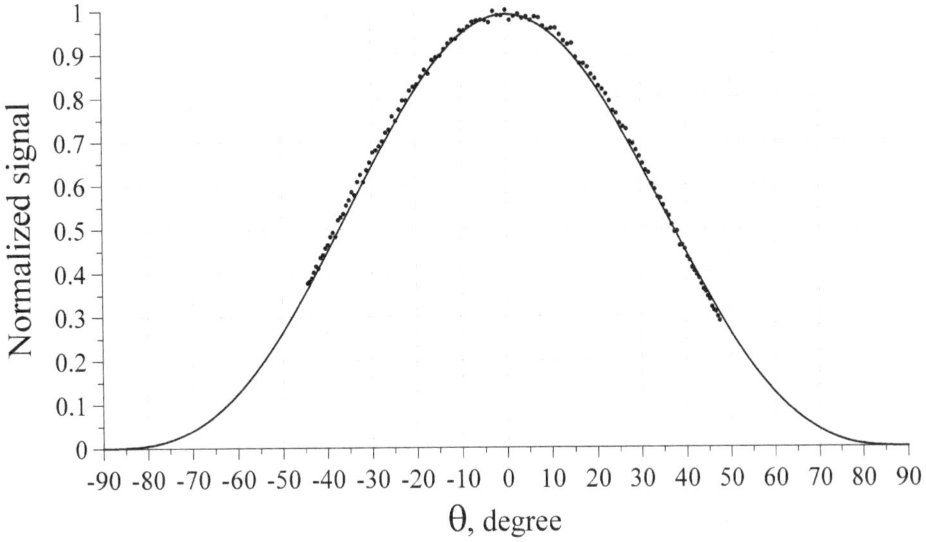

Fig. 7.61. Calibration curves for the Abel sensor. Signal values are plotted in dots and theoretical curve $\cos^3\theta$ in solid line.

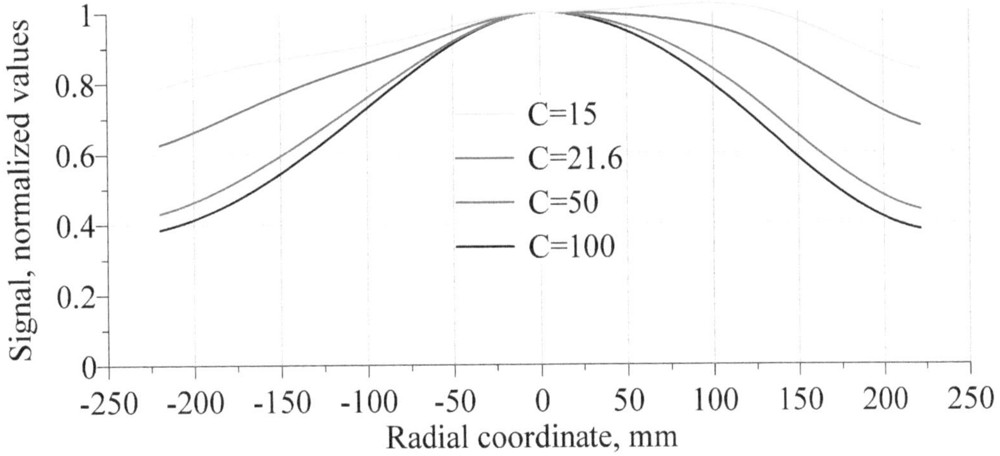

Fig. 7.62. Consecutive variation of plasma uniformity as it is measured by the Abel sensor with various values of adjustment capacitor C (in percent).

where k and k_0 are the ordinal and central numbers of the pixels, δ and h are explained in Fig. 7.54. The solid line represents theoretical function $\cos^3\theta$. The best fit gives actual values $h = 4.3$ mm, $\delta = 0.07$ mm, and central pixel $k_0 = 64.0$. Although the pixel width δ is always guaranteed by the photodetector specifications, the other two parameters are prone to variations during assembly process, and incorrect values lead to incorrect reconstruction. Therefore, calibration is always necessary for each newly assembled sensor.

Presenting measurements on real etch chambers, Fig. 7.62 summarizes results obtained on inductively coupled plasma with various values of coupling capacitor. Inductively coupled plasma is excited by two radio-frequency coils above the plasma volume, which are connected one with another by means of feedback capacitor C. This capacitor changes spatial distribution of electromagnetic field inside plasma volume, thus serving as an adjustment element to maximize uniformity of plasma. Without the Abel

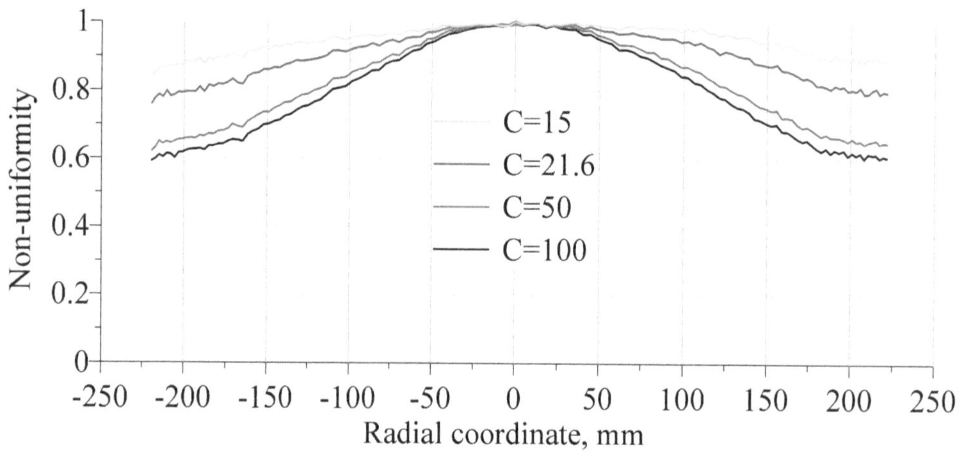

Fig. 7.63. Non-uniformity curves for the same processes as in Fig. 7.62.

sensor, process of adjustment is a lengthy procedure, requiring many cycles of loading-unloading test wafers and etching repeated one after another with various values of C. With the Abel sensor installed, the adjustment process is a matter of minutes.

Figure 7.62 reveals the fundamental problem of reconstruction: uncertainty of the reconstructed profile on axially asymmetrical spatial distributions. Mathematically, the Abel inverse transform and the entire algorithm are valid only for axially symmetrical distributions. If there is local axial asymmetry, then the result of reconstruction is not guaranteed, and reconstructed radial distribution may differ unpredictably from the real one. Figure 7.62 clearly shows radial asymmetry in red and green curves. It means that for these two cases ($C = 21.6$ and $C = 15$) reconstruction cannot be trusted. To cope pragmatically with such cases, it is possible to introduce another function: non-uniformity curves. If we denote the calibration signal recorded on the diffuser as calibr(k) and the signal measured on plasma as meas(k), then the non-uniformity curve is defined as

$$f(k) = \frac{\text{meas}(k)}{\text{calibr}(k)}.$$

The pixel number k can be easily recalculated into radial coordinate r:

$$r = L \cdot \sin\left[\arctan \frac{(k - k_0)\delta}{h}\right],$$

where L is the distance between the pinhole and the center of the chamber. Clearly, for the uniform plasma $f(r) = const$. Although function $f(r)$ does not have definite physical meaning and can be used only for general assessment of spatial non-uniformity, it is stable against axial asymmetry in a sense that small asymmetry causes equally small variations of $f(r)$. Figure 7.63 shows the results presented in Fig. 7.62 in terms of non-uniformity curves.

7.6. Two-Dimensional Sensor for Measuring Spatial Non-uniformity of Plasma

7.6.1. *The WDM WLS concept*

The Abel sensor, described in the previous section, delivers only one-dimensional result that does not explain two-dimensional (2D) distribution of plasma over the wafer. Its functionality relies on the assumption that plasma is distributed axially symmetrical. Although this assumption may be accepted as the initial approximation in most cases, developers of the semiconductor equipment need to know

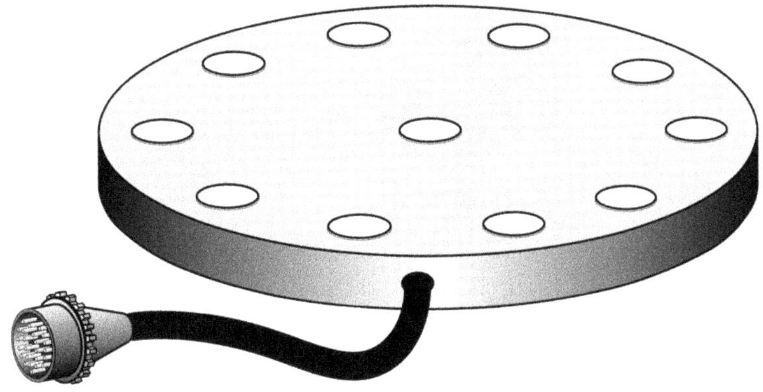

Fig. 7.64. Typical wafer-level sensor.

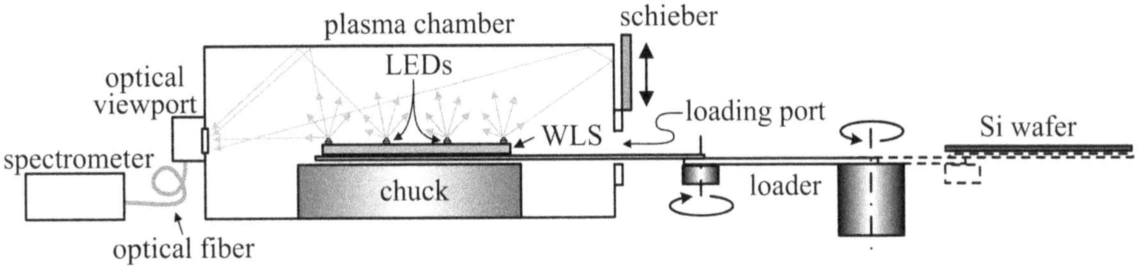

Fig. 7.65. The concept of WDM sensor.

more detailed information, especially when non-uniformity reaches extremely low levels of about several percent. One of the solutions is to install a 2D sensor inside the chamber instead of the silicon wafer itself. This type of sensors is called the wafer-level sensors (WLS), and they are freely available on the market. Typically, the WLS look like a 300-mm disk with several sensing areas grouped in a certain pattern over its upper surface (Fig. 7.64).

Such a sensor requires electrical power for functioning of its electronics, and electrical communication link to the computer, located outside the plasma chamber. Therefore, the sensor can be installed only into the opened chamber, connecting electrical cable to separate electronic module, situated outside the chamber, through optical viewport. In production line, however, opening the etching chamber is considered as emergency because the entire procedure requires much timc together with the necessity to restore vacuum after opening, not to say that it disturbs the entire carefully programmed and controlled technological process of the manufacturing line. Moreover, after installing such a sensor, the operator loses spectral information from the spectrometer, which must be disconnected from the chamber in order to use optical viewport for electrical cables. Therefore, the priority will always be given to those sensors that provide wireless communication and may be loaded into the process machine without opening it.

Among many concepts of wireless communication, a very attractive one is the use of the spectrometer itself as a mean of communication. If the WLS is equipped with light-emitting diodes (LEDs), radiating at different wavelengths, then the spectrometer can select these spectral components and measure their intensities, thus establishing a multi-channel optical link between the WLS inside the plasma chamber and the spectrometer outside. This concept is explained in Fig. 7.65 and can be named the Wavelength Division Multiplexing (WDM) WLS.

Among numerous other useful properties of this solution, which will be described in detail in the sections below, the most important for implementation in manufacturing line is the absence of any

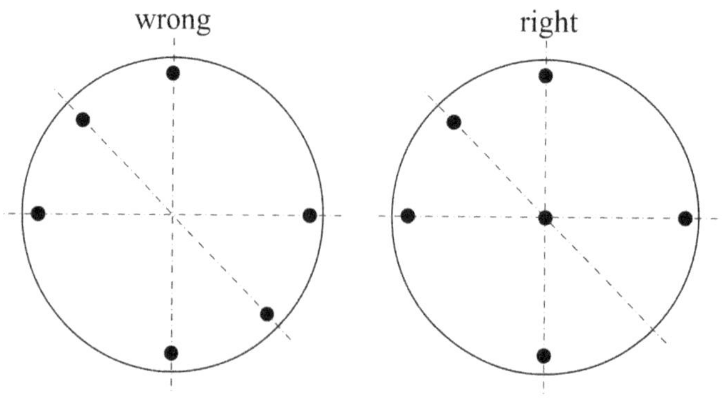

Fig. 7.66. Right and wrong configurations of measurement points on the WLS.

modifications to the process machine and its environment. Indeed, the WLS can be loaded into the chamber, using standard loader that is used for routine loading and unloading of silicon wafers, and the spectrometer — also a standard equipment of any process machine — is not disconnected. However, for this concept to be practically implemented, the weight of the sensor must be comparable to the weight of the silicon wafer itself, which is typically 150 g. Extracting pins, lifting the wafer from the chuck, are gentle in order to avoid jumping the wafer after it is released from the chuck, and cannot lift more than 250 g (typically). In addition, the sensor body must be thin enough to fit into narrow slit of loading port. These requirements impose strict limitation on the design of the WLS, which will be discussed below.

Another question is how many spectral channels must be established between the WLS and the spectrometer. It is not a matter of spectral resolution of the spectrometer — spectral resolution of a typical spectrometer far exceeds any practical needs of the WDM WLS technology (see Chapter 1). It is a matter of how many individual sensors are needed to reconstruct spatial distribution of plasma in two dimensions, and how many LEDs with different central wavelengths we can find on the market.

Axial symmetry of plasma chamber and configuration of exciting electro-magnetic fields determine dome-like distribution of plasma density above the wafer. The simplest approximation of such shapes is the second-order surface determined by the following equation:

$$F(x, y) = c_1x + c_2y + c_3x^2 + c_4y^2 + c_5xy + c_6,$$

where x and y are the Descartes coordinates in the plane of the sensor. Thus, six parameters c_1, c_2, c_3, c_4, c_5, c_6 must be measured in order to define the 2D distribution of plasma density. It means that plasma density must be measured in six points over the wafer surface: s_1, s_2, s_3, s_4, s_5, and s_6. Then, six equations can be composed and solved for c_1, c_2, c_3, c_4, c_5, and c_6. However, not every geometry of distributing the sensing points over the sensor is appropriate: some geometrical configurations do not produce consistent system of equations. As an example, Fig. 7.66 explains appropriate and inappropriate configurations:

Reconstruction begins from measuring signals s_1, s_2, s_3, s_4, s_5, and s_6 according to the scheme of Fig. 7.67.

With the radius of the circle of sensors being R, the following formulas give the solution:

$$c_1 = (s_2 - s_1)/2R; \quad c_2 = (s_3 - s_4)/2R; \quad c_3 = (s_2 + s_1 - 2s_5)/2R^2; \quad c_4 = (s_3 + s_4 - 2s_5)/2R^2;$$

$$c_5 = \frac{1}{R^2}\left(\frac{s_1 - s_2 + s_3 - s_4}{\sqrt{2}} + \frac{s_1 + s_2 + s_3 + s_4 - 4s_6}{2}\right); \quad c_6 = s_5.$$

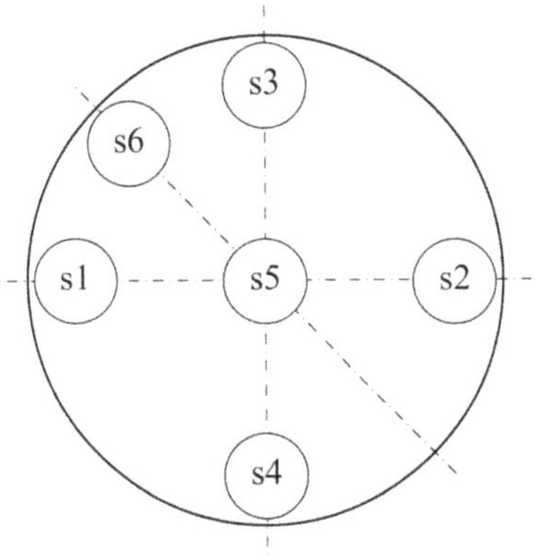

Fig. 7.67. Positions of measurement points on the WLS.

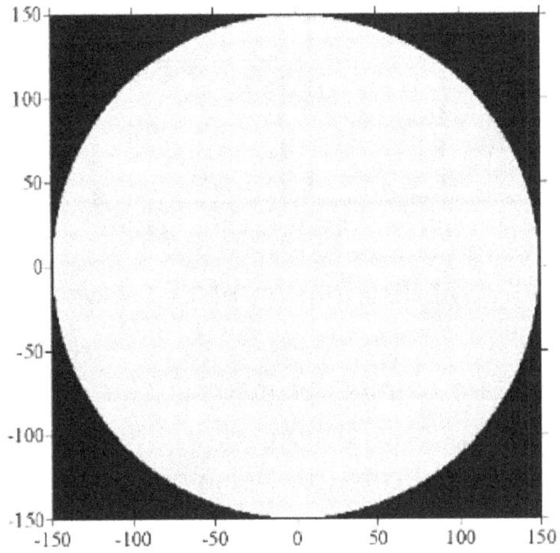

Fig. 7.68. Reconstructed uniform distribution.

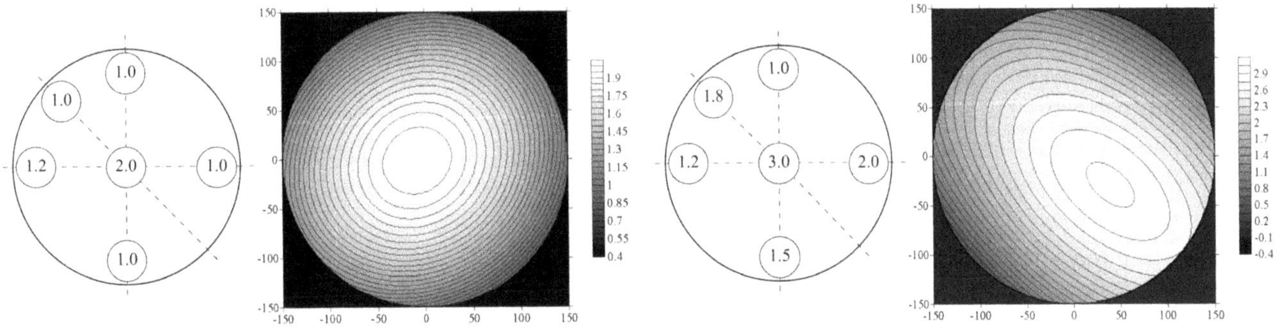

Fig. 7.69. Three different surfaces reconstructed from three different sets of measurements.

To understand how it works, assume for the beginning equality of all $s_i = 1$, $i = 1, 2, \ldots, 6$. Then, all c_i are zeros except for c_6, and as expected distribution of plasma is uniform:

If spatial distribution is not uniform, then the set of measurements s_i contains unequal components, which reconstruct to more complicated surfaces like those, shown in Fig. 7.69.

Table 7.5. Small package light-emitting diodes in visible domain.

Material	Visible color	Peak wavelength (nm)	Full spectral width (nm)	Direct voltage (Volt)
AlInGaN	Ultraviolet	365	24	3.6
InGaN	Blue	463	44	3.0
InGaN	Green	522	70	3.0
AlGaInP	Yellow green	565	30	1.9
AlGaInP	Orange	609	30	1.9
AlGaInP	Red	660	32	1.9
AlGaAs	Infrared	740	48	1.9

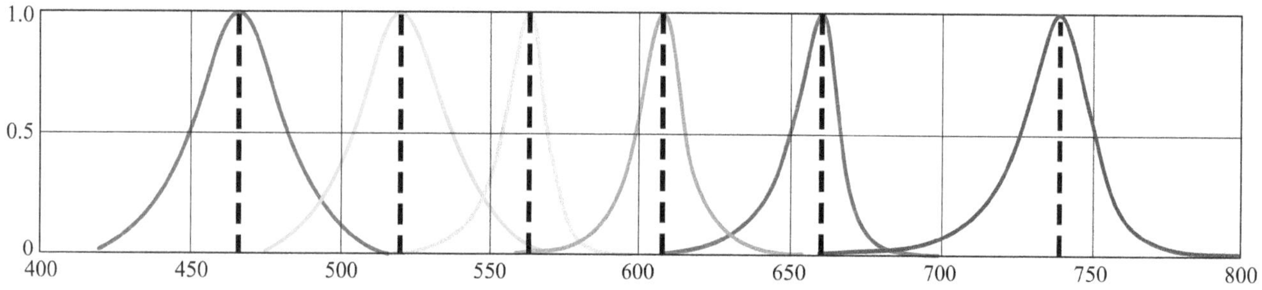

Fig. 7.70. Spectral distribution of LEDs wavelengths.

Clearly, the simple second-order surface cannot reconstruct sharp changes of plasma over the wafer. However, developers of process machines spent enormous efforts to reduce non-uniformity to only several percents, so that real distribution may be expected to be smooth, and the second-order approximation may produce acceptable results.

Thus, in order to optically relay information from the sensing elements, the WDM WLS must use six light-emitting diodes (LEDs) of different wavelengths, independently distributed within the spectral interval of the spectrometer. Table 7.5 summarizes possible choices for LEDs types, available in small packages.

From this list, six types must be chosen so that to provide minimum spectral overlapping as shown in Fig. 7.70. The ultra-violet LEDs are not suitable because their forward driving voltage reaches the limit of 4 Volt imposed by the battery (see Section 7.63). In order to minimize cross-talk between spectral channels, spectrometer must be programmed to record the integral signal only in narrow, about 1–2 nm, spectral interval around the peak of each LED. These narrow spectral intervals are shown in dashed lines.

Although some spectral overlapping does exist, its contribution may be considered insignificant, comparing to other sources of errors, such as second-order surface approximation or nonlinearity of electronics.

Optical signal, generated by the sensor inside the chamber filled with excited plasma, is always mixed with optical emission of the plasma itself. In order to separate them, optical flux generated by the sensor must be modulated. The frequency of this modulation cannot be high because time constant of the spectrometer is usually larger than 10 ms. Expecting about 20 spectral measurements during one period of modulation, the frequency of modulation may be set to 5 Hz. Then the signal, coming from the spectrometer, contains two components: the non-modulated one, slowly varying in time according to optical emission of plasma, and the modulated part, whose amplitude is determined by the sensor

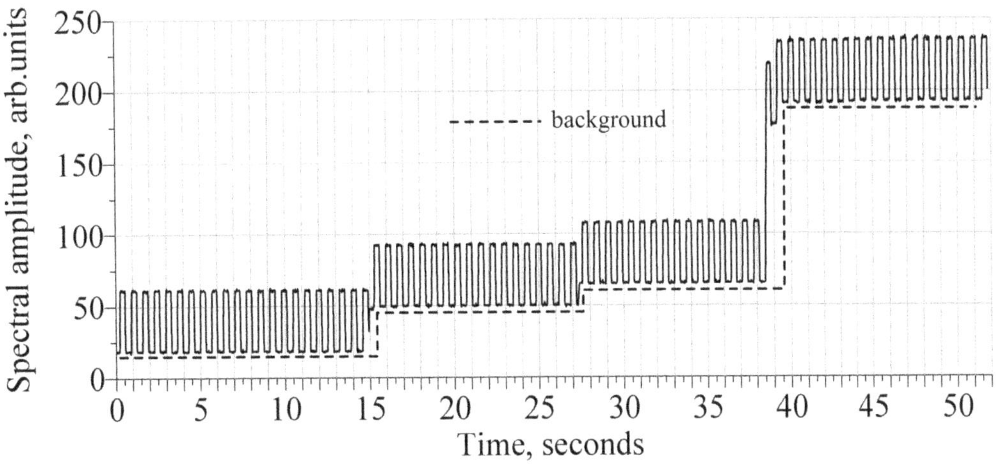

Fig. 7.71. Spectral trend with four levels of background.

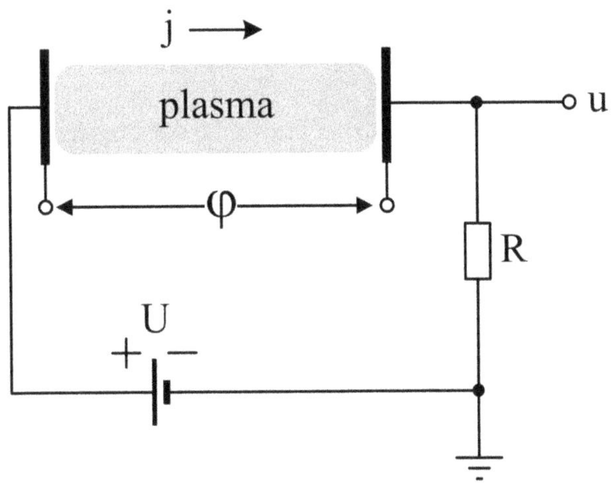

Fig. 7.72. Conceptual electrical scheme of measuring concentration of free charges in plasma.

alone. Amplification of the spectrometer is highly linear, making these two components independent of one another. This independence is clearly demonstrated by Fig. 7.71, showing temporal evolution (trend) of the spectrometer signal when optical emission of LED modulated at 2 Hz is mixed with the background illumination.

While strength of the background changes multifold, amplitude of modulated component remains the same.

7.6.2. *Theory of measuring plasma density*

The term plasma density means volumetric concentration of free charges in plasma, i.e. electrons and ions. Conductivity of plasma is proportional to concentration of free charges; therefore the scheme in Fig. 7.72 may be used to monitor variations of plasma density. Spatial variations of plasma density is the purpose of the WLS. Nonetheless, even absolute values can be measured after calibration against standard techniques for measuring plasma density, such as Longmuir probe, for instance.

The idea is to measure voltage variations at the output of the voltage divider formed by plasma, confined between two electrodes, and permanent resistor connected in series. For theoretical explanation, the following notations are used:

N — density of free charges;
e — electron charge;
m — mass of electrons;
M — mass of ions;
v — velocity of electrons;
w — velocity of ions;
S — area of electrodes;
R — scaling resistor;
u — control voltage;
j — current;
U — battery voltage;
ϕ — potential between electrodes.

Basic electrical equations are as follows:

$$u = jR; \quad U = \phi + jR.$$

The current between electrodes is formed by electrons and ions, coming from plasma. Electric potential between electrodes creates static electric field that deflects the charges towards electrodes. Having high velocity, these charges fly into the space between electrodes and may leave this volume without striking the electrodes. However, those charges, which has been deflected from their original trajectories and ended up on the electrodes, underwent definite variation of kinetic energy according to the following formulas:

single electron between electrodes

$$m\frac{v_1^2}{2} - m\frac{v_2^2}{2} = e\,\phi; \quad m\,v\,\Delta v = e\,\phi; \quad \Delta v = v_1 - v_2; \quad v = \frac{v_1 + v_2}{2};$$

single ion between electrodes :

$$M\frac{w_1^2}{2} - M\frac{w_2^2}{2} = e\phi.$$

Electron current:

$$Ne\Delta vS = \frac{e^2\phi NS}{mv}.$$

Ion current:

$$Ne\Delta wS = \frac{e^2\phi NS}{Mw}.$$

Total current:

$$j = e^2\phi NS\left(\frac{1}{mv} + \frac{1}{Mw}\right).$$

Then basic electrical equations give:

$$U = \phi + Ne^2S\phi\left(\frac{1}{mv} + \frac{1}{Mw}\right) \cdot R; \quad \phi = \frac{U}{1 + Ne^2SR\left(\frac{1}{mv} + \frac{1}{Mw}\right)}, \tag{7.1}$$

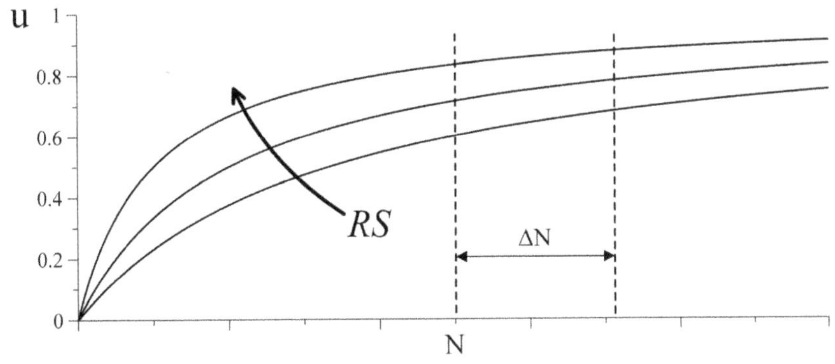

Fig. 7.73. Control voltage as a function of plasma density.

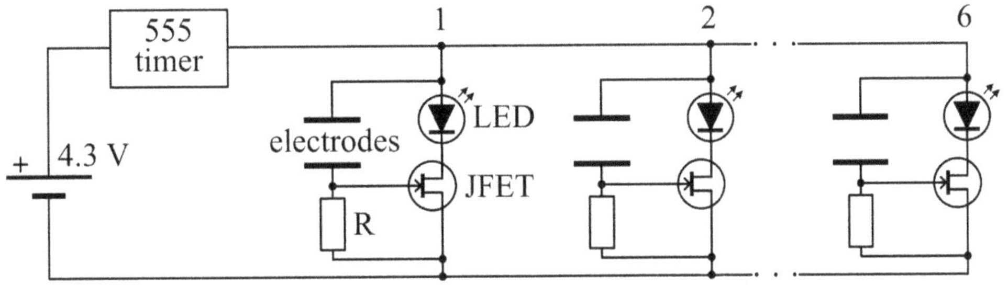

Fig. 7.74. Conceptual electronic circuit of the WDM WLS. The 555 timer produces modulation.

and total current is equal to

$$j = \frac{U}{\frac{1}{Ne^2S(\frac{1}{mv}+\frac{1}{Mw})} + R}.$$

This current develops control voltage u at the resistor R

$$u = \frac{U}{\frac{1}{Ne^2SR(\frac{1}{mv}+\frac{1}{Mw})} + 1}.$$

The control voltage is a function of many parameters, but we are interested in plasma density N. The function $u(N)$ is plotted in Fig. 7.73.

Due to negative feedback imposed by the resistor R, this function is not linear. However, since we are interested only in small variations ΔN of the order of several percent around N, the response within N and $N + \Delta N$ may be considered quite linear, as it is seen from Fig. 7.73.

7.6.3. *Electro-mechanical design and experimental results*

Plasma currents between electrodes in Fig. 7.72 are commonly small, unable to drive bi-polar transistors. On the other hand, these currents, being terminated by large resistors R of the order of tens or hundreds of kilo-Ohms, create voltage drop sufficient to drive field-effect transistors (FETs). Typical scheme that can be used is shown in Fig. 7.74. The 555 timer modulates optical flux from the LEDs to make measurements independent of plasma optical emission (see Section 7.6.1), and the spectrometer is programmed to measure only amplitude of this modulation at six spectral lines (Table 7.5).

Important comment should be made, regarding the nature of the field-effect transistors. There are basically two types of FET: the so-called junction FET (JFET), in which the current channel is insulated

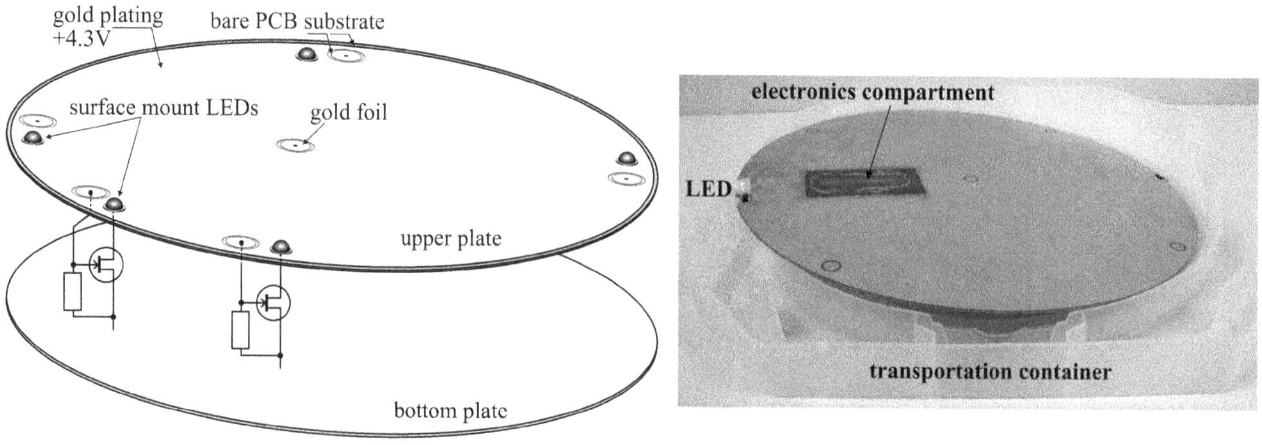

Fig. 7.75. Design concept (at left) and mass-dimensional model (at right) of the WDM WLS.

from the gate by a *p*–*n* junction, and the MOSFET, in which the current channel is insulated from the gate by an oxide layer. The MOSFET transistor is shut when the voltage at its gate is zero. Alternatively, the JFET transistor is partially opened at zero gate voltage, and this is exactly what is needed for operation of WDM WLS. Indeed, the individual LEDs have different luminosity, which may vary significantly from one to another. Therefore, before plasma is ignited, the sensor must be calibrated in order to equalize all its six channels. When there is no plasma in the chamber, there is no current between electrodes and all FETs have zero voltages at their gates. Hence, with MOSFETs, LEDs do not emit any light because the transistors shut their currents. As such, the spectrometer cannot sense the channels of the sensor and it is impossible to calibrate it. Alternatively, with JFETs, small currents flow through the transistors, keeping LEDs glowing, and the spectrometer can record these signals. The absence of plasma means its uniform distribution. Therefore, these six recorded signals s_1^0, s_2^0, s_3^0, s_4^0, s_5^0, s_6^0 should be taken as the reference that reconstructs uniform distribution. Obviously, it can be done by simple calibration: s_i/s_i^0, for all $i = 1,2,\ldots,6$.

The WDM WLS is made of two surface-to-surface connected printed circuit boards (PSBs) as explained in Fig. 7.75. Its total thickness is 1.9 mm and weight 208 g. All electronic components are chosen of surface-mount type with maximum thickness of about 0.5 mm, including 4.3 V ultra-thin rechargeable Li-ion NGK EnerCera EC382704P-C battery 0.5 mm thick.

The metallized surfaces of the upper and bottom PCBs are both connected to positive potential of the battery, thus shielding the internal electronic circuits against electromagnetic interference.

Experiments start with measuring six signals from the sensor when it is loaded into the chamber and plasma is not yet ignited. Amplitudes of modulation, recorded in a form of six trends, are presented in Fig. 7.76. Plasma is ignited at the 25th second.

The insert in the figure magnifies amplitudes of the signals before ignition of plasma. Clearly, relative amplitudes of the signals before and after ignition differ, signifying spatial non-uniformity of plasma. In order to reconstruct spatial distribution, the signals have to be normalized as was explained in the previous section. The normalized values are shown in the diagram in Fig. 7.77 together with the reconstructed 2D map.

Maximum variation of the 2D map over the sensor surface was measured to be only (151–141)/151≈6% — a reasonable value for high-quality 300 mm poly etcher.

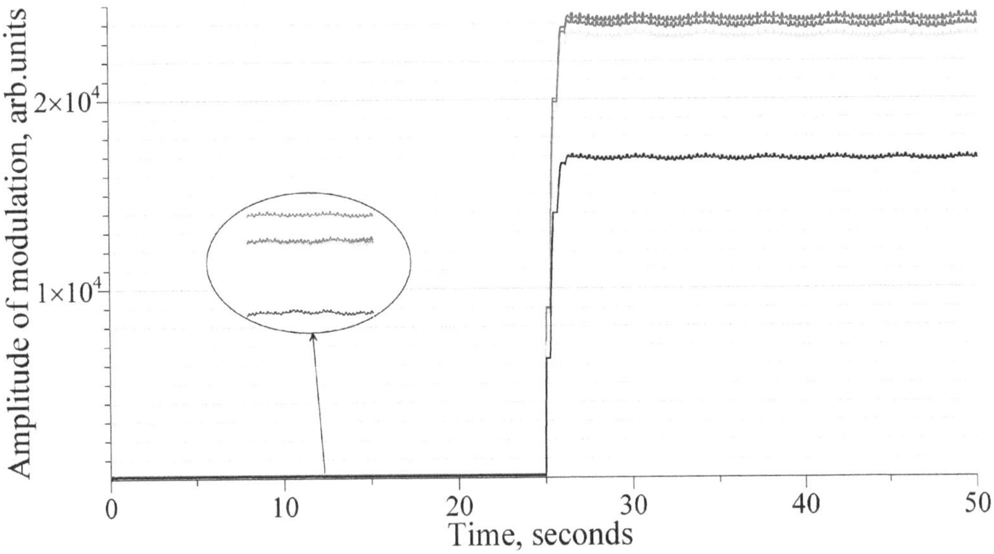

Fig. 7.76. Six trends of sensors signals as they are recorded by the spectrometer.

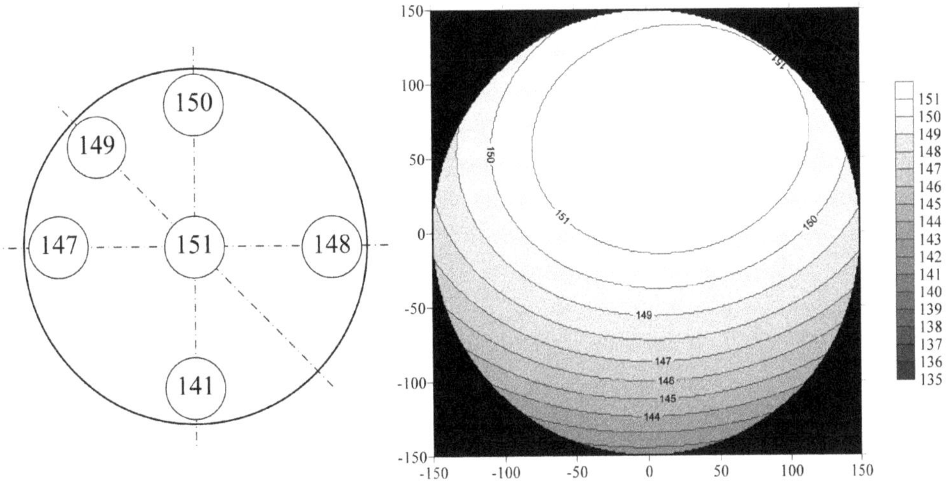

Fig. 7.77. Reconstructed spatial distribution of plasma density.

Supplemental Reading

B.M. Smirnov, *Plasma Processes and Plasma Kinetics*, WILEY-VCH Verlag, Weinheim, 2007.

G.S. Selwyn, *Optical Diagnostic Techniques for Plasma Processing*, AVS, New York, 1993.

I.P. Herman, *Optical Diagnostics for Thin Film Processing*, Academic Press, New York, 1996.

G.T. Herman, *Fundamentals of Computerized Tomography. Image Reconstruction from Projections.* 2nd edn., Springer Series in Advances in Pattern Recognition, Springer, 2009.

S.R. Deans, Radon and Abel transforms, Chapter 8 In: *Transforms and Applications Handbook*, 3rd edn., A.D. Poularikas (ed.), CRC Press, 2010.

H.A. Macleod, *Thin-Film Optical Filters*, 3rd edn., Institute of Physics Publishing, 2003.

Chapter 8

Spectral Reflectometry

8.1. Measurement of Thickness of Dielectric Films

8.1.1. *Basic phenomenology of spectral reflectometry*

Spectral reflectometry is a very simple and reliable tool for measuring thickness of thin films. This technique gained its popularity after commercialization of compact fiber-optic spectrometers (Chapter 1) — relatively inexpensive instruments, connected to a personal computer and easily operated through simple programs. Considering spectrometers and fiber-optic accessories as given components, a spectral interferometer can be assembled almost on a dinner table (Fig. 8.1). It requires minimum components: white-light source with fiber-optic adaptor, compact spectrometer connected to a computer, and bidirectional (bifurcated) optical fiber. In order to get strong reflected signal, the end of the fiber must

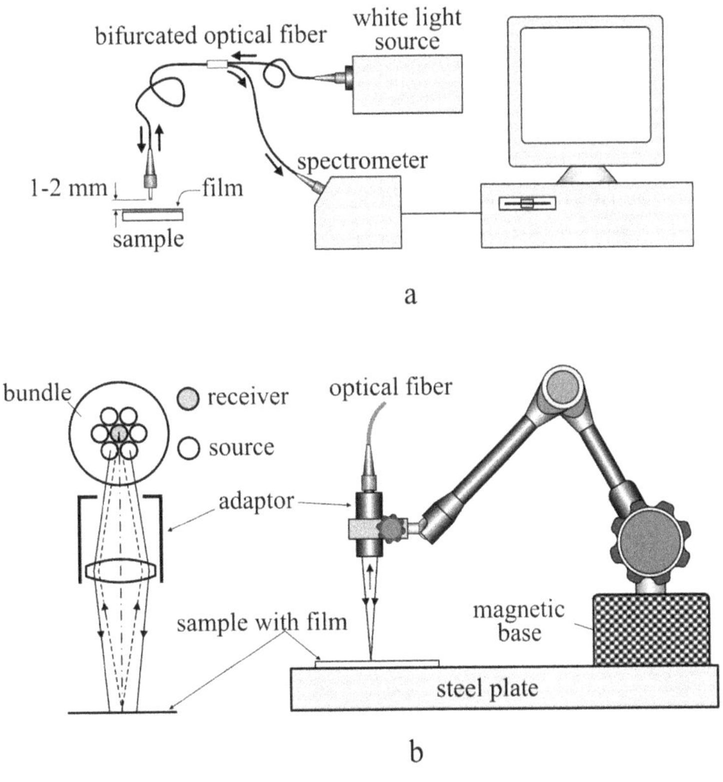

Fig. 8.1. (a) The simplest scheme of a spectral interferometer. (b) More reliable variant.

be kept firmly at roughly 1–2 mm above the surface of the film: it optimizes the angle at which the peripheral fibers in the bundle relay the light to the central core, connected to spectrometer. Almost professional system can be obtained by adding a focusing fiber-optic adaptor with a lens (Fig. 8.1(b)). The end of the fiber and the surface of the sample must be approximately conjugated relative to the lens, which means that the lens must project the image of the fiber onto the surface of the sample. To keep the adaptor in place, a standard triple-jointed arm with magnetic base on steel plate finalizes the system.

The concept is exceptionally simple: send a wave towards the transparent film on a substrate (on glass, for instance), and receive the wave reflected backwards. Since we are doing this through optical fiber, which is optically thin — typically 0.1–0.4 mm in diameter, the reflected wave can return back into the fiber only if the reflection is almost normal to the surface. Therefore, if we get some signal reflected back, then we may be sure that there was normal-incidence reflection. This feature makes the entire technique insensitive to angular adjustment and dramatically simplifies the theory of the phenomenon, making it possible to ignore spatial distribution of fields within the fiber core. As such, we may write down the returned wave E as a sum of two sinusoidal components, oscillating in time t with optical angular frequency ω, different amplitudes A_1 and A_2, and some phase shift between them:

$$E = A_1 \cos \omega t + A_2 \cos\left(\omega t + \frac{2\pi}{\lambda} 2hn\right).$$

The first component is reflected from the top of the film of thickness h and refractive index n, and the second — from its bottom, making the optical path difference equal to $2hn$. Photodetector responds to intensity E^2, averaging optical oscillations at frequency ω, therefore the electrical signal is proportional to

$$1 + \gamma \cos\left(\frac{2\pi}{\lambda} 2hn\right),$$

where γ is the contrast of the interference pattern, depending on A_1 and A_2. This holds true for every wavelength λ, and the spectrometer gives the signal for each λ separately, i.e. the spectrum. Therefore, the spectrum as a function of λ is modulated as $\cos(4\pi hn/\lambda)$. The first thing that must be emphasized is that this modulation is not periodic as a function of wavelength λ, but is periodic as a function of $1/\lambda$. Therefore, in order to determine film thickness h, the Fourier transform must be applied to the measured spectrum as a function of the inverse wavelength $x = 1/\lambda$.

Next, consider the simplest experiment shown in Fig. 8.2.

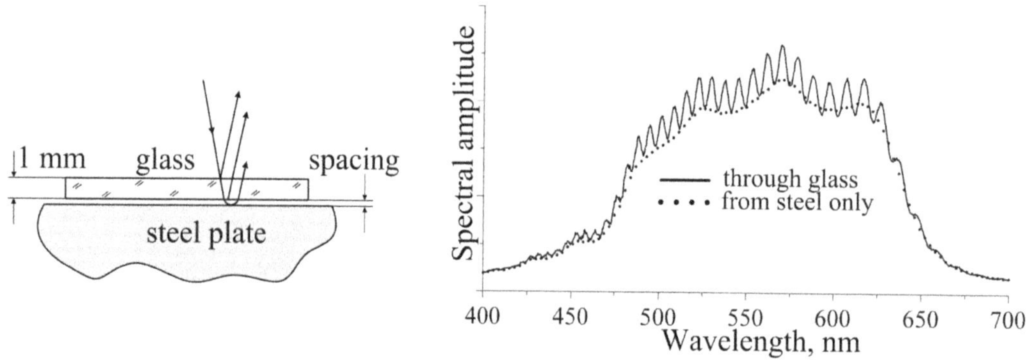

Fig. 8.2. Spectrum of white light on reflection from cover glass placed onto a steel plate. The dotted curve shows the spectrum without glass.

Clearly visible modulation is the result of interference because when the glass is removed there is no modulation. But the question is which waves produce this modulation? Indeed, there are three waves reflected from the steel base, from the bottom of the glass, and from the top of the glass. To answer this question, we have to explicitly determine spectral spacing of modulation $\Delta\lambda$:

$$\frac{2\pi}{\lambda_1}2hn - \frac{2\pi}{\lambda_2}2hn = 2\pi, \quad \Delta\lambda \approx \frac{\lambda^2}{2hn}.$$

At $\lambda = 500$ nm, the glass of $h = 1$ mm and refractive index $n = 1.5$ produces spacing $\Delta\lambda \sim 0.1$ nm — far below spectral resolution of a compact spectrometer that is typically ~ 1 nm. Thus, interference between the waves reflected from opposite surfaces of glass does exist, but it cannot be seen due to finite spectral resolution of the spectrometer. In general, spectral resolution δ of the spectrometer imposes the upper limit for the film thickness $h_{\max}$ that can be measured around the wavelength λ:

$$h_{\max} \approx \frac{\lambda^2}{2n\delta}.$$

What we do see in Fig. 8.2 is the interference between the waves reflected from the steel plate and bottom surface of glass. There is about six oscillations in between 525 nm and 575 nm, which gives the estimate $\Delta\lambda \approx 8$ nm and $h \approx$ 20 micron ($n = 1$ for air). It is a reasonable value because average size of a dust particle is about 20 micron, and the gap between the glass and the steel plate was due to dust, presumably.

It is not even necessary to perform a special mathematical analysis on data presented in Fig. 8.2 to notice the difference in periodicity between the blue (short wavelengths) and red (long wavelengths) wings of the spectrum. However, the same data presented as function of the inverse wavelength $x = 1/\lambda$ show exact periodical variations (Fig. 8.3).

To finalize this section, consider some very practical hints on how to analyze more complicated cases, using fast Fourier transform (FFT). The FFT can only be applied to uniformly distributed grids, i.e. to sets of points x_k with constant spacing:

$$x_{k+1} - x_k = const.$$

However, spectrometer produces data that are uniformly distributed in wavelength λ

$$\lambda_{m+1} - \lambda_m \approx const,$$

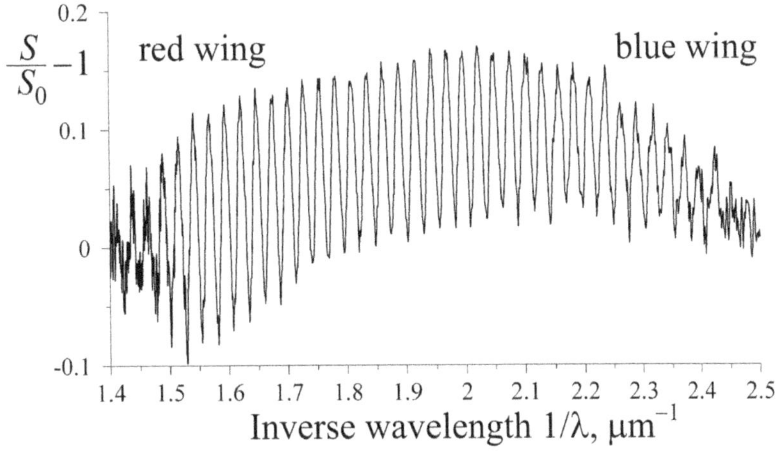

Fig. 8.3. The same spectrum $S(\lambda)$ as in Fig. 8.2 with subtracted non-interfering part $S_0(\lambda)$ caused by reflection from the steel plate. Horizontal axis shows inverse wavelength $1/\lambda$ in reciprocal microns.

and even this approximation is not accurate enough. Thus, the first task is to interpolate non-uniform grid $1/\lambda_m$ into uniform grid x_k, and there are many standard mathematical libraries that can do this job. The basic idea is to determine uniform grid x_k with as many points as we want, then to compute $\lambda_k = 1/x_k$, and after that to interpolate spectral points $S(\lambda_k)$, using the spectrum $S(\lambda_m)$ supplied by the spectrometer. Number N of interpolated points λ_k may far exceed the number of measured points λ_m.

Suppose this is done, and we have the set $S(x_k)$. Next, we need to apply some model to the spectrum, for example

$$S(x) = 1 + \sum_{j=1}^{M} \gamma_j \cos(2\pi x 2h_j n_j),$$

which represents a set of M films of thicknesses h_j and refractive indices n_j, producing fringe contrasts γ_j. Our goal is to determine γ_j and $b_j \equiv 2h_j n_j$. With the simplified presentation of $S(x)$ above, it is impossible to determine thicknesses h_j and refractive indices n_j separately — only in product. An exact theory is needed to extract h and n separately from measuring reflectivity of a thin film, and this topic will be explained in detail in the next section.

Now, apply the complex FFT to obtain the resultant set z_m as follows:

$$z_m = \sum_{k=1}^{N} S(x_k) e^{-2\pi i(k-1)(m-1)/N},$$

where N is the size of the grid. This formula shows how the FFT is computed: it works only with dimensionless integer numbers k and m, and does not want to know anything about our dimensional arguments x and h. Therefore, we need to establish a relation between the index m and $b = 2h \cdot n$. Doing this, we expect that the Fourier transform gives us a set of peaks — the harmonics — located at m_j, and each m_j identifies b_j. Relative amplitudes z_j of these peaks correspond to relative contrasts γ_j. Thus, substitute the model for $S(x)$ into the formula for FFT, expand cosine as the sum of complex exponents, and equalize the argument of the exponent to zero — this will show us locations of peaks:

$$x_k b = \frac{k-1}{N}(m-1).$$

Use

$$x_k = \frac{x_{\max} - x_{\min}}{N}(k-1)$$

to obtain

$$b = (m-1)\left(\frac{1}{\lambda_{\min}} - \frac{1}{\lambda_{\max}}\right)^{-1}.$$

Thus, every peak in the FFT, located at m_j, corresponds to a certain film with thickness

$$h_j = \frac{m_j - 1}{2n_j}\left(\frac{1}{\lambda_{\min}} - \frac{1}{\lambda_{\max}}\right)^{-1}.$$

It was already told that we do not know n_j, but for air-spaced gaps situation radically simplifies:

$$h_j = \frac{m_j - 1}{2}\left(\frac{1}{\lambda_{\min}} - \frac{1}{\lambda_{\max}}\right)^{-1}.$$

Thus, scaling the horizontal axis of the FFT according to this formula, we automatically measure thicknesses of all the spaces. Relative amplitudes of the peaks give contrasts γ_j. Typical result is shown in Fig. 8.4.

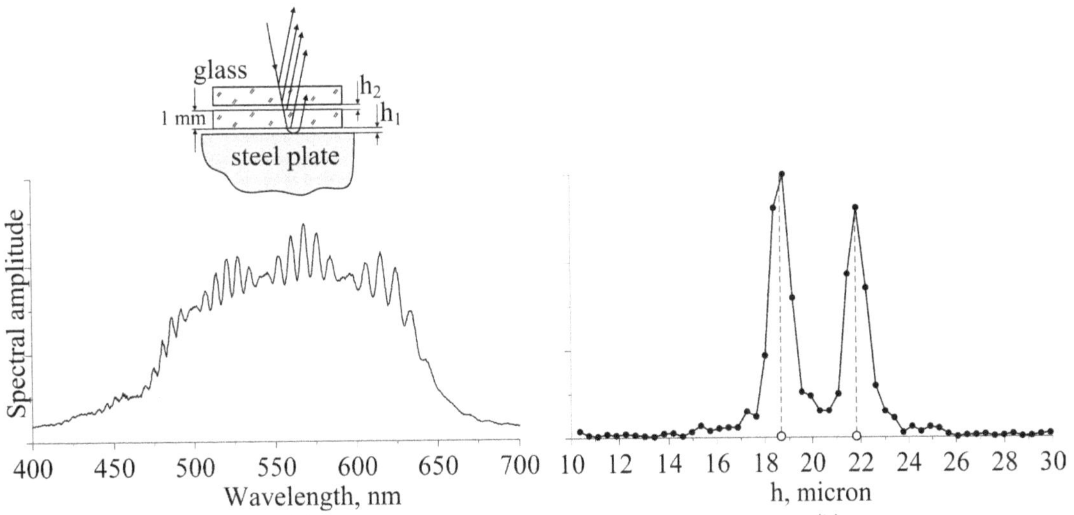

Fig. 8.4. An example of computing thicknesses of films.

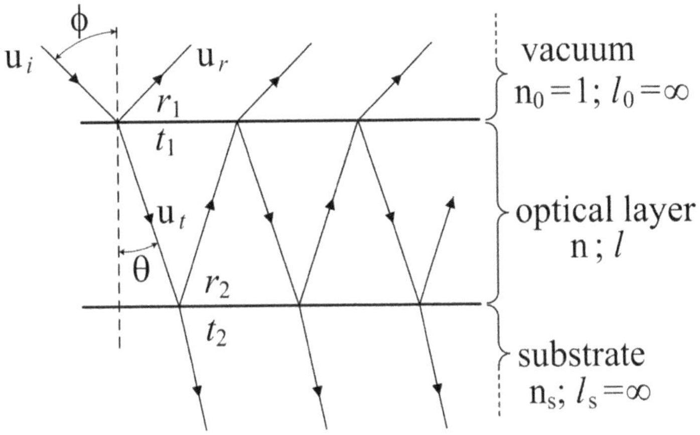

Fig. 8.5. Reflection from a single layer.

In this experiment, two glass plates of 1 mm thickness each were put one on top another, thus forming five reflecting surfaces, of which only four contribute in pairs to visible interference. Interference in two air spaces h_1 and h_2 produces modulated spectrum more complicated than in Fig. 8.3, and only experienced user can identify it as a beating curve produced by two sinusoids of different frequencies (Fig. 8.4(a)). However, the fast Fourier transform (FFT) not only readily identifies two gaps as two separate maxima, but being calibrated along the horizontal axis, gives the estimates of the spacings as 19 and 22 microns (Fig. 8.4(b)). Their vertical order in the stack cannot be identified exactly, but taking into consideration relative amplitudes, the plausible suggestion is that the higher peak comes from the steel plate.

8.1.2. *Theory of reflection from multiple layers*

To begin with, consider only one optical layer of thickness l and complex refractive index n deposited on an infinitely thick substrate of refractive index n_s (Fig. 8.5).

When a plane wave u_i with the wavelength λ in vacuum comes to the layer at the angle of incidence ϕ, the reflected wave u_r is composed of infinite number of partial waves reflected from the top and bottom surfaces of the layer. Since the substrate is considered of infinite thickness, reflection from its bottom surface may be neglected. According to the Snell law, the angles of incidence ϕ and transmission θ obey the sine law:

$$\sin\phi = n\sin\theta.$$

It means that the product $n\sin\theta$ within the layer is invariant of the layer refractive index n and is always equal to $\sin\phi$. As such, the phase delay δ that the transmitted wave u_t acquires on its way from the top to the bottom surfaces of the layer is equal to

$$\delta = \frac{2\pi}{\lambda}nl\cos\theta = knl\sqrt{1-n^{-2}\sin^2\phi} = kl\sqrt{n^2-\sin^2\phi},$$

where $k = 2\pi/\lambda$ is the wavenumber. The Fresnel reflection r and transmission t coefficients at the top interface are

$$r_s = \frac{n_0\cos\phi - n\cos\theta}{n_0\cos\phi + n\cos\theta}, \quad t_s = \frac{2n_0\cos\phi}{n_0\cos\phi + n\cos\theta}$$

for the s-polarized wave (electric vector perpendicular to the plane of incidence and parallel to the layer surface), and

$$r_p = \frac{n\cos\phi - n_0\cos\theta}{n\cos\phi + n_0\cos\theta}, \quad t_p = \frac{2n_0\cos\phi}{n\cos\phi + n_0\cos\theta}$$

for the p-polarized wave with the electric vector in the plane of incidence. The equivalent formulas hold true for the bottom interface and for the inverse direction of propagation with relevant changes of refractive indices and angles. Specifically, if we denote reflection coefficients for either s- or p-polarized wave coming from vacuum onto the layer at the angle of incidence ϕ as $r_{1\downarrow}$, and for the wave coming upward at the angle of incidence θ from the layer onto the top surface as $r_{1\uparrow}$, then from the above formulas follows the relation:

$$r_{1\downarrow} = -r_{1\uparrow}.$$

Also, for either s- or p-polarization, using the same notations $t_\downarrow$ and $t_\uparrow$ for the transmission coefficient of the waves coming downwards and upwards on the top surface, after some manipulations with the Fresnel formulas the following relation may be found:

$$1 - r^2 = t_\downarrow t_\uparrow.$$

The Fresnel coefficients satisfy boundary conditions at the interfaces, i.e. continuity of tangential components of electric and magnetic fields at the interface.

According to Fig. 8.5, the first partial reflected wave is formed by the first reflection of the incident wave u_i: $u_i \cdot r_{1\downarrow}$. The second partial reflected wave is formed by the first transmitted wave $u_t = u_i \cdot t_{1\downarrow}$ that reflects back from the bottom surface of the layer and passes through the top surface with the Fresnel transmission coefficient $t_{1\uparrow}$: $u_i \cdot t_{1\downarrow} \cdot r_{2\downarrow} \cdot t_{1\uparrow}$. The third partial reflected wave is formed by reflection of the second partial wave from the top surface, its transmission down to the bottom surface, reflection from it and passing through the top surface again: $u_i \cdot t_{1\downarrow} \cdot r_{2\downarrow} \cdot r_{1\uparrow} \cdot r_{2\downarrow} \cdot t_{1\uparrow} = u_i \cdot (t_{1\downarrow}t_{1\uparrow}r_{2\downarrow}) \cdot r_{1\uparrow}r_{2\downarrow}$. The next partial wave will be $u_i \cdot (t_{1\downarrow}t_{1\uparrow}r_{2\downarrow}) \cdot r_{1\uparrow}^2 r_{2\downarrow}^2$, and so on. Moreover, after each consecutive double pass of the layer, the wave acquires incremental phase delay 2δ. Eventually, the complex amplitude of

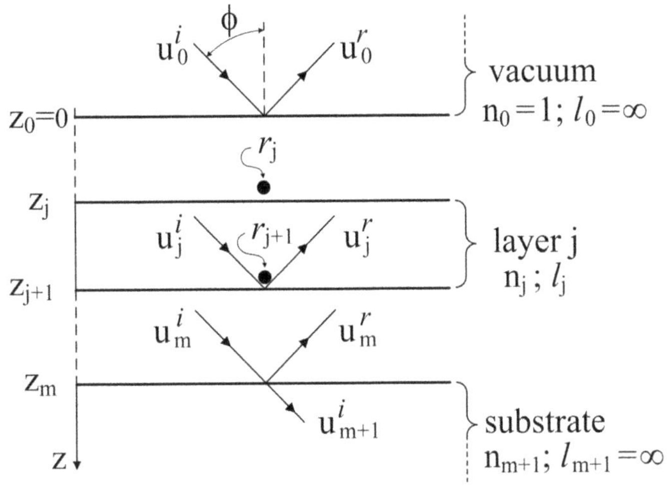

Fig. 8.6. Reflection of electromagnetic waves from a multilayer structure. The dots show the convention where the complex reflection coefficients are defined.

the reflected wave reduces to the following infinite sum:

$$u_r = u_i \left(r_{1\downarrow} + t_{1\downarrow} t_{1\uparrow} r_{2\downarrow} \sum_{j=0}^{\infty} r_{2\downarrow}^j r_{1\uparrow}^j e^{-2i\delta \cdot j} \right),$$

where $i = \sqrt{-1}$ and j is the ordinal number. The sum in this formula is the geometrical progression, which reduces to $(1 - r_{2\downarrow} r_{1\uparrow} e^{-2i\delta})^{-1}$, and the result transforms to

$$u_r = u_i \left(r_{1\downarrow} + \frac{t_{1\downarrow} t_{1\uparrow} r_{2\downarrow} e^{-2i\delta}}{1 - r_{2\downarrow} r_{1\uparrow} e^{-2i\delta}} \right).$$

The ratio u_r/u_i is the complex reflection coefficient of the layer r:

$$r = \frac{r_{1\downarrow} + r_{2\downarrow}(t_{1\downarrow} t_{1\uparrow} - r_{1\downarrow} r_{1\uparrow}) e^{-2i\delta}}{1 - r_{2\downarrow} r_{1\uparrow} e^{-2i\delta}}.$$

Instead of using separate reflection coefficients for the waves coming onto the top surface of the layer upward $r_{1\uparrow}$ and downward $r_{1\downarrow}$, it is convenient to accept only one — namely the $r_{1\downarrow}$, and denote it as r_1. The same notation will be accepted for $r_2 \equiv r_{2\downarrow}$. Then, using $r_{1\downarrow} = -r_{1\uparrow}$ and $1 - r^2 = t_\downarrow t_\uparrow$, the complex reflection coefficient for a single layer transforms to

$$r = \frac{r_1 + r_2 e^{-2i\delta}}{1 + r_1 r_2 e^{-2i\delta}}.$$

Now, consider a multilayer interfering structure, composed of m flat parallel uniform layers (Fig. 8.6).

Complex refractive index of the jth layer is n_j and its thickness is l_j. Assume that the multilayer structure is confined by vacuum from one side and by the substrate from another side. The substrate is the $(m+1)$th layer of infinite thickness. We also assume that the interfaces between the layers are smooth and accept the system of coordinate as in Fig. 8.6. Then, denoting the complex reflection coefficient on top interface of the j-th layer as r_j and r_{j+1} on its bottom interface, we obtain the recurrent procedure for computing all reflection coefficients within the multilayer, starting from the bottom:

$$r_j = \frac{r_j^F + r_{j+1} e^{-2i\delta_j}}{1 + r_j^F r_{j+1} e^{-2i\delta_j}}, \quad r_m = r_m^F, \quad r_{m+1} = 0,$$

where r_j^F are the Fresnel reflection coefficients, defined for both s- and p-polarizations in the following generalized way:

$$r_j^F = \frac{g_j\sqrt{n_j^2 - \sin^2\phi} - g_{j+1}\sqrt{n_{j+1}^2 - \sin^2\phi}}{g_j\sqrt{n_j^2 - \sin^2\phi} + g_{j+1}\sqrt{n_{j+1}^2 - \sin^2\phi}}, \quad g_j = \begin{cases} 1 \text{ for } s-\text{polarization}, \\ n_j^{-2} \text{ for } p-\text{polarization}, \end{cases}$$

$$\delta_j = kl_j\sqrt{n_j^2 - \sin^2\phi}.$$

Note that whereas the product $n\sin\theta$ is an invariant within the multilayer structure, the phase term δ_j does not require computations of refraction angles inside layers, so that only the angle of incidence ϕ on the front surface of the multilayer structure is needed.

The above recurrent procedure holds true for complex refractive indices as well, taking into account possible absorption in the layers. The reflectance R, i.e. the reflection coefficient as the ratio of energies of the reflected and incident waves, is equal to

$$R = |r_j|^2.$$

For the case of a single layer, this leads to the formula

$$R = \frac{|r_1|^2 + 2Re(r_2 r_1^* e^{-2i\delta_1}) + |r_2|^2}{1 + 2\text{Re}(r_1 r_2 e^{-2i\delta_1}) + |r_1|^2|r_2|^2},$$

where $r_{1,2}$ are the Fresnel reflection coefficients r_j^F, $j = 1,2$. For non-absorbing layers, this formula transforms to

$$R = \frac{r_1^2 + 2r_1r_2\cos 2\delta_1 + r_2^2}{1 + 2r_1r_2\cos 2\delta_1 + r_1^2r_2^2}.$$

Continuing the simplification, we may consider normal incidence, when

$$r_1 = \frac{n_0 - n_1}{n_0 + n_1}, \quad r_2 = \frac{n_1 - n_2}{n_1 + n_2}.$$

In this case, the formula for reflectance may be expressed in the form, directly containing refractive indices of the materials:

$$R = \frac{(n_0^2 + n_1^2)(n_1^2 + n_2^2) - 4n_0n_1^2n_2 + (n_1^2 - n_0^2)(n_2^2 - n_1^2)\cos 2\delta_1}{(n_0^2 + n_1^2)(n_1^2 + n_2^2) + 4n_0n_1^2n_2 + (n_1^2 - n_0^2)(n_2^2 - n_1^2)\cos 2\delta_1}.$$

As an example, Fig. 8.7 presents computations of spectral reflectance according to this formula with the layer thickness $l_1 = 5$ μm and values $n_0 = 1$ (the incident wave comes from air), $n_1 = 1.45$ (silicon oxide), and $n_2 = 3.88$ (silicon). Assumption of constant refractive indices in the entire spectral interval, i.e. ignoring spectral dispersion, helps to understand basic phenomenology, but is not true in reality. In practice, as it will be seen below in Section 8.1.3, dispersion introduces wobbling of the oscillating pattern.

Taking a closer look at the above formula for R, it is instructive to analyze at what wavelengths the reflectance reaches its maximum and minimum values R_{max} and R_{min}. The maximum of R is reached when $\cos 2\delta_1 = 1$, i.e. when

$$\frac{4\pi}{\lambda} n_1 l_1 = 2\pi m, \quad m = 0, 1, 2, \ldots$$

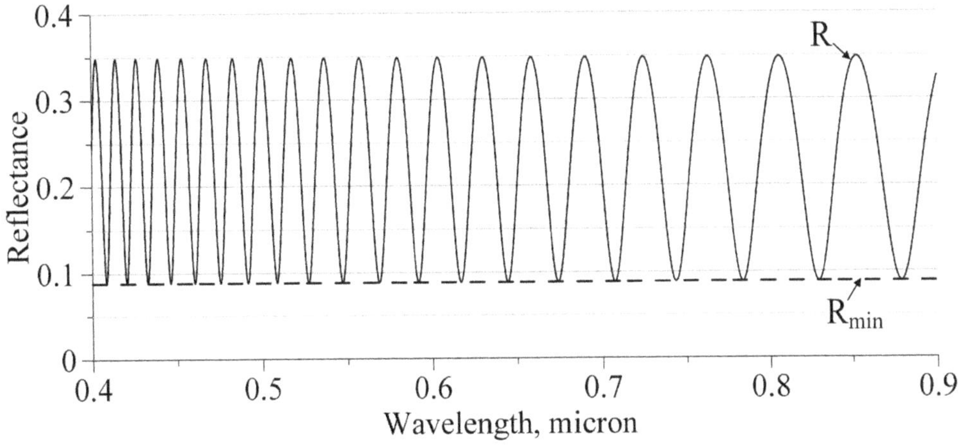

Fig. 8.7. Theoretical spectral reflectance curve. The dashed line shows the value given by the Abeles formula.

Substituting this value of the cosine, and performing obvious algebraic transformations, we obtain the formula

$$R_{\max} = \left(\frac{n_2 - n_0}{n_2 + n_0}\right)^2$$

which does not depend on n_1 and describes reflectance from the substrate alone. It means that, at the wavelengths where the reflectance reaches its maxima, the layer is fully transparent and its refractive index cannot be determined.

The minimum of R is reached when $\cos 2\delta_1 = -1$, i.e. at the wavelengths λ when

$$\frac{4\pi}{\lambda} n_1 l_1 = \pi + 2\pi m, \quad m = 0, 1, 2, \ldots$$

In this case, we have an equation, connecting the value $R_{\min}$ with n_1, which gives an opportunity to measure n_1 by measuring values of $R_{\min}$:

$$R_{\min} = \frac{x^2 + n_0^2 n_2^2 - 2n_0 n_2 x}{x^2 + n_0^2 n_2^2 + 2n_0 n_2 x}, \quad x = n_1^2.$$

Solution of this quadratic equation in x (bi-quadratic in n_1) gives the Abeles formula:

$$n_1^2 = n_0 n_2 \begin{cases} \dfrac{1 + \sqrt{R_{\min}}}{1 - \sqrt{R_{\min}}}, & n_1^2 > n_0 n_2, \\[2ex] \dfrac{1 - \sqrt{R_{\min}}}{1 + \sqrt{R_{\min}}}, & n_1^2 < n_0 n_2. \end{cases}$$

With these values of n_1,

$$R_{\min} = \left(\frac{n_0 n_2 - n_1^2}{n_0 n_2 + n_1^2}\right).$$

In Fig. 8.7, this value of reflectance is shown in straight dashed line.

For thinner layers $l_1 < \lambda/(2n) \sim 0.1\ \mu$m, multiple spectral oscillations of reflectance observable in visible domain for thick layers of the scale of microns (Fig. 8.7) degrade to smooth curviness, as computed in Fig. 8.8 for $l_1 = 0.1\ \mu$m and other parameters the same as in Fig. 8.7.

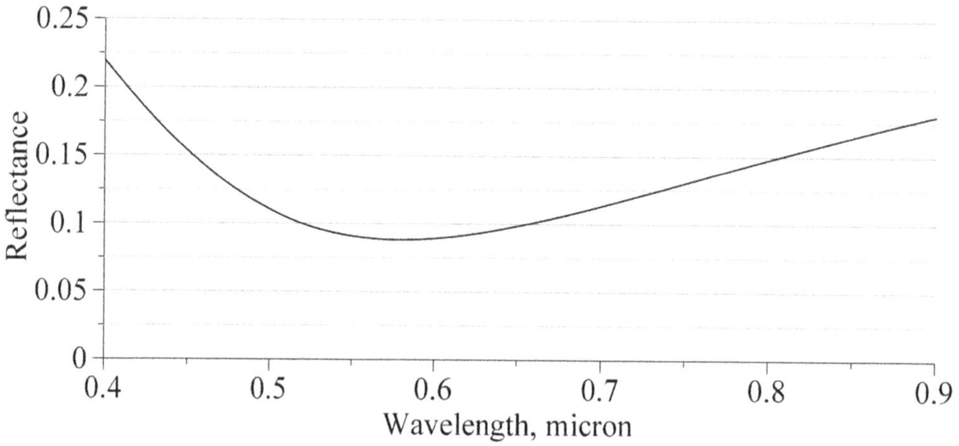

Fig. 8.8. Theoretical spectral reflectance for the layer 0.1 μm thick.

In such cases, the thickness and refractive indices also can be determined, but it must be done by mathematical fitting of theoretical computations into experimental data. Dispersion of the refractive indices of the film and substrate also can be included in these computations.

Up to this point, we considered common optical materials with low absorption. However, semiconductor manufacturing often deals with materials highly absorptive in visible domain, like silicon, or pure metals. Refractive index of an absorptive material may be mathematically presented in the complex form

$$\overline{n} = n \cdot e^{-i\alpha} = n\cos\alpha - in\sin\alpha \equiv n_r - in_i,$$

with n and α being real numbers. Then absorption of a plane wave with the wavelength λ, traversing the distance l in absorbing media with refractive index $\overline{n}$, is exponential (the Buger law), as it follows from the wave propagation term:

$$\exp[-ikl \cdot n(\cos\alpha - i\sin\alpha)] = \exp(-ikl \cdot n_r) \cdot \exp(-kl \cdot n_i), \quad k = 2\pi/\lambda.$$

Clearly, the non-absorbing material may be represented in the same form, assuming $\alpha = 0$. With this generalization, we are going to derive the formula for reflectance R at normal incidence for a plane wave, coming onto transparent film with refractive index n_1, deposited on an absorbing substrate with refractive index $n_2 e^{-i\alpha}$. This is a typical situation when thin film is formed on a silicon wafer.

According to the derived above basic formula for complex reflective coefficient r,

$$r = \frac{r_1 + r_2 e^{-2i\delta}}{1 + r_1 r_2 e^{-2i\delta}}, \quad r_1 = \frac{n_0 - n_1}{n_0 + n_1}, \quad r_2 = \frac{n_1 - \overline{n}_2}{n_1 + \overline{n}_2}, \quad \delta = k l_1 n_1.$$

The reflectance is then $R = |r|^2$, and performing some lengthy but primitive mathematical manipulations, the following formula may be obtained:

$$R = \frac{(n_0^2 + n_1^2)(n_1^2 + n_2^2) - 4n_0 n_1^2 n_2 \cos\alpha + (n_1^2 - n_0^2)(n_2^2 - n_1^2) \cdot \cos 2\delta + 2n_1 n_2 (n_1^2 - n_0^2)\sin\alpha \sin 2\delta}{(n_0^2 + n_1^2)(n_1^2 + n_2^2) + 4n_0 n_1^2 n_2 \cos\alpha + (n_1^2 - n_0^2)(n_2^2 - n_1^2) \cdot \cos 2\delta + 2n_1 n_2 (n_1^2 - n_0^2)\sin\alpha \sin 2\delta}.$$

Now, it is necessary to prove that this function reaches its local maxima and minima when $\cos 2\delta_1 = \pm 1$. For that, denote $\cos 2\delta = x$, $-1 < x < +1$, and find maxima and minima of the function $R(x)$:

$$R(x) = \frac{a - b + cx - d\sqrt{1 - x^2}}{a + b + cx - d\sqrt{1 - x^2}},$$

with the meaning of the constants a, b, c, and d easily deduced from the previous formula. This function does not have local extrema within the interval $-1 < x < +1$ and, therefore, the extrema are at the ends of this interval. To prove this, take the derivative of $R(x)$ and find that it does not have zeros within this interval: the numerator of the derivative is always $2b \equiv 8n_1^2 n_2 \cos\alpha \neq 0$. Hence, the maxima are reached when $\cos 2\delta = +1$ and the minima when $\cos 2\delta = -1$. Since δ depends only on optical properties of the film alone, it means that spectral positions of maxima and minima do not depend on optical properties of the substrate, no matter how absorptive it is.

The next question that must be answered, considering the effects of the substrate absorption, is whether the values of reflectance at its maxima $R_{\max}$ are equal to reflection from the substrate alone

$$\left|\frac{\overline{n}_2 - n_0}{\overline{n}_2 + n_0}\right|^2 = \frac{n_0^2 + n_2^2 - 2n_2 n_0 \cos\alpha}{n_0^2 + n_2^2 + 2n_2 n_0 \cos\alpha},$$

as it was in the case of the lossless substrate. From the above formula for R, we find:

$$R_{\max} = R|_{\cos 2\delta = 1} = \frac{n_0^2 + n_2^2 - 2n_2 n_0 \cos\alpha}{n_0^2 + n_2^2 + 2n_2 n_0 \cos\alpha}$$

which is exactly the reflectance from the substrate. This theoretical conclusion will be verified experimentally in the next section.

Finally, analyzing effects of absorptive substrates, it is worth mentioning the case of reflection from the film deposited onto highly conducting metals, like silver or aluminum. Absorption in that kind of materials is so high, that in the formula for complex refractive index $\overline{n}$ it is possible to approximately accept $\alpha \approx \pi/2$:

$$\overline{n} = n \cdot e^{-i\alpha} \approx -in_i.$$

Then $\cos\alpha \approx 0$ and from the formula for R we obtain $R \approx 1$, i.e. almost total reflectivity independently from the properties of the film.

The only theoretical topic that should be additionally considered in this section is the influence of angular divergence of the probe beam. Indeed, the formulas above were obtained for a single plane wave, and the question is how the result changes for a bundle of rays, coming onto the sample at different angles, particularly in case of a focused probe beam. In order to figure out basic phenomenology of this effect, we may again simplify the problem to a single layer and consider a conical bundle of non-polarized rays, coming from the vacuum ($n_0 = 1$) with the rotational axis of the cone perpendicular to the surface (Fig. 8.9).

Since the rays are not polarized, the energy of each ray is uniformly distributed between s- and p-polarizations and the energy reflection coefficient R for each ray, coming at the angle of incidence ϕ, is the arithmetic average of its s- and p-polarized components:

$$R(\phi) = 0.5(R_s + R_p),$$

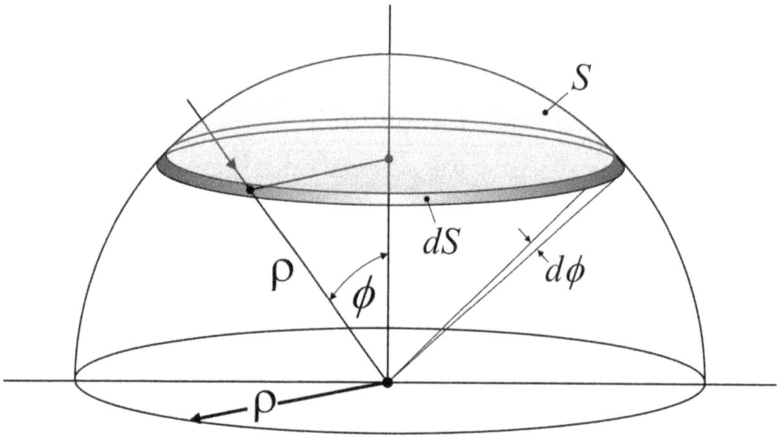

Fig. 8.9. Notations of the problem.

with

$$R_s = \frac{r_{1s}^2 + 2r_{1s}r_{2s}\cos(2\delta_1) + r_{2s}^2}{1 + 2r_{1s}r_{2s}\cos(2\delta_1) + r_{1s}^2 r_{2s}^2}, \quad R_p = \frac{r_{1p}^2 + 2r_{1p}r_{2p}\cos(2\delta_1) + r_{2p}^2}{1 + 2r_{1p}r_{2p}\cos(2\delta_1) + r_{1p}^2 r_{2p}^2},$$

$$r_{1s} = \frac{\sqrt{1-\sin^2\phi} - \sqrt{n_1^2 - \sin^2\phi}}{\sqrt{1-\sin^2\phi} + \sqrt{n_1^2 - \sin^2\phi}}, \quad r_{2s} = \frac{\sqrt{n_1^2-\sin^2\phi} - \sqrt{n_2^2 - \sin^2\phi}}{\sqrt{n_1^2-\sin^2\phi} + \sqrt{n_2^2 - \sin^2\phi}},$$

$$r_{1p} = \frac{\sqrt{1-\sin^2\phi} - n_1^{-2}\sqrt{n_1^2 - \sin^2\phi}}{\sqrt{1-\sin^2\phi} + n_1^{-2}\sqrt{n_1^2 - \sin^2\phi}}, \quad r_{2p} = \frac{n_1^{-2}\sqrt{n_1^2-\sin^2\phi} - n_2^{-2}\sqrt{n_2^2 - \sin^2\phi}}{n_1^{-2}\sqrt{n_1^2-\sin^2\phi} + n_2^{-2}\sqrt{n_2^2 - \sin^2\phi}},$$

and

$$\delta_1 = kl_1\sqrt{n_1^2 - \sin^2\phi}, \quad k = 2\pi/\lambda.$$

In this formula, the most influential angular contributor to the reflectance is the interference term with $\cos(2\delta_1)$, which oscillates the quicker the thicker the layer l_1 is. Thus, the primary consideration is that the effect of the finite angular width of the probe beam reveals itself in the strongest way for thick layers. In semiconductor manufacturing, the thickest layers are usually found in the so-called Chemical Mechanical Polishing (CMP) process, when a silicon wafer is being thinned from its original 0.8 mm thickness to about hundred microns or even less. Therefore, in CMP process, the effect of angular divergence of the probe beam is an important factor, which will be analyzed theoretically in full detail in Section 8.3. However, for thin optical layers of less than 1 micron, this effect may be expected to be much less important — the conclusion that is corroborated by numerical computations presented below.

The average reflectance within the cone of rays is equal to

$$\overline{R} = \frac{1}{S}\int_S R(\phi)ds = \frac{1}{2\pi\rho^2(1-\cos\phi_{\max})}\int_0^{\phi_{\max}} R(\phi)2\pi\rho^2\sin\phi d\phi = \frac{1}{1-\cos\phi_{\max}}\int_0^{\phi_{\max}} R(\phi)\sin\phi d\phi.$$

Note that the term $\sin\phi_{\max}$, determining angular dependence of the reflectance, is nothing else but numerical aperture NA of the objective lens that focuses light onto the sample. Therefore, we may analyze the effect of angular divergence in terms of the numerical aperture NA. In practice, standard objective lenses used in spectral reflectometry do not have high numerical apertures because not only

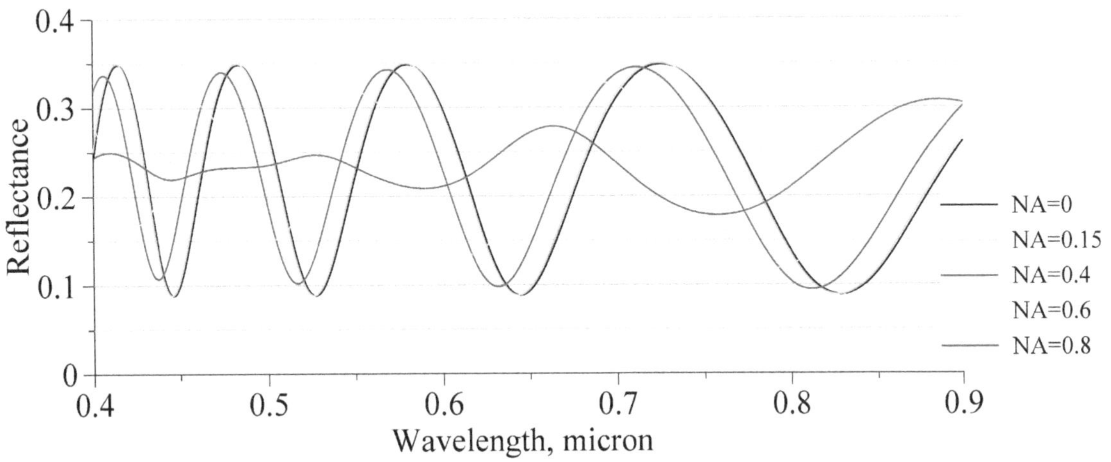

Fig. 8.10. Comparison of spectral reflectance of a single plane wave (solid line) with a bundle of plane waves at various angles of incidence around normal incidence (dashed line)

fine focal spot is needed but wide angular field of view as well. Hence, the values $NA \sim 0.15$ with magnifications around 5× may be considered as typical. Figure 8.10 compares spectra computed for 1 μm SiO_2 layer ($n_1 = 1.45$) on silicon ($n_2 = 0.388$) and several values of NA, including plane wave ($NA = 0$). Other parameters of computations are the same as those used in Fig. 8.7. Clearly, the results for the plane wave and $NA = 0.15$ may be considered as practically identical. However, larger numerical apertures, starting from $NA \approx 0.4$, degrade the spectrum, making spectral oscillations smaller and irregular. This result should be kept in mind, addressing the subject of Chapter 9 on angular-dependent reflectometry.

8.1.3. *Experimental techniques*

For practical implementation, it is important to focus the probe beam onto a certain small feature on the patterned semiconductor wafer. The primitive arrangement shown in Fig. 8.1 cannot tell where exactly the receiving optical fiber takes the reflected signal from. Therefore, another optical scheme with high-quality objective lens and beamsplitters is commonly used in combination with the imaging camera (Fig. 8.11). Typically, it contains two white light sources: one for illumination of the wafer (1) and the second one for spectral reflectometry (12). Also, two high-quality beamsplitters made of thin optical plates serve to relay the spectral reflectometer probe beam (beamsplitter 5) and distribute returning beams between the video-camera (11) and the spectrometer (10) (beamsplitter 8). In such a scheme, the operator can see on the computer display simultaneously the patterned area of the wafer (17) and the exact position of the reflectometer focal spot (18) on it. The wafer (14) fixed on the scanning table (15) can be transported to any desirable position for inspection.

In order to use theoretical considerations outlined in the preceding section, the raw signal from the spectrometer, shown in its typical form in Fig. 8.2 together with the background, must be separated from the background and normalized. Separation of the background is a simple procedure performed almost automatically before any measurement by measuring dark current of the photodetector and subtracting it from every subsequent measurement. As to the normalization, this procedure requires additional explanation and three different options exist.

The first normalization procedure. The first option is used when measuring the absolute value of the film thickness. In it, the true values of spectral reflectance $R(\lambda)$ must be computed and then either

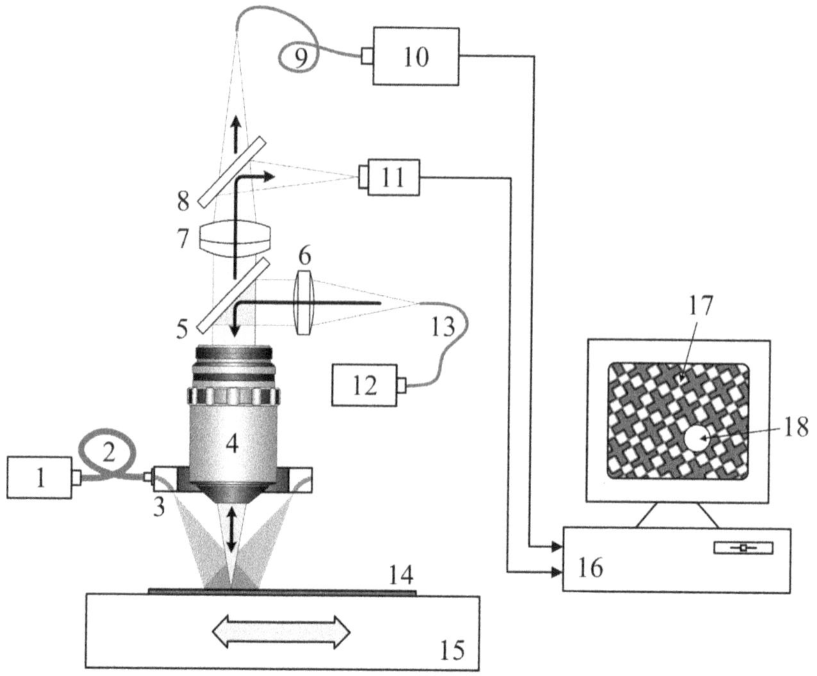

Fig. 8.11. Conceptual view of a spectral reflectometer. 1,12 — white light sources; 2,9,13 — optical fibers; 3 — illuminating ring; 4 — objective lens; 5,8 — beamsplitters; 6 — collimating lens; 7 — tube lens; 10 — spectrometer; 11 — video camera; 14 — wafer; 15 — scanning table; 16 — computer; 17 — image of the wafer; 18 — image of the focused light spot.

the Fourier transform applied, when the film is thick enough to produce multiple oscillations within the spectral range of the spectrometer, or a fitting procedure used when the film is thin and only smooth curviness is observed, as illustrated in Fig. 8.8. For that, the reference spectrum $S_0(\lambda)$ must be measured and stored in the computer memory when the high-quality mirror is placed instead of the wafer. Then, each time when the raw spectrum $S(\lambda)$ reflected from the sample is measured, the spectral reflectance $R(\lambda)$ may be computed as the ratio $S(\lambda)/S_0(\lambda)$. Of course, the dark current of the photodetector, measured and saved in the computer memory beforehand, must be subtracted from both $S(\lambda)$ and $S_0(\lambda)$. As an example, Fig. 8.12 presents spectral reflectance $R(\lambda)$ of the silicon wafer coated with approximately 6 μm silicon oxide layer, measured in a single local point on the wafer.

As it was explained in the previous section, the values $R_{\min}$ of spectral reflectance at the points of its minima determine the refractive index of the film n_1 if the refractive index of the substrate n_2 is known. The refractive index of a silicon wafer is well known at the wavelength of the He–Ne laser 0.6328 μm: it is complex and equal to $3.88 + i \cdot 0.02$. In Fig. 8.12, one minimum of oscillations, marked by a white circle, almost exactly corresponds to the wavelength of the He–Ne laser with the value of reflectance $R_{\min} = 0.10$. Thus, at this point, we may estimate the refractive index of the film, using only real part of the silicon refractive index:

$$n_1^2 = 3.88 \cdot \frac{1 - \sqrt{0.10}}{1 + \sqrt{0.10}} = 2.02, \quad n_1 = 1.42$$

which is a value close to the standard value 1.45 known for the silicon oxide.

The next step that we can make is to verify theoretical conclusion of the previous section, stating that the maxima of the oscillating spectral reflectance correspond to the values of reflectance from the

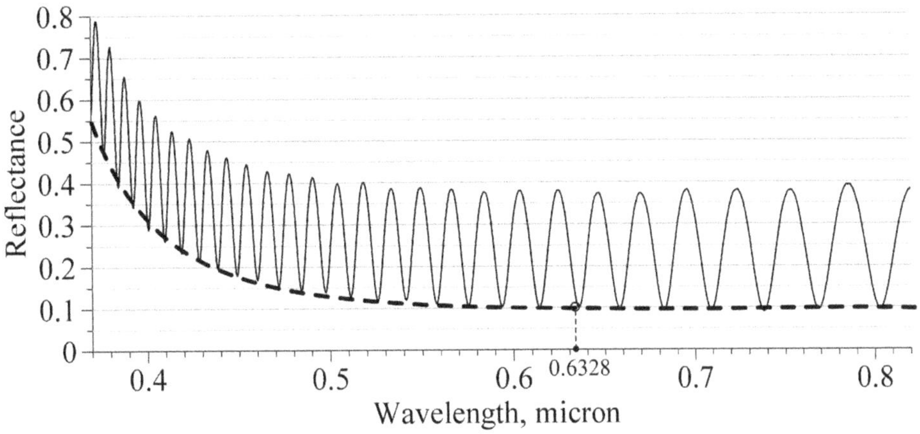

Fig. 8.12. Experimentally measured spectral reflectance. The dashed line approximately connects the points of minimum reflectance.

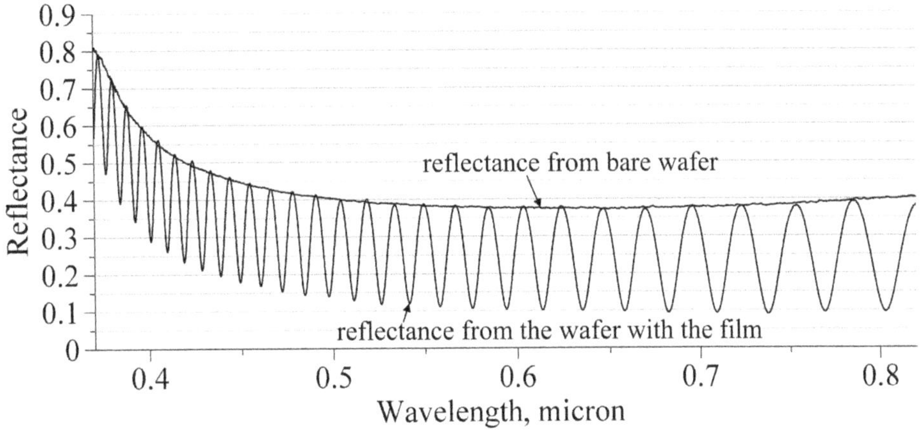

Fig. 8.13. Experimentally measured spectral reflectance of bare wafer goes through the maxima of oscillating spectral reflectance of the coated wafer as predicted by theory.

bulk substrate alone. For that, superimpose the curve in Fig. 8.12 with measured reflectance of the bare silicon wafer, as it is done in Fig. 8.13. Clearly, the theoretical prediction may be considered correct.

However, such good coincidence does not mean that it is possible to measure refractive index of silicon, using the theory of the previous section — those formulas were developed for transparent media, whereas silicon is strongly absorbing, especially in visible and ultraviolet domains.

The second normalization procedure. In semiconductor industry, the basic motivation for using spectral reflectometry is not the precise measurement of refractive indices — this job can be done calmly and with great precision in laboratory by other methods — but quick assessment of spatial non-uniformity of coatings over the area of the wafer. Therefore, the metrology tool must scan the wafer, producing a map of its uniformity. As such, wobbling of spectral reflectance, seen in Figs. 8.12 and 8.13 and caused mainly by dispersion of refractive index of silicon, is undesirable. Uniformity of spectral reflectance may be significantly improved, using the second normalization procedure: by measuring the reference spectrum $S_0(\lambda)$ not on the mirror but on bare silicon. Created in such a way ratio $S(\lambda)/S_0(\lambda)$ is not actually the reflectance, but this function does not show definite rise in the short-wavelength region of spectrum, and all the maxima of oscillations tend to be close to unity (100%). An example of this

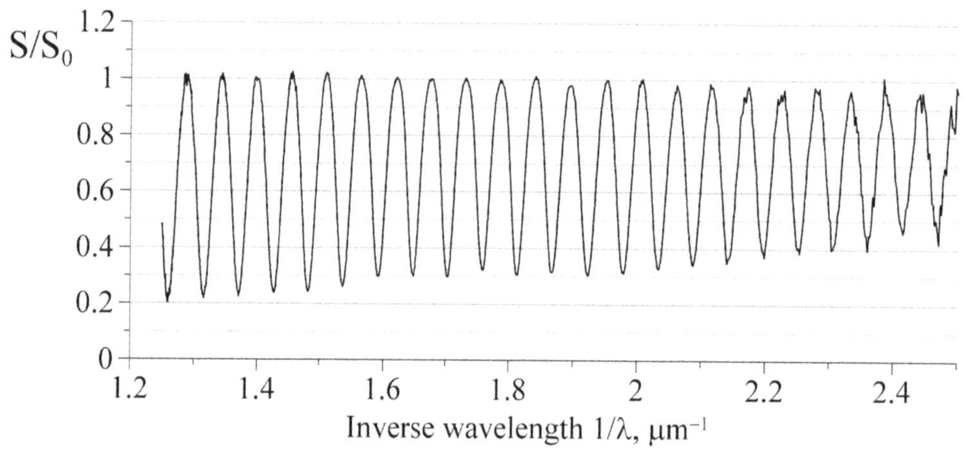

Fig. 8.14. Local spectral oscillations of a coated wafer after normalization to bare silicon.

type of normalization is shown in Fig. 8.14. In it, the ratio S/S_0 is plotted as a function of the inverse wavelength, making oscillations strictly periodical with the period $p = 0.055\ \mu\text{m}^{-1}$. Having determined the refractive index of the film previously as $n_1 = 1.42$, it is possible to estimate its thickness by formula

$$l_1 = \frac{1}{2n_1 p} = 6.4\ \mu\text{m}.$$

For better impression of how significant may spatial non-uniformity of the film thickness be, it is instructive to create the two-dimensional composition of a series of spectral oscillations recorded consecutively during one-dimensional scan along the surface of the wafer. Since periodicity of spectral oscillations directly depends on the thickness of a film, the natural waviness of the two-dimensional periodical picture is a good measure of spatial non-uniformity (Fig. 8.15).

The third normalization procedure does not use any reference sample at all — neither mirror, nor bare silicon. The reference spectrum $S_0(\lambda)$ is taken at an arbitrary point on the wafer, and from now on, this point becomes the reference point of measurements. Then all measurements at other points are being made relatively to the reference point by computing the ratio $S(\lambda)/S_0(\lambda)$. With this type of normalization, non-uniformity of the film thickness is encoded in the amplitude of the ratio $S(\lambda)/S_0(\lambda)$, which sometimes may be more convenient for quick analysis than measuring period of spectral oscillations like in Fig. 8.15. This statement requires additional explanation.

Consider the simplified model of spectral oscillations, used in the first section of this chapter, and write the reference spectrum at some chosen point of the wafer as

$$S_0(\lambda) = 1 + \gamma \cos(4\pi l_0 n \lambda^{-1}),$$

where l_0 and n are the thickness and refractive index of the film at the reference point, and γ is a constant. Moving mechanically to a neighboring point, the thickness changes to change to $l_0 + \delta$, and spectrum changes to

$$S(\lambda) = 1 + \gamma \cos[4\pi(l_0 + \delta) n \lambda^{-1}].$$

The normalized spectrum is

$$\frac{S(\lambda)}{S_0(\lambda)} = \frac{1 + \gamma \cos[4\pi(l_0 + \delta) n \lambda^{-1}]}{1 + \gamma \cos(4\pi l_0 n \lambda^{-1})}.$$

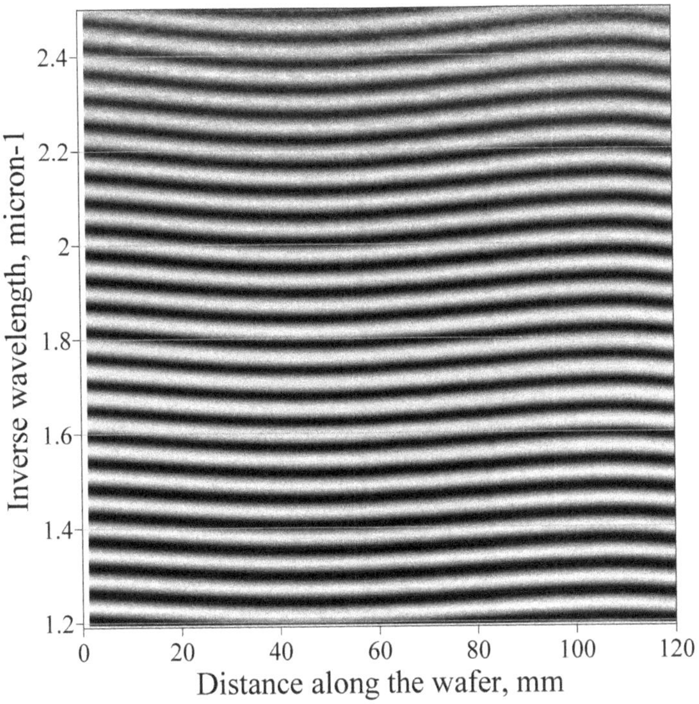

Fig. 8.15. Two-dimensional presentation of spectral oscillations recorded during one-dimensional scan along the 300 mm silicon wafer coated with silicon oxide layer.

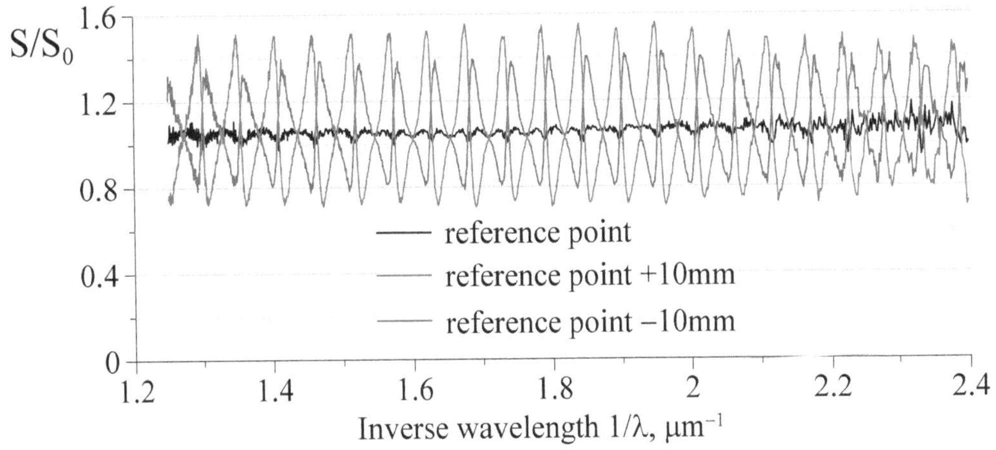

Fig. 8.16. Example of normalized spectra in two different points around the reference point.

Using the trigonometric identity $\cos(\alpha+\beta) = \cos\alpha\cdot\cos\beta - \sin\alpha\cdot\sin\beta$ and smallness of δ, this formula transforms to

$$\frac{S(\lambda)}{S_0(\lambda)} \approx 1 - \delta\cdot 4\pi\gamma n\lambda^{-1}\frac{\sin(4\pi l_0 n\lambda^{-1})}{1+\gamma\cos(4\pi l_0 n\lambda^{-1})}.$$

This formula describes spectral oscillations, the amplitude of which is proportional to spatial variation of the film thickness δ. An example of real measurements is shown in Fig. 8.16.

In this figure, the shape of spectral oscillations changes when the scanning spot crosses the reference point. It means that the reference point was chosen within the area where the film thickness varies

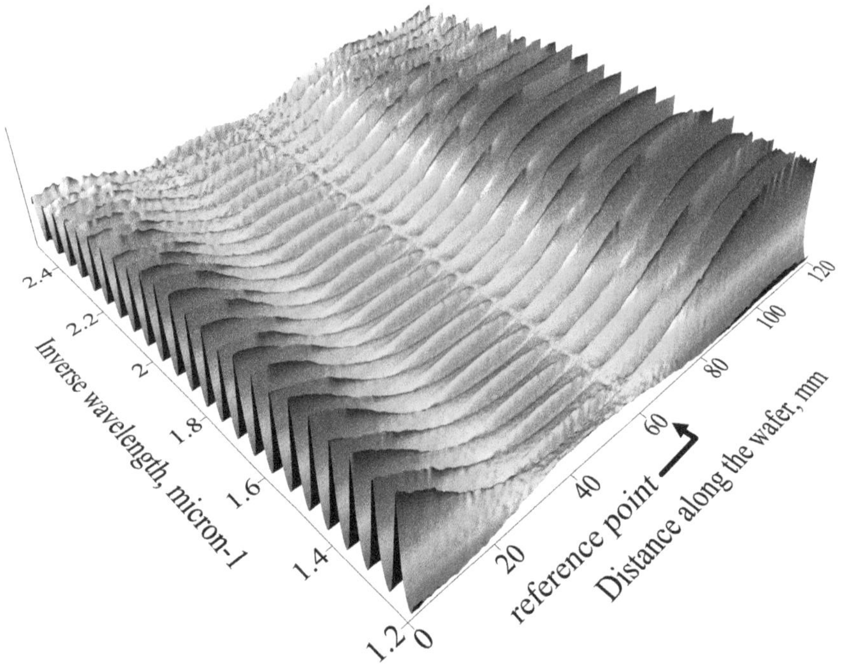

Fig. 8.17. Three-dimensional presentation of spectral oscillations during line scan over the wafer. The reference point was chosen in the middle of the scan.

monotonously in a wedge-like manner. In this case, variation of the film thickness δ changes its sign when the scan crosses the reference point, so that the function S/S_0 changes from

$$1 + |\delta| \cdot f(\lambda^{-1})$$

to

$$1 - |\delta| \cdot f(\lambda^{-1}),$$

where f is the periodical function that determines the shape of oscillations.

Using this technique, the entire variation of the film thickness along one 120 mm linear scan can be better visualized in the form of a 3D-map as in Fig. 8.17. Unlike Fig. 8.15, the third type of normalization readily displays non-uniformity of the film thickness in the form of amplitude variations of spectral oscillations.

8.2. Patterned Structures

8.2.1. *Phenomenology of spectral reflectometry on patterned structures*

Patterned surfaces are of the most interest in semiconductor manufacturing, especially in manufacturing memory devices, where relatively large areas are filled with sub-micrometer patterns with periods of the scale of tens to hundreds of nanometers. Metal interconnects form periodical sub-wavelength grids that act as polarizers in transmission mode or as diffraction gratings on reflection. As an example of memory structure manufactured on hundred-nanometer technology, Fig. 8.18 shows the microscopy image of the wafer as it is seen in the reflectometer described in Fig. 8.11.

The colors in the picture appear from diffraction of white light on microscopic diffraction gratings that form the structure of the memory device. Penetration depth of white light into such structures is

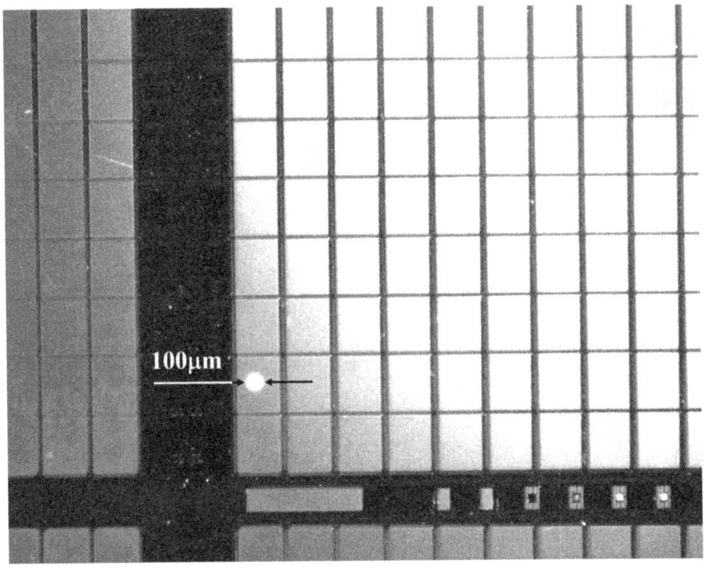

Fig. 8.18. A part of a memory structure with the probe beam focused into 100 μm spot within one of many rectangular memory banks of the device. The vertical and horizontal black strips contain read-out circuitry.

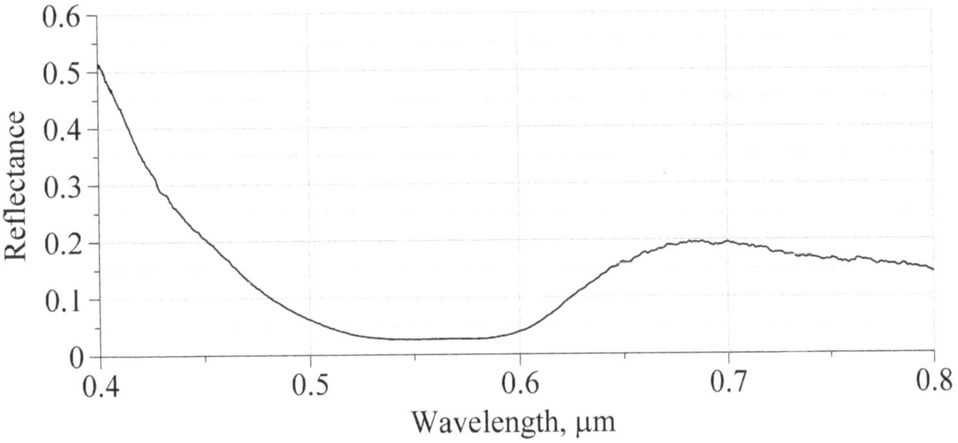

Fig. 8.19. Reflection-mode non-polarized spectrum returned from the spot marked by arrows in Fig. 8.18.

relatively small, of the order of one micron. Therefore, spectral variations of the reflected light do not reveal distinct oscillations, as shown in Fig. 8.19.

For the structures, containing micro- or nano-gratings, like the one shown in Fig. 8.18, the basic information about grating parameters may be encoded in difference between the spectra of the two orthogonal polarizations — with electrical vectors of the probe wave parallel and perpendicular to the grating grooves. As an example, Fig. 8.20 shows such polarization-dependent spectra, measured in the same point, where the non-polarized spectrum of the Fig. 8.19 was measured. The two lines present independently measured ratios $R_\mathrm{h}/R_\mathrm{v}$ and $R_\mathrm{v}/R_\mathrm{h}$ for the two reflected polarization components — with polarization vector in horizontal R_h and vertical R_v directions.

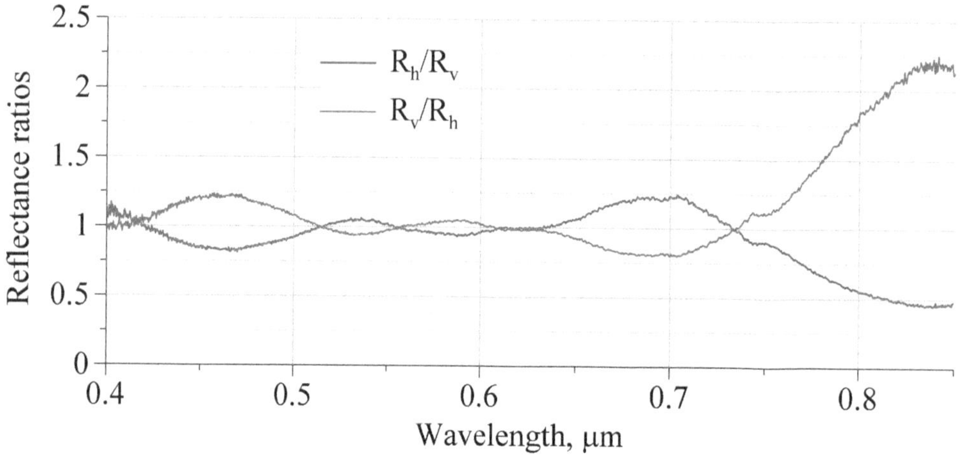

Fig. 8.20. The ratios of the two orthogonally polarized spectral components R_{h} and vertical R_{v} measured in the same spot as the spectrum in Fig. 8.19.

Clearly, in order to make any conclusions about the inner structure and defects in such layers, careful theoretical calculations of spectra must be made and fitted into experimental data. Thus, the problem of applying spectral reflectometry to patterned structures becomes more mathematical than technical, requiring reliable algorithms for calculating spectra of reflected waves.

8.2.2. *Theory of reflection from dielectric sub-wavelength patterned structures*

Nowadays, a variety of numerical methods and software packages are known that solve the Maxwell equations on the interface of a dielectric grating and compute reflection coefficients of diffraction orders: the coupled-wave method, the modal method, the C-method, and their modifications. All of them are equally accurate on dielectrics, but the most popular seems to be the coupled-wave method introduced in 1981 by Moharam and Gaylord. This method is widely known as the RCWA — an abbreviation for "rigorous coupled-wave analysis". Despite the ambitious word "rigorous" in its name, the RCWA is actually an approximation when applied to arbitrary grating shapes, and the better the approximation is, the longer the computation time is.

In 1993, Raguin and Morris introduced another algorithm that takes only milliseconds to compute the reflection from a patterned structure. It may be called the effective medium trapezoidal approximation (EMTA). Where the RCWA takes 20 seconds to compute the spectrum, the EMTA takes only 10 milliseconds. The basic assumption of this method is smallness of the pattern period relative to the wavelength of light. In recent years, the semiconductor technology came to nanometer-scale patterning — well below optical wavelengths in visible domain, making the EMTA method very accurate. Moreover, unlike all other methods, it is simple, stable, and may be finalized as a single analytical formula. Therefore, it deserves detailed explanation.

For the most of sub-wavelength structures, general trapezoidal model is applicable as it is shown in Fig. 8.21. The structure may be multilayered both in the patterned area and in the substrate area (Fig. 8.21(a)). However, in order to clarify explanation, a simplified model shown in Fig. 8.21(b) is accepted in this section. In it, only the substrate is allowed to be multilayered, while the patterned area is assumed to be made of a single material. This will reduce complexity of mathematical transformations to a minimum necessary for understanding of the concept.

When the period of a grating is much smaller than the wavelength, the grating, as well as any other periodically stratified medium, behaves as a continuous medium with two different dielectric constants

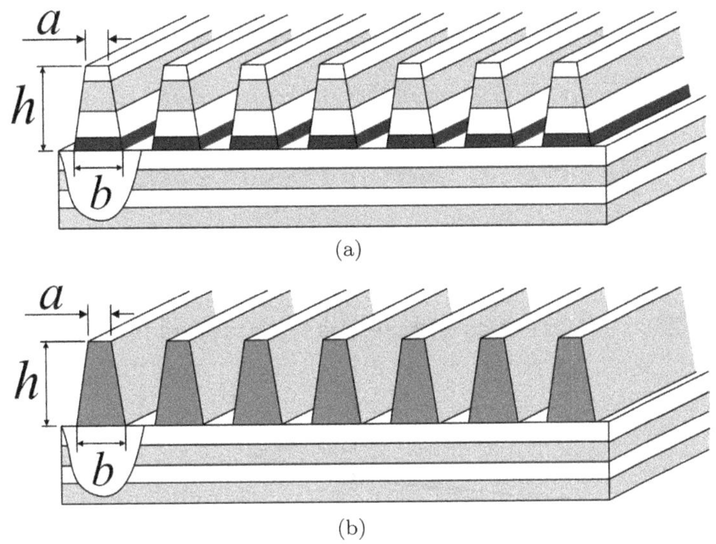

Fig. 8.21. Trapezoidal models: (a) the general model; (b) the simplified model accepted in computations below.

for the polarizations along and across to the grooves. In 1949, Rytov gave the zero and second order approximations for effective dielectric constants, having shown that the first-order terms vanish. Since then, his result was compiled without modification in numerous number of publications, and was proved experimentally correct. According to it, at normal incidence the effective dielectric constants are

$$\varepsilon_{||} = \varepsilon_{||}^{(0)} \left[1 + \frac{\pi^2}{3} \left(\frac{p}{\lambda} \right)^2 f^2 (1-f)^2 \frac{(\varepsilon_s - \varepsilon_i)^2}{\varepsilon_{||}^{(0)}} \right],$$

$$\varepsilon_{\perp} = \varepsilon_{\perp}^{(0)} \left[1 + \frac{\pi^2}{3} \left(\frac{p}{\lambda} \right)^2 f^2 (1-f)^2 (\varepsilon_s - \varepsilon_i)^2 \varepsilon_{||}^{(0)} \left(\frac{\varepsilon_{\perp}^{(0)}}{\varepsilon_i \varepsilon_s} \right)^2 \right],$$

$$\varepsilon_{||}^{(0)} = f \varepsilon_s + (1-f) \varepsilon_i,$$

$$\varepsilon_{\perp}^{(0)} = \left(\frac{f}{\varepsilon_s} + \frac{1-f}{\varepsilon_i} \right)^{-1},$$

where the notations "||" and "⊥" relate to the vector of electrical field being parallel and perpendicular to grooves, ε_s and ε_i are the dielectric constants of the grating material and the outer medium from where the wave comes, f is the filling factor defined as the ratio of the width of grating ridge to the period p, and λ is the wavelength. These formulas present the so-called effective medium approximation.

With today's semiconductor technology, the second-order corrections $(p/\lambda)^2$ in formulas for $\varepsilon_{||}$ and $\varepsilon_{\perp}$ are of the order of 10^{-2}. This means that the effective medium approximation works well for inspection of wafers, and that it is possible to use even zeroth-order approximations. In general, the filling factor f depends on vertical coordinate z (Fig. 8.22), so that for the trapezoidal model, the formulas for $\varepsilon_{||}$ and $\varepsilon_{\perp}$ may be rewritten as functions of z in the following forms:

$$\varepsilon_{||} = \varepsilon_1 \left[1 + \left(\frac{\varepsilon_2}{\varepsilon_1} - 1 \right) \frac{z}{L} \right], \quad \varepsilon_1 = \frac{a}{p} (\varepsilon_s - \varepsilon_i) + \varepsilon_i, \quad \varepsilon_2 = \frac{b}{p} (\varepsilon_s - \varepsilon_i) + \varepsilon_i,$$

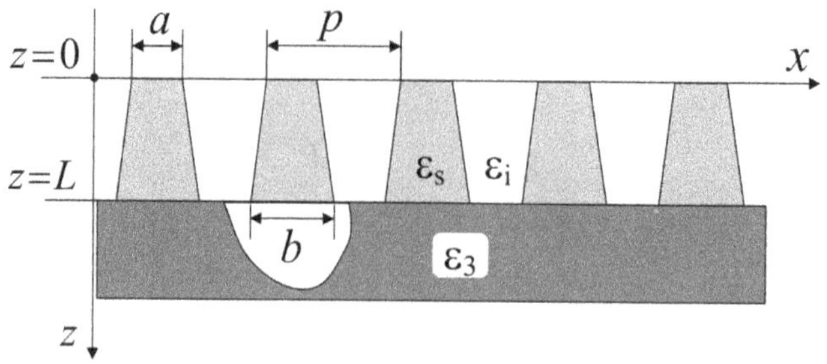

Fig. 8.22. System of coordinates in the trapezoidal model.

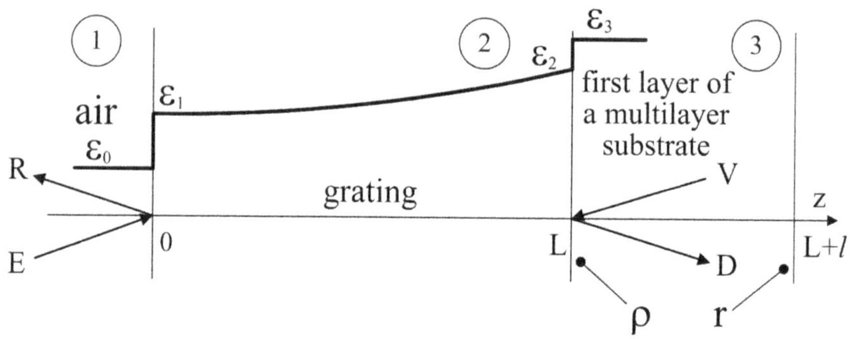

Fig. 8.23. Diagram of waves in consideration. Only the upper layer of a multilayer substrate is shown.

and

$$\varepsilon_{\perp} = \varepsilon_1 \left[1 + \left(\frac{\varepsilon_1}{\varepsilon_2} - 1\right)\frac{z}{L}\right]^{-1}, \quad \varepsilon_1 = \frac{1}{\varepsilon_i} - \frac{a}{p}\left(\frac{1}{\varepsilon_i} - \frac{1}{\varepsilon_s}\right), \quad \varepsilon_2 = \frac{1}{\varepsilon_i} - \frac{b}{p}\left(\frac{1}{\varepsilon_i} - \frac{1}{\varepsilon_s}\right).$$

These formulas can be presented in a general form

$$\varepsilon(z) = \varepsilon_1(1 + \delta_m z)^m, \quad m = \pm 1,$$

with δ_m being a constant parameter, which allows the exact solution of the Maxwell equations, using the approach proposed by Wallot in 1919. Wallot used a very peculiar mathematical transformation that presented the Maxwell equations as a set of the Bessel second order differential equations with analytical solution in terms of the Hankel functions. Using this approach, and considering the sub-wavelength patterned structure as continuous medium with depth-varying refractive index, it is possible to derive analytical formulas for complex reflection coefficients for linearly polarized waves reflected from such structures. The cross-sectional diagram of the problem is outlined in Fig. 8.23. Only normal incidence will be considered because it is the dominant optical scheme for metrology devices.

Then the Maxwell equations for electrical E and magnetic H fields can be written in a scalar form:

$$\begin{cases} \dfrac{\varepsilon}{c}\dfrac{\partial E}{\partial t} = -\dfrac{\partial H}{\partial z}, \\ \dfrac{1}{c}\dfrac{\partial H}{\partial t} = -\dfrac{\partial E}{\partial z}. \end{cases}$$

Here c is the speed of light, t is time. With the notations of Fig. 8.23, the electrical and magnetic fields in the first medium can be written as follows:

$$E_{\text{in}} = e^{i\omega t}(Ee^{-ik_1 z} + Re^{ik_1 z}),$$

$$H_{\text{in}} = e^{i\omega t}\sqrt{\varepsilon_0}(Ee^{-ik_1 z} - Re^{ik_1 z}),$$

where k_1 is the propagation vector in the first medium. Similarly, in the third medium:

$$E_{\text{out}} = e^{i\omega t}(D\,e^{-ik_3z'} + V e^{ik_3z'}),$$
$$H_{\text{out}} = e^{i\omega t}\sqrt{\varepsilon_3}(De^{-ik_3z'} - V\,e^{ik_3z'}),$$

where z' stands for $z-L$ and k_3 is the propagation vector in the third medium. Since at normal incidence there is a degeneracy in signs for reflected electrical and magnetic fields, in the formulas above, the signs of the reflected electrical fields are chosen same as those of the incident electrical fields.

In the second media, i.e. within the grating, Wallot gave the following solution:

$$E = e^{i\omega t}\left(\frac{\theta}{\theta_0}\right)^{1-q}[A_1H_{q-1}^{(1)}(\omega\theta) + A_2H_{q-1}^{(2)}(\omega\theta)],$$
$$H = -ie^{i\omega t}\left(\frac{\theta}{\theta_0}\right)^{q}\sqrt{\varepsilon_1}[A_1H_q^{(1)}(\omega\theta) + A_2H_q^{(2)}(\omega\theta)],$$

where $H_q^{(1)}$ and $H_q^{(2)}$ are the Hankel functions of the first and second kind, respectively,

$$\theta = \frac{2}{m+2}\frac{\sqrt{\varepsilon_0}}{\delta_m c}(1+\delta_m z)^{\frac{m+2}{2}} = \theta_0(1+\delta_m z)^{\frac{m+2}{2}}, \quad \theta_0 = \frac{2}{m+2}\frac{\sqrt{\varepsilon_0}}{\delta_m c}, q = \frac{m+1}{m+2},$$

and $A_{1,2}$ are the constants to be derived from the boundary conditions.

The continuity requirement at the interfaces between the first and the second medium, and between the second and the third medium give four equations for six unknowns:

$$E + R = [A_1H_{q-1}^{(1)}(\omega\theta_0) + A_2H_{q-1}^{(2)}(\omega\theta_0)],$$
$$\sqrt{\varepsilon_0}(E - R) = -i\sqrt{\varepsilon_1}[A_1H_q^{(1)}(\omega\theta_0) + A_2H_q^{(2)}(\omega\theta_0)],$$
$$D + V = \left(\frac{\theta_L}{\theta_0}\right)^{1-q}[A_1H_{q-1}^{(1)}(\omega\theta_L) + A_2H_{q-1}^{(2)}(\omega\theta_L)],$$
$$\sqrt{\varepsilon_3}(D - V) = -i\sqrt{\varepsilon_1}\left(\frac{\theta_L}{\theta_0}\right)^{q}[A_1H_q^{(1)}(\omega\theta_L) + A_2H_q^{(2)}(\omega\theta_L)].$$

Here the subscript "L" means that the function is taken at $z = L$. In order to make this system solvable, it is necessary to establish a relation between the direct- and back-propagating waves V and D. This can be done formally by introducing the complex reflection coefficient ρ at the interface between the second and the third medium (Fig. 8.23):

$$V = \rho D.$$

In Section 8.1.2, it was shown how to compute ρ, using standard multilayer recursion procedure. Eventually, the amplitude of the incident wave may be set equal to unity:

$$E = 1,$$

and this completes the system of equations that has to be solved for R.

For computational reasons, taking into consideration that not everyone mathematical software package provides efficient algorithms for the Hankel functions, the solution for complex reflection coefficient R may be expressed in terms of the Bessel functions. After some tedious but primitive algebraic transformations, using relations between the Hankel and Bessel functions

$$H_q^{(1)} = J_q + iY_q, \quad H_q^{(2)} = J_q - iY_q,$$

the solution for the complex reflection coefficient R can be written as follows:

$$R = \frac{M_+}{M_-},$$

$$M_\pm = \left[\sqrt{\frac{\varepsilon_2}{\varepsilon_3}}\left(\frac{1+\rho}{1-\rho}\right) J_q(\xi_L) - iJ_{q-1}(\xi_L)\right]\left[\pm\sqrt{\frac{\varepsilon_1}{\varepsilon_0}} Y_q(\xi_0) - iY_{q-1}(\xi_0)\right],$$

$$-\left[\sqrt{\frac{\varepsilon_2}{\varepsilon_3}}\left(\frac{1+\rho}{1-\rho}\right) Y_q(\xi_L) - i\,Y_{q-1}(\xi_L)\right]\left[\pm\sqrt{\frac{\varepsilon_1}{\varepsilon_0}} J_q(\xi_0) - iJ_{q-1}(\xi_0)\right]$$

and

$$\xi_0 = \frac{L}{\lambda}\cdot\frac{4\pi\sqrt{\varepsilon_1}}{(m+2)\left[\left(\frac{\varepsilon_2}{\varepsilon_1}\right)^{\frac{1}{m}} - 1\right]}, \quad \xi_L = \xi_0\cdot\left(\frac{\varepsilon_2}{\varepsilon_1}\right)^{\frac{m+2}{2m}}.$$

Here J_q and Y_q are the Bessel functions of the first and second kind, respectively.

Formulas for R and $M_\pm$ provide the exact solution to our problem in the trapezoidal model, and this solution is really "rigorous" unlike that of the RCWA. And the most important advantage of this solution is that the speed of computations is incomparably higher than that with the RCWA. However, computation of the Bessel functions (or Hankel functions) requires some additional resources from the computer, and due to it, in some cases, computational time increases significantly. Fortunately, the formulas for R can be asymptotically simplified for utmost computational speed, preserving at the same time necessary accuracy.

For the asymptotic expansion, it is better to present the result in terms of the Hankel functions as it was originally done by Wallot rather than in the Bessel functions as it was done by Raguin and Morris. The above result is equivalent to

$$R = -1 + \frac{2b}{bc+ad} H^{(1)}_{q-1}(\xi_0) + \frac{2a}{bc+ad} H^{(2)}_{q-1}(\xi_0) \quad ,$$

$$a = H^{(1)}_q(\xi_L) - i\frac{n_3}{n_2}\left(\frac{1-\rho}{1+\rho}\right) H^{(1)}_{q-1}(\xi_L) \quad ,$$

$$b = i\frac{n_3}{n_2}\left(\frac{1-\rho}{1+\rho}\right) H^{(2)}_{q-1}(\xi_L) - H^{(2)}_q(\xi_L) \quad ,$$

$$c = H^{(1)}_{q-1}(\xi_0) - i\frac{n_1}{n_0} H^{(1)}_q(\xi_0) \quad ,$$

$$d = H^{(2)}_{q-1}(\xi_0) - i\frac{n_1}{n_0} H^{(2)}_q(\xi_0) \quad ,$$

where $n_m = \sqrt{\varepsilon_m}$, $m = 0,1,2,3$ are the complex refractive indices.

The most important feature of semiconductor wafers' structures that makes it possible to further enhance the speed of computations, is that the wall angle of a ridge is typically small, i.e. the ratio $a/b \sim 1$. It is quite unlike the antireflection coatings, where ridges has basically the triangular shape and consequently $a/b \sim 0$. If the wall angle is small then, according to the formulas for $\varepsilon_{||}$ and $\varepsilon_\perp$, approximately $\varepsilon_1 \approx \varepsilon_2$, and $|\xi_0| \gg 1$. As such, it is possible to use very efficient asymptotic expansion for Hankel functions which is valid when $x \gg q$:

$$H^{(1)}_q(x) = \sqrt{\frac{2}{\pi x}} e^{i\left(x - \frac{\pi}{4} - \frac{q\pi}{2}\right)}, \quad H^{(2)}_q(x) = \sqrt{\frac{2}{\pi x}} e^{-i\left(x - \frac{\pi}{4} - \frac{q\pi}{2}\right)}.$$

Semiconductor wafer can be characterized by the typical values $L = 300$ nm, $a = 50$ nm, $b = 70$ nm, and the refractive index of silicon $3.88 + i\,0.02$. With these values, the argument of Hankel functions is of the order of $\xi_0 \sim\ 10 + i\ 0.001$, while

$$q \equiv \frac{m+1}{m+2} \leq 1.$$

Thus, in this case, the asymptotic expansion provides good approximation. Then complexity of the result reduces dramatically:

$$R = -1 + 2\frac{e^{-2i\Delta\xi}\left[1 - \frac{n_2}{n_3}\left(\frac{1+\rho}{1-\rho}\right)\right] - \left[1 + \frac{n_2}{n_3}\left(\frac{1+\rho}{1-\rho}\right)\right]}{e^{-2i\Delta\xi}\left[1 - \frac{n_2}{n_3}\left(\frac{1+\rho}{1-\rho}\right)\right]\left(1 - \frac{n_1}{n_0}\right) - \left[1 + \frac{n_2}{n_3}\left(\frac{1+\rho}{1-\rho}\right)\right]\left(1 + \frac{n_1}{n_0}\right)},$$

with

$$\Delta\xi = \xi_L - \xi_0 = \frac{2\pi}{\lambda} L n_1 \frac{2}{m+2} \frac{\left(\frac{n_2}{n_1}\right)^{\frac{m+2}{m}} - 1}{\left(\frac{n_2}{n_1}\right)^{\frac{2}{m}} - 1}.$$

With smaller pattern height L, of less than 50 nm, the asymptotic version may be not enough correct, and then the formulas, containing the Hankel or Bessel functions, have to be used. Multilayer substrate influences the result through complex reflection coefficient ρ. The mathematical formalism of computing complex reflection coefficients of multilayer structures is described in Section 8.1.2.

The results in Figs. 8.24 and 8.25 present validation of the EMTA algorithm on two tasks with definite theoretical solutions. The first one (Fig. 8.24) requests calculation of the spectral reflectivity of a single

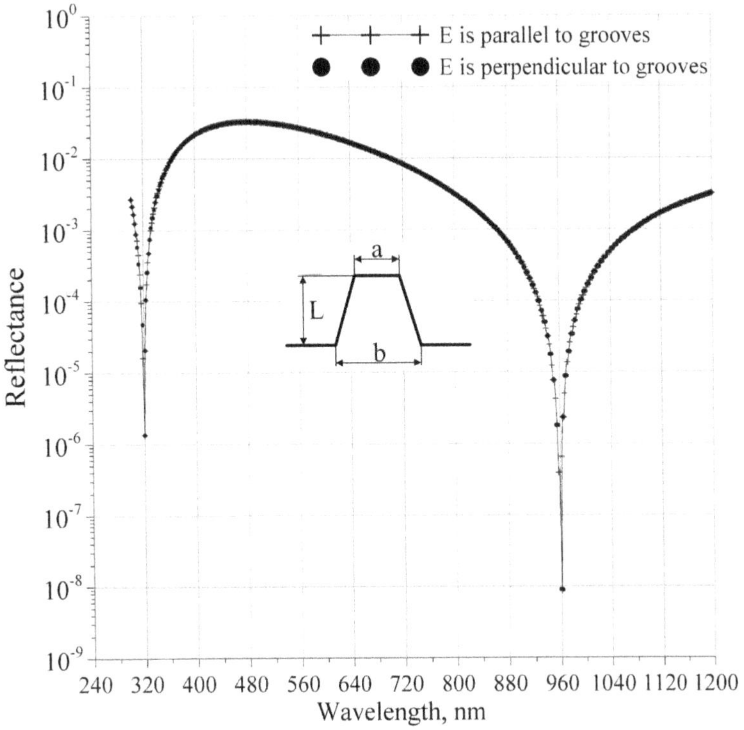

Fig. 8.24. Validation of the algorithm. No dispersion. Thickness $L = 200$ nm; period $p = 120$ nm; $b = 119.999$ nm; $a = 119.99$ nm; refractive indices $n_{\text{substrate}} = 1.44$; $n_{\text{grating}} = 1.2$. With these values, theory predicts reflectance minima at 320 nm and 960 nm, reflectance maximum at 480 nm, and reflectance values in minima must be zero. Both perpendicular and parallel configurations must have the same reflectances.

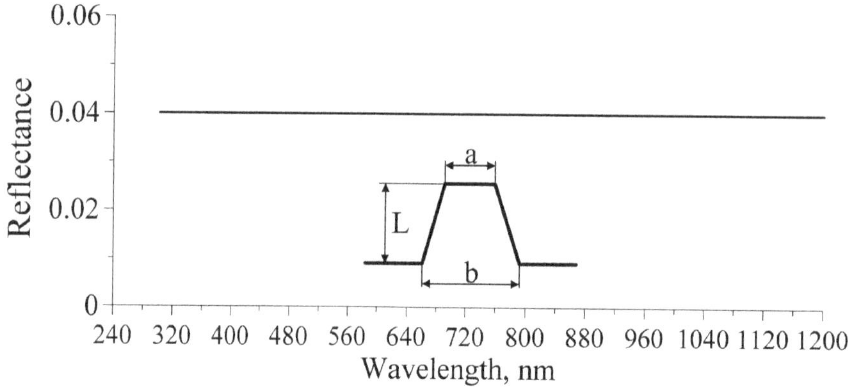

Fig. 8.25. Validation of the algorithm. No dispersion. Thickness $L = 0.5$ nm; period $p = 120$ nm; $b = 60$ nm; $a = 59.999$ nm; $n_{\text{substrate}} = 1.5$; $n_{\text{grating}} = 1.2$. With these values, theory predicts reflectivity to be equal to 0.04 regardless of the grating pattern and refractive index. Both perpendicular and parallel configurations must have same reflectances.

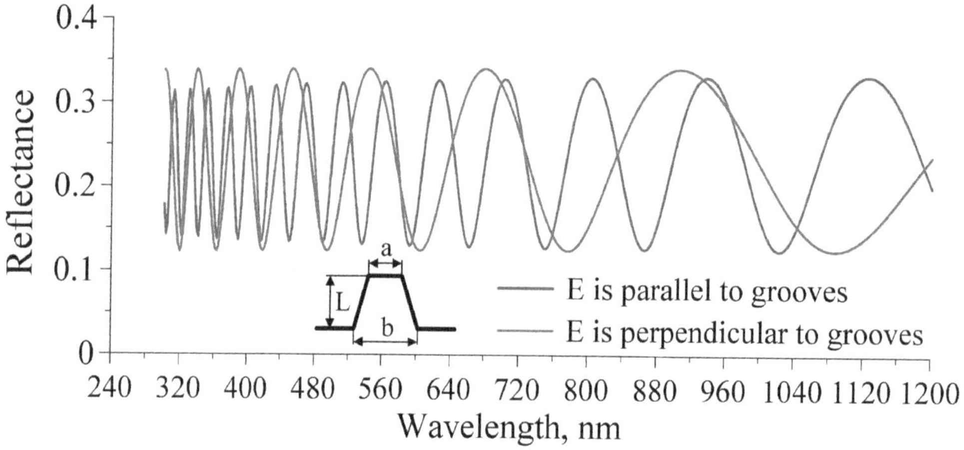

Fig. 8.26. General case. No dispersion. Thickness $L = 1.0$ μm; Period $p = 120$ nm; $b = 60$ nm; $a = 58$ nm; $n_{\text{substrate}} = 3.88 + i0.02$ (silicon); $n_{\text{grating}} = 1.2$.

solid (non-patterned) quarter-wave layer on a glass substrate. This problem was modeled by streaming the filling factor of the grating to a unity. The numerical solution delivered by the EMTA proved to be correct within values close to machine precision on complex variables. Lines, representing the parallel and perpendicular cases, coincide. The second task requires computation of the spectral reflectance of an arbitrary patterned grating when its thickness tends to zero. In this case, theory predicts the Fresnel reflectivity on the substrate alone regardless of the grating pattern and its refraction index. The EMTA delivers the result that complies with the theory, as it can be seen from Fig. 8.25. Lines, representing the parallel and perpendicular cases, coincide.

In general, spectral reflectances of the parallel and perpendicular polarizations may differ significantly, as exemplified in Fig. 8.26.

8.3. Chemical Mechanical Polishing

8.3.1. *Principle of optical control of silicon thickness*

Before packaging, silicon wafer undergoes process of thinning — chemical mechanical polishing (CMP). During this process, the bottom side of the wafer is being grinded and polished, reducing its thickness from the original 0.8 mm to a residual thickness of silicon of only hundreds of microns, or even tens of microns. Such thinned wafers look more like sheets of paper, rather than solid plates (Fig. 8.27).

During the entire CMP process, the thickness of silicon is being carefully monitored by means of infrared spectral reflectometers, as explained schematically in Fig. 8.28.

Silicon is transparent in infrared domain, starting from 1 μm and up to 10 μm. The probe beam created by infrared light source, coming upright through optical window in the platen, focuses on the rear side of the wafer and reflects back to the spectrometer. The two partial waves — the one reflected from the polished side of the wafer and the second one reflected from the device structure formed in the

Fig. 8.27. 300 mm silicon wafer after the CMP process.

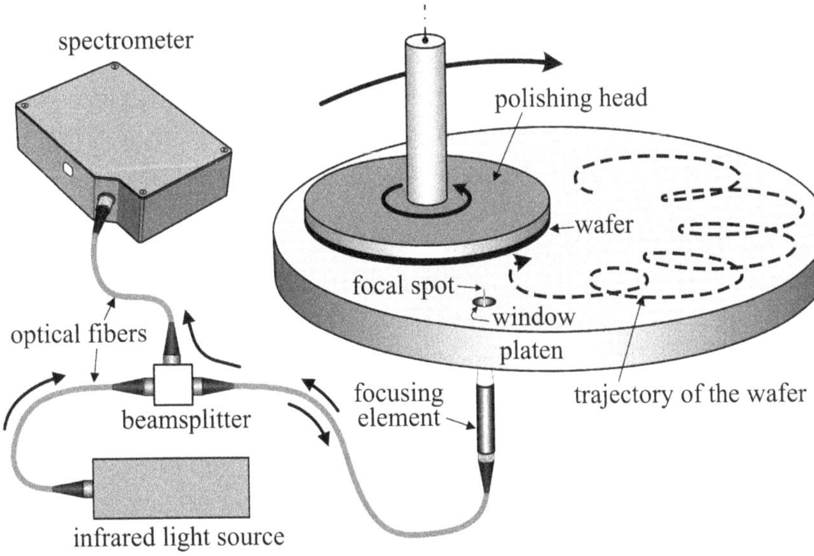

Fig. 8.28. The concept of monitoring thickness of residual silicon during CMP process.

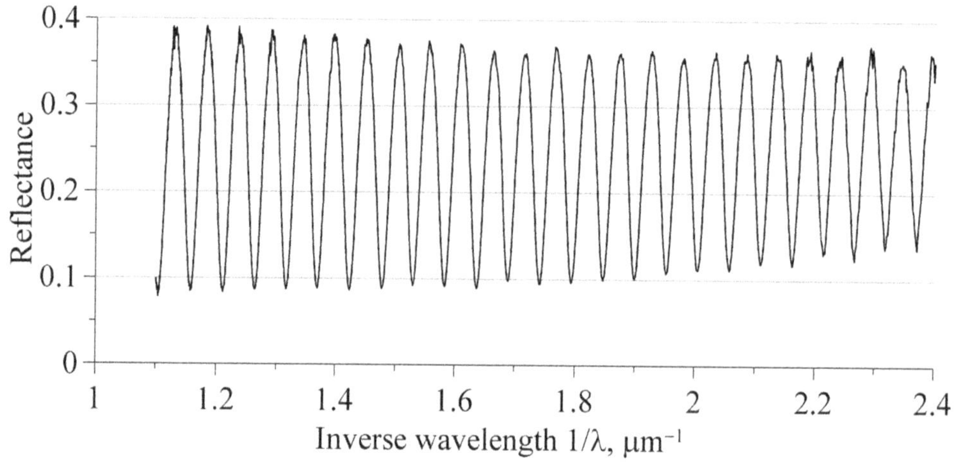

Fig. 8.29. Typical oscillating spectrum, delivered by spectral reflectometry in visible domain.

wafer — interfere, producing spectral oscillations with all the physics of this phenomenon explained in the beginning of this chapter.

Roughness of interfaces inside the wafer, large refractive index of silicon, and finite numerical aperture of the focused optical beam significantly deviate phenomenology of spectral reflectometry in CMP process from what is known for visible optics. Traditionally, theory of spectral reflectometry is developed on the models of smooth interfaces and plane waves, as was discussed in the preceding sections of this chapter. However, experimental results observable in CMP applications cannot be explained, using traditional optical theories. Therefore, the sections below present typical experimental results of spectral reflectometry on rough optical films, identify problems that must be explained, and derives theoretical corrections that explain the observable phenomenology.

8.3.2. *Basic phenomenological differences from visible optics*

We already know from the preceding sections of this chapter that reflection from a thin optical film produces oscillations in spectrum of the reflected light — the effect of spectral interferometry. As a typical example, Fig. 8.29 shows spectrum of light reflected from a several micron thick photoresist on a silicon wafer. Silicon is opaque in visible domain, therefore major portion of light reflects from the bulk silicon and interferes with a smaller portion of light, reflected from the photoresist, whose refractive index may be roughly taken 1.5. Therefore, the spectrum has a distinct form of sinusoid as two waves interfere. On such smooth surfaces, roughness is typically less than 100 Å, and as a common practice, it is taken into account as a narrow interface layer, producing no noticeable changes to theoretical shape of the spectrum, computed for ideally smooth surfaces.

Theoretically, reflection at normal incidence of a linearly polarized plane wave from a layer on a bulk substrate is described by the following formula (Section 8.1.2):

$$R = \left| \frac{r_{12} + r_{23}e^{i\psi}}{1 + r_{12}r_{23}e^{i\psi}} \right|^2,$$

where r_{12} and r_{23} are the Fresnel reflection coefficients on the interfaces air-layer and layer-substrate, and $\psi = 2knt$ with $k = 2\pi/\lambda$, n, and t being, respectively, the wavenumber, refractive index of the layer and its thickness. For non-polarized light, this formula must be averaged over orthogonal polarizations. For smooth layers, like the one that produced the spectrum in Fig. 8.29, theoretical formula above gives good qualitative and quantitative agreement with experiments, as shown in Fig. 8.30.

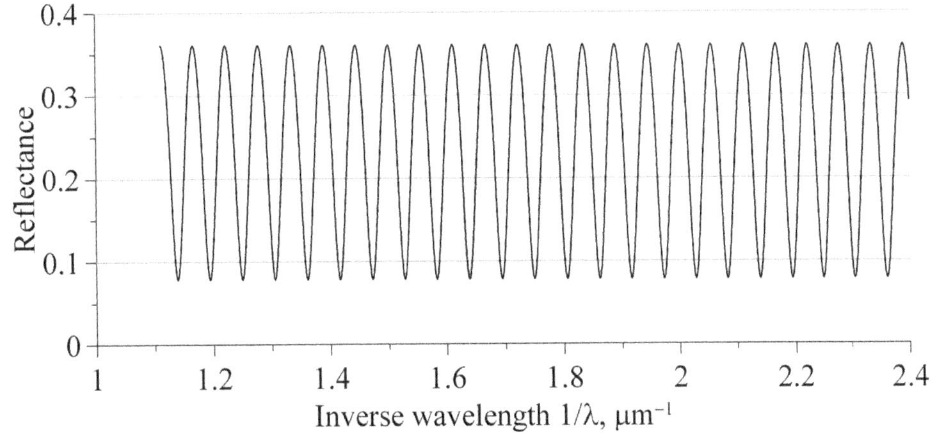

Fig. 8.30. Theoretical reflection spectrum, computed for non-polarized light with constant refractive index over the spectral interval (no dispersion).

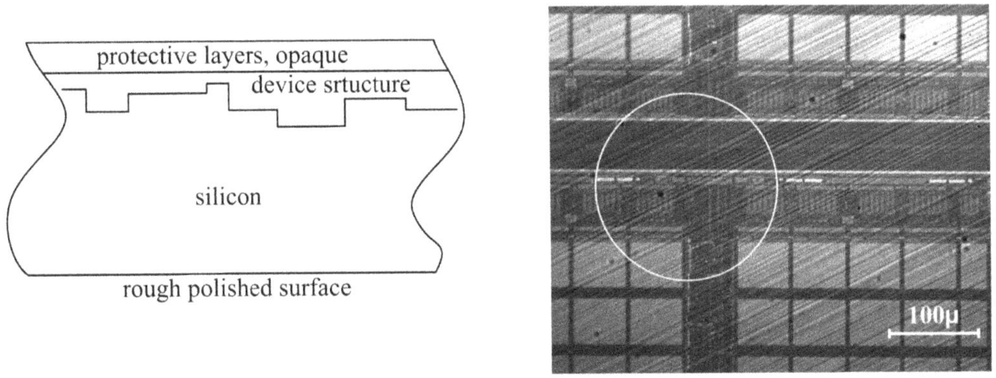

Fig. 8.31. Schematic cross section of the wafer (at left) and its image through polished silicon (at right). The white circle shows approximate area, covered by a focused light spot of 200 μm in diameter.

However, in CMP, there are two very important factors, making measurements more complicated than in visible domain: high refractive index of silicon (for standard wafers, 3.56 at the wavelength 1 μm) and high roughness of the reflecting layers. Figure 8.31 shows schematically the cross section of a thinned wafer and the real image of the patterned device structure, taken through the silicon layer at the wavelength around 1 μm.

Diagonal scratches on silicon surface, left after grinding and polishing, are clearly seen in the image. Typical optical arrangement, used in spectral reflectometry, is explained in detail in Section 8.1.3. In it, the focal spot of a probe beam is commonly created by projecting the image of, an optical fiber core onto the sample plane. Therefore, typical diameter of the focal spot is about 200 μm, as shown by the white circle in Fig. 8.31. Obviously, the reflected light contains reflections from numerous elements of the device structure with various phases of optical waves. Moreover, the refractive index of silicon at the wavelength 1 μm is much higher than refractive indices of typical layers, used in visible domain, making the roughness more important than in thin films in visible domain. Due to all the aforementioned effects, typical reflection spectra, obtained from such rough structures, differ significantly from the examples considered above, as shown in Fig. 8.32.

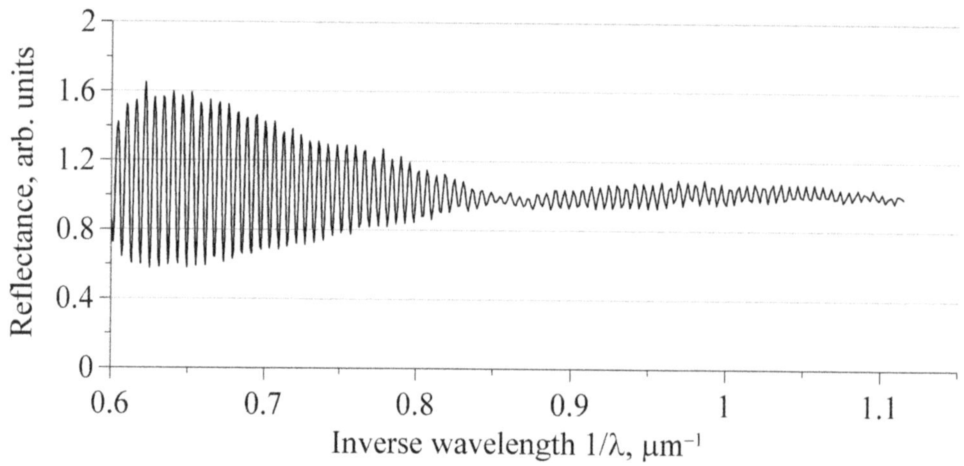

Fig. 8.32. Example of a spectrum obtained from a memory chip with silicon layer thinned to 25 μm.

Here are four main features that cannot be explained by the existing theory:

- sinusoidal shape of oscillations;
- bigger oscillating amplitude at longer wavelengths than at the shorter ones;
- waists in oscillations;
- periodicity of waists.

Below, we shall discuss all these phenomena.

8.3.3. *Sinusoidal shape of oscillations*

Consider visible domain with smooth films. As a rule, refractive index n of such films is around 1.5. At the interface air-film, at normal incidence, the Fresnel reflection coefficient is

$$|r_{12}| = \left|\frac{1-n}{1+n}\right| \sim 0.2,$$

i.e. several times less than unity. As such, the product $r_{12}r_{23}$ in the formula for reflectance R is significantly less than unity, and the formula itself can be approximated as

$$R \approx |r_{12} + r_{23}e^{i\psi}|^2 = A + B \cdot \cos(2knt)$$

describing a pure sinusoidal oscillation as the function of the wavenumber. This nearly sinusoidal oscillation is seen in Figs. 8.28 and 8.29.

With silicon in infrared domain, the situation is quite different. Refractive index of silicon at 1 μm is 3.56, and the refractive index of the device (Fig. 8.31) is even bigger. Therefore, $r_{12}r_{23} \sim 1$ and sinusoidal approximation for R is invalid. Theoretical reflection curve, computed according to exact formula for R, is shown in Fig. 8.33.

The theoretical spectrum is not sinusoidal in shape, as expected, whereas the experimental one is quite sinusoidal. This difference can be explained from the point of view of finite numerical aperture of the focused beam.

The exact formula for reflectance can be rewritten in the form

$$R = \left|\frac{r_{12} + r_{23}e^{i\psi}}{1 + r_{12}r_{23}e^{i\psi}}\right|^2 = \frac{r_{12}^2 + r_{23}^2 + 2r_{12}r_{23} \cdot \cos\psi}{1 + r_{12}^2 r_{23}^2 + 2r_{12}r_{23} \cdot \cos\psi} \equiv F(\cos\psi).$$

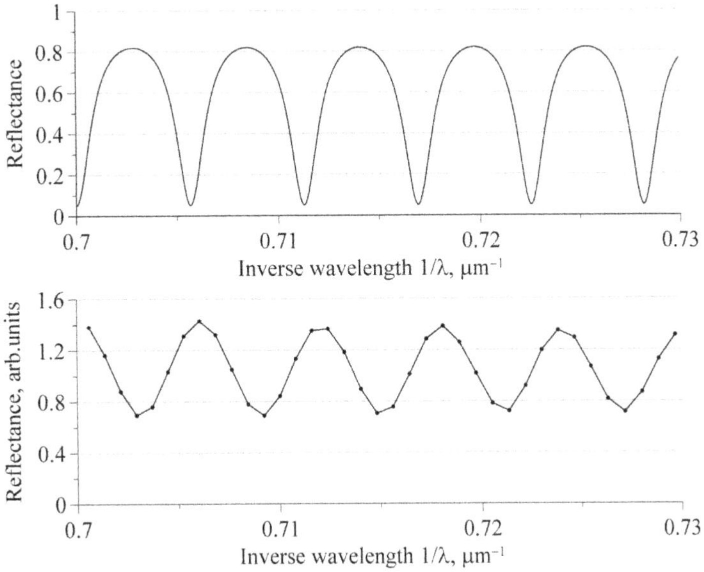

Fig. 8.33. Reflected spectrum from silicon layer 25 μm thick: calculated (above) and experimental (below), copied from Fig. 8.32.

Function $F(x)$ is the monotonous one in the domain of the argument $-1 < x < +1$, and can be expanded in the McLaurin series:

$$F(x) = \sum_{m=0}^{\infty} \frac{1}{m!} F^{(m)}(0) \cdot x^m,$$

where $F^{(m)}$ is the mth derivative of $F(x)$. Substituting $x = \cos\psi$, and using trigonometric identity $\cos\alpha \cdot \cos\beta = \frac{1}{2}\cos(\alpha+\beta) + \frac{1}{2}\cos(\alpha-\beta)$, it is easy to derive the powers of the cosine:

$$\cos^2\psi = \frac{1}{2} + \frac{1}{2}\cos 2\psi,$$

$$\cos^3\psi \equiv \cos^2\psi \cdot \cos\psi = \frac{1}{2}\cos\psi + \frac{1}{2}\cos 2\psi \cdot \cos\psi = \frac{3}{4}\cos\psi + \frac{1}{4}\cos 3\psi,$$

$$\cos^4\psi = \frac{3}{4} + \frac{1}{2}\cos 2\psi + \frac{1}{8}\cos 4\psi,$$

and so on. Substituting the powers of cosine into the McLaurin expansion and combining terms with $\cos m\psi$, the reflectance can be rewritten as

$$R = \sum_{m=0}^{\infty} A_m \cdot \cos m\psi.$$

This formula is an exact identity with all the coefficients A_m explicitly defined by proper algebraic manipulations. Now we need to average it over all angles of incidence that the focused conical beam contains.

For waves with the angle of incidence θ inside the film, the phase term is equal to

$$\psi = 2knt \cdot \cos\theta,$$

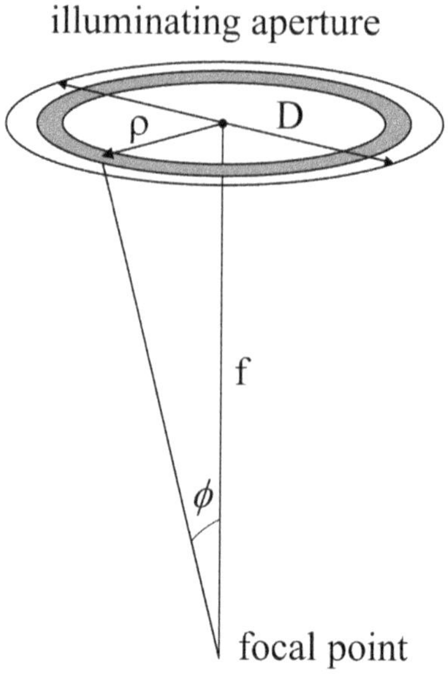

Fig. 8.34. Geometry of the focused beam.

and for the angle of incidence ϕ in air (the Snell law)

$$\psi = 2kt \cdot \sqrt{n^2 - \sin^2 \phi}.$$

Consider Fig. 8.34. The relative number of rays, coming at the angle of incidence ϕ, is the ratio of the area covered by thin ring $2\pi\rho d\rho$ to the area of the illuminating aperture $\pi D^2/4$:

$$\frac{8\pi\rho d\rho}{\pi D^2} \approx \frac{2\phi d\phi}{NA^2},$$

where $NA = D/(2f)$ is approximately the numerical aperture. Typical values of numerical apertures, used in practice, are less than 0.5. For such small values, the maximum value of ϕ is

$$\phi_{\max} \approx NA.$$

Coefficients A_m and the phase ψ depend on ϕ. But A_m are slowly varying functions, and their values are approximately the same for all coming rays. Therefore, averaging R over ϕ takes the form

$$\overline{R} = \sum_{m=0}^{\infty} A_m \cdot \frac{1}{NA^2} \int_0^{NA} \cos(m2kt\sqrt{n^2 - \sin^2 \phi})2\phi d\phi,$$

where the dash over R denotes its average value. In the approximation of small NA, when for all the angles $\sin \phi \approx \phi$, this integral can be computed analytically. Indeed,

$$\sqrt{n^2 - \sin^2 \phi} \approx \sqrt{n^2 - \phi^2} = n\sqrt{1 - \frac{\phi^2}{n^2}} \approx n\left(1 - \frac{\phi^2}{2n^2}\right).$$

Denoting

$$u \equiv \frac{\phi^2}{2n^2}$$

and realizing that

$$2\phi d\phi = 2n^2 du,$$

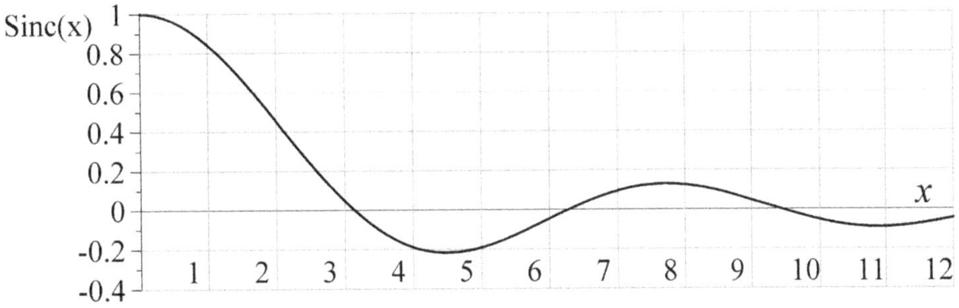

Fig. 8.35. Function Sinc(x). Zeroes are distributed as πj, $j = 1, 2, \ldots$.

the integral transforms to

$$2n^2 \int_0^{u_{\max}} \cos[m2ktn(1-u)]du, \quad u_{\max} = \frac{NA^2}{2n^2}.$$

With another new variable

$$v = 1 - u,$$

the integral becomes of a standard type

$$-2n^2 \int_0^{v_{\max}} \cos(m2ktnv)\, dv, \quad v_{\max} = 1 - \frac{NA^2}{2n^2}.$$

Integrating, and using trigonometric identity

$$\sin\alpha - \sin\beta = 2\sin\left(\frac{\alpha-\beta}{2}\right) \cdot \cos\left(\frac{a+\beta}{2}\right),$$

it is easy to come to a final result:

$$\overline{R} = \sum_{m=0}^{\infty} A_m \cdot \mathrm{Sinc}\left(\frac{NA^2 mkt}{2n}\right) \cdot \cos\left[m2knt\left(1 - \frac{NA^2}{4n^2}\right)\right],$$

where the well-known function

$$\mathrm{Sinc}(x) = \frac{\sin x}{x}$$

is presented, for the convenience of readers, in numerically computed form in Fig. 8.35.

Consider formula for $\overline{R}$ in detail. To begin with, the quadratic term in the argument of the cosine, correcting spectral frequency of oscillations, is small and may be neglected in all practical cases. For instance, for $NA = 0.5$ and $n = 3.56$

$$\left(\frac{NA}{2}n\right)^2 = 4.9 \times 10^{-3}.$$

Next, we shall analyze the role of numerical aperture NA. With $NA = 0$, the averaged spectral reflectance is the same as the non-averaged one, i.e. the non-sinusoidal one, due to higher order spectral frequencies with $m > 1$. With $NA > 0$, amplitudes of higher order spectral frequencies, as well as the amplitude of the fundamental spectral frequency at $m = 1$, decrease due to function Sinc(x). However, amplitudes of the higher orders decrease quicker due to the factor m in the argument of Sinc(x). This is the first reason why real measurements reveal almost sinusoidal spectral modulation, like in Fig. 8.33.

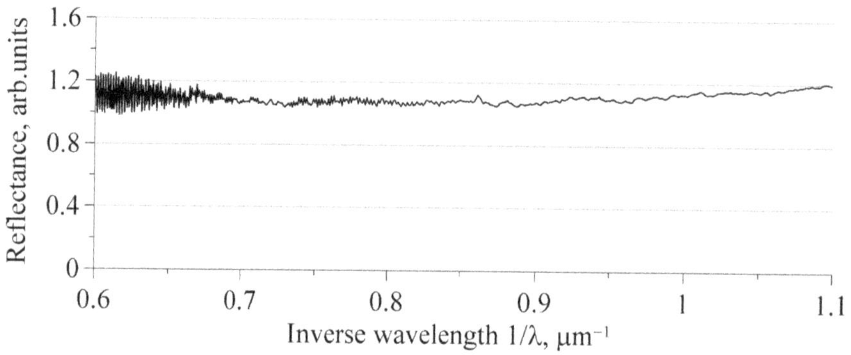

Fig. 8.36. Example of a spectrum obtained from a memory chip with silicon layer thinned to 84 μm. Both the amplitude of the fundamental modulation is smaller than in Fig. 8.32, and position of the first waist is shifted to a smaller value around 0.72 μm^{-1}.

Particularly, it is instructive to analyze, at what thickness t of silicon layer the amplitude of the second harmonic $m = 2$ turns to zero for the first time:

$$NA^2 2t = n\lambda.$$

For the wavelength 1 μm, refractive index 3.56, and practical numerical aperture 0.5, this formula gives $t = 7\,\mu$m. From the formula for $\overline{R}$, it follows that all higher-order harmonics are also zeros at this thickness.

Finite numerical aperture is also one of the reasons, why spectral modulation gets smaller with higher wavenumbers k and exhibits almost zero amplitudes — the waists — at some values of k, like in Fig. 8.32. It happens because the argument of the Sinc function is proportional to $k = 2\pi/\lambda$, and the average amplitude of spectral reflectivity $\overline{R}$ follows the shape of the Sinc function. This is also the reason, why the waists are distributed almost equidistantly.

Formula for $\overline{R}$ also predicts, that the amplitude of the fundamental harmonic with $m = 1$ depends on the thickness of silicon layer t as the Sinc function: the thicker the silicon layer, the smaller the amplitude, and it falls quicker as a function of the wavenumber k. Figure 8.36 confirms this prediction.

If the finite numerical aperture were the only factor, causing decrease of spectral oscillations, then formula for $\overline{R}$ could produce good estimate of the thickness t of the remaining silicon: measure the wavelength λ_1 of the first waist of the fundamental oscillation with $m = 1$ and compute

$$t = \frac{n\lambda_1}{NA^2}.$$

For instance, in Fig. 8.32, $\lambda_1 \approx 1.17$ μm. Then with $NA = 0.2$ and $n = 3.56$, this formula gives $t = 104$ μm. However, the value computed from the period of spectral oscillations gives $t = 25$ μm – roughly four times smaller. In Fig. 8.36, $\lambda_1 \approx 1.4$ μm, which gives $t = 124$ μm, whereas the value computed from the period of spectral oscillations gives $t = 84$ μm. The reason for this discrepancy is the second factor — roughness of layers, which is discussed in the next section.

8.3.4. *The role of roughness*

In the previous section, it was explained why, in practice, the shape of spectral oscillations is sinusoidal. Mathematically, it means that higher order spectral harmonics are small, and it is possible to present average reflectivity as

$$\overline{R} = a \cdot \mathrm{Sinc}\left(\frac{NA^2 kt}{2n}\right) \cdot \cos(2knt).$$

The Sinc term describes slowly varying amplitude of oscillations. The term "slowly" must be verified. For that, consider, how many oscillations of the cosine term take place within the first lobe of the function Sinc. Using

$$\psi = 2knt \cdot \cos\theta,$$

the above formula may be rewritten for normal incidence as

$$\overline{R} = a \cdot \mathrm{Sinc}\,(\alpha\psi) \cdot \cos(\psi), \quad \alpha \equiv \left(\frac{NA}{2n}\right)^2.$$

One period of oscillations corresponds to $\psi_p = 2\pi$, and the first lobe of Sinc is defined by $\alpha\psi_1 = \pi$. Thus, the number of oscillations

$$\eta = \frac{\psi_1}{\psi_p} = \frac{1}{2\alpha}.$$

With $NA = 0.5$ and $n = 3.56$ this formula gives $\eta = 100$. It proves that in $\overline{R}$, the Sinc term is indeed a slowly varying function.

From Fig. 8.31 it follows that, within the focused spot of light, many elements of the device structure contribute to reflection, so that the thickness of silicon t is not constant:

$$t = \bar{t} + \delta,$$

where $\bar{t}$ is the average thickness within the focused spot and δ is a random variation. Since nothing is known about δ, it is a common practice to assume it being a random variable uniformly distributed within $\pm\Delta/2$. Therefore, it is necessary to average $\overline{R}$ over δ. Whereas we have already proved that the number of spectral oscillations within the lobe of the function Sinc is big ($\eta >> 1$), the averaging may be applied only to the cosine:

$$\overline{\overline{R}} = a \cdot \mathrm{Sinc}\left(\frac{NA^2k\bar{t}}{2n}\right) \cdot \frac{1}{\Delta}\int_{-\Delta/2}^{+\Delta/2} \cos\left[2kn(\bar{t}+\delta)\right]\,d\delta = a \cdot \mathrm{Sinc}\left(\frac{NA^2k\bar{t}}{2n}\right) \cdot \mathrm{Sinc}(kn\Delta) \cdot \cos(2kn\bar{t}).$$

Here the double dash over R means double averaging over both roughness and the cone of rays. The formula for $\overline{\overline{R}}$ shows that the amplitude of spectral oscillations falls more rapidly than predicted by the formula for $\overline{R}$, and it explains why our earlier theoretical estimates of silicon thickness t proved to be inconsistent with experiments. Since the previous estimates turned out to be far from reality, it is now logical to assume that the first waist λ_1 in Fig. 8.32 is caused by roughness, and using the formula for $\overline{\overline{R}}$, it is possible to estimate the scale of roughness Δ:

$$\Delta = \frac{\lambda_1}{2n}.$$

For $\lambda_1 \approx 1.17\ \mu$m in Fig. 8.32 and $n = 3.56$, this formula gives $\Delta \approx 0.17\ \mu$m. In Fig. 8.36, $\lambda_1 \approx 1.4\ \mu$m, which gives $\Delta \approx 0.2\ \mu$m. Parameter Δ was introduced as a combined roughness of the top and bottom interfaces, i.e. interfaces air-silicon and silicon-device. Therefore, the individual roughness of any one of the two interfaces should be less than these estimates.

Supplemental Reading

O.E. Heavens, *Optical Properties of Thin Solid Films*, Dover Publications Inc., New York, 1991, 261p.

F. Abelès. La détermination de l'indice et de l'épaisseur des couches minces transparentes, *J. Phys. Radium*, **11**(7), pp. 310–314 (1950). (https://hal.archives-ouvertes.fr/jpa-00234262).

E.D. Palik, *Handbook of Optical Constants of Solids*, vol. 1, Academic Press, 1998.

R. Petit (ed.), *Electromagnetic Theory of Gratings*, Springer-Verlag, Heidelberg, 1980.

M. Neviere, E. Popov, *Light Propagation in Periodic Media*, Marcel Dekker Ltd., New York, 2002.

J. Chandezon, D. Maystre, G. Raoult, A new theoretical method for diffraction gratings and its numerical application, *J. Optics, Paris*, **11**, pp. 235–241 (1980).

I.C. Botten, M.S. Craig, R.C. McPhedran, J.L. Adams, J.R. Andrewartha, The dielectric lamellar diffraction grating, *Optica Acta*, **28**(3), pp. 413–428 (1981).

M.G. Moharam, T.K. Gaylord, Rigorous coupled-wave analysis of planar grating diffraction, *J. Opt. Soc. Am.*, **71**(7), pp. 811–818 (1981).

S.M. Rytov, Electromagnetic properties of a finely stratified medium, *Sov. Phys. JETP*, **2**(3), pp. 456–475 (1956).

D.H. Raguin, G.M. Morris, Analysis of antireflection-structured surfaces with continuous one-dimensional surface profiles, *Appl. Opt.*, **32**(14), pp. 2582–2598 (1993).

J. Wallot, Der senkrechte Durchgang elektromagnetischer Wellen durch eine Schicht raumlich veranderlicher Dielektrisitatskonstante, (Vertical passing of electromagnetic waves through a layer with space-variable dielectric constant) *Ann. Physik*, **60**(4), (v. 365 — total publication volume number), pp. 734–762 (1919).

Chapter 9

Related Non-spectroscopic Techniques

9.1. Angular Reflectometry

9.1.1. *The concept of angular reflectometry*

Angular reflectometry is used for measuring thickness and optical constants of thin transparent optical layers — the same task that can be solved by spectral reflectometry (Chapter 8). The basic positive feature of angular reflectometry, comparing to spectral reflectometry, is much finer focal spot of the probe beam that can be localized within small test area on a semiconductor wafer. These test areas are specially organized microscopic pitches on the wafer that are not part of any functional devices but serve only for monitoring deposition or etch processes. Such fine optical spots can be formed only by high numerical aperture objective lenses that cannot be used in spectral reflectometry for the reasons explained in Chapter 8. Another positive feature of angular reflectometry is the use of monochromatic light — the property that eliminates errors associated with spectral dispersion of the refractive index. However, there are also strong disadvantages that make angular reflectometry much more cumbersome, costly, and tricky technology: necessity of very accurate autofocusing and poor performance on thick layers of micrometer scale. All the aforementioned requires detailed explanation.

Consider a plane monochromatic wave with the wavelength λ, coming at the angle of incidence ϕ from vacuum to a film of thickness t and refractive index n_1, deposited on the infinitely thick substrate of refractive index n_2 (Fig. 9.1).

Reflection coefficients are governed by the following formulas derived in Chapter 8:

$$r_{1s} = \frac{\sqrt{1-\sin^2\phi} - \sqrt{n_1^2 - \sin^2\phi}}{\sqrt{1-\sin^2\phi} + \sqrt{n_1^2 - \sin^2\phi}}, \quad r_{2s} = \frac{\sqrt{n_1^2-\sin^2\phi} - \sqrt{n_2^2 - \sin^2\phi}}{\sqrt{n_1^2-\sin^2\phi} + \sqrt{n_2^2 - \sin^2\phi}},$$

$$r_{1p} = \frac{\sqrt{1-\sin^2\phi} - n_1^{-2}\sqrt{n_1^2 - \sin^2\phi}}{\sqrt{1-\sin^2\phi} + n_1^{-2}\sqrt{n_1^2 - \sin^2\phi}}, \quad r_{2p} = \frac{n_1^{-2}\sqrt{n_1^2-\sin^2\phi} - n_2^{-2}\sqrt{n_2^2 - \sin^2\phi}}{n_1^{-2}\sqrt{n_1^2-\sin^2\phi} + n_2^{-2}\sqrt{n_2^2 - \sin^2\phi}},$$

$$R_s \equiv |r_s|^2 = \frac{r_{1s}^2 + 2r_{1s}r_{2s}\cos(2\delta) + r_{2s}^2}{1 + 2r_{1s}r_{2s}\cos(2\delta) + r_{1s}^2 r_{2s}^2}, \quad R_p \equiv |r_p|^2 = \frac{r_{1p}^2 + 2r_{1p}r_{2p}\cos(2\delta) + r_{2p}^2}{1 + 2r_{1p}r_{2p}\cos(2\delta) + r_{1p}^2 r_{2p}^2},$$

and

$$\delta = kt\sqrt{n_1^2 - \sin^2\phi}, \quad k = 2\pi/\lambda.$$

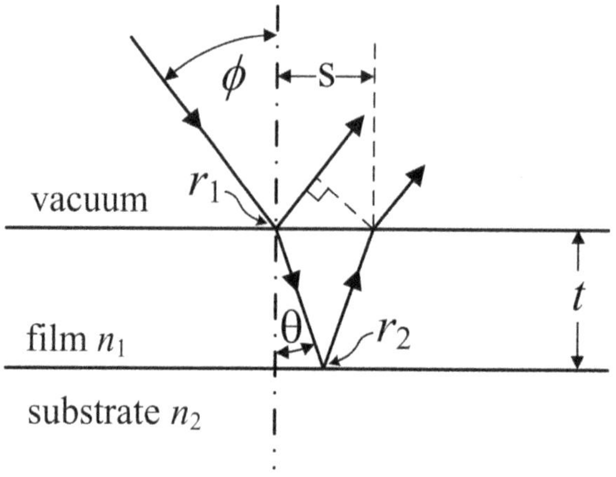

Fig. 9.1. Notations of the task.

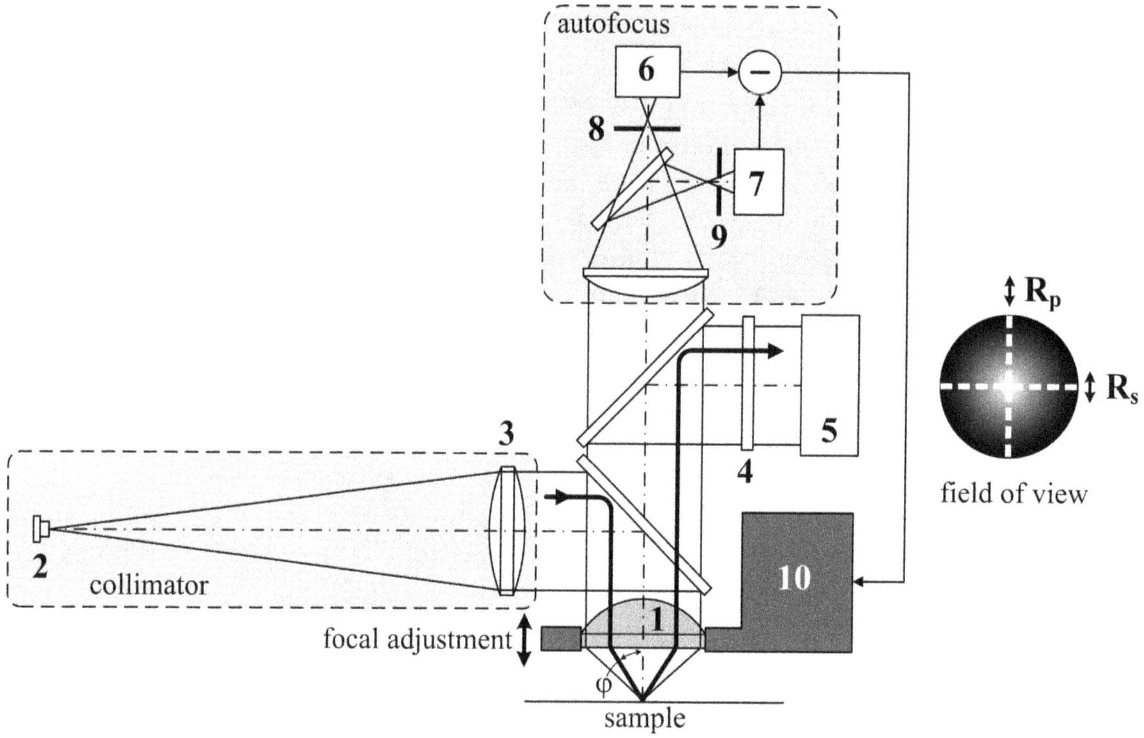

Fig. 9.2. Generalized optical scheme of the angular reflectometer. 1 — high numerical aperture objective lens; 2 — light-emitting diode; 3 — collimating lens; 4 — polarizer; 5 — imaging camera; 6,7 — photodetectors; 8,9 — pinholes; 10 — servo-motor.

Indices "s" and "p" describe polarizations: the perpendicular and parallel to the plane of incidence, respectively. Thus, changing the angle of incidence ϕ, it is possible to measure functions $R_s(\phi)$ and $R_p(\phi)$, and estimate t, n_1, and n_2 by fitting theoretical formulas into experimental data. This is the general idea of angular reflectometry, and its typical practical implementation is outlined in Fig. 9.2.

Instead of changing angle of incidence ϕ of a single ray at each step of measurement, the high numerical aperture objective lens is used, directing simultaneously a plurality of rays at different angles of incidence ϕ onto the sample, and the entire picture is recorded at once in the plane of the imaging camera. Nowadays, the majority of objective lenses on the market are designed infinity-conjugated, i.e. to produce least possible aberrations at the focus when the incoming rays are parallel (coming from infinity). Therefore, the rays reflected from the sample in focal plane and passing backwards through the lens are also parallel, so that in any cross section of the reflected beam the radial coordinate r and the angle of incidence ϕ are connected:

$$r = f \tan \phi,$$

where f is the focal length of the objective. Thus, any pixel in the imaging camera corresponds to a certain angle of incidence ϕ, and electrical signal from this pixel is proportional to reflectance R. In order to measure separately reflectances of s- and p-polarized waves, the polarizer should be installed in front of the camera, or the light from the source must be definitely polarized. In particular geometry of Fig. 9.2, when polarization axis of the polarizer is in the plane of the figure, the pixels in the central vertical axis of the camera produce signals proportional to $R_p(\phi)$, while the orthogonal axis of the camera gives $R_s(\phi)$. Explaining this case, the field of view of the camera and polarization directions are shown in the right part of Fig. 9.2.

The bigger the numerical aperture is, the wider the interval of angles of incidence ϕ_{max} is, according to formula:

$$NA = \sin \phi_{\text{max}}.$$

For instance, NA=0.8 gives $\phi_{\text{max}} = 53.1°$; NA= 0.9 corresponds to $\phi_{\text{max}} = 64.1°$. The bigger ϕ_{max}, the better the fitting result and precision of measurements are. However, objective lenses with high numerical apertures require extremely precise focusing onto the sample in order to maintain parallel beam. The depth of focus for the objective lens with NA= 0.9 is less than a micrometer, which requires nano-scale positioning and autofocusing system. As an example, Fig. 9.2 presents one possible autofocusing scheme with two photodetectors and two pinholes in front of them. The pinholes are positioned strictly coaxial with the optical beam and at equal distances from the planes of the best focusing in two shoulders of the system: the first pinhole closer to the first photodetector, another pinhole farther from the second one. When the two signals from the photodetectors are equal, the objective is accurately focused. When not, the sign of the difference between these signals gives the direction of vertical adjustment, and the absolute value of this difference — the initial speed of motion. Numerous other autofocusing schemes and algorithms may be used, which, however, go beyond the scope of this book.

Angular reflectometry makes it easy to measure refractive index of bulk materials by determining the Brewster angle ϕ_{B}, i.e. the angle of incidence at which $R_p(\phi_{\text{B}}) = 0$. When no film is on the substrate with refractive index n,

$$R_p(\phi) = \left| \frac{\sqrt{1 - \sin^2 \phi} - n^{-2}\sqrt{n^2 - \sin^2 \phi}}{\sqrt{1 - \sin^2 \phi} + n^{-2}\sqrt{n^2 - \sin^2 \phi}} \right|^2,$$

which leads to the formulas

$$\phi_{\text{B}} = \arcsin \frac{n}{\sqrt{n^2 + 1}}, \qquad n = \tan \phi_{\text{B}}.$$

Typical theoretical curves of $R_p(\phi)$ and $R_s(\phi)$, computed for silicon ($n = 3.88$) and quartz ($n = 1.45$), are shown in Fig. 9.3. The Brewster angles are respectively 76° and 56°.

Using angular reflectometry, it is also possible to elegantly measure refractive index of a film deposited on the substrate. Indeed, when the angle of incidence ϕ equals the Brewster angle of the film material, then p-polarized rays do not reflect from the upper surface of the film. The condition for that is

$$r_{1p} = 0, \quad \text{or} \quad \sqrt{1 - \sin^2 \phi} = n_1^{-2}\sqrt{n_1^2 - \sin^2 \phi}.$$

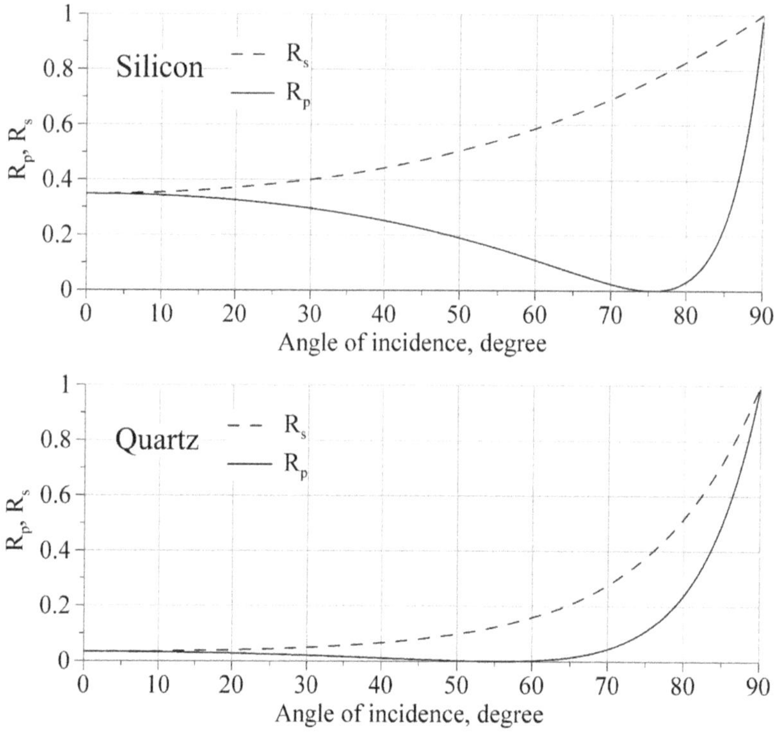

Fig. 9.3. Angular reflection coefficients for bare silicon and quartz.

Then

$$R_p = |r_{2p}|^2 = \frac{\sqrt{1-\sin^2\phi} - n_2^{-2}\sqrt{n_2^2 - \sin^2\phi}}{\sqrt{1-\sin^2\phi} + n_2^{-2}\sqrt{n_2^2 - \sin^2\phi}}$$

exactly the value of reflection coefficient from bulk substrate at this angle of incidence. Thus, the method formalizes as follows: (1) measure the reflectance curve $R_p(\phi)$ of the bare substrate, (2) measure the reflectance curve $R'_p(\phi)$ with film on the substrate, (3) superimpose $R_p(\phi)$ with $R'_p(\phi)$ and find the angle ϕ_0, at which $R_p(\phi_0) = R'_p(\phi_0)$. Then refractive index of the film

$$n_1 = \tan\phi_0.$$

The idea of this method becomes clearer from theoretical curves presented in Fig. 9.4. Judging visually on the curves, the intersect point is approximately at $\phi_0 = 55°$, thus giving $n_1 = 1.43$ — a value very close to the one used for computations.

9.1.2. *Theoretical restrictions*

The critical parameter of angular reflectometry is spectral purity of the probe beam. When the wavelength changes by $\Delta\lambda$, the entire angular distribution also changes, and angular maxima at one wavelength may superimpose with angular minima at another wavelength (Fig. 9.5). This effect not only makes measurement unreliable, but also leads to lower contrast of the picture, making at worst case interference pattern completely undetectable.

Therefore, it is instructive to analyze requirements to spectral width of the probe beam. The solution can be found in two steps: first, deriving the formula, connecting $\Delta\lambda$ with the angular shift $\Delta\phi$ of the entire pattern $R(\phi)$, and then deriving the formula for angular separation between maxima (or minima)

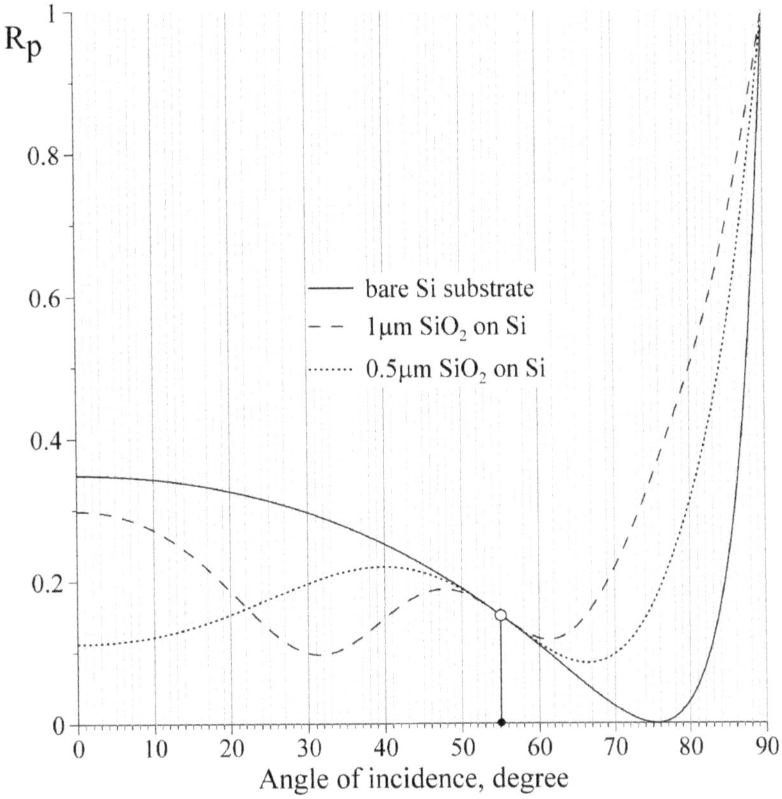

Fig. 9.4. Computed reflection coefficients R_p for bare silicon substrate ($n_2 = 3.88$) and two SiO_2 ($n_1 = 1.45$) films of different thickness on silicon substrate. Wavelength 0.6 μm.

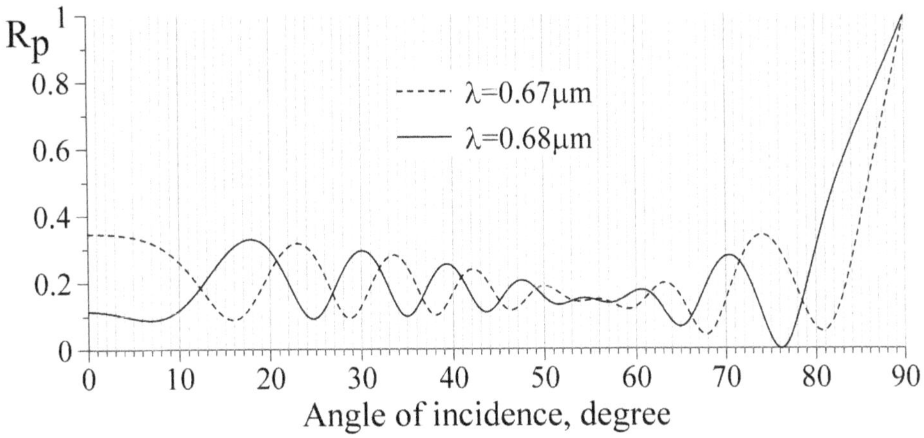

Fig. 9.5. Small variations of the wavelength of the probe beam lead to significant changes in the interference pattern ($n_2 = 3.88$, $n_1 = 1.45$, $t = 6$ μm).

in $R(\phi)$. As a result, the requirement to spectral width will be obtained as the request for $\Delta\phi$ to be much smaller than the angular separation of two consecutive maxima (or minima) in angular oscillations.

Angular oscillations of the reflection curves are caused by interference between the waves reflected from the film and from the substrate (Fig. 9.1). Applying not entirely rigorous physical interpretation, it is possible to say that when the wavelength changes from λ to $\lambda + \Delta\lambda$, the angular pattern $R(\phi)$

shifts to a new location $R(\phi + \Delta\phi)$, at which the phase difference between interfering waves remains the same. This phase difference is determined by the optical path difference δ. From pure geometrical considerations of Fig. 9.1, it follows that

$$\delta = \frac{2tn}{\cos\theta} - s \cdot \sin\phi = 2tn \cdot \cos\theta = 2t\sqrt{n^2 - \sin^2\phi},$$

and the phase difference between the interfering waves is equal to $2\pi\delta/\lambda$. Here n is the refractive index of the film (subscript "1" is dropped for the sake of conciseness). Then $\Delta\lambda$ and $\Delta\phi$ are connected through the following formulas:

$$\delta_1 = 2t\sqrt{n^2 - \sin^2\phi}; \quad \delta_2 = 2t\sqrt{n^2 - \sin^2(\phi + \Delta\phi)}; \quad \frac{\delta_1}{\lambda} = \frac{\delta_2}{\lambda + \Delta\lambda}.$$

Taking into consideration smallness of $\Delta\lambda/\lambda$ and $\Delta\phi$, the subsequent algebraic manipulations are straightforward:

$$n^2 - \sin^2(\phi + \Delta\phi) = \left(\frac{\lambda}{\lambda + \Delta\lambda}\right)^2 (n^2 - \sin^2\phi);$$

$$\sin(\phi + \Delta\phi) = \sin\phi \cdot \cos\Delta\phi + \cos\phi \cdot \sin\Delta\phi \approx \sin\phi + \cos\phi \cdot \Delta\phi;$$

$$\left(\frac{\lambda}{\lambda + \Delta\lambda}\right)^2 = \left(\frac{1}{1 + \Delta\lambda/\lambda}\right)^2 = \left(1 + \frac{\Delta\lambda}{\lambda}\right)^{-2} \approx 1 - 2\frac{\Delta\lambda}{\lambda};$$

$$\Delta\lambda \approx \lambda \cdot \frac{\sin\phi \cdot \cos\phi}{n^2 - \sin^2\phi} \cdot \Delta\phi.$$

The second step of our computations is to derive angular separation between oscillations in $R(\phi)$, i.e. the angular interval between maxima (or minima) of oscillations. Angular positions of consecutive maxima ϕ_0,ϕ_1,ϕ_2, ... ,ϕ_p is determined by a set of equations:

$$\frac{2\pi}{\lambda} \cdot 2t\sqrt{n^2 - \sin^2\phi_0} = m \cdot 2\pi;$$

$$\frac{2\pi}{\lambda} \cdot 2t\sqrt{n^2 - \sin^2\phi_1} = (m - 1) \cdot 2\pi;$$

$$\vdots$$

$$\frac{2\pi}{\lambda} \cdot 2t\sqrt{n^2 - \sin^2\phi_p} = (m - p) \cdot 2\pi; \quad p = 0, 1, 2, 3, \ldots.$$

Here m is an unknown natural number — the ratio of the optical path difference to the wavelength. The same sequence of equations, with adding π in the right-hand side, holds true for the minima. Subtracting the first equation from the others, eliminate m:

$$\frac{2t}{\lambda}\sqrt{n^2 - \sin^2\phi_p} = \frac{2t}{\lambda}\sqrt{n^2 - \sin^2\phi_0} - p.$$

Assume $\phi_0 = 0$. Then

$$\sin^2\phi_p = n^2 - \left(n - \frac{\lambda p}{2t}\right)^2 = \left(n - n + \frac{\lambda p}{2t}\right)\left(2n - \frac{\lambda p}{2t}\right).$$

Definite oscillations in $R(\phi)$ occur only for relatively thick films $t \gg \lambda$. In this case, $\frac{\lambda p}{2t} \ll 2n$, and approximately

$$\sin^2\phi_p \approx \frac{\lambda np}{t}, \quad \phi_p \approx \arcsin\sqrt{\frac{\lambda np}{t}}, \quad p = 0, 1, 2, 3, \ldots.$$

From this formula, we see that angular maxima (and minima) in $R(\phi)$ are populated denser as the angle increases, which is consistent with computations presented in Fig. 9.5. Thus, the most restrictive area of angles is close to the maximum available angle of incidence $\phi_{\max}$, for which $\sin\phi_{\max} = NA$ — numerical aperture of the objective lens. Thus,

$$\sin^2\phi_{\max} \equiv NA^2 \approx \frac{\lambda n}{t} p_{\max}, \quad \text{and} \quad p_{\max} = \frac{NA^2 t}{\lambda n}.$$

We must address our request to the minimal angular separation $\Delta\phi$, which occurs near $\phi_{\max}$:

$$\Delta\phi = \phi_{p_{\max}} - \phi_{p_{\max}-1} = \arcsin NA - \arcsin\sqrt{NA\left(1 - \frac{\lambda n}{NA^2 t}\right)}.$$

Assuming $\lambda n/(NA^2 t) << 1$, and applying first-order approximation to the arcsine, we obtain

$$\Delta\phi \approx \frac{\lambda n}{2t \cdot NA\sqrt{1 - NA^2}}.$$

Substituting this result into formula for $\Delta\lambda$ and using $\sin\phi_{\max} = NA$, the final restriction for the probe beam spectral width comes as the request for it to be much smaller than

$$\Delta\lambda \approx \frac{\lambda^2}{2nt(1 - NA^2/n^2)}.$$

From here, two significant qualitative conclusions come: the bigger the refractive index and the thickness of the film are, the narrower the spectral width of the probe beam must be. Nowadays, with great advance of semiconductor memory V-NAND technology, thickness of dielectric films to be monitored increases from nanometers to several micrometers. As a quantitative example, consider $t = 6$ μm, $n = 1.5$, $NA = 0.8$, and $\lambda = 0.6$ μm. Then the above formula gives $\Delta\lambda = 28$ nm, and spectrum of the light source must be much narrower than this value, not wider than several nanometers only. Typical spectral characteristics of the most popular sources are presented in Fig. 9.6. It looks as if the laser diode and, of course, He–Ne laser (Chapter 3) would be the best choice for angular reflectometry.

However, there is also the lower limit for spectral width of the light source, dictated by speckles. The He–Ne lasers, whose coherence length is of the scale of kilometers, are well known for strong speckles in cross section of their beams. The laser diodes, although having orders of magnitude wider spectrum, also produce speckles in the form of concentric circles already at their output window, which can be directly seen in a microscope focused on the output window of a laser diode (Fig. 9.7).

The light-emitting diodes, whose spectral width is around 30 nm, do not produce noticeable speckle but may be suitable as light sources only for nanometer thick films. As to the thick micrometer-scale films, the superluminescent diodes with spectral width of about 10 nm may be considered as the only choice.

9.1.3. *Experimental examples*

In order to better realize basic features of angular reflectometry, first consider typical experimental results that may be obtained on low-refractive index substrates like glass, and then compare them to the results obtained on silicon substrates. In both cases, the superlumiscent diode with 9 nm spectral width at 0.68 μm central wavelength and Nikon objective lens with 0.8 numerical aperture ($\phi_{\max} = 53°$) were used.

Figure 9.8 presents cross-sectional image of the beam reflected from the glass substrate alone. Noticeable speckle and circular interference fringes, produced by finite coherence of the superluminiscent diode, is an unavoidable part of the experiment.

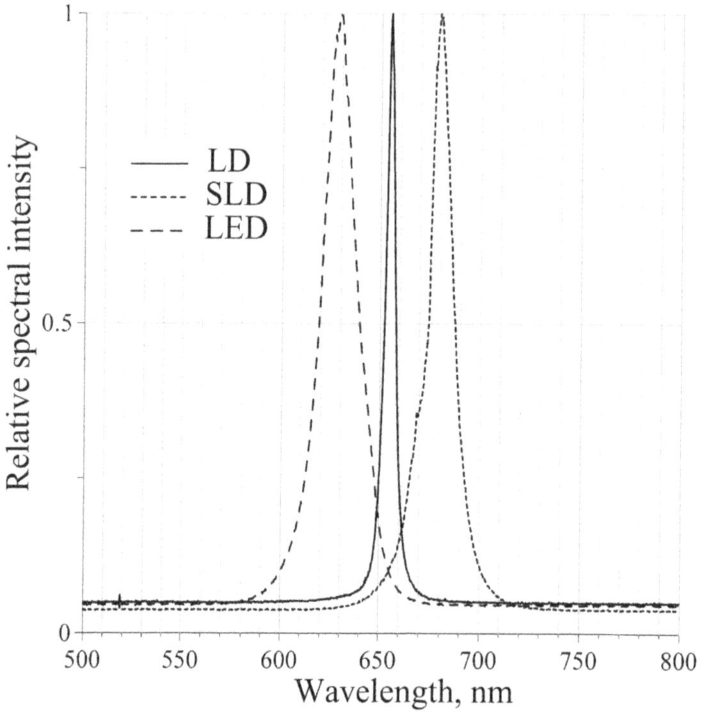

Fig. 9.6. Comparison of spectra of three types of light sources: laser diode (LD), light-emitting diode (LED), and superluminiscent diode (SLD).

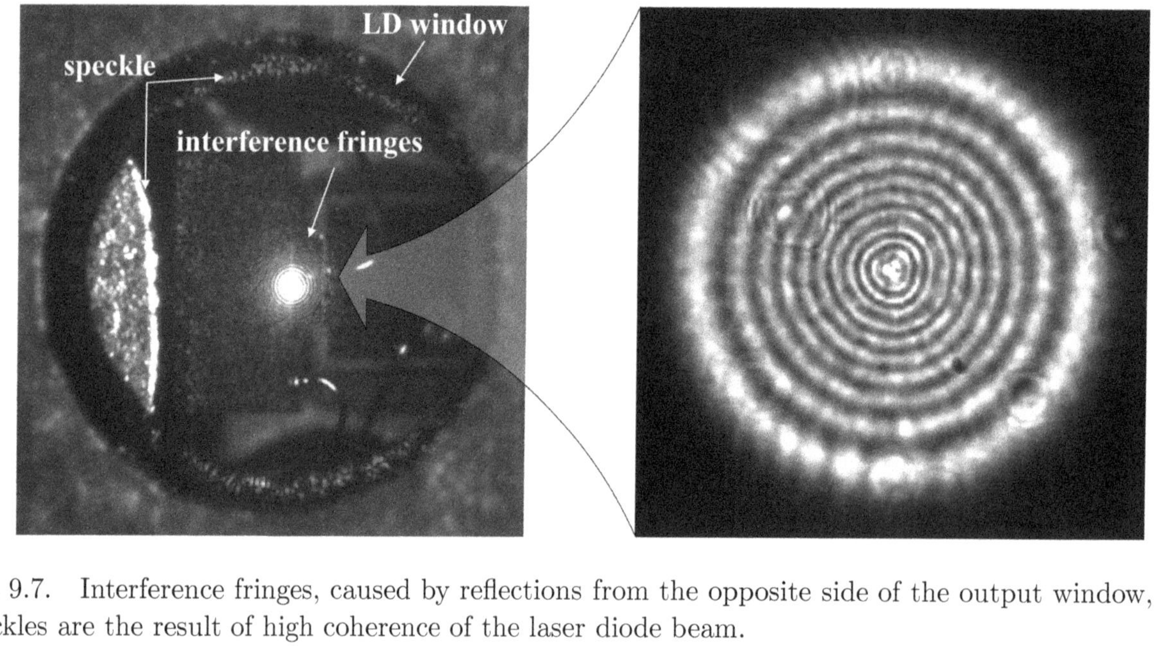

Fig. 9.7. Interference fringes, caused by reflections from the opposite side of the output window, and speckles are the result of high coherence of the laser diode beam.

The Brewster angle for quartz is 56° (Fig. 9.3), which is almost equal to $\phi_{\max}$. Therefore, horizontal borders of the circle, determined by R_p, are almost black. On the contrary, vertical borders, brightness of which is proportional to R_s, are brighter than the central part. With the film on the glass, wide dark circle appears in the picture, indicating interference in the film (Fig. 9.9).

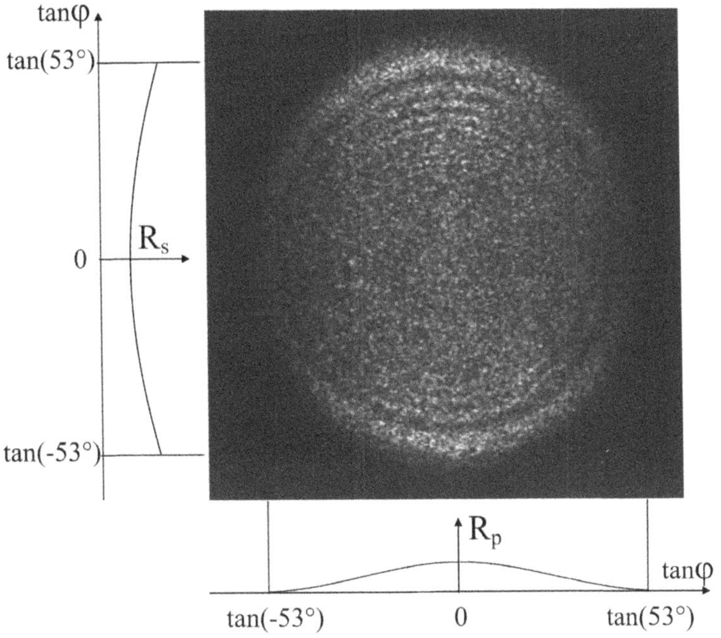

Fig. 9.8. Picture of angular reflectance from bare glass.

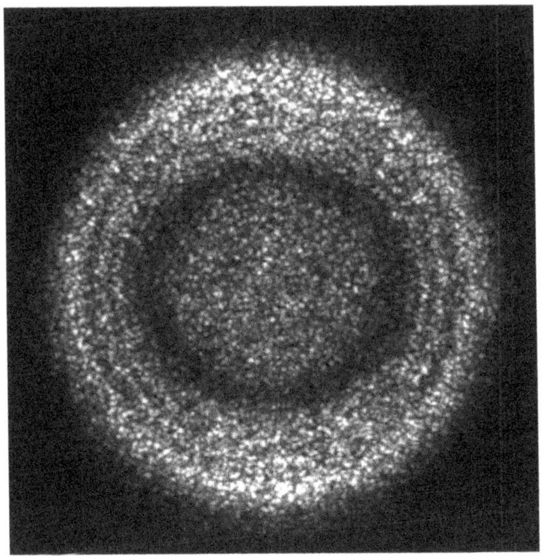

Fig. 9.9. Picture of angular reflectance from film deposited on glass.

Real part of the refractive index of silicon at the wavelength around 0.6 μm is 3.88, more than 2 times higher than that of glass. Therefore, the Brewster angle $\arctan(3.88) = 76^\circ$ significantly exceeds $\phi_{\max}$, making intensity distribution in the cross section of the beam almost uniform (Fig. 9.10(a)). Finally, consider relatively thick 6 μm silicon oxide layer deposited on silicon wafer (Fig. 9.10(b)). As the layer thickness increases, the number of interference fringes also increases. It must be emphasized that, for such big number of fringes to be observed, coherence of the light source must be high, as discussed above in Section 9.1.2.

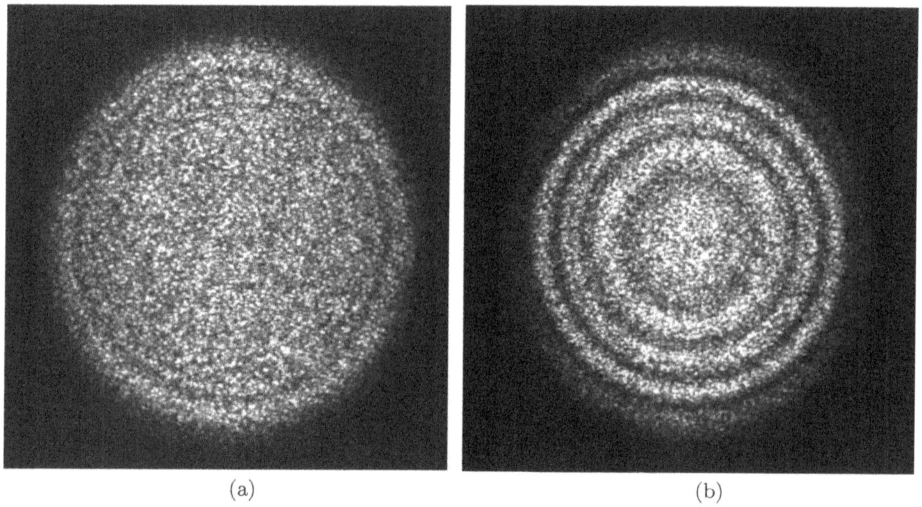

Fig. 9.10. Pictures of angular reflectance from (a) bare silicon and (b) 6 μm oxide layer on silicon.

9.2. Surface Polarimetry

9.2.1. *Principle of surface polarimetry*

Working area of lithography masks for manufacturing memory chips is composed of many parallel reflecting sub-wavelength stripes (metallic or multilayered) spaced by air gaps. Therefore, a mask may be considered as a wire grid polarizer for visible optics, and it is reasonable to expect that polarimetry techniques may be efficient for detecting tiny variations of the mask parameters. This is the basic idea, and the heterodyne polarimetry concept is shown in Fig. 9.11. The details of the heterodyne technique are explained in Chapter 3.

The output beam of the Zeeman laser consists of two orthogonal linearly polarized components E_1 and E_2 with the frequencies ω_1, ω_2 and real amplitudes a_1 and a_2:

$$E_1 = a_1 e^{i\omega_1 t + i\phi_1}; \quad E_2 = a_2 e^{i\omega_2 t + i\phi_2}.$$

Suppose the wave E_1 is polarized along the sample structure, while the wave E_2 — perpendicularly to it. Then the waves reflected from the grating exhibit different amplitude and phase changes:

$$E_{||} = E_1 r_{||} = |r_{||}| a_1 e^{i(\omega_1 t + \phi_1 + \phi_{||})}; E_{\perp} = E_2 r_{\perp} = |r_{\perp}| a_2 e^{i(\omega_2 t + \phi_2 + \phi_{\perp})};$$

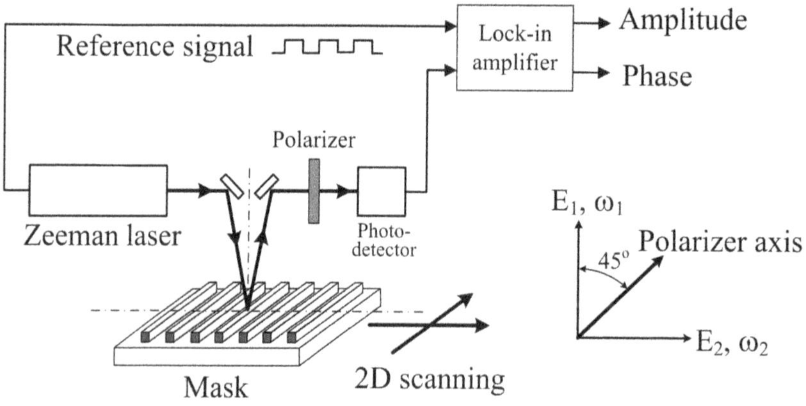

Fig. 9.11. The concept of heterodyne polarimetry.

where $r_{||}$ and $r_\perp$ are the complex reflection coefficients for the waves polarized parallel and perpendicularly to the structure, and $\phi_{||}$, and $\phi_\perp$ are additional phase shifts acquired after reflection. The two orthogonally polarized waves do not interfere. Therefore, in order to obtain interference beatings at intermediate frequency, a polarizer is always installed in front of the photodetector as shown in Fig. 9.11. The intermediate frequency component of the detector photo-current is proportional to

$$j = |r_{||}||r_\perp|a_1 a_2 \cos[(\omega_1 - \omega_2)t + \phi_1 - \phi_2 + \phi_{||} - \phi_\perp].$$

While a_1, a_2, $\omega_1 - \omega_2$, and $\phi_1 - \phi_2$ are the parameters that depend on laser operation, the other parameters — $|r_{||}|$, $|r_\perp|$, $\phi_{||}$, and $\phi_\perp$ — depend on the surface structure. Using synchronous detector (lock-in amplifier) and the reference signal from the laser, it is possible to obtain the two output electrical signals proportional to the amplitude $|r_{||}||r_\perp|$ and phase $\phi_{||} - \phi_\perp$ of the intermediate frequency component. The quantity $R = |r_{||}||r_\perp| = \sqrt{|r_{||}|^2|r_\perp|^2} = \sqrt{R_{||} \cdot R_\perp}$ represents the geometrical average of the power reflection coefficients $R_{||}$ and $R_\perp$ of the parallel and perpendicularly polarized laser components.

Consider the main physical reasons, affecting the amplitude and the phase outputs. It is commonly agreed that polarization properties of a wire greed polarizer, i.e. the values of $|r_{||}|$ and $|r_\perp|$, are determined by different energy losses of the electrons, moving along and perpendicularly to the wires. This simple explanation is applicable, of course, only to conducting structures, while the polarization effects are also observable on dielectric structures. However, lithography masks always contain metal layers, so that we may formally apply the aforementioned explanation to our problem. Since the period of the structure is constant, the average reflectivity of a sample is basically determined by the average amount (mass) of metal per square unit of the surface, or by the filling factor. The phase channel provides combined information about both the filling factor and phase uniformity of the mask at its working wavelength. Although this information is macroscopic, i.e. averaged over the cross section of the probe beam (about 50 μm), it is supposedly closely related to the microscopic phase shifts which are the main concern for the manufacturers.

To finalize this section, it is necessary to emphasize that heterodyne polarimetry is sensitive to macroscopic lateral variations of average parameters of periodical structures rather than to single microscopic defects. The single microscopic defects would not contribute to the signal because of their negligibly small area with respect to the cross section of the probe beam.

9.2.2. *Principle of measuring critical dimension*

The only way to measure critical dimensions with polarimetry tools is to compare experimental results with the data provided by mathematical simulation. Therefore, the calibration should be carried out, finding particular numerical values for the parameters of simulating software that deliver accurate simulation for any newly introduced mask. For an arbitrary mask with known both the period and average filling factor, the process of calibration means determining two coefficients c_1 and c_2 in the linear approximation of the reflectivity $R = |r_{||}||r_\perp|$:

$$R = c_1 f + c_2,$$

in a narrow interval around the average value of the filling factor f. This can be done, using simulating software, prior to measurements. Then simple renormalization of signal amplitudes gives the map calibrated in the filling factor. It is important that the renormalization is many orders of magnitude faster than the execution of complicated simulation routines, and therefore, the map of the filling factor spatial distribution can be obtained online. Thus, the simulation software only performs a transfer of calibration coefficients from a calibration mask to any new mask with different parameters.

Nowadays, a variety of so-called "rigorous" numerical approaches is used to solve the Maxwell equations directly in order to compute the parameters of the waves reflected by gratings. All of them work well with the dielectric gratings but fail to describe quantitatively the metallic ones because in metals the electron mean-free-path (several Angstroms) is of the order of the penetrating depth, therefore, macroscopic electrodynamics cannot be applied.

For the subject this chapter deals with, it is only important to understand that application of any of the "rigorous" numerical approaches to metal gratings is not a physically correct procedure but rather an approximate assessment of what it is possible to expect qualitatively. The most widely used computational routine is known as the RCWA — an abbreviation for "rigorous coupled-wave analysis". The experimental results described below were compared with the RCWA. Originally, the RCWA routine was designed for lossy dielectrics, and later adopted as a *de-facto* standard for metals, ignoring the physical difference between these two types of solids. Therefore, using the RCWA, we have to use the complex refractive index as a parameter of the metal media. Although the basic mathematical routine of the RCWA can be considered as the exact numerical solution to the diffraction problem, in reality the RCWA is only an approximation because of the finite number of terms in the modal expansion and the approximate representation of the grating profile by bar sections. In all computations below we have been retaining 41 term in the modal expansion and 10 profile sections, reasonably balancing between the speed and computational accuracy.

In 1990, Petit and Tayeb introduced very compact and, at the same time, efficient algorithm based on the thin metal layer approximation (TMLA). In this method, it is assumed that the metal is equivalent to a lossy dielectric with relative permittivity $1 + i\sigma/\varepsilon_0\omega$ where σ is the real conductivity of the metal, ε_0 is the permittivity of vacuum, and ω is the angular frequency of the field. Thus, in this model, the metal is characterized by only one parameter — its conductivity. It was shown that when the thickness of metal h_0 tends to zero, the reflected field tends to a certain limit field, which is determined entirely by the profile of the grating $h(x)$ and the dimensionless parameter $s = h_0\sigma\eta_0$, where $\eta_0 = \sqrt{\mu_0}/\varepsilon_0$ is the vacuum impedance. The basic formulas of the TMLA are summarized below in the form suitable for practical applications.

Consider an infinitely thin metal grating on a dielectric substrate with permittivity ε. The grating lines are directed along the z-axis, the surface normal is along y-axis, and the boundary lies in the $x - z$ plane. In the notations of the original work by Petit and Tayeb, the boundary conditions for the TE-wave $u(x, y)$, coming at the angle of incidence θ, are

$$u^+ = u^-, \quad \left(\frac{\partial u}{\partial y}\right)^+ - \left(\frac{\partial u}{\partial y}\right)^- = -ikshu^+,$$

and for the TM-wave:

$$\left(\frac{\partial u}{\partial y}\right)^+ = \frac{1}{\varepsilon}\left(\frac{\partial u}{\partial y}\right)^-, \quad u^+ - u^- = \frac{ish}{k}\left(\frac{\partial u}{\partial y}\right)^+,$$

where $k = 2\pi/\lambda$ is the wavenumber, $s = h_0\sigma\eta_0$, and $h(x)$ is the grating profile. As usual, notations "+" and "−" correspond to the functions above and beneath the boundary. Applying then the Rayleigh expansion to the field $u(x, y)$

$$u(x, y) = \sum_{m=-\infty}^{+\infty} u_m \exp(ik_{xm}x + ik_{ym}y), \quad k_{xm}^2 + k_{ym}^2 = k^2,$$

and expanding the profile $h(x)$ into the Fourier series, the boundary equations can be rewritten in terms of the n-th diffraction-order components of the reflected r_n and transmitted t_n waves, and the Fourier coefficients of the profile h_n:

$$\begin{cases} r_n + \delta_{n0} = t_n \\ \sum\limits_{m=-\infty}^{+\infty} [(\gamma_n + \beta_n)\delta_{nm} + ksh_{n-m}]r_m = -ksh_n - (\gamma_n - \beta_n)\delta_{n0} \end{cases} \quad \text{for the TE-wave,}$$

$$\begin{cases} r_n - \delta_{n0} = -\frac{\gamma_n}{\beta_n \varepsilon} t_n \\ \sum\limits_{m=-\infty}^{+\infty} \left[\frac{k}{s}\left(1 + \frac{\beta_n}{\gamma_n}\varepsilon\right)\delta_{nm} + \beta_m h_{n-m}\right] r_m = -\beta_0 h_n - \frac{k}{s}\left(1 - \frac{\beta_n}{\gamma_n}\varepsilon\right)\delta_{n0} \end{cases} \quad \text{for the TM-wave.}$$

Here $\alpha_n = k \cdot \sin\theta + n \cdot 2\pi/d$ with d being the grating period, $\beta_n^2 = k^2 - \alpha_n^2$, $\gamma_n^2 = k^2\varepsilon - \alpha_n^2$, $n = 0, 1, 2, \ldots$, and δ_{nm} is the Kronecker symbol. These are the two infinite systems of equations for r_n and t_n. The solution starts with finding r_n from the truncated systems of equations

$$\sum_{m=-M}^{+M} [(\gamma_n + \beta_n)\delta_{nm} + ksh_{n-m}]r_m = -ksh_n - (\gamma_n - \beta_n)\delta_{n0} \quad \text{for the TE-wave,}$$

$$\sum_{m=-M}^{+M} \left[\frac{k}{s}\left(1 + \frac{\beta_n}{\gamma_n}\varepsilon\right)\delta_{nm} + \beta_m h_{n-m}\right] r_m = -\beta_0 h_n - \frac{k}{s}\left(1 - \frac{\beta_n}{\gamma_n}\varepsilon\right)\delta_{n0} \quad \text{for the TM-wave.}$$

Typically, $M = 100$ terms in modal expansion is enough for good convergence. Any available routine for solving linear systems of equations is suitable, for instance, the LSLCG from the IMSL package. After the set of r_n is obtained, the t_n can be found from the rest two equations:

$$r_n + \delta_{n0} = t_n, \quad \text{for the TE-wave,}$$

$$r_n - \delta_{n0} = -\frac{\gamma_n}{\beta_n \varepsilon} t_n, \quad \text{for the TM-wave.}$$

Now, we have to determine analytical model for grating profile $h(x)$. The basic consideration for this is simplicity of expansion into the Fourier series. As such, the following approximation may suffice:

$$h(x) = 1 - e^{v(|x| - 0.5fd)}, \quad 0 \le |x| \le 0.5fd,$$

where d is the grating period, v determines sharpness of the edges, and f is the formal filling factor as it is determined at the metal-substrate interface. In order to work with the dimensionless parameters, this formula can be rewritten in the form

$$h(x) = 1 - e^{0.5w(|t| - f)}, \quad 0 \le |t| \le f$$

with $t = 2x/d$, $w = vd$. In the limit $w \to \infty$ the profile is rectangular (Fig. 9.12).

With this model, it is easy to compute analytically the Fourier coefficients of $h(x)$:

$$h_n = \frac{\sin(\pi n f)}{\pi n} - \frac{2w}{w^2 + (2\pi n)^2}\left[\cos(\pi n f) + \frac{2\pi}{w}\sin(\pi n f) - e^{-0.5wf}\right], \quad w = vd.$$

The free parameter $s = h_0\sigma\eta_0$ requires calibration. In order to determine it from experimental measurements, consider homogenous metal layer, which corresponds to $w = \infty$ and $f = 1$. In this case, the zeroth-order term $h_0 = 1$. Then, at normal incidence $\alpha_0 = 0$, boundary equations for the TE and TM waves are equivalent, and we get

$$r_0 = \pm\frac{s + \sqrt{\varepsilon} - 1}{s + \sqrt{\varepsilon} + 1}.$$

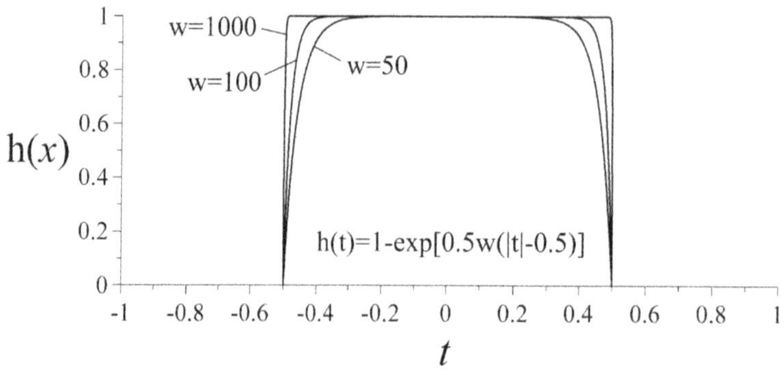

Fig. 9.12. The model shapes of the grating profile.

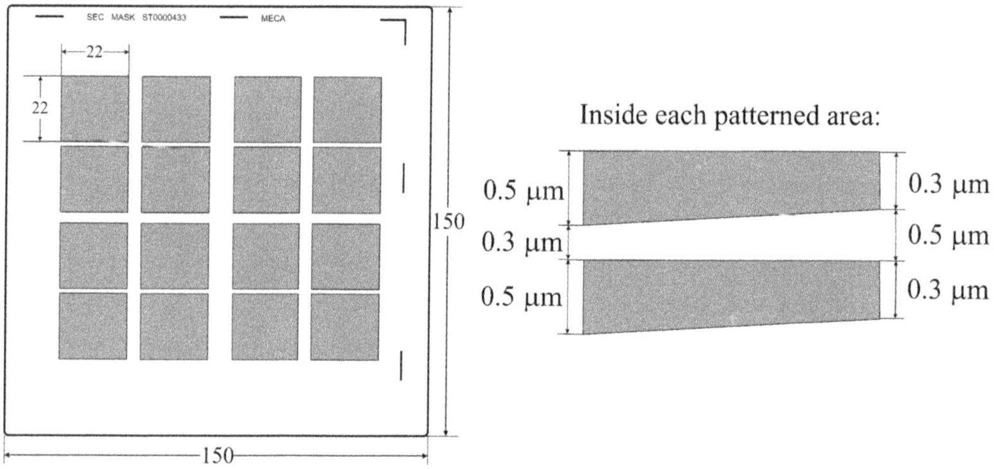

Fig. 9.13. Calibration mask.

The sign "±" here merely reflects the degeneracy of the normal incidence case with respect to the sign of the reflected field. If the energy reflectance at normal incidence $R = |r_0|^2$ is known, then the appropriate value of s follows:

$$s = \frac{1+\sqrt{R}}{1-\sqrt{R}} - \sqrt{\varepsilon}.$$

Another way of calibration is the direct measurement on special sample with precisely measured pattern dimensions. Configuration of the calibration mask, used in the experiments described below, is shown in Fig. 9.13.

The chromium layer 55 nm thick was etched to the pattern shown at the right of the picture. This pattern is the same inside each of the sixteen 22×22 mm^2 square patterned areas. The period is kept constant all over the sample, while the filling factor varies linearly from 0.375 to 0.625.

Figure 9.14 presents the results obtained with the amplitude channel. The line scan shown in dots is calibrated in geometrical average reflectivity $R = |r_{||}| \cdot |r_{\perp}|$, showing that the filling factor of 0.5 corresponds to the reflectivity of about 0.2. Calibration of reflectivity was done on the reference mirror of known reflectance.

The level of the noise in the amplitude channel establishes a limit for detecting the smallest possible variations of the filling factor. This limit was calculated as 0.5% of the filling factor when the time

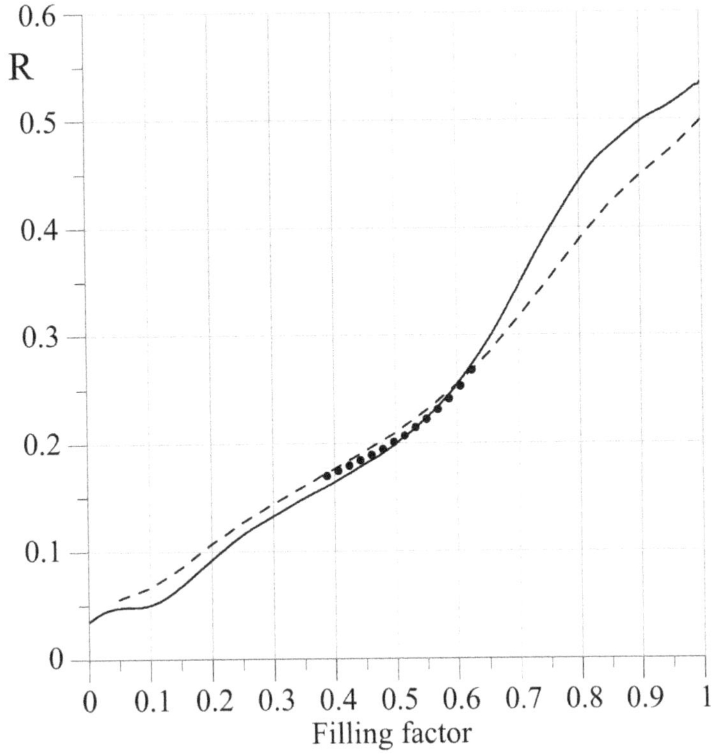

Fig. 9.14. Fitting of the amplitude data. Solid line — the RCWA, dashed line — the TMLA. Experimental data are shown in dots.

constant equals 3 ms. It means that if the period is equal to 0.8 μm and the average filling factor equals 0.5, then the technique is capable of detecting as small line width variations as 2 nm.

Experimental line scan shown in Fig. 9.14 was compared to the RCWA and TMLA simulation. The RCWA simulation could be fitted into the experimental curve, varying the real and imaginary parts of the chromium refractive index. The TMLA simulation was fitted without variation of any parameter, just computing s by the last formula with $R = 0.50$ measured on the non-patterned part of the mask. Both the RCWA and TMLA simulations do not depend much on the exact value of the sharpness parameter w if it is taken large enough, about 10^4. That large value of the sharpness parameter means that the fitted profile is practically a rectangle. It can be seen also that RCWA gives the reflection coefficient on uniform chromium layer, i.e. with the filling factor equal to unity, larger than the experimental value of 0.50.

The amplitude fitting with the TMLA turned out to be more accurate and much simpler than that with the RCWA. Note that both the uniform chromium layer reflectivity and the experimental data from the patterned area were fitted equally well. However, the phase curve computed with the TMLA appeared to be much less accurate than that fitted with the RCWA (Fig. 9.15). It is not a surprise: the RCWA operates with the two variables (real and imaginary parts of a refractive index), while the TMLA with only one (parameter s).

In principle, it is possible to use for calibration both amplitude and phase channels. However, since the phase curve was practically flat around the filling factor of interest, the amplitude channel was chosen for calibration. According to the principle of measuring critical dimensions, coefficients c_1 and c_2 were determined and used to compute the filling factor variations. Some experimental results on real lithography masks are summarized in the next section.

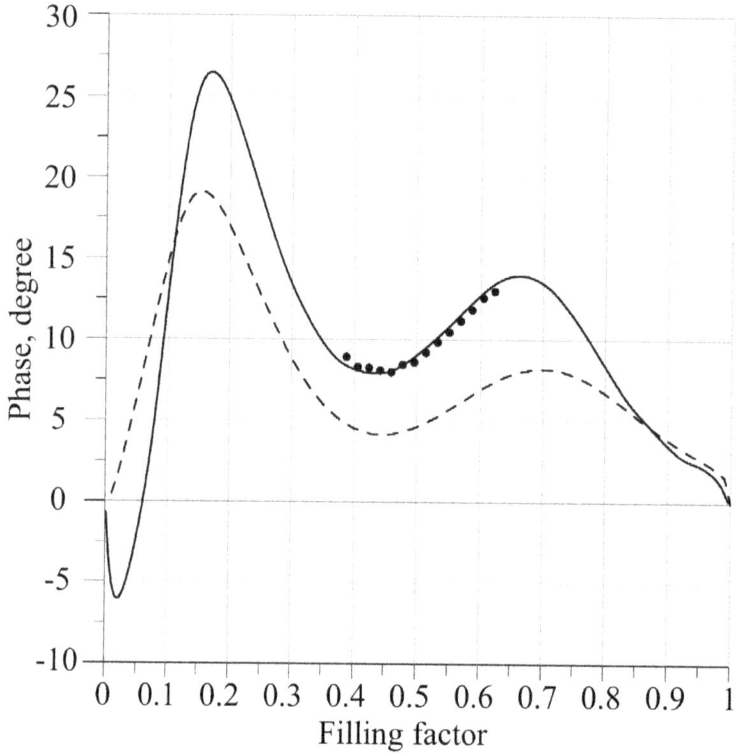

Fig. 9.15. Fitting of the phase data. Solid line — the RCWA, dashed line — the TMLA. Experimental data are shown in dots.

9.2.3. *Experimental results*

Figure 9.16 presents design concept of the experimental installation. The Wavetronics WT307 Zeeman laser 1 with the frequency split of about 2.13 MHz is the key component. Stability of the laser output is determined by the length of the laser cavity which is permanently controlled to produce equal intensities of the two independent output waves. Therefore, the laser is mounted on the heavy brass table 9.2, floating on rubber shock absorbers 3, in order to minimize the effect of outer vibrations. The diameter of the beam emerging from the laser cavity is of about 0.3 mm. With 40 mm focal length of the lens, focusing the beam onto the mask, the laser spot on the mask has the full width of 60 μm.

The two-dimensional orthogonal scanning unit was designed for the $150 \times 150 \times 6$ mm^3 sample masks and simplest line-by-line scanning algorithm. The scanning speed was limited by the signal-to-noise ratio in the phase channel. Typically, the data were acquired with either 3 ms or 10 ms time constants, depending on the necessary spatial resolution, but in some cases, when exceptionally clean phase measurements were necessary, the 30 ms time constant was used. With 1 ms time constant, a map, containing 800×600 pixels, has been routinely acquired in 50 min. For more precise measurements, say, for maps of 800×800 pixels and 10 ms time constant it took 4 hrs.

Figure 9.17 portrays the amplitude and phase maps of real phase-shifted mask for memory chips. In the amplitude map, the overall variation of the filling factor was calculated to be equal to 0.022. With the period equal to 460 nm, this value corresponds to the variation of the critical dimension of about 10 nm. The resolution is believed to be equal to 0.1 of the entire peak-to-peak variation, which corresponds to 1 nm. The difference between the amplitude and phase maps is substantial: although the phase map does follow trend of the amplitude map in the middle, it has definitely different peripheral distribution (a diagonal one).

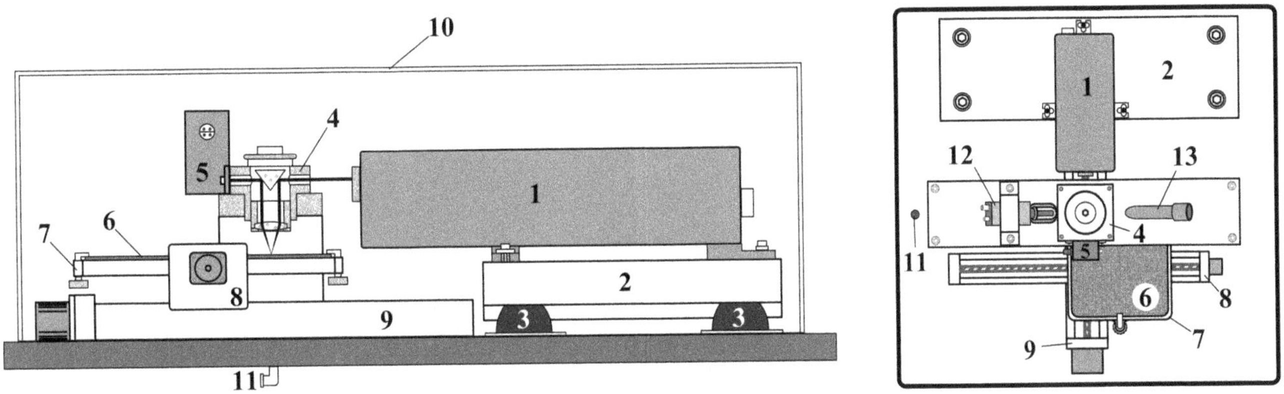

Fig. 9.16. The plan and side views of the experimental installation. 1 — Zeeman laser; 2 — anti-vibration table; 3 — rubber shock absorbers; 4 — optical module; 5 — photo-detector with pre-amplifier; 6 — mask; 7 — mask holder; 8, 9 — two-dimensional scanning stage; 10 — plastic box; 11 — dry air inlet; 12 — camera for visual tracking. The footprint of the system measures 0.8×0.7 m^2.

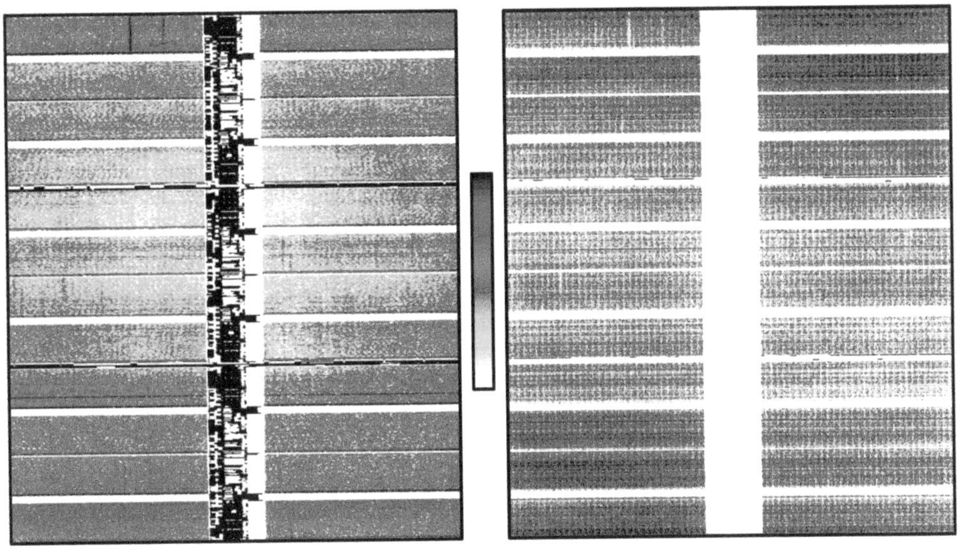

Fig. 9.17. The amplitude (left) and phase (right) maps of a phase-shifted lithography mask. Color scale bar applicable to both maps is located between them. In the amplitude map, the filling factor value of 0.489 corresponds to white color and the value of 0.511 — to black color. In the phase map peak-to-valley variation of phase equals to 0.344°.

Finally, Fig. 9.18 compares the results of the heterodyne polarimetry to other optical inspection techniques available on the market: Atlas-M of Nanometrics, n&k 5700 of n&k Technology, and Aera 193 of Applied Materials. In these experiments, the inspected mask was composed of 8×8 vertically elongated rectangular areas of grating structures with the micro-period equal to 460 nm and the filling factor of 0.5.

It is interesting to compare the pictures with respect to both global (long-scale) and local variations, keeping the scanning electron microscope (SEM) image as a reference. First, consider the strongly localized imperfection marked with green color in the SEM map in the upper left corner. This area was marked by only two techniques: the heterodyne polarimeter and n&k 5700 (blue regions in the

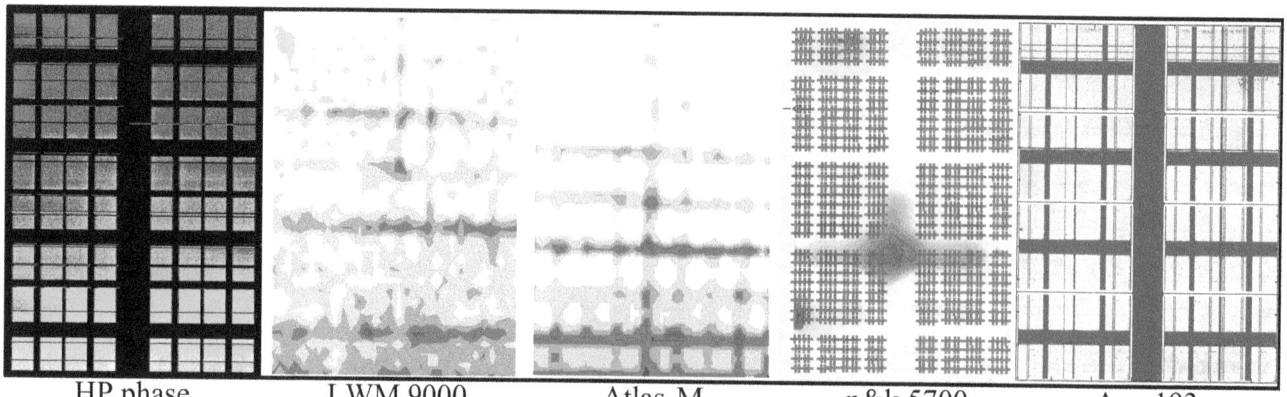

Fig. 9.18. Mask uniformity portraits. From left to right: heterodyne polarimeter (HP) phase map; scanning electron microscope LWM 900 map; optical critical dimension inspector Atlas-M map; spectral reflectometer n&k 5700 map; Aera 193 at 193 nm.

both maps). However, the n&k 5700 shows deep red area in the middle, which does not correlate with the SEM. On the contrary, the SEM shows smooth transition from the left upper corner to the lower right corner, which correlates strongly with the heterodyne polarimeter and Atlas-M maps, and not so obviously with the Aera 193 map.

9.3. Phase-Resolved Heterodyne Microscopy

9.3.1. *Principle of heterodyne microscopy*

Almost eight decades ago, phase contrast microscopy introduced by Zernike in 1942 revolutionized biological research by making it possible to study transparent objects. The success was so astonishing that Zernike was awarded the Nobel prize in 1953. The next great step in phase-contrast microscopy was accomplished in 1969 by Zeiss company, which in collaboration with Nomarski introduced the first commercial differential interference contrast microscope. This technique has great advantage over the Zernike phase-contrast scheme because it produces high-contrast images of the edges of objects and fine structural details within transparent specimens without bright diffraction halos, interference fringes, and other artifacts. Today, with all its minor modifications, differential interference contrast technology, also commonly referred to as the Nomarski microscope, remains the basic instrument in biological research. As time elapsed, advances of the semiconductor industry towards nano-scale technology produced new challenges in spatial resolution and contrast of images of extremely thin layers less than 100 nm thick, which gave birth to new solutions. Let us consider them in more detail.

In traditional phase-contrast microscopy, including the Nomarski technique, the phase contrast is visualized indirectly, by means of interference of two waves with different phases. The phase-contrast picture is in fact the intensity variation of the sum of the two spatially coherent waves, and the contrast of this picture is a quadratic function of the phase shift when the latter is small. Thus, the contrast quadratically tends to zero with the phase shift, making the Nomarski technique practically insensitive to phase shifts less than 5°. The solution to this problem is to measure phase shift directly, by means of heterodyning. The idea of the scanning heterodyne microscope was first introduced and experimentally verified by Sawatari in 1973. He suggested splitting a laser beam into probe and reference beams, introducing a frequency shift into one of the beams, scanning the object under the focused probe beam, and recombining the beams onto the photo-detector. Then, according to basic methodology of laser heterodyning, the output electrical signal contains full information about the amplitude and phase of the laser beams, which can be used for imaging. In comparison with classical microscopes, this scheme has several advantages: strong suppression of incoherent background illumination, strong axial selectivity (a feature of confocal scanning microscope), and phase sensitivity linearly proportional to phase variations of the object. Its main disadvantage turned out to be exceptional sensitivity to any vibrations, which made the entire concept impractical. It can be theoretically proved that, despite coherent illumination, spatial resolution of the Sawatari scheme is exactly the same as that of the incoherent microscope.

To overcome vibrational sensitivity, the differential heterodyne scanning microscope was proposed by See in 1985. In it, the lateral position of the laser spot focused onto the object is harmonically modulated within the spot diameter, making the system sensitive to only the gradient of the object surface structure. Differential heterodyne microscope has extremely high sensitivity to surface relief, capable of detecting the Angstrom-level height variations. Its spatial resolution, however, is not as good as that of a conventional optical microscope, and its imaging is anisotropic. These drawbacks are the consequences of fine jitter of the focal spot along the direction of scanning. It can be shown theoretically that the spatial resolution of the differential heterodyne microscope is about the same as that of an ideal coherent imaging system, which is poorer than that of an incoherent one. Strictly speaking, this microscope is not a phase-contrast technique but rather a gradient-contrast one because the amplitude and phase of the reflected optical wave are mixed in the output electrical signal.

Considering spatial resolution, it is worth mentioning that with the Nomarski microscope the image is also formed by a differential technique, using two sheared light spots, separated by a fraction of the focused beam diameter. For this reason, the image is also anisotropic and spatial resolution is somewhat poorer than that of an ideal incoherent imaging system.

Further development of phase-resolved microscopy was driven by requirements of semiconductor industry to overcome limitations of spatial resolution, inherent to traditional microscopes, and to build three-dimensional images. One type of such multi-functional inspection tools is described in this section — a dual-channel heterodyne microscope. This technology may be considered unique in a sense that it incorporates several most important features in a single device:

- simultaneous amplitude and phase-contrast imaging;
- phase sensitivity of better than 0.1°;
- highest spatial resolution in the amplitude mode equal to that of an ideal incoherent imaging system;
- super-resolution in the phase-contrast mode;
- isotropic imaging;
- insensitivity to any background or scattered radiation.

As can be seen from the experimental results presented below, comparative analysis of the amplitude and phase-contrast images of the same area of an object captured at the same moment provides unprecedented opportunities for detection of image features of different nature. It is especially important for recognition of various defects in the semiconductor industry.

In realization, the idea is similar to the aforementioned heterodyne microscopy techniques but based on a completely new principle. Consider a two-frequency cross-polarized laser beam focused onto a vertical wall of a patterned surface as it is shown in Fig. 9.19.

A linearly polarized component of light with its polarization vector $\mathbf{e}_1$ directed along the wall has angular frequency ω_1, while a second component with its polarization vector $\mathbf{e}_2$ directed across the wall has angular frequency ω_2. Since the boundary conditions for these two waves are different, they will be reflected with different phase shifts, which can be measured at the intermediate frequency $\omega_1 - \omega_2$, using the standard heterodyne technique. By moving the sample or scanning the beam in two directions,

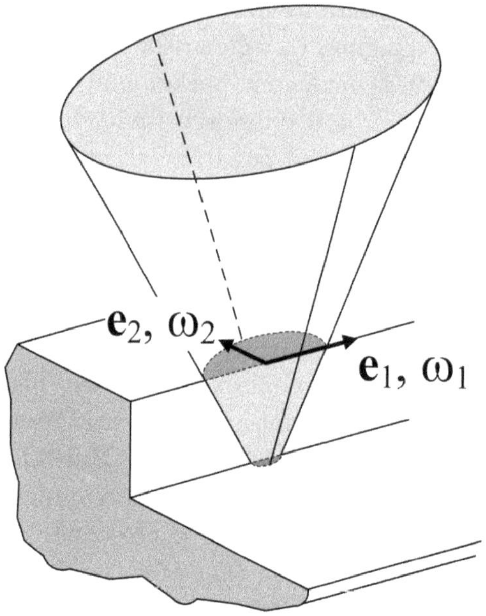

Fig. 9.19. Different boundary conditions for the two orthogonal polarizations create phase shift between reflected waves.

it is possible to get simultaneously two different types of images: the traditional amplitude image that uses only amplitude information, and the phase-contrast image built on phase information. Since spatial variations of phase are not directly associated with the size of the focused beam, it can be expected that phase-contrast mode of operation can resolve closely positioned objects in those cases when they cannot be resolved by the traditional microscopes.

9.3.2. *Optical scheme and instrumentation*

Proper design is crucial for stable operation of the proposed technology. Although the overall optical arrangement has very few components and, therefore, looks rather simple (Fig. 9.20), there are two fundamental design problems: isolation of the laser from the back-reflected beam, and independence of the phase of the receiver output signal from amplitude variations. Isolation is vital because any reflecting surface positioned in the focal plane of an objective acts as a perfect retro-reflector, sending the reflected beam directly back into the laser cavity, thereby disturbing its performance. Traditional isolation techniques such as Faraday rotators cannot be used in this case because the laser beam is composed of two orthogonally polarized components. Another simple but ineffective solution could be a neutral optical attenuator in the laser beam so that the back-propagating wave would be attenuated by a factor squared with the respect to directly propagating wave. The best solution proved to be a path-propagation attenuation scheme, which is explained in Fig. 9.21 for the particular case of the Zygo laser (Chapter 3).

In it, the output laser beam is additionally expanded to such an extent that the magnification of the additional expander is bigger than the ratio of the output and input beam diameters. A certain part of the beam's energy is thus wasted, and due to this, the back-reflected beam is de-magnified to such an extent that its diffraction divergence becomes much bigger than that of the original laser beam. Consequently, the back-reflected beam becomes uncoupled to the laser cavity, while the quality of the probe beam remains high.

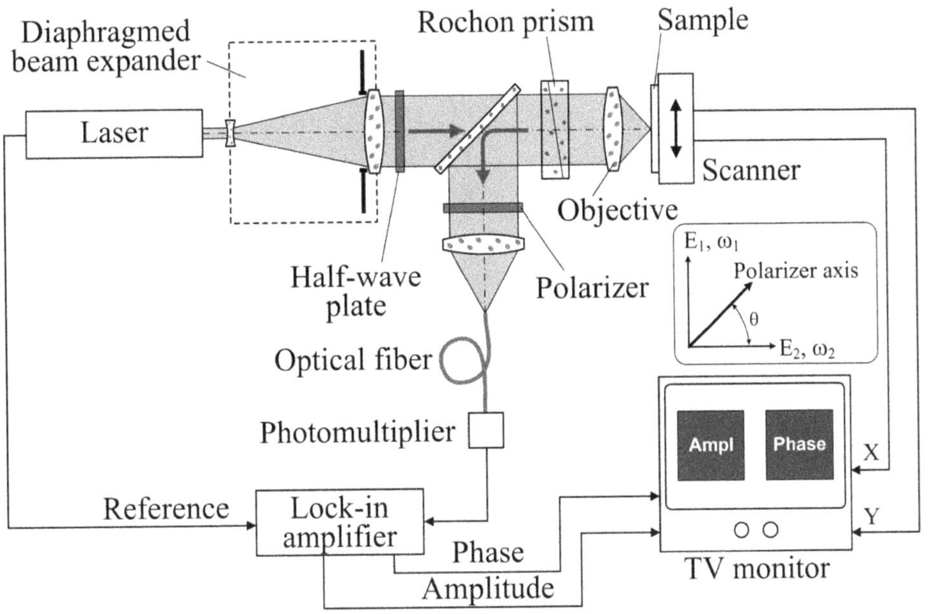

Fig. 9.20. Optical scheme of the system.

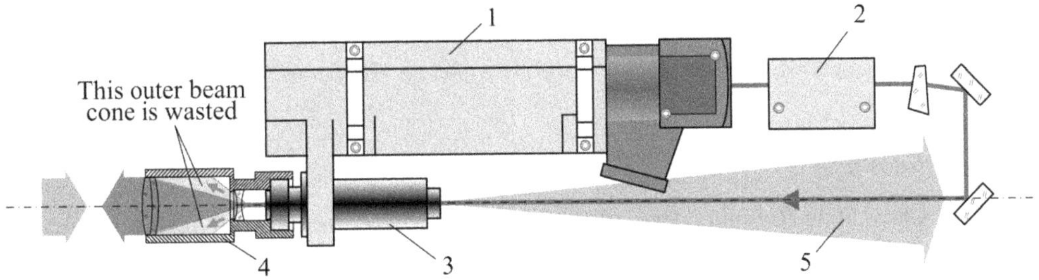

Fig. 9.21. The path-propagation optical isolation. 1 — gas-discharge module; 2 — acousto-optical module; 3 — inner beam expander; 4 — additional beam expander; 5 — back-reflected wave.

There are three commercially available types of the two-frequency He–Ne lasers ($\lambda = 0.63\ \mu$m) widely used in nano-positioning and polarimetry: Hewlett-Packard (Agilent) Zeeman lasers, Wavetronics Zeeman lasers, and Zygo lasers. All of them were experimentally tested and it was found that Zygo lasers are least sensitive to back-propagating waves, ensuring perfectly stable operation with the isolation technique described above. Another reason for choosing Zygo laser was an order of magnitude higher frequency split (20 MHz), which enabled an image acquisition rate of better than 1 μs per pixel.

For isotropy and spatial resolution of images it is important that the wavefronts of the two partial waves, composing the laser output beam, coincide. This condition may be verified by measuring the phase shift over the cross section of the laser output beam. The result presented in Fig. 9.22 shows that across the 8 mm aperture maximum phase shift of the Zygo laser is about 26°, which corresponds to an angular offset of the wavefronts equal to 5.6 μrad.

This value is much smaller than the diffraction divergence at this aperture equal to 75 μrad. Therefore, the spatial separation of the two partial focal spots is negligible. As a matter of fact, the Hewlett-Packard laser showed the smallest phase mismatch of about 2° among the three aforementioned lasers.

The intensity of the reflected light may vary substantially over the sample, making it necessary to guarantee independence of the phase of the signal from its amplitude. Extensive experimentation showed

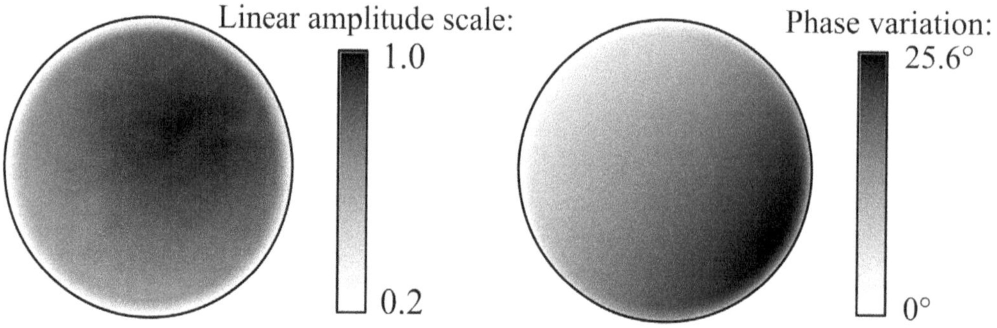

Fig. 9.22. Amplitude (left) and phase (right) spatial distributions in the output cross section of Zygo laser.

that it could not be achieved using circuits with photodiodes: inner photodiode resistance connected in parallel with the output capacitance inevitably introduces a phase delay that depends on the intensity of the light. Therefore, the only choice is a photomultiplier. Hamamatsu developed a very compact sealed photomultiplier H6780 with maximum sensitivity at the wavelength 0.63 μm. The low saturation threshold inherent to photomultipliers can be neutralized by setting a suitable gain by means of a trimmer resistor. Since the photomultiplier is essentially a source of current, the fast trans-impedance operational amplifier AD844 was used at the output, providing a flat frequency response up to 30 MHz. Any wide-angle background or back-scattered radiation is blocked by inserting an optical fiber in front of the photomultiplier. In this configuration, the fiber coupled to the focusing lens acts as a spatial filter.

9.3.3. *Qualitative theory*

Consider an ideal objective lens focusing two plane waves with frequencies ω_1 and ω_2 to two identical focal spots whose radial distributions, normalized to unity at their maximum, are described by

$$h(r) = 2\frac{J_1\left(3.83\frac{r}{\Delta_R}\right)}{3.83\frac{r}{\Delta_R}},$$

as illustrated in Fig. 9.23.

Formula for $h(r)$ represents the well-known Airy function, with J_1 being the Bessel function of the first order and Δ_R being the Rayleigh resolution parameter specific for the particular lens and wavelength. At a fixed moment of observation, the optical axis of the lens, which changes its position according to scanning program, intersects the sample, located in the focal plane, at the point ρ. Suppose there are two point-like objects in the focal plane positioned at r_1 and r_2 and separation between them δ. These two point-like objects generate reflected spherical waves with amplitudes proportional to $h(r_1 - \rho)$ and $h(r_2 - \rho)$ and frequencies ω_1 and ω_2, respectively. After passing back through the objective lens, the spherical waves become plane waves again, whose wavefronts subtend angles α_1 and α_2 relative to the lens plane p. Thus, the reflected complex fields with the frequencies ω_1 and ω_2 are identical (except for polarization, which is of no importance for the subject of this section) in the input pupil of the objective, and may be represented as

$$E_{1,2}(p) = h(r_1 - \rho) \cdot e^{ik\alpha_1 p - i\omega_{1,2}t} + h(r_2 - \rho) \cdot e^{ik\alpha_2 p - i\omega_{1,2}t},$$

where k is the wavenumber and t is time.

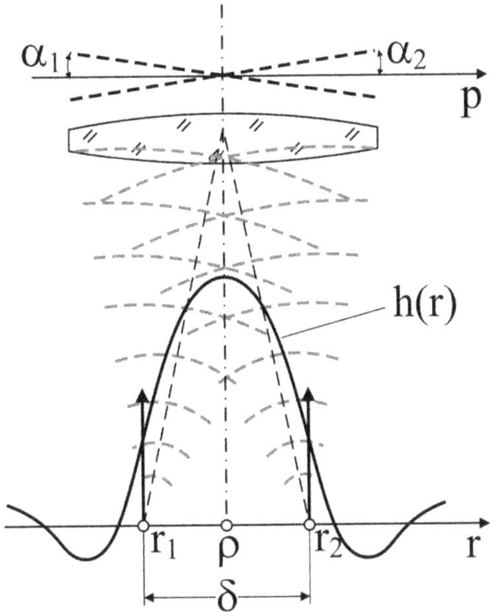

Fig. 9.23. Theoretical model of spatial resolution.

Complex amplitude $A(\rho)$ of the electrical signal of the heterodyne receiver at the intermediate frequency $\omega_1 - \omega_2$ is proportional to

$$\int E_1(p)E_2^*(p)d^2p,$$

where integration is taken over the plane p. For the sake of simplicity of presentation, we write the variables in a scalar form. However, one must have in mind that p and $\alpha_{1,2}$ have certain directions, and integration in should be performed accordingly, taking into account that for the lens with the focal length f paraxial angles are equal to $\alpha_{1,2} = r_{1,2}/f$. Then it is easy to show that

$$\int e^{ik\frac{\mathrm{r}}{f}\mathrm{p}}d^2p = S \cdot 2\frac{J_1\left(3.83\frac{r}{\Delta_R}\right)}{3.83\frac{r}{\Delta_R}} = S \cdot h(r),$$

where S is the area of the lens input aperture. With it,

$$A(\rho) = h^2(r_1 - \rho) + h^2(r_2 - \rho) + 2h(r_1 - \rho)h(r_2 - \rho)h(r_1 - r_2).$$

This non-trivial formula will be used for estimating spatial resolution in the amplitude channel. The point-spread function (PSF), normalized to unity at its maximum, follows from this formula by setting $r_1 = r_2$ and recalling that $h(0) = 1$:

$$\mathrm{PSF} = h^2(\rho).$$

Thus, the point-spread function of the amplitude channel is equal to that of an ideal incoherent imaging system. This conclusion, however, should not be confusing: spatial resolution in our case is somewhat different from that of an ideal incoherent imaging system. Indeed, traditional incoherent imaging system is linear with respect to intensity and, therefore, two point-like sources will produce the signal

$$A_{\mathrm{incoherent}}(\rho) = h^2(r_1 - \rho) + h^2(r_2 - \rho).$$

In the case of coherent imaging system

$$A_{\mathrm{coherent}}(\rho) = h(r_1 - \rho) + h(r_2 - \rho).$$

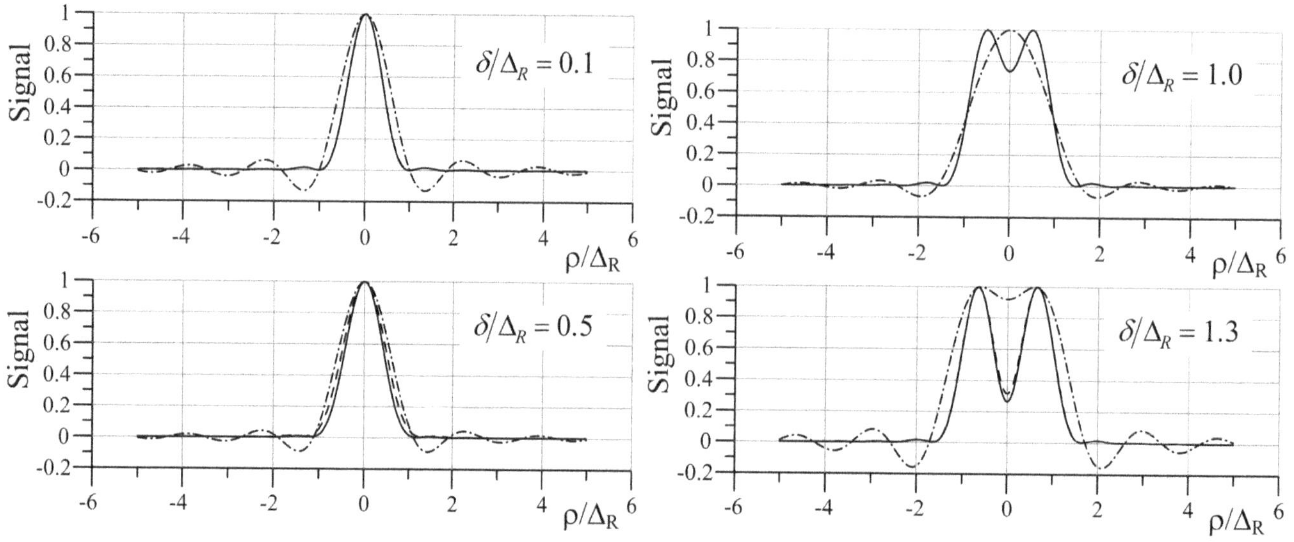

Fig. 9.24. Signals of two point-like sources separated by δ. Solid line — heterodyne amplitude channel; dashed line — ideal incoherent system (mostly overlap with the heterodyne); dashed-solid line — ideal coherent system.

The difference between the spatial resolutions of coherent and incoherent imaging systems is well known and we shall not dwell on it. But it is interesting to compare them with our system.

Figure 9.24 presents a series of pictures, showing computed signals $A(\rho)$, $A_{\text{incoherent}}(\rho)$, and $A_{\text{coherent}}(\rho)$ for several values of δ.

The lines, showing heterodyne amplitude signal, in most cases overlap with those of the ideal incoherent signal, showing that spatial resolution is approximately same as that of traditional microscopes. A more precise comparison can be performed by analyzing the contrast of the normalized signals $1-A(0)$ as a function of δ. Figure 9.25 shows that any noticeable difference between the heterodyne and incoherent systems exists only for small separation $\delta < 2\Delta_R$.

For semiconductor manufacturing, the most interesting topic is separation of long narrow lines. Therefore, the qualitative theory of heterodyne phase-contrast imaging should be presented using the physical model of imaging the edges, i.e. step-height borders that are relatively long with respect to the focused laser spot. Therefore, it is appropriate to speak about the edge-spread function (ESF) rather than the point-spread function (PSF), as in traditional theory of optical imaging.

Consider now the cross section of a circular focused laser beam when its center at O is offset by r from the edge line (Fig. 9.26). The two cross-polarized laser waves have respectively orthogonal polarizations $\mathbf{e}_1$ and $\mathbf{e}_2$, and frequencies ω_1 and ω_2.

When there is no edge within the focused spot, then the phase shift between the reflected waves is the same as that of the incident waves. With the edge inside, small phase variations are introduced proportional to the area of a narrow stripe of the width ε along the edge. This phase variation is different for the polarization components parallel and perpendicular to the edge: $\phi_{||}(r)$ and $\phi_{\perp}(r)$. The magnitude of these variations is proportional to the integral of the wave amplitude over narrow rectangular area of the width ε and infinite length along the edge line:

$$p(r) = \varepsilon \int_{-\infty}^{+\infty} h\left(\sqrt{y^2+r^2}\right)\, dy = \varepsilon \int_{-\infty}^{+\infty} 2\frac{J_1\left(3.83\frac{\sqrt{y^2+r^2}}{\Delta_R}\right)}{3.83\frac{\sqrt{y^2+r^2}}{\Delta_R}} dy.$$

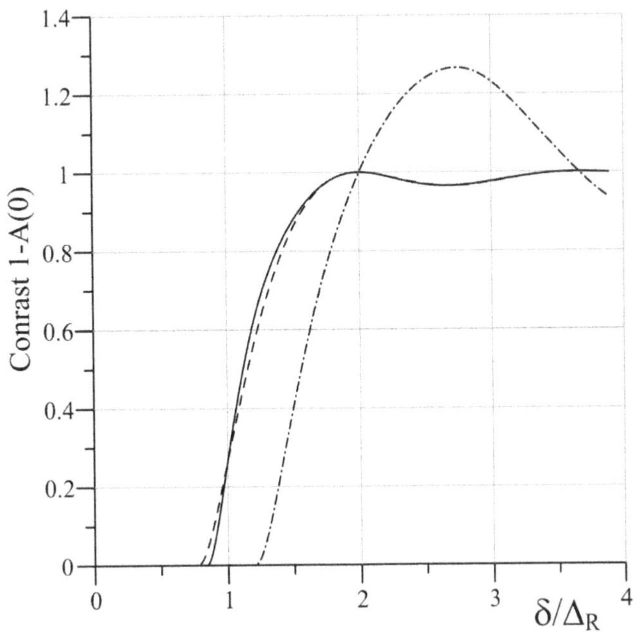

Fig. 9.25. Contrast as a function of separation. Solid line — heterodyne amplitude channel; dashed line — ideal incoherent system; dashed-solid line — ideal coherent system.

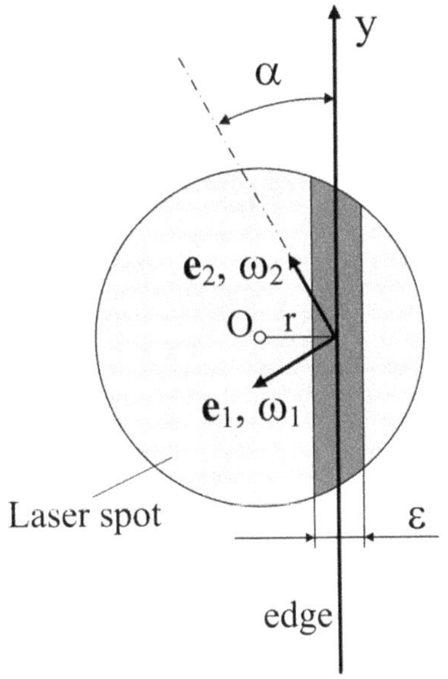

Fig. 9.26. Geometrical derivation of the ESF.

The effect itself, i.e. the phase shift per unit edge length and unit field amplitude, may be supposedly characterized by the constants $\phi_{||}$ and $\phi_{\perp}$. Thus,

$$\phi_{||,\perp}(r) = \phi_{||,\perp} \cdot p(r).$$

Depending on the polarization direction, the wave field may be either parallel or perpendicular to the edge, so that it is necessary to introduce trivial projection coefficients $|\cos\alpha|$ and $|\sin\alpha|$ to account for the different magnitude of the effect in the intermediate configuration. These considerations lead to the following qualitative formulae for the phase shifts in the reflected waves with frequencies ω_1 and ω_2:

$$\omega_1 : \quad \phi_{||}(r) \cdot |\sin\alpha| + \phi_{\perp}(r) \cdot |\cos\alpha|,$$
$$\omega_2 : \quad \phi_{||}(r) \cdot |\cos\alpha| + \phi_{\perp}(r) \cdot |\sin\alpha|.$$

Hence, the phase shift between the waves with frequencies ω_1 and ω_2 may be written in the form

$$\Delta\phi(r,\alpha) = p(r) \cdot (\phi_{||} - \phi_{\perp}) \cdot (|\sin\alpha| - |\cos\alpha|).$$

This formula expresses the ESF of the phase channel and is the basic formula of the qualitative theory.

Radial dependence of the ESF is of a primary importance. It is determined by $p(r)$. Being normalized as $p(0) = 1$, this function takes the form

$$p(r) = \int_0^\infty \frac{J_1\left(\sqrt{a^2r^2 + z^2}\right)}{\sqrt{a^2r^2 + z^2}} dz, \quad a = \frac{3.83}{\Delta_R}$$

because

$$\int_0^\infty \frac{J_1(z)}{z} dz = 1.$$

Figure 9.27 shows this function.

For functional comparison only, the dashed line in this figure shows the PSF of an ideal coherent imaging system $h(r)$. It is seen that the first zero of $p(r)$ is positioned closer to the origin than that of $h(r)$. Although, strictly speaking, this fact does not signify any advantage in spatial resolution since the ESF and PSF describe completely different physical situations, we still may hope that spatial resolution in the phase-contrast mode will not be inferior to that of an ideal coherent imaging system. Experimental results presented below show that, in fact, it is even better than that of an ideal incoherent system.

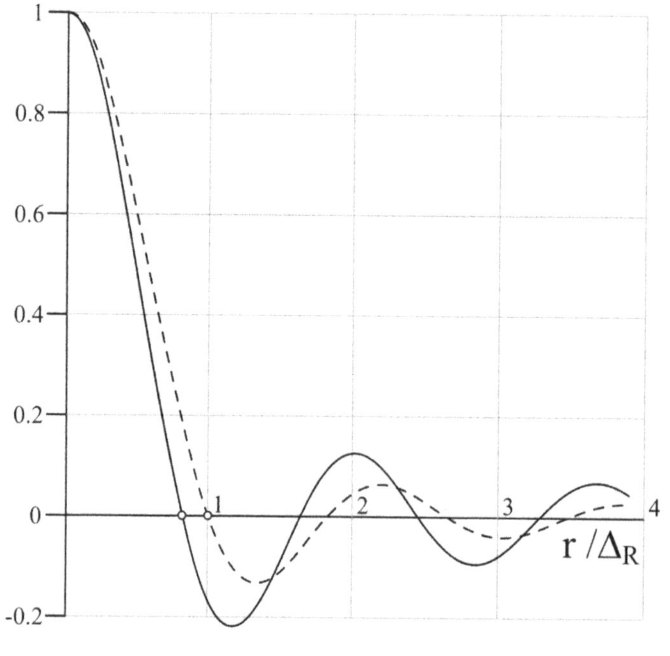

Fig. 9.27. ESF curve (solid line). Dashed line shows PSF of the ideal coherent imaging system.

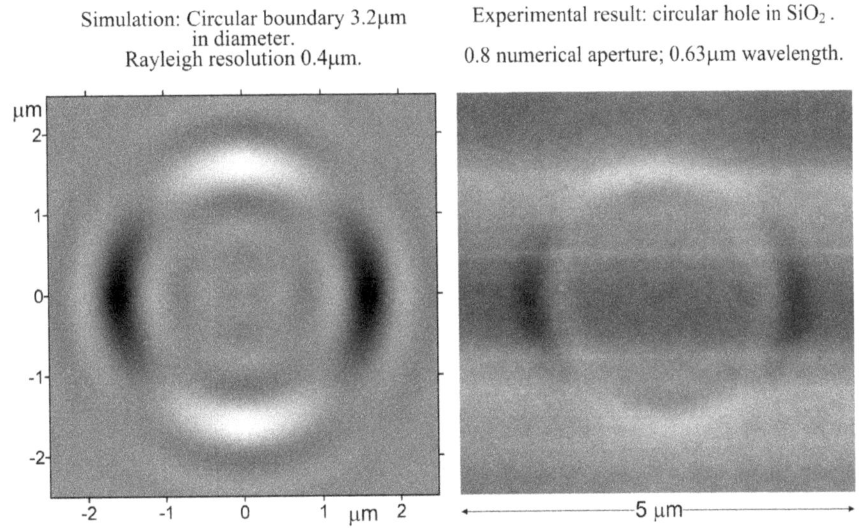

Fig. 9.28. Simulated phase-contrast image of the circular hole 3.2 μm in diameter with the Rayleigh resolution parameter $\Delta_R = 0.4$ μm (left). Experimental phase-contrast image of the hole in SiO_2 layer 20 nm thick on silicon wafer (right). Polarizations of the two laser waves are directed vertically and horizontally.

Since the qualitative form $\Delta\phi(r, \alpha)$ of the ESF in the phase channel have been derived, it is logical to apply it for assessment of spatial resolution as was done for the amplitude channel (Fig. 9.24). For that, it would be necessary to simulate scanning over two closely positioned narrow edge walls as it was done for the two closely positioned point-like sources for the amplitude channel in Fig. 9.24. But alas: we cannot do that because we do not know how the phase shift changes when the width of the wall becomes narrower than ε (see Fig. 9.26). Therefore, with the current limitation of theory, assessment of spatial resolution in the phase channel can only be performed experimentally.

However, we can simulate images of simple objects such as circular pillar or hole with a diameter much bigger than that of the focal spot, and compare them with the experimental images. In this simulation, the image signal is determined by $\Delta\phi(r, \alpha)$, while the touch-point between the laser spot and the pillar's circumference determines the angle α between the tangent line and directions of polarization in formula for $\Delta\phi(r, \alpha)$. The result is shown in Fig. 9.28 together with the experimental image of the circular hole in SiO_2 layer 20nm thick on silicon wafer. Qualitative agreement is clear.

9.3.4. *Experimental results*

Figures 9.29–9.33 present experimental results for simple samples. For each sample, two images may be obtained simultaneously, using amplitude and phase channels. In order to facilitate comparison, these types of images are grouped into pairs, and the pairs are arranged in two sections, corresponding to non-uniformly and uniformly reflecting samples. There are no scale bars in the figures but all the images have the same width of 80 μm, so it is straightforward to calculate the size of features. Simultaneous imaging in both the traditional amplitude and phase-contrast modes may be considered as a useful feature for inspection applications. From the point of view of visual perception, phase-contrast images provide better rendering and an illusion of three-dimensional vision.

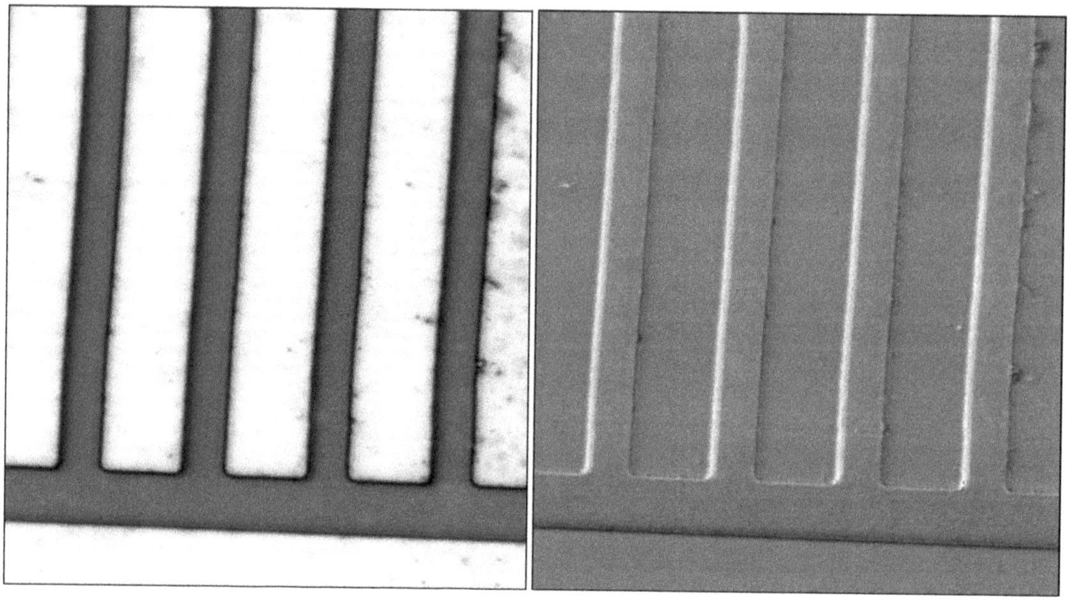

Fig. 9.29. Non-uniformly reflecting sample: optical encoder disk with 20 μm period. Left picture: amplitude image; right picture: phase-contrast image. Numerical aperture of the objective 0.6. Image width 80 μm.

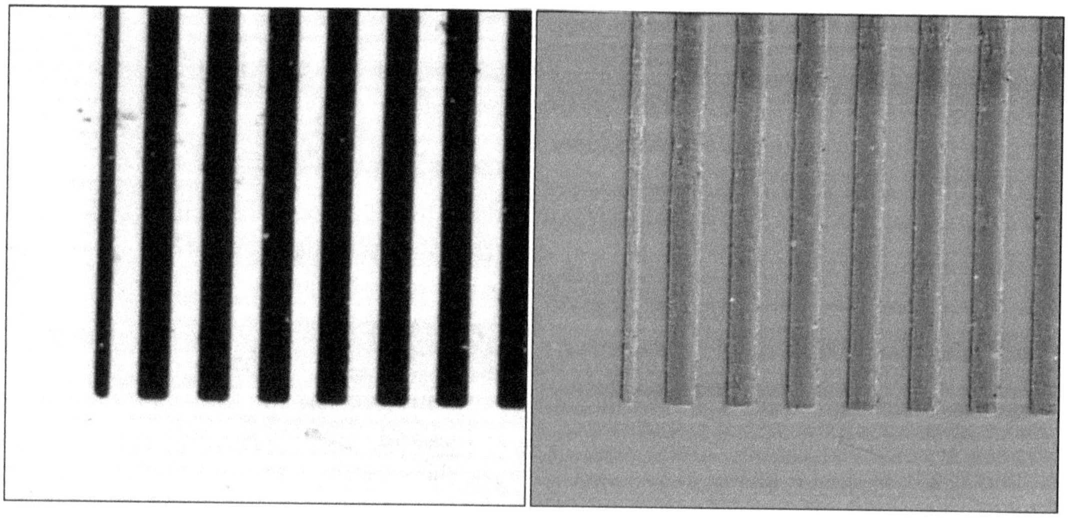

Fig. 9.30. Non-uniformly reflecting sample: chromium grating with 9 μm period on bare glass. Left picture: amplitude image; right picture: phase-contrast image. Numerical aperture 0.8. Image width 80 μm.

9.3.5. *Super-resolution*

Both the Rayleigh criterion and Abbe formula give approximately the same estimate for the spatial resolution of an ideal incoherent imaging system:

$$\Delta r = (0.5 - 0.6) \cdot \frac{\lambda}{NA},$$

where λ is the wavelength and NA is the numerical aperture of the objective lens. The Rayleigh criterion states that the contrast in the traditional amplitude image of the two point-like sources separated by

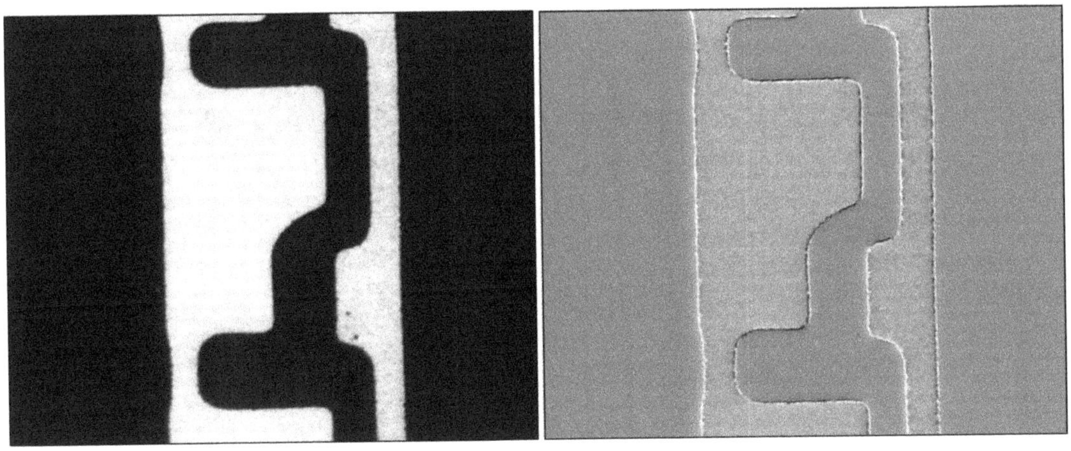

Fig. 9.31. Non-uniformly reflecting sample: liquid crystal display pattern of a mobile phone. Left picture: amplitude image; right picture: phase-contrast image. Numerical aperture 0.8. Image width 80 μm.

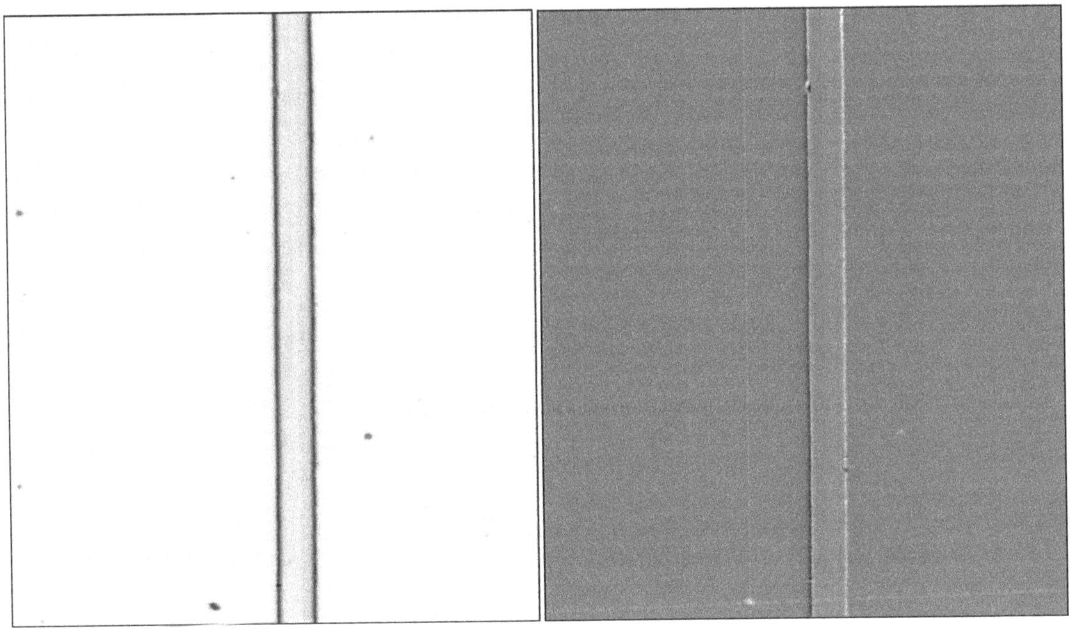

Fig. 9.32. Uniformly reflecting sample: single trench in silicon. Reflectivity 40%. Left picture: amplitude image; right picture: phase-contrast image. Numerical aperture 0.8. Image width 80 μm.

a distance Δr is about 0.2 (see Fig. 9.24). For linear structures, such as gratings, the contrast would be somewhat different from this value but this is not the point of our discussion. What matters is whether or not the contrast in the phase channel exceeds that in the amplitude channel. There are good theoretical reasons for the contrast of features of wavelength dimension to be higher with phase-contrast imaging than with traditional incoherent amplitude imaging. This phenomenon is commonly referred to as super-resolution. For the He–Ne laser with a wavelength 0.63 μm and an objective lens with $NA = 0.8$ the Rayleigh–Abbe formula gives the value $\Delta r \approx 0.4$ μm. In order to verify super-resolution capabilities of the phase channel, a 0.5-μm period trench grating was manufactured in silicon (the NTT-AT company product AS100P-D). The result is presented in Fig. 9.34.

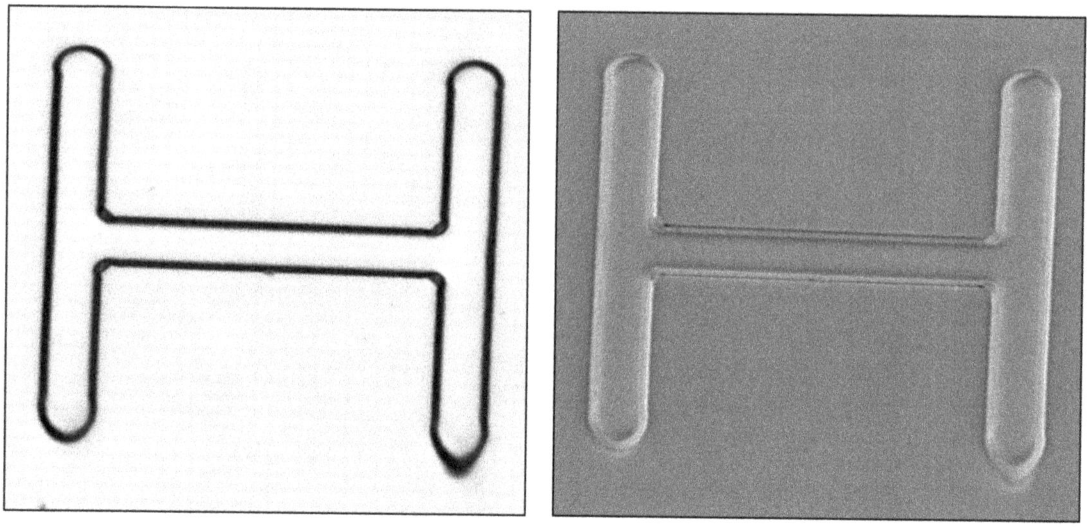

Fig. 9.33. Uniformly reflecting sample: letter "H" imprinted into highly transparent polymer on bare glass. Reflectivity 5%. Left picture: amplitude image; right picture: phase-contrast image. Numerical aperture 0.8. Image width 80 μm.

In the amplitude channel, the contrast defined as $\frac{a_{\max}-a_{\min}}{a_{\max}+a_{\min}}$ is equal to less than 5%, while in the phase channel it is equal to 23%.

Figures 9.35 and 9.36 clearly show the effect of phase-contrast super-resolution on the single-trench structures. In the amplitude mode, narrow rectangular trenches with two walls look exactly like single step-height borders that form the frame of the image in Fig. 9.33, whereas the phase-contrast mode explicitly shows the trench with two narrowly separated walls.

9.3.6. *Dark-field and bright-field modes of operation*

Consider Fig. 9.37, which shows orientation of polarization vectors in the microscope.

The key component of the microscope is the dual-frequency cross-polarized laser, producing two orthogonal linearly polarized waves $E_1 = a_1 e^{i\omega_1 t}$ and $E_2 = a_2 e^{i\omega_2 t}$ with the frequency difference $(\omega_1 - \omega_2)/2\pi = 20$ MHz. Polarization vectors of the both waves are mutually orthogonal to each other, so $\mathbf{e}_1 \cdot \mathbf{e}_2 = 0$, and therefore, when entering the photo-detector, these waves produce the output current proportional to the sum of intensities:

$$|\mathbf{e}_1 a_1 e^{i\omega_1 t} + \mathbf{e}_2 a_2 e^{i\omega_2 t}|^2 = |a_1|^2 + |a_2|^2.$$

Such waves are called decoupled because they do not produce interference and no phase information is available at the output of the photodetector. To couple these waves, and hence to obtain the intermediate frequency signal, containing phase information, one needs a polarizer installed at some angle θ between $0°$ and $90°$. The amplitude of this signal is proportional to the product of geometrical projections of vectors $\mathbf{e}_1$ and $\mathbf{e}_2$ onto the polarizer axis (shown in dashed lines in Fig. 9.37). The signal reaches its maximum when $\theta = 45°$, which corresponds to the bright-field mode of operation. In this mode, both amplitude and phase channels provide valuable information.

The signal is zero when $\theta = 0°$ or $90°$. However, if a depolarization occurs after reflection, and a third wave with the amplitude v and polarization vector along $\mathbf{e}_1$ or $\mathbf{e}_2$ appears, the situation changes, and an oscillating component appears in the photo-current even if $\theta = 0°$ or $90°$:

$$|\mathbf{e}_1 a_1 e^{i\omega_1 t} + \mathbf{e}_2 v e^{i\omega_1 t + i\phi} + \mathbf{e}_2 a_2 e^{i\omega_2 t}|^2 = |a_1|^2 + |a_2|^2 + 2a_2 v \cos[(\omega_1 - \omega_2)t + \phi].$$

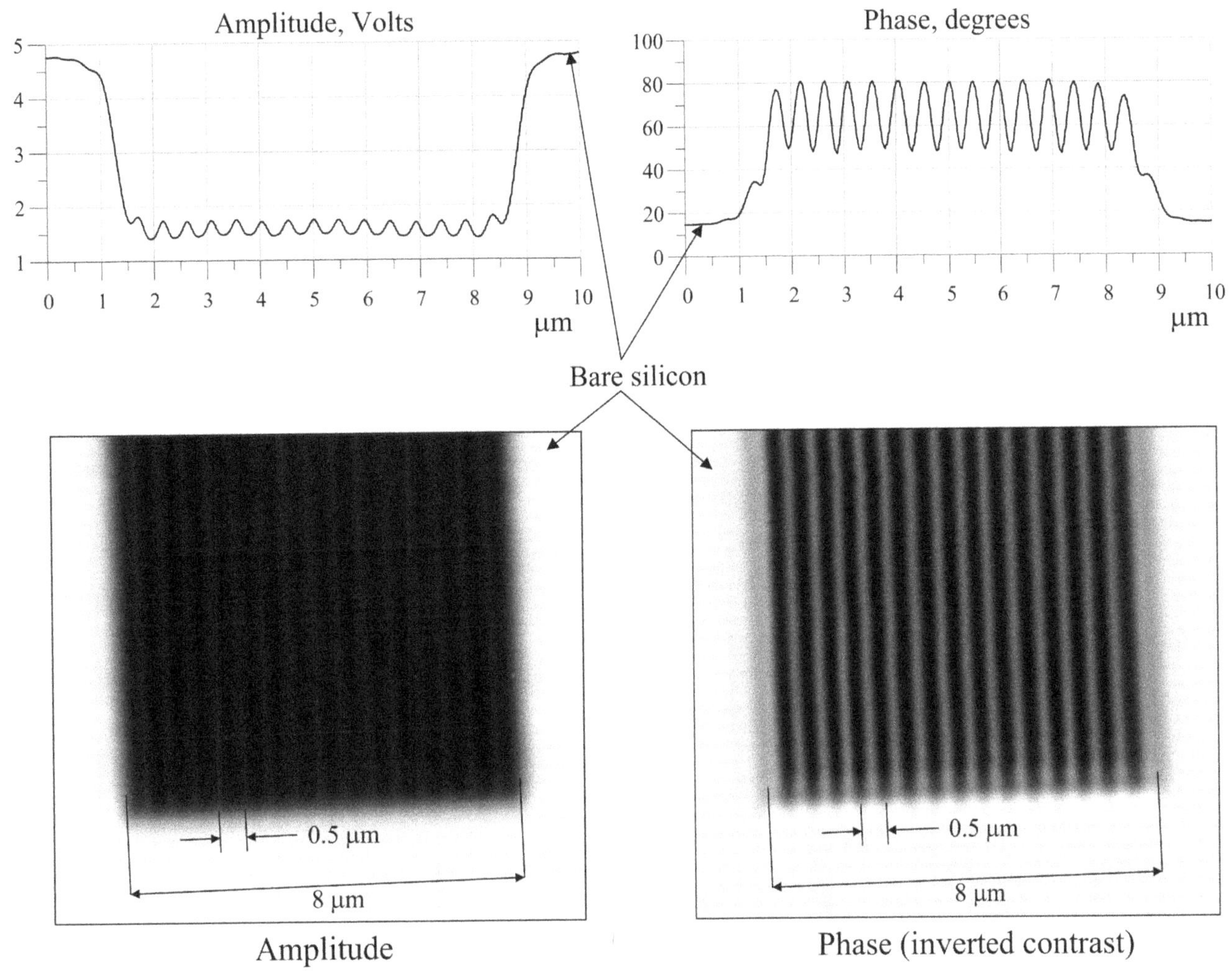

Fig. 9.34. Periodical rectangular trench grating in silicon. For the sake of comparison, the phase-contrast image is presented in inverted contrast.

Thus, if the polarizer is originally set to zero or ninety-degree orientation, reflection from clean non-patterned surface produces a uniformly black image. However, if some dust or imperfections exist on the surface, they depolarize the reflected wave, and a bright spot will be observable at this point of the image. This is the idea of the dark-field mode of the microscope. In this mode, the phase channel does not provide any valuable information because depolarization does not have any significant influence on the phase of the signal when the polarizer is directed along $\mathbf{e}_1$ or $\mathbf{e}_2$.

What the dark-field mode can do for inspection of real semiconductor devices is shown in Fig. 9.38 that presents pictures of a fragment of an integrated circuit in the amplitude, phase-contrast, and dark-field modes. Each mode of operation delivers its own useful information for inspection. For example, the phase-contrast image clearly shows two parallel bars that are not visible in other pictures (Fig. 9.38(b), the area marked with white dashed circle). The phase-contrast and the dark-field $\theta = 90°$ images are the only ones that show 45° interconnects (the area marked with white dashed rectangles in Fig. 9.38(b) and Fig. 9.38(d)). The dark-field $\theta = 0°$ image better shows point-like defects as bright spots (the area marked with white dashed circle in Fig. 9.38(c)). It also seems that the dark-field mode is more sensitive

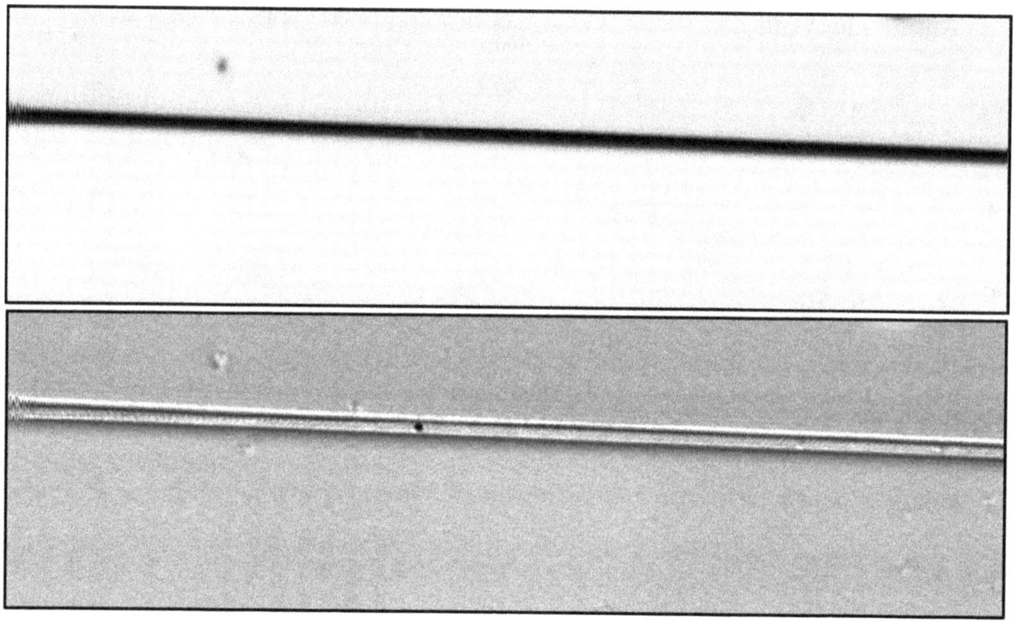

Fig. 9.35. Super-resolution: single horizontal trench (along scanning) 1 μm wide in highly transparent polymer on bare glass. Reflectivity 5%. Upper picture: amplitude image; bottom picture: phase-contrast image. Numerical aperture 0.8. Image width 80 μm.

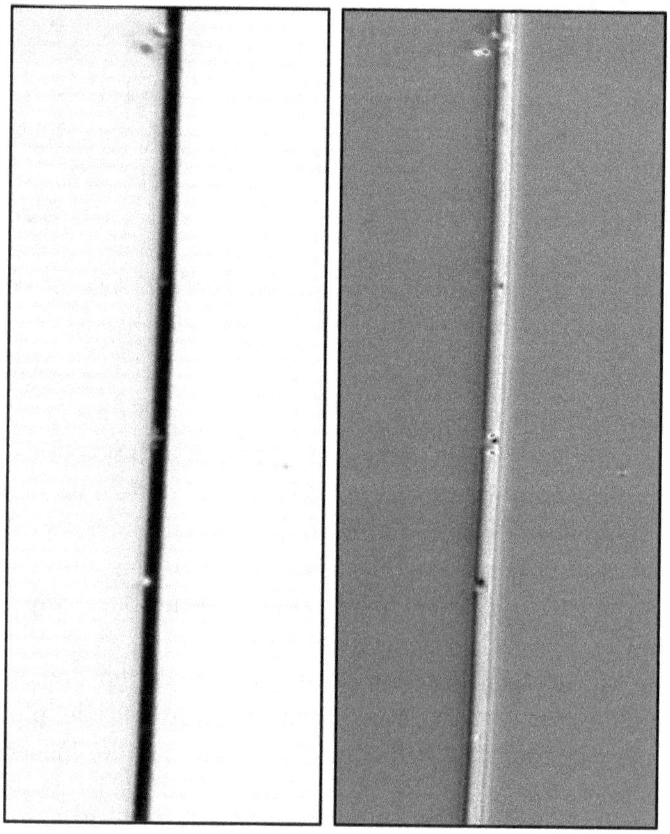

Fig. 9.36. Super-resolution: single vertical trench (across scanning) 1 μm wide in highly transparent polymer on bare glass. Reflectivity 5%. Left picture: amplitude image; right picture: phase-contrast image. Numerical aperture 0.8.

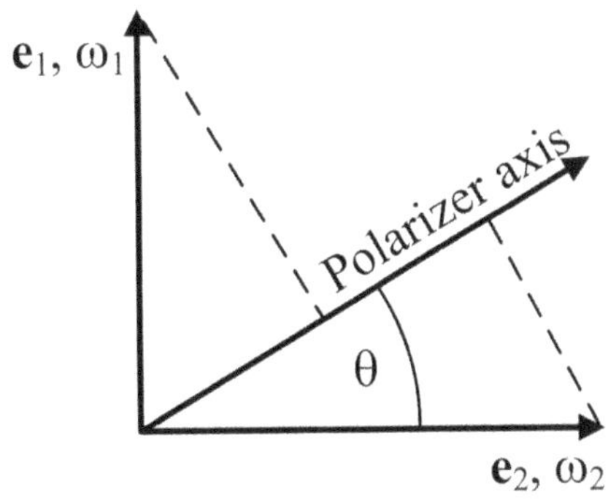

Fig. 9.37. Polarizer couples the two orthogonal polarizations to produce useful signal at the output of the photodetector.

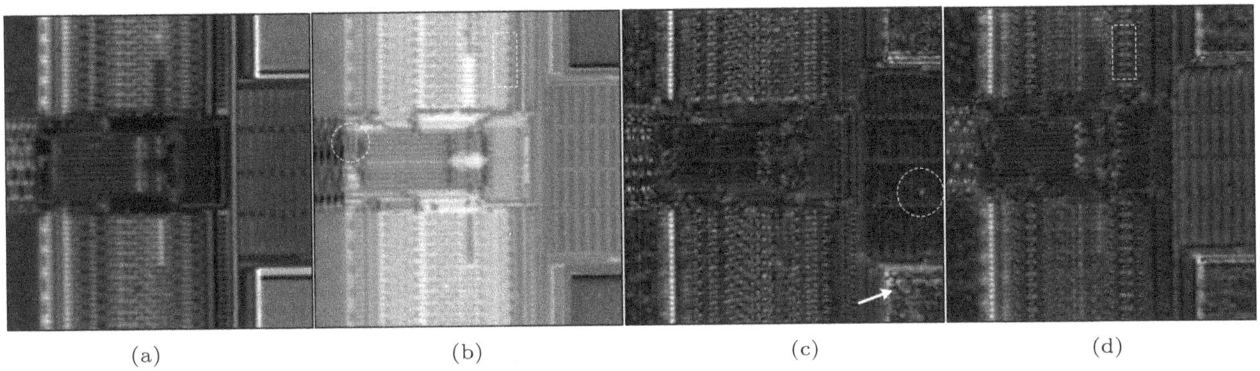

Fig. 9.38. Fragment of an integrated circuit. (a) Amplitude image; (b) phase-contrast image; (c) dark-field image $\theta = 0°$; (d) dark-field image $\theta = 90°$. Image width 30 μm.

to what may be considered as smooth surface variations, to which the amplitude and phase-contrast modes are practically insensitive (white arrow in Fig. 9.38(c)).

9.3.7. *Non-patterned anisotropic surfaces*

Some non-patterned surfaces may still be anisotropic due to inner stress or molecular structuring. The heterodyne microscope can effectively recognize such surfaces by inverse phase contrast. Consider Fig.9.39, explaining mutual orientation of the partial laser waves with polarization vectors $\mathbf{e}_1$, $\mathbf{e}_2$ and the ellipsoid of refractive indices with principal axes directed along n_1, n_2.

The simplest case of $\mathbf{e}_1$, $\mathbf{e}_2$ coinciding with the principal axes n_1, n_2 is sufficient to understand the idea. After reflecting from the sample, the partial waves E_1 and E_2 with initial phases ϕ_1, ϕ_2 and wavenumbers $k_1 \approx k_2 = k$ acquire additional phases proportional to the thickness of the layer h:

$$E_1 = \exp(i\omega_1 t + i\phi_1 + ik2hn_1); \quad E_2 = \exp(i\omega_2 t + i\phi_2 + ik2hn_2).$$

Then the photocurrent $j = E_1E_2^* + E_2E_1^*$ acquires additional phase $\delta = k2h(n_1 - n_2)$, so that the total phase is equal to

$$\phi = \phi_0 + \delta,$$

where $\phi_0 = \phi_1 - \phi_2$. If the sample is rotated by 90° then the total phase will be

$$\phi = \phi_0 - \delta,$$

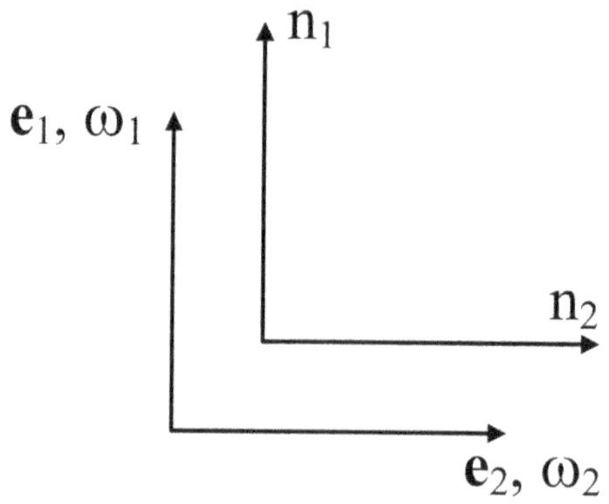

Fig. 9.39. In anisotropic medium, the refractive indices n_1, n_2 are different for the two orthogonal polarizations $\mathbf{e}_1$, $\mathbf{e}_2$.

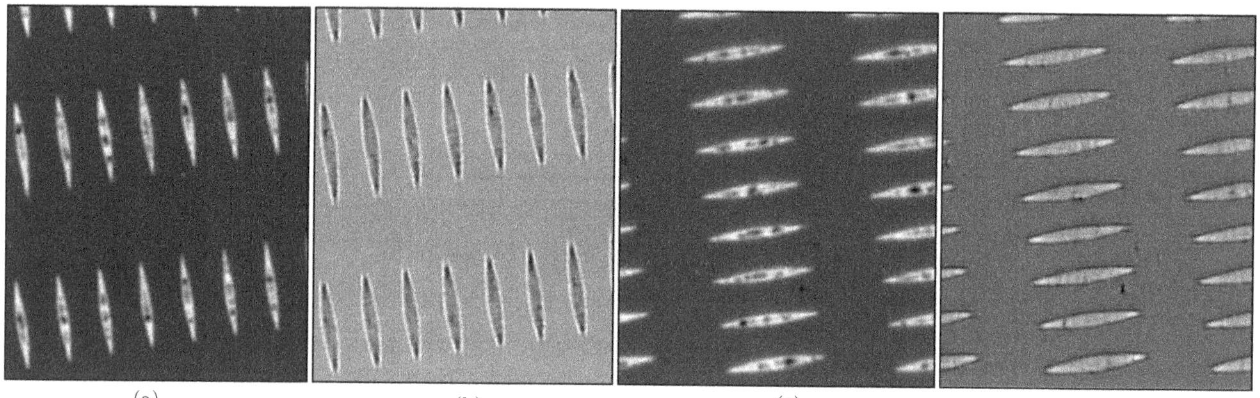

Fig. 9.40. Anisotropic surface. (a) and (c): Amplitude images; (b) and (d): Phase-contrast images. Image width 30 μm.

which means the inverse phase contrast. Note that the amplitude contrast, i.e. the amplitude of the photocurrent, does not change.

Figure 9.40 presents a series of two pairs of images of the calibration sample for scanning near-field optical microscopes (NT-MDT, SNG-01) rotated by 90° with respect to each other. This sample is composed of elongated rhomboidal vanadium islands, 30 nm thick, on a quartz substrate. The short period of this structure is 4.5 μm. Whereas the amplitude images are the same, the phase-contrast images show inversed contrast within the vanadium segments for 90° rotated orientations: dark areas of rhombuses in Fig. 9.40(b) change to white ones in Fig. 9.40(d). This clearly indicates the presence of anisotropy. To make such a comparison, it is not even necessary to rotate the entire sample, as was the case for Fig. 9.40: the half-wave plate in Fig. 9.20 makes this switching. It can be quickly done even during image acquisition, as shown in Fig. 9.41.

9.3.8. *Three-dimensional profiling of opaque and transparent samples*

The cross-polarized laser heterodyning makes it possible to measure profiles of surfaces very accurately. Particularly in the dual-channel heterodyne microscope, it can be done by inserting a Rochon quartz prism (4 mm thick) to split the cross-polarized partial laser waves by 4 μm in the focal plane (Fig. 9.20). This option was added for inspection of phase-shifting lithography masks in semiconductor industry,

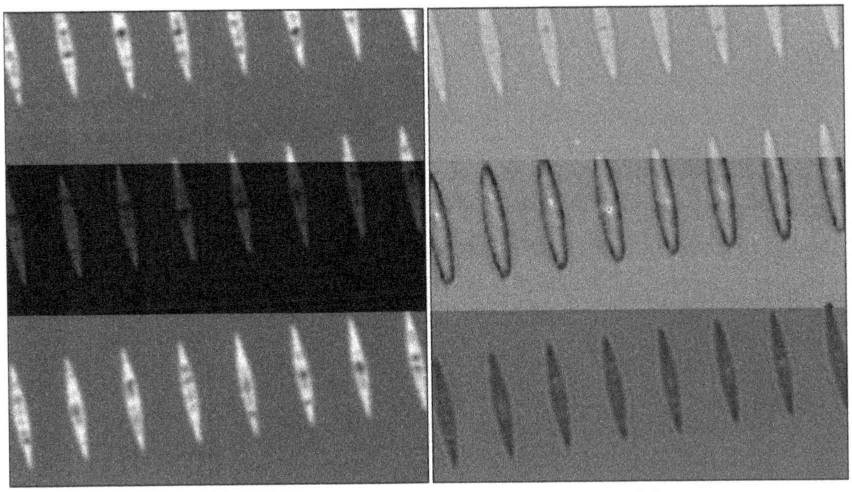

(a) (b)

Fig. 9.41. During image acquisition, the polarization was switched from 45° at the first part of the image, to 0° in the middle, and to −45° in the end of the measurement. The middle part of the image corresponds to the dark-field mode.

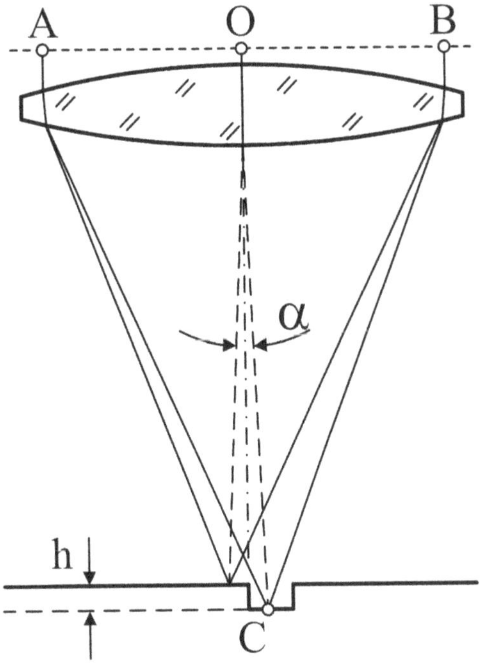

Fig. 9.42. Optical path length is the same for all the rays, focusing in C.

where step-height relief variations are of a primary importance. By placing one focal spot onto the mask surface and the other one into the phase-shifting trench, it is possible to measure the phase shift and the trench depth at the laser wavelength, and then to recalculate the phase shift to the working wavelength. This is explained in Fig. 9.42.

According to the basic principle of a lens, optical paths for all the rays AC, OC, and BC are the same. Therefore, the phases of all rays focusing to the point C are the same, and the phase difference between the two orthogonally polarized laser waves is merely $\phi_0 + 2hk$, where h is the trench depth, k

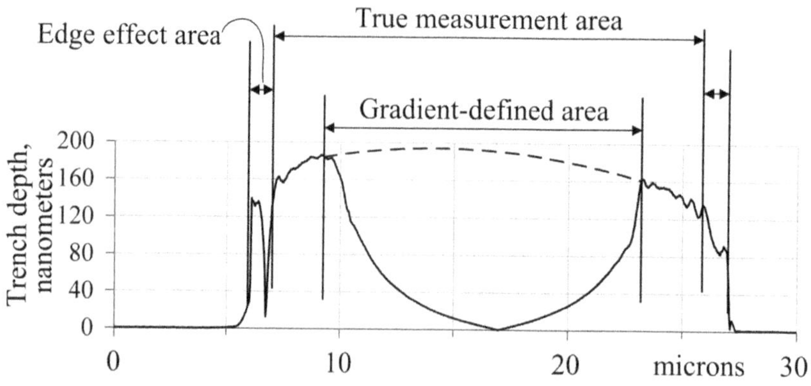

Fig. 9.43. The depth scan over a trench in silicon. At the left side of the trench, the depth is more than 160 nm, whereas at its right side it is less than 160 nm.

is the wavenumber, and ϕ_0 is unimportant constant phase shift, cumulatively introduced by the laser, optics, and electronics.

In the scanning microscope, it is possible not only to measure a single value of phase at one point of the surface, but to obtain the entire line scan of the surface profile. For example, Fig. 9.43 shows the cross-scan of the trench in silicon.

While the scan progresses from left to right, there are three clearly defined areas that appear in the graph. As the right focal spot approaches the left wall of the trench, the outer parts of this focal spot begin to sense the wall. This is a rather narrow area of the edge effect, where no definite conclusion can be reached about the trench depth, and the phase rapidly varies, according to the unknown spatial phase distribution within the focal spot. When the right focal spot entirely resides in the trench, the true measurement area begins. In it, the depth is measured as

$$h = \frac{\phi}{2k},$$

where ϕ is the phase of the signal. The trench is wider than the focal separation, therefore the left (second) focal spot soon resides inside the trench together with the first one. At this point, the phase of the signal is determined by the gradient of the trench bottom. The gradient-defined area spreads until the right (first) focal spot leaves the trench. Here the true measurement area continues, but the phase is of the opposite sign because the two focal spots swapped their positions.

Not only totally reflecting, but also transparent layers can be measured in this way. In order to estimate the height of the layer, its refractive index should be known. Consider this in more detail with reference to Fig. 9.44.

The wave, after passage through the transparent layer, interferes with the wave reflected from the front surface of this layer to produce

$$E_1 = r_1 e^{i\omega t} + t_1 r_2 t_1 e^{i(\omega t + 2hkn_1)},$$

where

$$r_1 = \frac{n_1 - 1}{n_1 + 1}, \quad r_2 = \frac{n_2 - n_1}{n_2 + n_1}, \quad \text{and} \quad t_1 = \frac{2}{n_1 + 1}$$

are the Fresnel reflection and transmission coefficients, and $n_{1,2}$ are the complex refractive indices of the medium (at normal incidence). The wave reflected from the second focal spot is

$$E_2 = r_3 e^{i(\omega t + 2hk)}$$

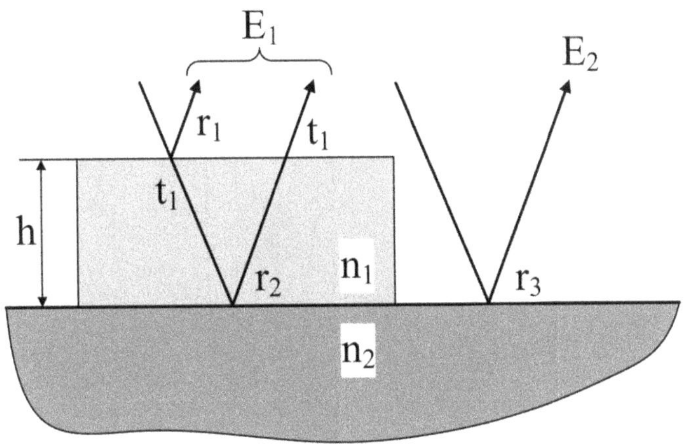

Fig. 9.44. Reflected wave E_1 is composed of two interfering waves.

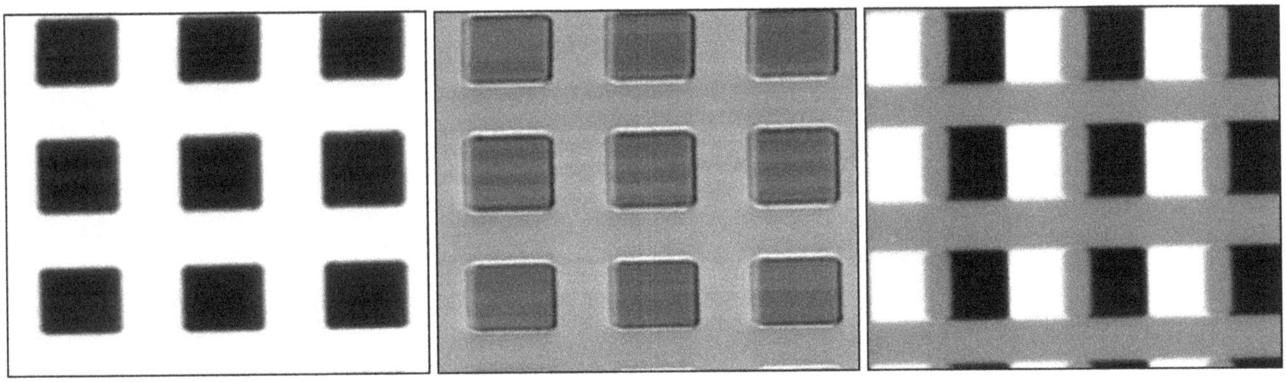

Fig. 9.45. Images of 100 nm thick SiO_2 vertical square pillars on silicon with 10 μm periodicity (Budget Sensors, HS-100MG). Image width 30 μm. (a) Amplitude image, no Rochon prism; (b) phase-contrast image, no Rochon prism; (c) phase channel image, Rochon prism is inserted and splits the focal spot by 4 μm.

with

$$r_3 = \frac{n_2 - 1}{n_2 + 1}.$$

Using computer simulation, it is possible to calculate the phase difference between the two complex variables E_1 and E_2 for any known $n_{1,2}$, and to find the best fit h for the experimental data. This is the idea, and the results below show how it works in practice.

When the Rochon prism is inserted into the beam (Fig. 9.20), imaging does not make great sense because there are two spatially separated focal spots that blur the image. Nonetheless, if a 2D scan is made in this mode, then the result will be two overlapping pictures, each obtained by one of the two focal spots. This is illustrated by Fig. 9.45.

In Fig. 9.45(a), the pillars look black due to interference in transparent SiO_2 layer. For the same reason, the pillars look dark in the phase picture (Fig. 9.45(b)). For 3D profiling, a line scan of the picture (c) provides all the necessary information about height (Fig. 9.46).

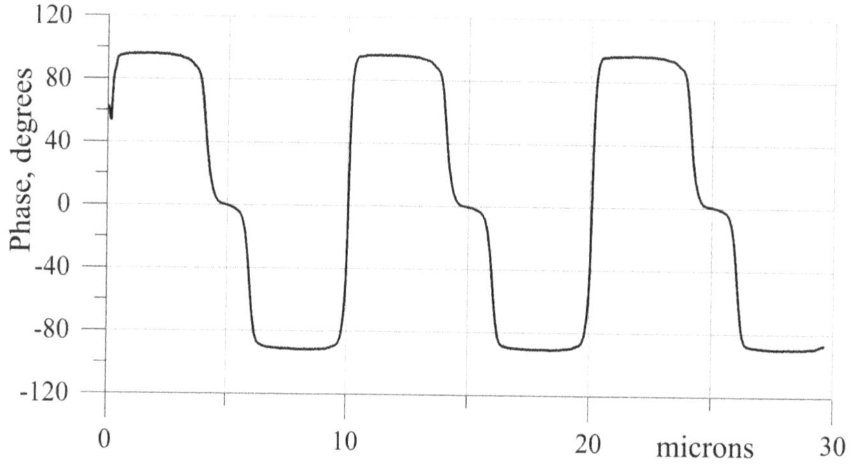

Fig. 9.46. One line scan of the image in Fig. 9.45(c).

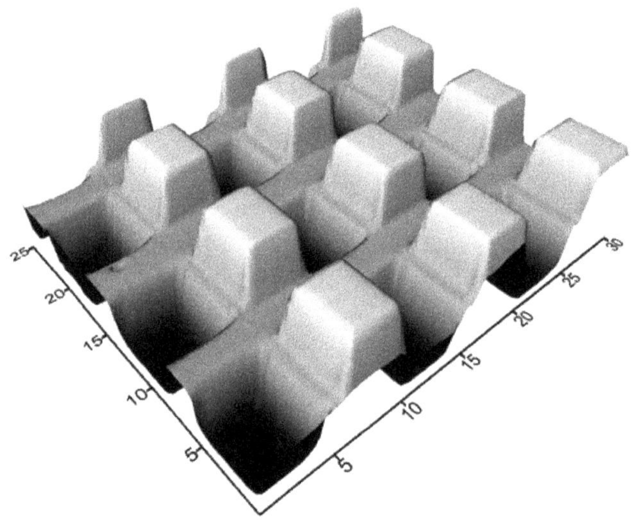

Fig. 9.47. Three-dimensional representation of the image in Fig. 9.45(c).

The positive and negative swings of the phase must be exactly the same as the pair of focal spots sweeps the pillar from left to right. This makes it easy to eliminate any possible uncertainty in defining the zero phase level ϕ_0 by computing the amplitude of phase variation as

$$\Delta\phi = \frac{\phi_{\max} - \phi_{\min}}{2}.$$

For the scan in Fig. 9.46, this gives the value $\Delta\phi = 93.8^\circ \pm 0.1^\circ$. With it, the above three-wave simulation, with the values of the refractive indices of SiO_2 and Si being respectively 1.542 and $3.88 + i\ 0.02$, gives the height of the structure 106.94 ± 0.005 nm. The certainty of this fitting is astonishing, and the only factor of possible errors may be incorrect values of refractive indices. On the other hand, the manufacturer calibrated the particular sample as 116 nm by means of atomic-force microscopy. The most probable reason for this discrepancy is an inaccurate value of the refractive index of silicon. It is also possible that the SiO_2 layer is slightly birefringent (anisotropic), which would account for the effects described in the Section 9.3.7.

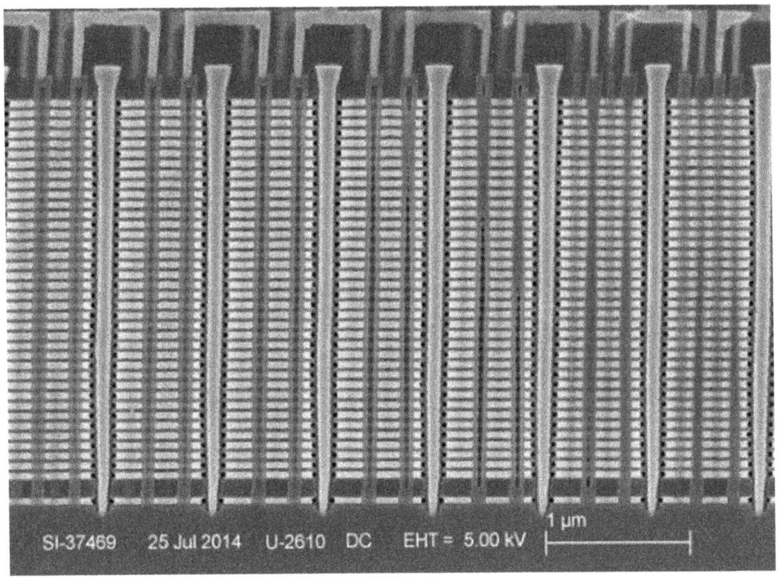

Fig. 9.48. Typical structure of the V-NAND memory.

Since the phase channel picture in Fig. 9.45(c) relates to the height of the pillar, it is almost irresistible to draw a surrogate of a "3D picture" of the structure, using the phase values as the third coordinate. The result is shown in Fig. 9.47.

9.4. Optical Arbitrary Waveform Generator

9.4.1. *Problem statement*

In recent years, the progress in electronic memory devices was associated mainly with the development of the so-called V-NAND technology — semiconductor devices, in which dimensional limitations of functional elements, imposed by diffraction, were overcome by means of their vertical stacking in a form of a multilayer structure. An example of such a vertical structure is shown in Fig. 9.48.

This technology consists of about 500 successive operations, starting with deposition of a multilayer structure, composed of a 32 or 64 alternating SiO_2–SiN layers. At the next stages, a series of extremely narrow (40 nm) and relatively deep (up to 5 μm) holes (channels, seen as pins in Fig. 9.48) is formed by means of plasma-chemical etching. These channels, being afterwards filled with tungsten, play the role of electrodes. The critical operation in the entire V-NAND production is manufacturing of these channels. The process of etching is monitored by recording optical emission of plasma at characteristic lines, associated with the etch bi-products (Chapter 7). The etching process must be stopped very accurately in time in order to prevent over-etching. As the etching propagates through alternating layers of SiO_2 and SiN, intensities of the chosen spectral lines oscillate, and the number of oscillations gives the precise depth of channels. However, the amplitudes of oscillations decrease in time, as the coherence of etching at various points on the 300 mm silicon wafer decrease. This is clearly seen in the typical spectroscopic trend recorded at the wavelength 387 nm (CN line), which is portrayed in Fig. 9.49.

This phenomenon creates significant difficulties to counting the number of oscillations, necessitating development of special highly sensitive spectroscopic techniques and algorithms. In turn, such a development requires simulating tools that generate arbitrary optical trends as in Fig. 9.49, without expensive experiments on real plasma machines. However, optical modulation techniques cannot provide linear optical response to driving electrical signal. As such, the optical output is always different

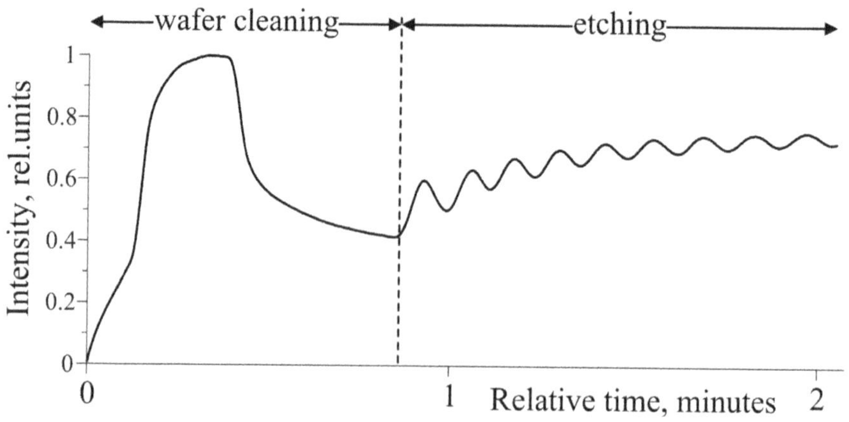

Fig. 9.49. Real optical etching trend, recorded at the wavelength 387 nm.

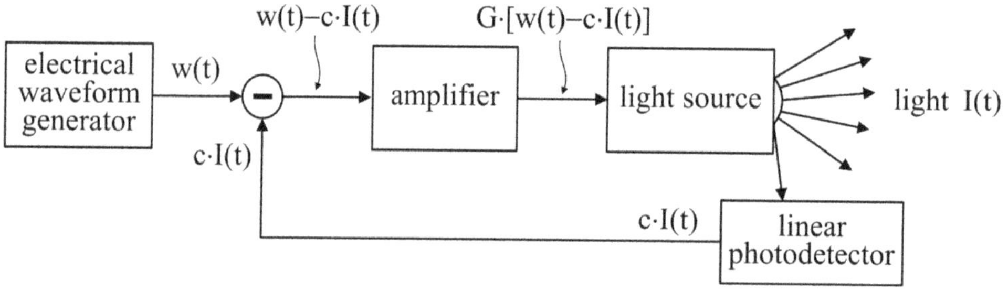

Fig. 9.50. Generalized layout of an OAWG. The photodetector is coupled to electrically controlled light source and provides negative feedback for it.

from the electrical input, and fitting one to another requires tedious adjustment, which works only for a particular shape of the signal and cannot be automatically applied to other shapes. The concept of optical arbitrary waveform generator (OAWG) presented in the current section solves this problem.

9.4.2. *The concept*

The generalized concept of OAWG is presented in Fig. 9.50.

It is necessary to begin with mathematical explanation why the waveform of light $I(t)$ is a good approximation of the given waveform $w(t)$, whatever the nonlinearity of the light source is. After that, the quantitative criterion will be derived. In the end of this section, the importance of linearity of a photodetector response will be explained.

The basic assumption and the requirement for the high quality of approximation is linearity of a photodetector: its output electrical signal $v(t)$ must be proportional to $I(t)$:

$$v(t) = c \cdot I(t),$$

where t is time. As such, the driving signal that is applied to the light source is

$$s(t) = G \cdot [w(t) - cI(t)],$$

where G is gain of the amplifier.

Whatever the nonlinearity of the light source is, its intensity may be expanded in a Taylor series:

$$I(t) = a_1 s + a_2 s^2 + a_3 s^3 + \cdots = \sum_{n=1}^{\infty} a_n s^n.$$

Practically, there is always a dominant coefficient in this expansion: for fairly linear sources it is a_1; for parabolic source it is a_2; for cubic source it is a_3; etc. It means that if the dominant coefficient is a_m then

$$|a_n| < |a_m| \quad \text{for all}\, n \neq m.$$

Since the system is in equilibrium, we have an equation:

$$I(t) = \sum_{n=1}^{\infty} a_n G^n \cdot [w(t) - cI(t)]^n.$$

In practice, there is always a limit to the output intensity:

$$I(t) < I_{\text{sat}},$$

which is commonly understood as the saturation value of the output optical flux.

We are now interested in what happens when G tends to infinity: $G \to \infty$. Since the last inequality always holds true, we can write down clear mathematical relation that holds true for each and everyone term in the expansion series of $I(t)$:

$$|a_n G^n \cdot [w(t) - c\,I(t)]^n| < I_{\text{sat}}, \quad n = 1, 2, 3, \ldots \quad ,$$

or

$$|w(t) - cI(t)| < \sqrt[n]{\frac{I_{\text{sat}}}{|a_n|}} \cdot \frac{1}{G}, \quad n = 1, 2, 3, \ldots \quad .$$

When $G \to \infty$

$$\lim_{G \to \infty} |w(t) - cI(t)| = 0,$$

and we have

$$w(t) = cI(t)$$

or

$$I(t) = \frac{w(t)}{c}.$$

This relation means that, with $G \to \infty$, OAWG according to Fig. 9.50 generates optical waveform $I(t)$ exactly the same as the desired waveform $w(t)$, independently of nonlinearity of the light source.

An important feature of the OAWG is that intensity of the output light $I(t)$ depends on the conversion gain c of the photodetector: the smaller the gain, the bigger the intensity. Therefore, there are two options for varying intensity of light in OAWG:

- by varying amplitude of the driving waveform $w(t)$;
- by varying conversion gain c of the photodetector.

Next, we are going to figure out what happens in reality when $G < \infty$. Recalling that there is always a dominant nonlinearity of the order m, inequality for the error $|w(t) - cI(t)|$ can be limited from above:

$$|w(t) - cI(t)| < \sqrt[m]{\frac{I_{\text{sat}}}{|a_m|}} \cdot \frac{1}{G}.$$

This relation means that the maximum decline Δ of the optical flux $I(t)$ from the given waveform $w(t)$ is less than

$$\Delta = \sqrt[m]{\frac{I_{\text{sat}}}{|a_m|}} \cdot \frac{1}{G}.$$

In terms of saturation voltage V_{sat} of the light source, according to the expansion of $I(t)$,

$$I_{\text{sat}} \sim |a_m| \cdot V_{\text{sat}}^{\text{m}},$$

and the last formula for Δ transforms to an estimate

$$\Delta < \frac{V_{\text{sat}}}{\text{G}}$$

or

$$\frac{\Delta}{V_{\text{sat}}} < \frac{1}{\text{G}}.$$

For example, if we want to have less than 1% waveform error in the output light, then the gain of the amplifier must be more than 100.

Our next step in understanding the principle of OAWG is to analyze the importance of linearity of the photodetector. Our previous results show that, if the photodetector provides linear response, the error in the waveform of light can be made as small as necessary by increasing gain G of the amplifier, whatever the nonlinearity of the light source is. However, if the photodetector is not linear, then this is not true. Indeed, let the photodetector be nonlinear, quadratic for example. Quadratic response means that

$$v(t) = cI^2(t).$$

Then, repeating previous manipulations, we obtain

$$\lim_{G\to\infty} |w(t) - c\,I^2(t)| = 0$$

or

$$I(t) = \sqrt{\frac{w(t)}{c}},$$

which means that light intensity $I(t)$ is distorted as the square root of the given waveform $w(t)$. Therefore, the photodetector must be linear as much as possible. Nonlinearity of the photodetector cannot be amended by high gain of the amplifier. One possible but not the only solution is to use a photomultiplier tube (PMT), or briefly a photomultiplier. It is also known that, in order to obtain the largest possible diapason of linearity, a PMT must be connected to an operational amplifier with a large feedback resistor, and that, choosing properly the gain of PMT and the gain of operational amplifier, it is possible to adjust the photoreceiver to the maximum necessary intensity of light. This is the reason why, in practical implementations of the OAWG, the photodetector may be composed of a PMT and an operational amplifier connected in series.

9.4.3. *Variants of practical realization*

The waveform generator in Fig. 9.50 may possibly be implemented either as a stand-alone function generator, or as an integrated computerized unit capable of producing any mathematically conceivable function. The first case may be considered as a low-cost solution, whose practical implementation is shown in Fig. 9.51.

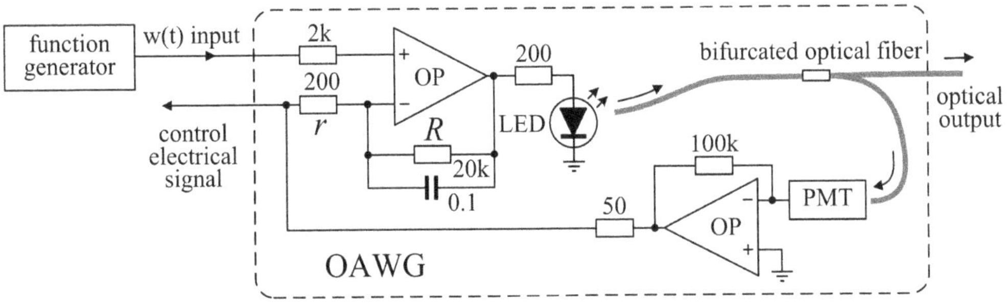

Fig. 9.51. OAWG with a function generator. LED — light-emitting diode; OP — operational amplifiers; PMT — photomultiplier tube.

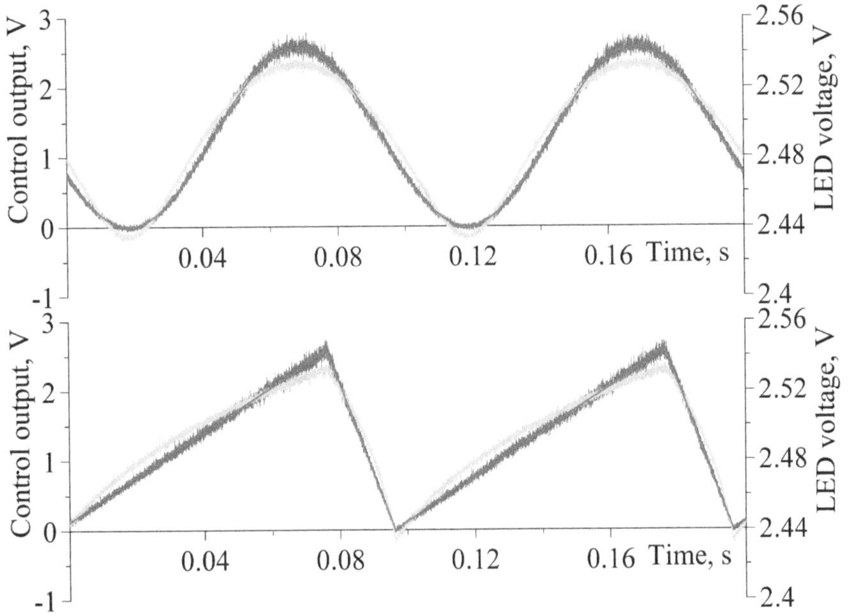

Fig. 9.52. Examples of sinusoidal and ramp signals generated by OAWG. The input waveform $w(t)$ is shown in red line on the background of a wider noisy blue trace of the control signal that represents light intensity $I(t)$. The green trace shows the driving voltage at LED (right axis).

Here, operations of subtraction and amplification are performed by a single operational amplifier. In order to guarantee responses to quickly altering waveforms on the scale of microseconds, the operational amplifier must have the gain-bandwidth product more than 100 MHz. However, for slower signals, quick response is not needed, and then a small damping capacitor may be installed in the feedback to sever spontaneous oscillations that may happen through the air on two operational amplifiers connected in series. The control electrical signal from the inverting input of the operational amplifier may be used to monitor actual waveform of light at the output of the OAWG. This voltage follows the intensity $I(t)$ of the output light because the photomultiplier with the second operational amplifier create highly linear electrical response to light.

Figure 9.52 illustrates how accurately the OAWG converts input waveform $w(t)$ into light intensity $I(t)$. Two standard waveforms are presented: the sinusoid and the ramp signal with 80% duty ratio. The difference between the input waveform $w(t)$ and driving voltage at light-emitting diode (LED) is clearly seen. For example, to produce a sinusoidal waveform, the LED driving signal must not be a sinusoid in order to keep sinusoidal light output. Excellent performance of OAWG near very low levels of light, practically near zero levels, is even better seen on the traces of a ramp signal.

Figure 9.53 illustrates the concept of changing intensity of light by varying conversion gain c, which is proportional to PMT gain M.

Here, intensity of light is presented as an electrical signal from an auxiliary photodetector. The signal at the control output of the OAWG would not show variation of amplitude when the gain of the PMT is changed because this signal is always approximately equal to the waveform $w(t)$ — the basic concept of the OAWG.

Figure 9.54 confirms the effect of the amplifier gain G on the discrepancy between the waveform $w(t)$ and intensity of light $I(t)$. The amplifier gain was changed by varying the resistor r in Fig. 9.51:

$$G = \frac{R}{r}.$$

Clearly, $G = 100$ may suffice many practical applications.

For waveforms more complicated, than an ordinary function generator can produce, a computerized system must be used. An example of a fully computerized OAWG is shown in Fig. 9.55. Here, not only any arbitrary waveform shape can be programmed, but the intensity of light also can be set from the computer, using the digital PMT concept (Chapter 7).

With a computerized OAWG, it is possible to modulate light by any mathematical function. Suppose we want to generate light with intensity, evolving initially like a semi-circle and then like a triangle — a house roof (Fig. 9.56).

For that, define the functions:

$$f_1(i) = \sqrt{1 - \frac{(i - i_0)^2}{r^2}}, \quad f_2(i) = p \cdot \left(1 + \frac{i - i_1}{d}\right), \quad f_3(i) = p \cdot \left(1 - \frac{i - i_1}{d}\right).$$

Start the program, and the light at the output of OAWG will evolve as necessary (Fig. 9.57).

In reality of semiconductor manufacturing, any spectral line may be chosen for the experiment and numerous lines may be traced simultaneously. Therefore, the light-emitting diodes (LED), which were discussed in the previous sections, are not suitable for this task because they work only at a predefined wavelength. Instead, the white light sources with wide spectrum, like tungsten halogen lamps, should be used as shown in Fig. 9.58.

For demonstration, a commercially available variable white light source (Luminar Ace, LA-100USW) with 100 W tungsten halogen lamp (KLS, JCR 12V 100WH10) was chosen. The time constant of the lamp was about 0.5 seconds — the value completely suitable for the experiments with the periodicity of light oscillations about seconds (Fig. 9.49).

The informative part of the optical trend, marked as "etching" in Fig. 9.49, was simulated as a command voltage in the following way:

$$3 \cdot \{0.5(1 - e^{-0.05t}) + 0.1[1 + \sin(2t - 1.75)e^{-0.05t}]\},$$

with the time t in seconds and the amplitude of this signal in Volt. The result is presented in Fig. 9.59.

The lower trend is composed of two, almost coinciding, lines: the command voltage (shown in light grey line) and the optical intensity, measured at the output of the PMT (solid black line). It shows that the output optical signal follows closely the command signal, simulating the real optical trend during etch. Besides, it is very instructive to look at the voltage on the lamp — the upper trend in the figure. In order to mimic the command signal in optical intensity, the voltage on the lamp differs noticeably from the latter, reacting to variations of the lamp optical emissivity. The insert in the figure, showing magnified in time sharp responses of control loop, displays millisecond-scale bursts generated

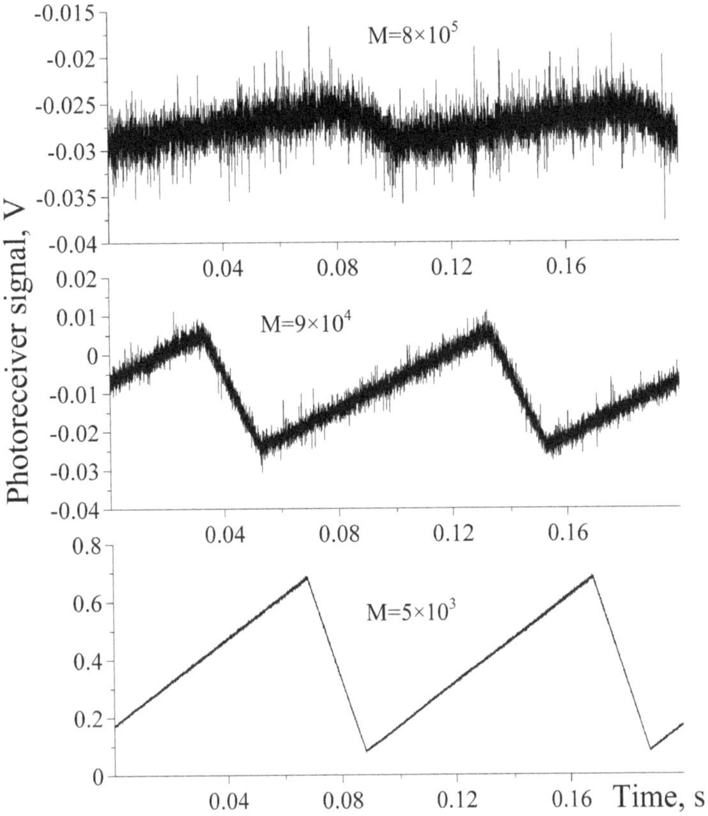

Fig. 9.53. In OAWG, optical intensity is inversely proportional to PMT gain M.

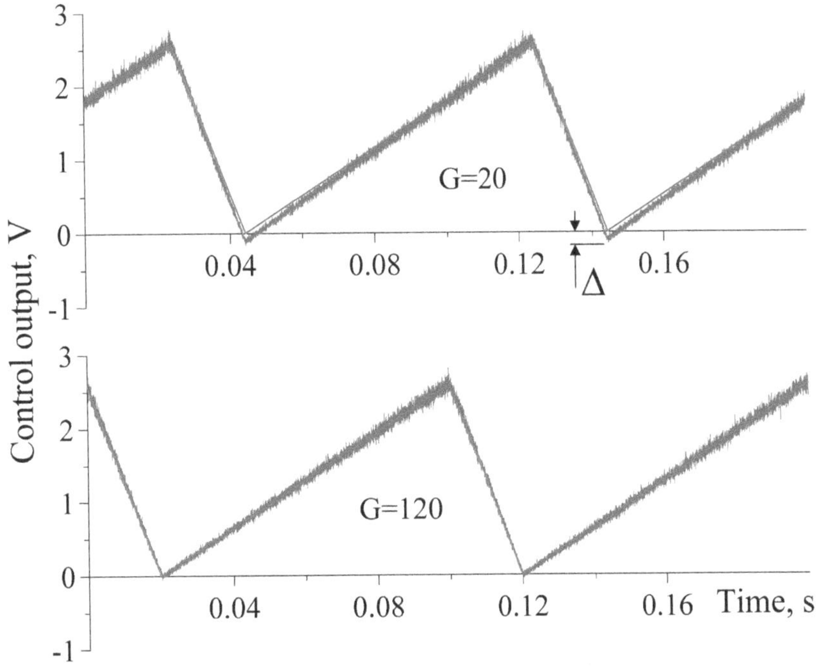

Fig. 9.54. Discrepancy between intensity of light (blue lines) and the waveform (red lines) decreases with the amplifier gain G.

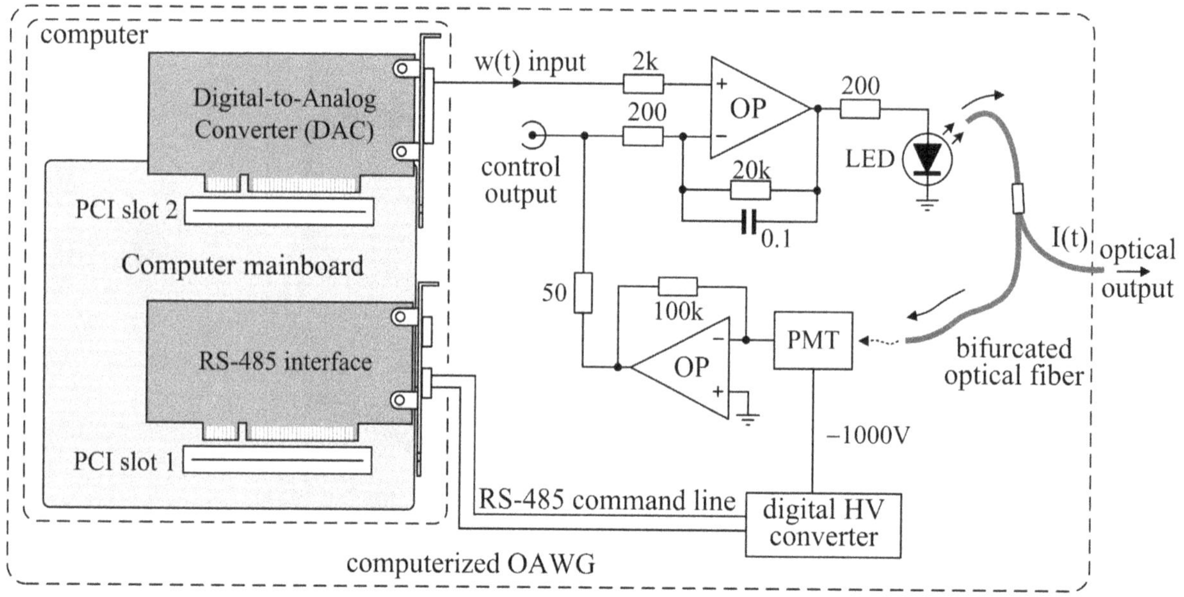

Fig. 9.55. One possible configuration of a computerized OAWG. The only principle difference from Fig. 9.51 is the computer instead of a function generator as a source of the waveform $w(t)$.

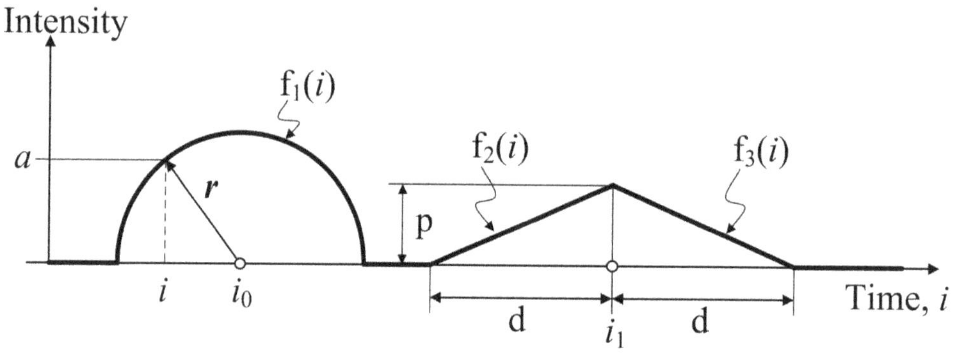

Fig. 9.56. Example of an arbitrary waveform that can be generation by a computerized OAWG.

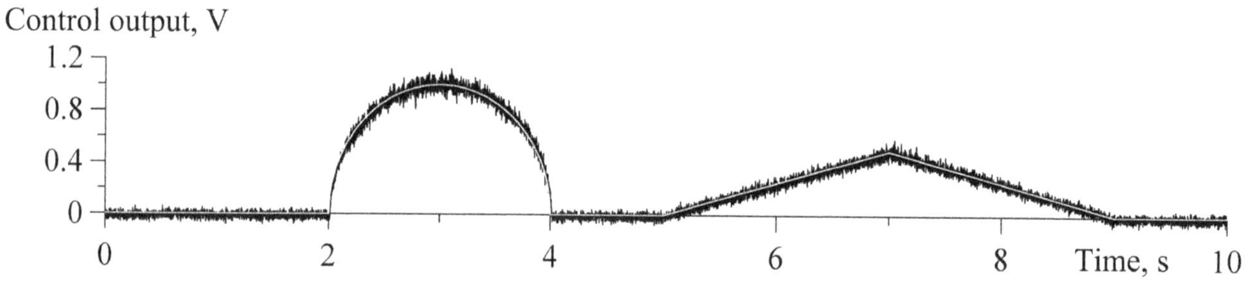

Fig. 9.57. Evolution of light intensity, measured at the control output (noisy black curve) superimposed with the mathematical function (white line). Parameter p is set to 0.6 V.

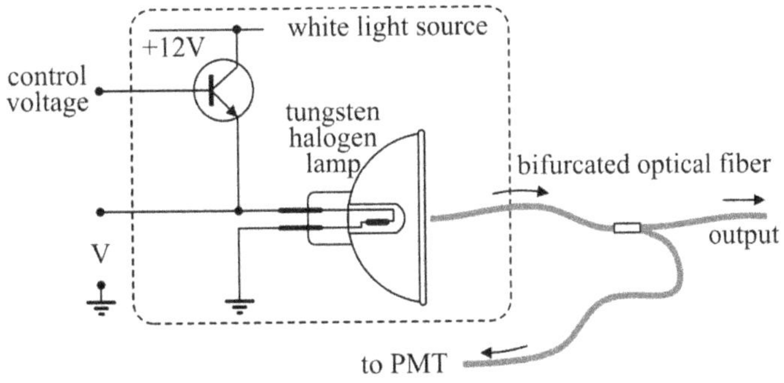

Fig. 9.58. Inside a variable white light source, the tungsten-halogen lamp is connected to +12 V power supply in an emitter-follower scheme.

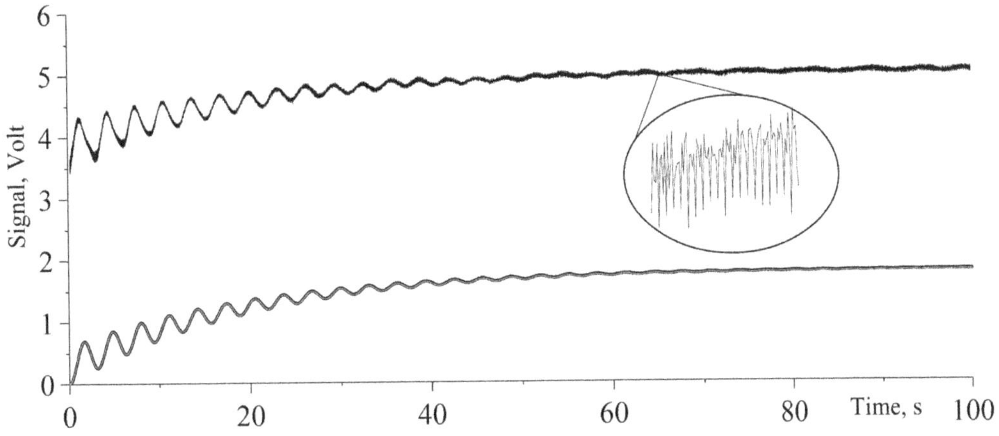

Fig. 9.59. Simulation of the etch trend in white light. Three trends: the command signal and the optical intensity, taken from the PMT (Fig. 9.55) — in the lower part of the figure, and the voltage V on the lamp — above them.

by operational amplifiers in order to compensate for the differences between the command signal and real optical intensity. These bursts are not visible in the output light, being smoothed by the lamp thermal inertia.

Supplemental Reading

O.E. Heavens, *Optical Properties of Thin Solid Films*, Dover Publications Inc., New York, 1991, 261p.

F. Abelès, La détermination de l'indice et de l'épaisseur des couches minces transparentes, *J. Phys. Radium*, **11**(7), pp. 310–314 (1950) (https://hal.archives-ouvertes.fr/jpa-00234262).

J.C. Dainty (ed.), *Laser Speckle and Related Phenomena*, Springer Topics in Applied Physics, Vol. 9, 1975.

L.D. Landau, E.M. Lifshitz, *Electrodynamics of Continuous Media*, 2nd edn., Chapter 10, The Propagation of Electromagnetic Waves, Pergamon Press, 1984.

M. Born, E. Wolf, *Principles of Optics*, Cambridge University Press, 7th edn., 1999.

S.A. Kovalenko, Optical properties of thin metal films, *Semiconductor Physics, Quantum Electron. and Optoelectron.*, **2**(3), pp. 13–20 (1999).

M.G. Moharam, T.K. Gaylord, Rigorous coupled-wave analysis of planar grating diffraction, *J. Opt. Soc. Am.*, **71**(7), pp. 811–818 (1981).

R. Petit, G. Tayeb, Theoretical and numerical study of gratings consisting of periodic arrays of thin and lossy strips, *J. Opt. Soc. Am. A*, **7**(9), pp. 1686–1692 (1990).

R.D. Allen, G.B. David, G. Nomarski, The Zeiss–Nomarski differential interference equipment for transmitted-light microscopy, *Z. Wiss. Mikrosk.*, **69**(4), pp. 193–221 (1969).

V.A. Andreev, K.V. Indukaev, The problem of sub-Rayleigh resolution in inteference microscopy, *J. Russian Laser Res.*, **24**(3), pp. 220–236 (2003).

I.S. Gradshteyn, I.M. Ryzhik, *Table of Integrals, Series, and Products*, 7th edn., Art 6.561.17. Academic Press, 2007.

E. O'Neil, *Introduction to Statistical Optics*, 3rd edn., Dover Publications, Mineola, NY (2003).

T. Sawatari, Optical heterodyne scanning microscope, *Appl. Opt.*, **12**(11), pp. 2768–2772 (1973).

C.W. See, M.V. Iravani, H.K. Wickramasinghe, Scanning differential phase contrast optical microscope: application to surface studies, *Appl. Opt.*, **24**(15), pp. 2373–2379 (1985).

C.W. See, M.V. Iravani, Differential amplitude scanning optical microscope: theory and application, *Appl. Opt.*, **27**(13), pp. 2386–2792 (1988).

F. Zernike, Phase contrast: A new method for microscopic observation of transparent objects. Part I, *Physica*, **9**(7), pp. 686–698 (1942).

F. Zernike, Phase contrast: A new method for microscopic observation of transparent objects. Part II, *Physica*, **9**(10), pp. 674–686 (1942).

Index

A

Abbe formula, 286–287
Abel, 204
  eigenfunctions, 205
  transform, 204, 206
  sensor, 205, 210–212
Abeles formula, 231
Airy
  formula, 54
  function, 54, 75, 88, 280
angle of reflection, 2
astigmatism, 14, 16

B

band-pass filter, 66, 148
beamsplitter, 35, 37–38, 41,43, 86–87, 93–94, 96, 100–101, 146, 149, 235–236
  polarizing (PBS), 72
  non-polarizing (NPBS), 72
Bessel
  function, 75, 88, 245–247, 280
  differential equations, 244
Brewster angle, 61, 261, 266–267
buried structures, 48–49

C

calibration, 16
  source, 9–10, 18, 29–30
  module, 17
  spectrum, 17–18
chemical mechanical planarization (CMP), 41, 234, 249–251
coherence, 91, 106, 110, 265–267, 297
  length, 91, 265
  spatial, 68
  temporal, 92, 110
confocal, 57–59, 277
conjugated point, 92
corner-cube reflector (CCR), 87–94, 96–99, 102–103, 106–107, 110, 112
correlation function, 68–69

D

decay time, 151–153, 156, 194
Descartes principle, 58
diffraction, 4
  Fraunhofer, 4
  order, 3–5, 7–9, 29
    positive, 3, 7
    negative, 3
    sorting filter, 29
diffraction grating, 1, 4, 12
  blazed, 2, 4–5
    blaze angle, 5
  flat, 1, 2, 12
  curved, 2, 12
  grooves, 2, 5–6, 12–14
  equation, 1, 3, 7, 13, 15
  spacing, 5
  dispersion,
    angular, 5–7
    linear, 6, 16, 121
  curvature, 13–14
  field curvature, 15–16, 83
  meridional rays, 14–16

sagittal rays, 14–16, 28
VLS, 15
holographic, 83
anamorphism, 83
Dirac delta-function, 66, 69, 78
dispersion
spectral, 35, 43, 143, 230, 232, 237, 247–248, 251, 259
statistical, 116

E

endpoint, 18, 169, 171–176, 182, 184–186
exposure time, 1, 21, 24–25, 102–103, 107–109, 115, 134–137, 139–140, 179

F

Fermat principle, 92
filter
neutral-density, 110, 137
interference, 32, 53, 92, 112, 176, 182–184, 190–192, 195
finesse, 55–57, 60
fluorescence, 126–128, 147
focal length, 5, 7
Fourier
transform, 35, 48, 50, 66, 69, 71, 73, 79, 94, 97–98, 105, 107–108, 122, 143–144, 152, 165, 224, 226, 236
fast Fourier transform (FFT), 35, 225, 227
series, 270–271
coefficients, 271
free spectral range (FSR), 55, 60–62
Fresnel
formulas, 48
reflection, 51
transform, 80–81

G

geometrical progression, 54
Glan-Taylor prism, 61

H

Hankel functions, 244–247
He–Ne laser, 37, 60, 71–72, 91, 98–99, 112, 122, 236, 265, 279, 287
heterodyne, 61, 64–68, 70–74, 76–79, 268–269, 275–277, 281–283, 291–292

I

image intensifier, 113–115, 118, 120, 129–130, 132, 134, 136–139
IMSL Fortran library, 123, 271
instrument function, 97
integration time, 1
interference fringes, 87–88, 90–91, 94
localization, 94
depth, 95
interferogram, 38–39, 51, 96–97, 103, 107–109
interferometer, 35, 40, 44, 89–94
Michelson, 35–36, 87–88, 90–91, 93–94
scanning, 53
Fabry-Perot (IFP), 53–55, 63
scanning IFP, 53–55, 59, 61
flat mirror IFP, 55
spherical mirrors IFP, 57
confocal IFP, 57–58
IFP cavity, 53, 55, 57

L

Lambertian law, 210
laser mode analyser, 53
laser-induced break-down spectroscopy (LIBS), 121, 129, 133–135
Levenberg-Marquardt algorithm, 122–124, 206
light-emitting diode (LED), 84, 112, 127, 129, 148, 151, 210, 213, 216, 260, 265–266, 301–302
Lorentz spectra, 71

M

Maxwell
equations, 242, 244, 270
distribution, 155
mercury doublet, 17–19, 29, 103–104, 108–109, 111, 121–124
micro-channel plate (MCP), 113–115, 120, 137
monochromator, 9–10, 31
uncrossed Czerny-Turner scheme, 31
Fastie–Ebert configuration, 31, 33
multi-beam interference, 53

N

Nomarski, G., 276
numerical aperture, 14–15, 27, 95, 108–109, 168, 234–235, 250, 252, 254–256, 259–261, 265, 286–288, 290

O

objective
 Mireau, 86
 Michelson, 86
operational amplifier, 187–188, 280, 300–301, 305
optical fiber, 27, 29, 44, 92, 124–125
 bundle, 27–28, 106, 132, 138, 159, 184
  imaging, 105–107
 core, 45
order-sorting filter, 8, 10, 17, 83
 thin-film, 8–9, 11–12, 29

P

parabolic mirror, 41, 45
Petit and Tayeb algorithm, 270
Photodetector, 1, 20
 single-channel, 20
 multi-element, 20
 CCD, 16–18, 20–22, 25, 27–28, 31, 83–84, 96, 102–103, 106–108, 113, 116, 118–120, 132, 138, 148, 195
 MOS, 20
 dark current, 25–26
 cooling, 25, 30
photo-multiplying tube (PMT), 96, 100, 108, 124–125, 146, 149, 151–152, 177–182, 186–189, 194–195, 300–303, 305
photoresist, 49
piezo-transducer (PZT), 56, 59, 91, 96–98, 102, 106–107, 145–146
plasma
 diagnostics, 118, 128, 137, 143, 154, 168, 190
 fluctuations, 176, 184–186
point-spread function, 95, 120, 122, 124, 281–282

Q

quantum efficiency, 10, 17, 20, 25–26, 65, 74, 77, 79, 100–101, 103, 108, 115, 117, 177, 179
quarter wave plate (QWP), 72

R

Raguin and Morris algorithm (EMTA), 242, 246–248
Rayleigh
 resolution criterion, 18–19, 121, 280, 285–287
 expansion, 270
reflectometer, 41–43, 49, 96, 103, 105, 235–236, 240
 Fourier-transform, 41–42
 imaging, 105–106
 spectral, 41, 235–236, 249
refractive index, 48–49, 83
 of silicon, 48
resolution
 spectral, 1, 5, 7, 18–19, 27, 29–32, 36–37, 40, 43, 46, 48–51, 53, 55–56, 59–60, 62, 64, 71, 74, 97–98, 103, 105, 108–109, 120–146, 182, 214, 225
 spatial, 45–46, 83, 94, 113, 274, 276–277, 279, 281–282, 284–286
rigorous coupled-wave analysis (RCWA), 242, 246, 270, 273–274
Rowland circle, 12, 14–16

S

sensitivity, 1, 9, 27, 30
shrinkage, 49
Siegman's antenna theorem, 77, 79
signal-to-noise ratio (SNR), 102–103, 110, 112, 115–118, 121–122, 129–135, 173–175, 179–180
silicon wafer, 46, 48–50
slit, 1, 4, 6–7, 13, 27, 86, 108
Snell law, 228, 254
spatial-spectral picture, 83, 85–86
spectral
 components, 3
 interferometry, 84–86
 oscillations, 38, 46, 48–49, 51, 85–86, 97, 231, 235, 238–240, 250, 256–257
 plane, 1, 83
 range, 1, 11
 reflection, 9
 transmission, 9
spectrometer, 1
 grating spectrometer, 1
  fiber-optic spectrometer, 8, 32

wide-range spectrometer, 1, 7, 27
compact spectrometer, 5, 8, 27
Czerny–Turner scheme, 12
crossed, 12, 29
uncrossed, 12
Rowland scheme, 15, 29–30
Wadsworth scheme, 14–16
Fourier-transform spectrometer, 35, 37–38, 40–41, 43, 87, 103, 112, 154
absorption type, 37–39
emission type, 40–41
infrared (FTIR), 41
imaging, 83–87, 93
gated, 86, 113, 119, 127–129, 133, 135–136, 149, 154, 194–196, 200
spectrum, 1
Stokes formula, 51
synchronous detector, 143–146, 148, 158, 269

T

Taylor expansion, 55
telecentric lens, 197–200
thin metal layer approximation (TMLA), 270, 273–274

W

Wallot, J., 244
wavefront, 4, 58, 74
matching condition, 77, 79–80
wavelength range, 6
wavenumber, 46, 49–50, 88
Whittaker–Kotelnikov sampling theorem, 73, 96, 103
Wiener filter, 122–123

X

xenon lamp, 8

Z

Zeeman laser, 61, 71–73, 98, 268, 274–275, 279
Zernike, 276
Zygo laser, 60–64

CPSIA information can be obtained
at www.ICGtesting.com
Printed in the USA
JSHW040713071222
33860JS00004B/19

9 789811 257599